CELL BIOLOGY

A Comprehensive Treatise

Volume 4

Gene Expression: Translation and the Behavior of Proteins

CONTRIBUTORS

Richard G. W. Anderson
D. Banerjee
S. J. Chan
Boyd Hardesty
John W. B. Hershey
Samuel B. Horowitz
Gisela Kramer
Aldons J. Lusis
Michael D. McMullen
J. Marsh

Terence E. Martin
Philip L. Paine
C. Patzelt
James M. Pullman
P. S. Quinn
Rudolf A. Raff
C. M. Redman
Peter J. Roach
D. F. Steiner
Richard T. Swank

H. S. Tager

ADVISORY BOARD

Wolfgang Beermann
Donald Brown
Joseph G. Gall
G. P. Georgiev
Paul B. Green
George Klein
George Lefevre
A. W. Linnane
Daniel Mazia
Brian McCarthy

Lee Peachey
Robert P. Perry
Keith R. Porter
Frank H. Ruddle
Robert T. Schimke
S. J. Singer
Tracy M. Sonneborn
Noboru Sueoka
Hewson Swift
George J. Todaro

Gordon Tomkins*

* Deceased

CELL BIOLOGY
A Comprehensive Treatise

Volume 4

Gene Expression: Translation and
the Behavior of Proteins

Edited by

DAVID M. PRESCOTT

LESTER GOLDSTEIN

Department of Molecular, Cellular and Developmental Biology
University of Colorado
Boulder, Colorado

1980

Academic Press

A Subsidiary of Harcourt Brace Jovanovich, Publishers
New York   London   Toronto   Sydney   San Francisco

ACADEMIC PRESS, INC.
111 Fifth Avenue, New York, New York 10003

United Kingdom Edition published by
ACADEMIC PRESS, INC. (LONDON) LTD.
24/28 Oval Road, London NW1 7DX

Library of Congress Cataloging in Publication Data
Main entry under title:

Gene expression: translation and the behavior of proteins.

(Cell biology, a comprehensive treatise ; v. 4)
Includes bibliographies and index.
1. Genetic translation. 2. Protein biosynthesis.
3. Gene expression. I. Goldstein, Lester.
II. Prescott, David M. , Date III. Series.
[DNLM: 1. Translation, Genetic. 2. Proteins—
Physiology. 3. Proteins—Genetics. QH574 C393
1977 v. 4]
QH574.C43 vol. 4 [QH450.5] 574.87s [574.19'245]
ISBN 0-12-289504-5   80-16454

PRINTED IN THE UNITED STATES OF AMERICA

80 81 82 83   9 8 7 6 5 4 3 2 1

Contents

1 The Translational Machinery: Components and Mechanism

John W. B. Hershey

2 Regulation of Eukaryotic Protein Synthesis

Gisela Kramer and Boyd Hardesty

3 Masked Messenger RNA and the Regulation of Protein Synthesis in Eggs and Embryos

Rudolf A. Raff

         Urchin Eggs 114
    IV. Masking and the Control of Protein Synthesis in Eggs
         and Embryos 116
    V. Concluding Remarks 128
        References 129

4 Structure and Function of Nuclear and Cytoplasmic
   Ribonucleoprotein Complexes

   Terence E. Martin, James M. Pullman, and
   Michael D. McMullen

    I. Introduction 137
    II. Nuclear RNP 139
    III. Cytoplasmic RNP Containing mRNA 160
        References 167

5 Proteolytic Cleavage in the Posttranslational
   Processing of Proteins

   D. F. Steiner, P. S. Quinn, C. Patzelt, S. J. Chan,
   J. Marsh, and H. S. Tager

    I. Introduction 175
    II. Early Cleavages—The Presecretory Proteins 177
    III. Posttranslational Cleavages—Intracellular
         (The Proproteins) 187
        References 194

6 Principles of the Regulation of Enzyme Activity

   Peter J. Roach

    I. Introduction 203
    II. Some Basic Concepts of Biochemical Regulation 205
    III. Mechanisms of Controlling Enzyme Activity 218
    IV. Integration of Metabolic Controls 260
    V. Conclusion 284
        References 285

7 The Movement of Material between Nucleus and Cytoplasm

   Philip L. Paine and Samuel B. Horowitz

    I. Introduction 299
    II. Determinants of Nucleocytoplasmic Movements 301

8   Regulation of Location of Intracellular Proteins

Aldons J. Lusis and Richard T. Swank

9   The Biogenesis of Supramolecular Structures

Richard G. W. Anderson

10   Protein Secretion and Transport

C. M. Redman and D. Banerjee

List of Contributors

Numbers in parentheses indicate the pages on which the authors' contributions begin.

Richard G. W. Anderson (393), Department of Cell Biology, Health Science Center, The University of Texas, Dallas, Texas 75235

D. Banerjee (443), Lindsley F. Kimball Research Institute of the New York Blood Center, New York, New York 10021

S. J. Chan (175), Department of Biochemistry, The University of Chicago, Chicago, Illinois 60637

Boyd Hardesty (69), Clayton Foundation Biochemical Institute, Department of Chemistry, The University of Texas, Austin, Texas 78712

John W. B. Hershey (1), Department of Biological Chemistry, School of Medicine, University of California, Davis, California 95616

Samuel B. Horowitz (299), Cellular Physiology Laboratory, Department of Biology, Michigan Cancer Foundation, Detroit, Michigan 48201

Gisela Kramer (69), Clayton Foundation Biochemical Institute, Department of Chemistry, The University of Texas, Austin, Texas 78712

Aldons J. Lusis (339), Department of Medicine, Division of Hematology-Oncology, School of Medicine, University of California, Los Angeles, California 90024

Michael D. McMullen (137), Department of Biology, The University of Chicago, Chicago, Illinois 60637

J. Marsh (175), Department of Biochemistry, The University of Chicago, Chicago, Illinois 60637

Terence E. Martin (137), Department of Biology, The University of Chicago, Chicago, Illinois 60637

Philip L. Paine (299), Cellular Physiology Laboratory, Department of Biology, Michigan Cancer Foundation, Detroit, Michigan 48201

C. Patzelt (175), Department of Biochemistry, The University of Chicago, Chicago, Illinois 60637

James M. Pullman (137), Department of Biology, The University of Chicago, Chicago, Illinois 60637

P. S. Quinn (175), Department of Biochemistry, The University of Chicago, Chicago, Illinois 60637

Rudolf A. Raff (107), Program in Molecular, Cellular and Developmental Biology, Department of Biology, Indiana University, Bloomington, Indiana 47401

C. M. Redman (443), Lindsley F. Kimball Research Institute of the New York Blood Center, New York, New York 10021

Peter J. Roach* (203), Department of Pharmacology, University of Virginia School of Medicine, Charlottesville, Virginia 22908

D. F. Steiner (175), Department of Biochemistry, The University of Chicago, Chicago, Illinois 60637

Richard T. Swank (339), Department of Molecular Biology, Roswell Park Memorial Institute, Buffalo, New York 14263

H. S. Tager (175), Department of Biochemistry, The University of Chicago, Chicago, Illinois 60637

* Present address: Department of Biochemistry, Indiana University School of Medicine, Indianapolis, Indiana 46223.

Preface

The four volumes of this treatise are devoted to cell genetics. Volumes 1 and 2 covered cell inheritance and its molecular basis. Volume 3 extended the subject into the area of the molecular and cytological basis of gene expression, focusing particularly on the synthesis and processing of RNA molecules and their regulation. Volume 4 completes the theme with discussions of the translation of genetic information into proteins and the final stage of gene expression—namely, the activities and behaviors of proteins.

Each volume is designed to serve as a comprehensive source of primary knowledge at a level suitable for graduate students and researchers in need of information on some particular aspect of cell biology. Thus we have asked contributors to avoid emphasizing up-to-the-minute reviews with the latest experiments, but instead to concentrate on reasonably well-established facts and concepts in cell biology.

David M. Prescott
Lester Goldstein

Contents of Other Volumes

1

The Translational Machinery: Components and Mechanism

John W. B. Hershey

I. INTRODUCTION

Protein synthesis is a complex metabolic process whereby amino acids are polymerized into long linear polypeptides. The chemistry of the reaction is basically quite simple. The carboxyl group of one amino acid is linked to the amino group of another amino acid to form a peptide bond. Energy is required and is supplied by prior activation of the carboxyl

1

group. The reaction is repeated by condensation of the activated carboxyl group of the peptide with a new amino acid until the protein is completed. Thus the nascent polypeptide grows from its amino terminus toward its carboxy terminus. Were polymerization per se the only consideration, the enzymology of protein synthesis would surely be relatively simple, perhaps comparable to proteolysis. However, proteins are rich in information; they are composed of 20 different kinds of amino acids arranged in specific sequences. The amino acid sequence of a protein is specified by a sequence of nucleotides in DNA. Therefore, protein synthesis is also the process whereby information is *translated* from nucleotide sequences in DNA into amino acid sequences in proteins.

A broad overview of protein synthesis follows. Amino acids are chemically activated in an ATP consuming reaction by forming esters with specific transfer RNA's (tRNA's). Polymerization of the activated amino acids occurs on small organelles, called ribosomes, which are composed of RNA and numerous proteins. Information for the order of polymerization is specified by messenger RNA (mRNA) which is transcribed from DNA that comprises the gene for the protein. The polymerization process on the ribosome is divided conceptually into three phases: initiation, elongation, and termination. Initiation is the process by which the first aminoacyl-tRNA interacts with a precise initiator region of the mRNA and forms a complex with the ribosome. Elongation involves the sequential binding of specific aminoacyl-tRNA's as determined by the mRNA and incorporation of their aminoacyl derivatives into the growing polypeptide. Termination is the hydrolysis of the completed protein from the final tRNA as specified by special signals in the mRNA. Each of the phases of protein synthesis is promoted by protein factors, called initiation, elongation, or release factors, which transiently bind to ribosomes and catalyze the process. Energy is supplied through the hydrolysis of GTP. In all, more than 150 macromolecules are involved in the translation of a simple mRNA. Thus, amino acids are polymerized into information-rich proteins with the consumption of energy. The machinery, i.e., ribosomes, factors, tRNA's, and mRNA's, is unaltered and may be reutilized repeatedly to synthesize more proteins.

In this chapter, the macromolecular components of the translational machinery are identified and their structures, biogenesis, and cellular levels are described. The pathway for interaction of the components is then considered and attempts are made to describe the process in precise molecular or enzymological terms. Special attention is given to protein synthesis in eukaryotic cells, but prokaryotic components and mechanisms are described also. This is appropriate because the broad features of protein synthesis are remarkably similar in the two cell types, and because many bacterial structures and mechanisms are better understood at this time. The review develops the broad outlines and fundamental concepts of the translational process. Fuller experimental details and exhaustive references are available in the numerous extensive reviews cited in the text.

As the mechanism of protein synthesis is considered, three questions are of paramount importance: (1) How is the synthesis of proteins initiated at the correct place in the mRNA? (2) How does protein synthesis proceed with a very low frequency of error? (3) How may the translational process be modulated and its rate controlled? Ultimately we hope to explain all of the steps of protein synthesis in terms of well-understood chemical forces. The material which follows documents our progress towards this goal.

II. MACROMOLECULAR COMPONENTS

A. Aminoacyl-tRNA's

Aminoacyl-tRNA's are directly involved as precursors in the polymerization of amino acids on ribosomes. Transfer RNA's constitute a class of small single-stranded RNA molecules, each composed of 74 to 94 nucleotides with a mass of about 25,000 daltons. An amino acid is covalently attached to a tRNA by an ester linkage between the carboxyl group of the amino acid and a ribose hydroxyl group at the 3'-terminal adenosine of the tRNA. The reaction involves the hydrolysis of ATP and the formation of a "high energy" ester bond, thus activating the amino acid for subsequent peptide bond formation. Attachment is catalyzed by a specific enzyme, called aminoacyl-tRNA synthetase (or amino acid tRNA ligase). All cells contain at least one specific synthetase for each of the 20 amino acids. In eukaryotic cells, a complete set of these enzymes is present in the cytoplasm, while another set is found within mitochondria or chloroplasts. For each amino acid there are at least one, but frequently a number of, different tRNA molecules (called isoacceptor families) to which a given amino acid can be attached.

Aminoacyl-tRNA's perform two major functions in protein synthesis: (1) the amino acid is "activated" for peptide bond formation, and (2) the tRNA portion acts as an "adaptor" between the mRNA and the amino acid on the ribosome. The fidelity of proteins synthesis depends on the synthetase mechanism, which attaches an amino acid specifically to its cognate tRNA, and on the mechanism of mRNA-directed binding of aminoacyl-tRNA's to ribosomes. Thus tRNA's play a critical and central role in these important reactions. tRNA's also participate in a variety of cellular functions besides protein synthesis. These special functions, which are outside the scope of this chapter, have been outlined by Rich and RajBhandary (1976).

1. Transfer RNA's

a. **Multiplicity of Species.** The cytoplasm of eukaryotic cells may contain more than 100 different species of tRNA. It follows that for a given amino acid, there may be numerous specific tRNA's, called isoacceptor tRNA's. Isoacceptor tRNA's are denoted by the amino acid abbreviation in superscript (e.g., tRNA^{Leu}) and different members of the family are indicated by subscript numerals (e.g., tRNA$_1^{Leu}$, tRNA$_2^{Leu}$, etc.). The different species can be separated and purified by such fractionation procedures as ion exchange and reverse phase chromatography (Cantoni and Davies, 1971; Pearson *et al.*, 1971). Fractionation of isoacceptor species is based on differences in the number, composition, sequence, and extent of modification of the bases. For example, from the bacterium *Escherichia coli* B, five species of tRNA^{Leu} can be isolated; the presence of three to four species per amino acid is common in mammalian cells. Members of an isoacceptor family often differ appreciably in base composition and sequence, and most differ in the anticodon region. They may be present in different amounts in a given cell type, some being relatively abundant and others relatively scarce.

Degeneracy of the genetic code is consistent with the existence of families of isoacceptor tRNA's; 61 codons specify 20 amino acids. The number of code words for an amino acid varies from one each for methionine and tryptophan to six each for leucine, arginine, and serine. It is thought that generally there is one tRNA for each codon. However, examples are known where different isoacceptor species recognize a single, specific codon and some tRNA's recognize up to four codons, each of which differs in the third letter of the code word. Such interactions, explained by the "wobble" concept (Crick, 1966), are discussed below in Section IV,B,4. The multiplicity of tRNA species is described in greater detail by Ofengand (1977).

b. **Structure of tRNA's.** Transfer RNA's are single stranded polyribonucleotides containing 74 to 94 nucleotides that vary greatly in base composition. The first nucleotide sequence was obtained for yeast tRNA^{Ala} by Holley and co-workers (1965). Particularly striking was the large number (10 out of 77) of unusual bases such as pseudouridine, ribothymidine, methylguanosine, dihydrouridine, and inosine. Recent technical advances in RNA sequencing methodology permit tRNA sequences to be determined with relative ease (for a recent review, see Rich and RajBhandary, 1976). Primary structures have been reported for more than 100 different tRNA's obtained from a variety of organisms ranging from bacteria to mammals (Gauss *et al.*, 1979). A large number of recent detailed reviews on tRNA chemistry and structure are available (Rich and RajBhandary, 1976; Clark, 1977; Ofengand, 1977; Rich, 1978). An outline of the basic structures and principles is given below.

Transfer RNA's are not random coils of single-stranded RNA, but are folded into active globular conformations which can be denatured. All of the known tRNA sequences may be organized conceptually into a planar secondary structure which resembles a cloverleaf pattern. The cloverleaf, as shown in Fig. 1A for yeast tRNA^{Phe}, is composed of a stem and three arms; each arm consists of a stem and loop. Stems are antiparallel double-strained helical regions, while loops are non-base-paired regions. Comparison of different tRNAs indicates that the cloverleaf structure has both constant and variable features (see Fig. 1B). The following generalizations apply to nearly all known tRNA's.

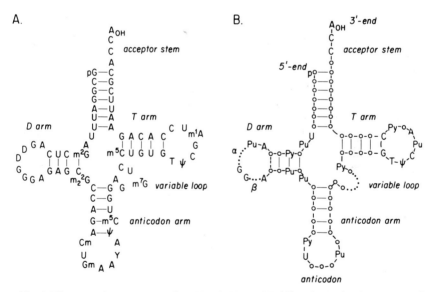

Fig. 1. The secondary structure of tRNA. (A) Yeast tRNA^{Phe}. The nucleotide sequence is organized into the planar cloverleaf pattern typical of all tRNA's. Symbols used for modified nucleosides are Ψ, pseudouridine; T, ribothymidine; Y, modified purine; D, dihydrouridine; m, methylated bases. (B) Constant and variable features of tRNA. Nucleotide residues are shown by circles except for invariant bases, which are identified by letters. Regions with variable numbers of nucleotides are shown by dotted lines and are discussed in the text. Symbols used are Py, pyrimidine; Pu, purine; Ψ, pseudouridine; T, ribothymidine. Adapted from Rich (1978).

1. The 3'-terminus is CpCpA, where the amino acid is attached to the 2'- or 3'-hydroxyl of the terminal adenosine. There is a phosphate at the 5'-terminus, usually pG. Nucleotidyltransferase, an enzyme present in all cells, can add or remove the CCA sequence from tRNA's (Deutscher, 1974).

2. The acceptor stem contains 7 base pairs and 4 non-base-paired nucleotides at the 3'-terminus.

3. The T arm comprises a 5 base pair stem and a loop of exactly 7 nucleotides containing the sequence TΨC.

4. The anticodon arm contains a stem of 5 base pairs and a loop of 7 nucleotides. In the middle of the loop are 3 bases, called the anticodon, which interact during protein synthesis with mRNA by base-pairing to a suitable triplet sequence, called the codon. At the 5'-side of the anticodon are two pyrimidines, the closest being invariably U. Adjacent to the anticodon at the 3'-side is a hypermodified purine.

5. The D arm is more variable than the other arms and consists of a stem of 3 to 4 base pairs and a loop containing 7 to 12 nucleotides. Constant features of the loop are the presence of A bases next to the stem and two adjacent G bases. Between the constant G's and A's are regions of variable length, denoted α and β in Fig. 1B. The variable regions generally contain dihydrouridine, from which the arm obtained its name.

The variability in nucleotide number is strictly confined to three regions of the molecule. The α and β regions of the D loop mentioned above together account for differences of only 3 nucleotides. The greatest variability occurs in a region between the T arm and the anticodon arm called the variable loop. This extra arm can vary in length from 3 to 21 nucleotides.

The three-dimensional structure of yeast tRNA[Phe] has been determined from X-ray diffraction studies of crystals by Rich and co-workers (Kim et al., 1974) and by Klug, Clark, and colleagues (Robertus et al., 1974). The molecule is L- shaped, with the acceptor stem at one end and the anticodon loop at the other (Fig. 2). All of the hydrogen bonding interactions predicted by the cloverleaf structure are found in the tertiary structure. The acceptor stem and the T stem are stacked together to form one arm of the L structure, while the D stem and anticodon stem are stacked to form the other arm of the L. The D and T loops are located at the "elbow" and interact with each other. Stabilization of the globular structure of tRNA is derived primarily from hydrogen bonding interactions, but base stacking is important also. It appears likely that the conformation of the tRNA in the crystal is essentially the same as that in solution. The crystal structure is consistent with studies, such as chemical

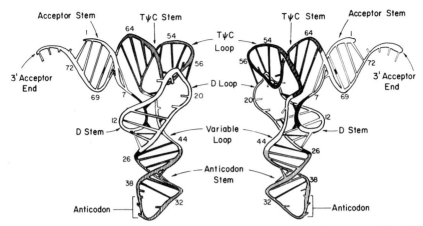

Fig. 2. A schematic diagram of the three-dimensional structure of yeast tRNA^Phe. Two side views are shown of the model based on X-ray crystallographic analyses. The ribose phosphate backbone is depicted as a coiled tube, and base pair interactions are shown as cross-ranges. Taken from Rich (1977).

modification and nuclear magnetic resonance (nmr) spectroscopy, of tRNA in solution [see Rich and RajBhandary (1976) for details].

Many of the constant features of tRNA's are critically involved in interactions which stabilize the tertiary structure. The two G's in the D loop bind to the ΨC sequence in the T loop. The pyrimidine between the T stem and the variable loop binds with the constant purine in the D loop. An A base in the D loop interacts with the U at the base of the acceptor stem. The T and A bases of the T loop bind to each other. Most of these hydrogen bonding interactions are not of the standard Watson–Crick type, but are more specialized. A detailed description of the structure of yeast tRNA^Phe is available (Rich and RajBhandary, 1976).

c. Initiator tRNA's. Initiation of protein synthesis in eukaryotic cells involves a unique initiator tRNA, $tRNA_i^{Met}$. These initiator tRNA's are functionally distinct from all other tRNA's (reviewed by Ofengand, 1977), in that they alone bind to the small ribosomal subunit in a reaction promoted by initiation factors, and they cannot form a ternary complex with elongation factors and GTP. In bacteria and mitochondria, the Met-$tRNA_i^{Met}$ is formylated by a transformylase enzyme, producing fMet-$tRNA_i^{Met}$. Transformylase is absent from the cytoplasm of eukaryotic cells and the initiator species is not formylated. However, eukaryotic Met-$tRNA_i^{Met}$ can be formylated *in vitro* by the bacterial enzyme.

What structural features do initiator tRNA's possess which account for these differences? Mammalian cytoplasmic $tRNA_i^{Met}$ lacks the invariant sequence TΨ in the T loop and contains AU instead. The T loop appears to be of critical importance, since its sequence is essentially identical in all eukaryotic initiator species examined. In the anticodon loop, C is usually found in place of the constant U on the 5'-side of the anticodon. Initiator tRNA's from prokaryotes are different from other tRNA's in that they lack the base pair interaction of the 5'-terminal nucleotide with the fifth base from the 3'-terminus. Crystals of *E. coli* $tRNA_i^{Met}$ recently have been analyzed by X-ray diffraction techniques, and the results indicate that basically the same kinds of tertiary folding occur in the initiator tRNA as in yeast $tRNA^{Phe}$ (N. Woo and A. Rich, personal communication). A satisfactory structural explanation of the functional differences of initiator tRNA's remains to be elucidated.

2. Aminoacyl-tRNA Synthetases

There appears to be one aminoacyl-tRNA synthetase for each amino acid and its cognate isoacceptor tRNA family. (In eukaryotic cells, a second set of synthetase molecules has been identified which are limited to organelle protein synthesis.) Thus, a single synthetase enzyme can attach its amino acid to all members of the isoacceptor family. This is even true for the methionine synthetase, which recognizes both the initiator $tRNA_i^{Met}$ and the $tRNA_m^{Met}$ used to insert methionine internally into proteins.

The synthetases differ considerably in structure, even though each catalyzes essentially an identical reaction (Kisselev and Favorova, 1974; Söll and Schimmel, 1974). Bacterial synthetases vary in molecular weight from 44,000 to 262,000 and their subunit structures include monomers (α), dimers (α_2 or $\alpha\beta$), and tetramers ($\alpha_2\beta_2$). Eukaryotic synthetases are similarly constituted, but appear to be present in the cytoplasm as high molecular weight complexes containing all 20 synthetases, tRNA's, and cholesterol esters (Bandyopadhyay and Deutscher, 1973). The functional significance of these fragile complexes is unknown. Most monomeric synthetases fall in the size range 70,000–120,000, while the subunit size of most dimeric enzymes is exactly half, 35,000–60,000. This finding and other evidence suggest that the larger monomeric polypeptides may have arisen by gene duplication and fusion of the subunit genes of dimers (Ofengand, 1977). Thus all tRNA synthetases may have evolved from a basic gene for a 35,000–60,000 molecular weight polypeptide.

3. The Aminoacylation of tRNA

a. **Mechanism.** The highly specific aminoacylation or charging of tRNA is shown by the following general equation

Amino acid + tRNA + ATP $\rightleftarrows$ aminoacyl-tRNA + AMP + pyrophosphate

The reaction is catalyzed by a specific synthetase enzyme and requires the presence of Mg^{2+}. The enzymes are specific for ATP, although ATP analogues such as $AMP-P(CH_2)P$ are functional with some synthetases. The amino acid must possess a free amino group and be in the L configuration. During the reaction, the $\alpha-\beta$ pyrophosphate bond of ATP is broken and an ester bond between the amino acid and the tRNA is formed. The reaction is reversible and its equilibrium constant is near or slightly less than 1. Thus the ester bond is "high energy," and the amino acid is "activated" for subsequent peptide bond formation. The product, pyrophosphate, is hydrolyzed to phosphate by a pyrophosphatase, which prevents the back reaction and subsequently shifts the reaction equilibrium to fully charged tRNA. In effect, two high energy pyrophosphate bonds are cleaved in order to aminoacylate one tRNA.

The mechanism of aminoacylation of tRNA has been studied in considerable detail. For recent reviews, see Ofengand (1977), Söll and Schimmel (1974), and Eigner and Loftfield (1974). Synthetases possess one or two active sites, depending on their subunit structure: monomers (α) and heterodimers ($\alpha\beta$) possess one site, while homodimers (α_2) and tetramers ($\alpha_2\beta_2$) possess two sites which apparently act independently. The order of addition of the three reaction components and the order of release of products have been studied by kinetic methods and appear to depend on the specific synthetase used. K_m values for ATP range from 40 to 2300 μM, for amino acids from 5 to 150 μM, and for tRNA's from 0.03 to 0.5 μM. Maximal velocities of aminoacylation range from 10 to 2200 molecules/synthetase/min (Loftfield, 1972). An intermediate, aminoacyl-AMP, has been identified and isolated. Its presence suggests that the aminoacylation reaction proceeds in two steps.

$$\text{Amino acid} + \text{ATP} \rightleftarrows \text{aminoacyl-AMP} + \text{pyrophosphate} \qquad (1)$$
$$\text{Aminoacyl-AMP} + \text{tRNA} \rightleftarrows \text{aminoacyl-tRNA} + \text{AMP} \qquad (2)$$

Alternatively, aminoacyl-AMP may be a dead-end product and the aminoacylation reaction may involve all three substrates in a concerted reaction (Loftfield and Eigner, 1969). It has not yet been possible to prove which alternative is correct and whether all synthetases act by the same mechanism.

The site of aminoacylation of tRNA depends on the specific synthetase examined (Ofengand, 1977). Some synthetases catalyze ester bond formation at the 2'-hydroxyl of the 3'-terminal adenosine of tRNA, while others attach the amino acid to the 3'-hydroxyl. However, at neutral pH an amino acid migrates rapidly from one hydroxyl to the other. The result is an equilibrium mixture, usually with about 70% attachment to the 3'-hydroxyl and 30% to the 2'-hydroxyl. Approach to equilibrium is very rapid, requiring only milliseconds.

b. **Specificity.** Synthetases attach the correct amino acid to their cognate tRNA's with an error rate of only 10^{-3} to 10^{-4}. The problem is to explain how a particular synthetase is able to discriminate so accurately between the specific amino acid or tRNA and structurally similar noncognate amino acids or tRNA's. As a class, synthetases exhibit very high affinity for their cognate amino acid, but in some cases the K_m for a noncognate amino acid is only 50- to 100-fold greater. Studies with homologous and heterologous synthetases and tRNA species indicate that affinities for noncognate tRNA's may be as high as for cognate tRNA's, but that reaction velocities are much greater with the correct tRNA. The results suggest that specificity is determined in part by the relative K_m's of the reaction substrates, but more critically during the catalytic step (V_{max} effect). With respect to the amino acid, two reactions are involved if aminoacyl-AMP is a true reaction intermediate. Thus discrimination could be made during binding of both the amino acid and the aminoacyl-AMP (K_m effects) and/or in the rate of formation of the aminoacyl-AMP and again in the rate of transfer of the amino acid to tRNA. Cognate tRNA could act as a positive allosteric effector during the first step. The chemical forces involved in amino acid binding are not known precisely, but are thought to be similar to those generally involved in the recognition of small molecules by enzymes.

The ability of synthetases to discriminate between tRNA's is even less well understood. The interactions of tRNA's and synthetases have been studied extensively in order to determine which part of the tRNA is recognized by the synthetase. Sequence comparisons between members of an isoacceptor family and between different families have not revealed a consistent pattern of recognition. tRNA modifications obtained either chemically or genetically have led to changes in specificity due to the substitution of a single base for another. Such studies (among numerous others) indicate that three separate tRNA loci are important in their interaction with synthetases: the acceptor stem, the D stem, and the anticodon. Some synthetases interact at all three loci, while others may recognize only one or two. The three loci are found on one side of the tRNA structure shown in Fig. 2 and are separated by about 78 Å. Thus a synthetase molecule which recognizes the anticodon must span this distance, but this is greater than the diameter of a globular protein the size of a common synthetase. Synthetase enzymes could be elongated, nonglobular proteins, but the known three-dimensional structures of synthetases do not support this notion. Alternatively, the conformation of tRNA complexed with synthetases may differ from the L-shaped structure found in crystals (Reid, 1977). It is necessary to compare crystallographic analyses of synthetase–tRNA complexes with analyses of the components alone to resolve this difficult problem. In addition, such analyses must be performed with a number of synthetases before generalizations can be made for all.

B. Ribosomes

Ribosomes were first observed as ribonucleoprotein particles in extracts of animal and bacterial cells in the 1940's, and then shown to function in protein synthesis by Zamecnik's group in the mid-1950's (Littlefield *et al.*, 1955). These complex organelles provide the stage or surface on which the process of protein synthesis occurs. Ribosomes catalyze the formation of covalent peptide bonds, but it is primarily the noncovalent interactions of their structural components with each other and with the various components of the translational machinery which are characteristic of ribosome function.

Ribosomes are divided into two classes: 80 S and 70 S ribosomes. The 80 S ribosomes are found in the cytoplasm of eukaryotic cells and contain about 70 different proteins and 4 RNA molecules. The smaller 70 S class are found in prokaryotic cells and in the eukaryotic organelles, mitochondria and chloroplasts, and are composed of fewer and smaller molecules: only 3 RNA's and about 55 proteins. The two classes of ribosomes are also distinguished by their sensitivity to antibiotics: the 70 S class is sensitive to chloramphenicol but not to cycloheximide; the 80 S class is sensitive to cycloheximide but not to chloramphenicol. All ribosomes dissociate into nonidentical subunits: 80 S particles produce 40 S and 60 S subunits; 70 S particles give rise to 30 S and 50 S subunits. Substantial progress has been made in defining the components and elucidating the three-dimensional structure of the subunits of 70 S and 80 S ribosomes, especially those from *Escherichia coli*. These results, together with the methods used and the general principles obtained, are described at some length below. Greater details about the structure of prokaryotic ribosomes and comprehensive references to original work are available in recent extensive reviews (Nomura *et al.*, 1974; Kurland, 1977; Stöffler and Wittmann, 1977; Brimacombe *et al.*, 1978). Work on eukaryotic ribosomes is only just beginning; progress to date is outlined here and is reviewed in detail by Bielka and Stahl (1978) and Wool (1979).

1. Escherichia coli Ribosomes

a. **Composition and Properties.** The 30 S subunit contains a single RNA molecule, called 16 S rRNA and 21 different proteins, named S1 to S21. The 50 S subunit is comprised of two RNA molecules, a 23 S rRNA and a 5 S rRNA and 34 proteins, L1 to L34. For both subunits, the RNA constitutes about two-thirds of the particle mass. Small-angle X-ray scattering

studies (Van Holde and Hill, 1974) indicate that 30 S particles are oblate
ellipsoids or flat discs (55 × 220 × 220 Å), while 50 S subunits are more
symmetric and appear as triaxial ellipsoids (130 × 170 × 260 Å). In contrast, electron microscopic analysis (Stöffler and Wittmann, 1977) of
dried, stained subunits show more complex shapes: the 30 S subunit (Fig.
3A) possesses two major domains, a smaller "head" separated by a
constriction from a larger "body," with overall dimensions of
80 × 100 × 200 Å; the 50 S subunit (Fig. 3B) is more globular, with dimensions 160 × 200 × 230 Å. Both particles are highly hydrated, containing 0.4 gm/gm of internal water (Van Holde and Hill, 1974). This suggests
that the particles may be quite "loose" and may be able to assume a
number of different conformations. Some of the properties of 70 S ribosomes and subunits are summarized in Table I.

The RNA components have been purified and characterized extensively. Some nucleosides of rRNA are covalently modified: 16 S rRNA
contains 10 methylated nucleosides and 23 S rRNA contains 14; two
pseudouridylic acid residues have been reported as well. RNA sequencing
procedures have been used to determine the primary structures of 5 S
rRNA (120 nucleotides) (Brownlee et al., 1968) and 16 S rRNA (1542
nucleotides) (Ehresmann et al., 1975; Carbon et al., 1978), while substantial progress has been made on the sequence of 23 S rRNA (3200 nucleotides) (Branlant et al., 1977). Recently the primary structure of 16 S
rRNA was determined by sequencing the DNA from a cloned fragment
containing the gene for 16 S rRNA (Brosius et al., 1978). The rapid DNA
sequencing techniques also are being used to determine the primary structure of 23 S rRNA. Analyses of the primary sequences of 5 S and 16 S
rRNA suggest that double-stranded regions of RNA may be present, and

TABLE I

Physical and Chemical Properties of Ribosomes

Particle	Mass (daltons)	RNA species	Number of proteins	Protein mass total mass
70 S	2.7 × 10⁶	3	53	0.33
30 S	0.9 × 10⁶	16 S	21	0.33
50 S	1.8 × 10⁶	23 S	34	0.33
		5 S		
80 S	4.3 × 10⁶	4	~75	0.49
40 S	1.4 × 10⁶	18 S	~33	0.56
60 S	2.9 × 10⁶	28 S	~45	0.41
		5.8 S		
		5 S		

physical studies confirm extensive secondary structure. Numerous models for rRNA secondary structure have been proposed, but there is little evidence available to judge the validity of these proposals.

All 55 ribosomal proteins have been purified to near-homogeneity. Most are small, basic proteins, the majority of which have molecular weights of less than 15,000. The weight average molecular weight determined by sodium dodecyl sulfate–polyacrylamide gel electrophoresis is 21,000 for both 30 S and 50 S proteins. The complete amino acid sequences of 33 proteins are known, and work on the remaining proteins is in progress (Brimacombe et al., 1978). The results to date indicate that each of the ribosomal proteins is distinctly different except that S20 and L26 are identical proteins; L7 and L12 are nearly identical, differing only in that L7 is acetylated at the N-terminus; and L8 is a complex of L7, L12, and L10. The uniqueness of the proteins means that ribosomal subunits are structurally dissymmetric, in sharp contrast to the high protein symmetry found in RNA virus particles. Hydrodynamic and small-angle X-ray scattering studies on individual ribosomal proteins purified by nondenaturing methods indicate that many of the proteins are elongated rather than globular in shape. The secondary structures for isolated proteins have been predicted on the basis of their primary sequences (Dzionara et al., 1977) and are being analyzed by circular dichroism and nmr techniques.

It is likely that all 53 of the identified ribosomal proteins are present in 70 S particles in one copy each, except for the proteins L7/L12, which are found in four copies (Hardy, 1975). However, a rigorous, unambiguous definition of the composition and stoichiometry of the 70 S ribosome has not yet been made. When ribosomes are isolated from cell lysates by differential centrifugation, a large number of proteins are found associated with the particles but can be removed by subsequent centrifugation in high salt buffers. This washing procedure also removes some ribosomal proteins, resulting in a heterogeneous population of ribosomes partially deficient in protein. Thus, a continuum of loosely to tightly associated protein exists, making the distinction between ribosomal proteins, ribosomal factors, and gratuitously associated proteins arbitrary by this criterion alone. Functional studies with proteins suspected of participating in protein synthesis are required to resolve these questions.

   b. **Reconstitution.** The *in vitro* reconstitution of active ribosomal subunits from their dissociated RNA and protein components was first accomplished in Nomura's laboratory for 30 S particles from *Escherichia coli* (Traub and Nomura, 1968) and for 50 S particles from *Bacillus stearothermophilus* (Nomura and Erdmann, 1970). Reconstitution of *E. coli* 50 S subunits has been reported more recently (Nierhaus and Dohme, 1974; Amils et al., 1978). These experiments establish that the information required for the assembly of ribosomal subunits is contained in the structures of the RNA and protein components.

Reconstitution techniques are also useful in studying the structure of the ribosome and the function of its components (Nomura and Held, 1974). When 16 S rRNA and 30 S ribosomal proteins are separated by urea–LiCl extraction procedures, only six or seven of the purified 21 proteins bind to 16 S rRNA alone. Some of the remaining proteins are then able to bind to 16 S rRNA complexes containing one or more of the "primary binding proteins." Such specific dependencies led Nomura to propose an "assembly map" which likely reflects protein–protein interactions in the particle (Mizushima and Nomura, 1970). Reconstitution experiments also have been used to construct 30 S or 50 S particles deficient in a single component. Loss of functional activity sometimes can be correlated with the specific involvement of the omitted protein, but this approach has not been so useful as expected, possibly because most ribosomal functions appear to involve the cooperative interaction of a number of different proteins. It is noteworthy that maximal functional activity requires the addition of one equivalent of most of the proteins, even those which are found in fractional amounts in salt-washed ribosomes.

c. **Structure.** In order to understand the molecular mechanisms of protein synthesis, it is necessary to elucidate in detail the three-dimensional structure of the ribosome. Ideally, resolution of the structure at the 2–3 Å level is required, as has been obtained for many enzymes. Since crystals of ribosomes have not yet been prepared which are satisfactory for X-ray diffraction studies, other methods have been used to elucidate the RNA and protein topography of the ribosomal subunits. Most progress to date has been made in defining the spatial relationship of ribosomal proteins by four different techniques: immune electron microscopy, protein crosslinking, fluorescence energy transfer and neutron scattering. These methods are briefly described below and have been reviewed in greater detail recently (Brimacombe et al., 1978).

Immune electron microscopy provides the most dramatic evidence for ribosome structure and the spatial arrangement of its proteins. This approach has been pursued vigorously in the laboratories of Stöffler and of Lake, and their results have been reviewed in detail recently (Brimacombe et al., 1978). The technique relies on two basic facts: suitably stained ribosomal subunits have readily discernable shapes when viewed by electron microscopy, and antibody molecules specific for individual ribosomal proteins are able to react with their cognate proteins in the intact particles. Subunit–antibody complexes are formed and analyzed. The position of reacting antibody is determined in relation to the shape of the subunit and a map of specific antigenic determinants is obtained. Locations of all of the 30 S protein and most of the 50 S proteins have been determined and are shown in Fig. 3. The results indicate that many of the proteins are greatly elongated in the particles, some antigenic

(A)   (B)

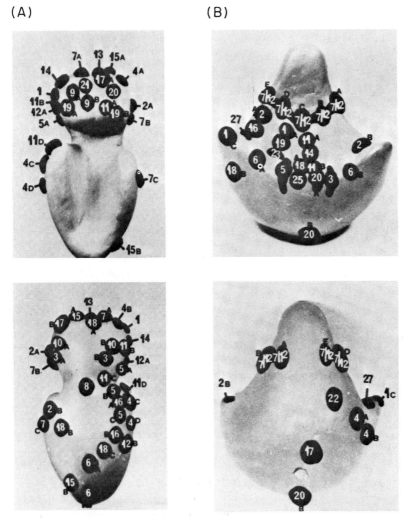

Fig. 3. Three-dimensional models of 30 S and 50 S ribosomal subunits. (A) The 30 S subunit. Front and back views are shown of a model based on images obtained by electron microscopy. The numbers indicate the locations of the centers of antibody binding sites for the 21 proteins. (B) The 50 S subunit. Front and back views, and the locations of antibody binding sites for 19 proteins, are shown. Taken from Brimacombe *et al.* (1978); reproduced, with permission, from *Annu. Rev. Biochem.* **47.** © 1978 by Annual Reviews Inc.

determinants of a single protein being separated by over 100 Å. The technique is particularly satisfying because the location of proteins can be visualized directly. Three difficulties are apparent, however: (1) it is difficult to evaluate the effect on shape that sample preparation imposes; (2) discerning the shapes of the subunits at different orientations is not obvious, resulting in differences in interpretation of similar data; and (3) the purity and specificity of the antibodies is critical. A recent report indicates that there are no exposed antigenic determinants for protein S4 and that the extended map positions for S4 are due to contaminating antibodies specific for other proteins (Winkelmann and Kahan, 1978). Nevertheless, this exciting technique has given the most extensive data concerning the topography of ribosomal proteins. Work is in progress to refine the analysis by relating specific antigenic determinants to regions in the primary structure of the proteins.

Protein–protein cross-linking with bifunctional reagents is being used to define the neighborhoods of ribosomal proteins. Subunits are treated with the reagent; dimers or higher oligomers are isolated, and the components are identified by immunochemical or electrophoretic techniques. More than 35 pairs of cross-linked proteins have been characterized for the 30 S subunit, and three-dimensional models have been proposed based primarily on cross-linking results (Traut et al., 1974; Sommer and Traut, 1976). Extensive progress in defining the protein topography of the 50 S subunit also is being made (Kenny and Traut, 1979). In general, the results obtained by these methods are consistent with the models based on immune electron microscopy. The cross-linking approach is appealing for its directness and versatility: bifunctional reagents differing in length or specificity can be employed. Interpretation of the results is complicated by the fact that the yield of cross-linked products is a function not only of the nearness of the proteins, but also of the reactivity of the amino acid side chains. If ribosomal particles are flexible rather than rigid, some crosslinks could arise between highly reactive proteins not normally located near each other. The formation of misleading crosslinks could also occur in protein-deficient subunits, since the ribosomes used in structural studies are heterogeneous. That cross-linked protein pairs are still functional has been demonstrated in only two cases, by reconstituting S5–S8 and S13–S19 dimers into active 30 S particles (Lutter and Kurland, 1973). Another difficulty is due to the elongated structures of the proteins themselves. Three proteins of the 30 S subunit, S4, S8, and S13, each have been crosslinked to eight other ribosomal proteins. For greater precision, analyses will need to identify where the cross-link occurs in the primary structures of the linked proteins.

Energy transfer between fluorescence labeled proteins has been used to measure the distance between proteins in ribosomal subunits (Cantor *et al.*, 1974). Two proteins are covalently labeled, each with a different fluorescence dye and the derivatized proteins are incorporated into particles by reconstitution techniques. The efficiency of transfer of energy from one dye to the other is measured, and the distance separating the dye molecules is calculated. If each protein is randomly derivatized, the measurements reflect the distance separating their centers of mass. Fourteen pairs in the 30 S subunit have been analyzed, and the results are generally consistent with the other methods described above.

The neutron scattering method (Moore *et al.*, 1977) is used to measure the distance separating the centers of mass of two deuterated proteins incorporated into 30 S particles by *in vitro* reconstitution. Eighteen pairs have been studied, and the results compare favorably with those of the other techniques. Analysis of the data provides information not only about distances but also about protein shape. Nearly all of the proteins appear to have axial ratios deviating significantly from unity. The centers of mass of RNA and protein within the particles can be determined also. RNA and protein appears rather uniformly distributed in the 30 S particle, but their centers of mass are displaced in the 50 S subunit. The results indicate that the 50 S particle is composed of an RNA core surrounded by a shell of protein (Stuhrmann *et al.*, 1977).

Little is known at this time about the secondary and tertiary structure of rRNA, which constitutes two-thirds of the ribosome's mass. Four general kinds of investigations are being made: (1) Specific regions in the sequence of rRNA which interact with individual ribosomal proteins are being identified (Brimacombe *et al.*, 1978). Nuclease-protected regions of RNA specific for single protein complexes are identified, or the composition of RNA and proteins is elucidated in subparticle fragments produced by limited nuclease digestion. (2) Regions of rRNA which are accessible to modification are determined in intact particles. Kethoxal attacks non-base-paired guanine residues and the location of reactive guanine bases in the 16 S rRNA sequence have been identified (Noller, 1974). (3) RNA–RNA cross-linking experiments (e.g., with psoralens) indicate that regions of 16 S rRNA are neighbors in the 30 S subunit but are greatly separated in the sequence. (4) Ribosomal proteins are being cross-linked to RNA and the region in the RNA sequence identified. Thus work on the structure of rRNA and how it is incorporated into ribosomal subunits is only just beginning.

From the sections above, it is clear that substantial progress has been made in elucidating the structure of the *E. coli* ribosome. The primary structures of all of the proteins and rRNAs will soon be known, and the spatial arrangement of the components is being determined at a resolution of 30–40 Å. Further details about ribosomal structure and function are given in Sections IV,A,4 and IV,B,3 below. These studies suggest that in general the ribosome functions by the cooperative interaction of its components; individual steps of protein synthesis involve parts of a number of structural components of the ribosome rather than, for example, a single ribosomal protein. In order to understand the molecular mechanisms of these reactions, considerable refinement of the models of ribosome structure is needed. More precise structural determinations are a formidable challenge for researchers in this field.

2. Other 70 S Class Ribosomes

Progress in the elucidation of the composition and structure of the ribosome is not nearly so extensive in species other than *Escherichia coli*. Limited studies indicate that the size and composition of most, if not all, prokaryotic ribosomes are similar. The small and large subunits have sedimentation coefficients of approximately 30 S and 50 S, and each contains 5 S, 16 S, and/or 23 S rRNA. The ribosomal proteins from different bacteria have been compared by polyacrylamide gel electrophoresis and are similar in number and molecular weight (Sun *et al.*, 1972), but differ significantly in charge (Geisser *et al.*, 1973). Proteins also have been tested for immunological cross-reactivity with antibodies against *E. coli* ribosomal proteins (Stöffler, 1974). Strong cross-reactivity was observed within the family Enterobacteriaceae, while significantly weaker cross-reactions were observed with other families, such as Bacillaceae. On the other hand, the degree of amino acid sequence homology is quite high between some of the *E. coli* and *Bacillus stearothermophilus* ribosomal proteins (Higo and Loertscher, 1974; Yagushi *et al.*, 1974), and the proteins are functionally interchangeable (Higo *et al.*, 1973). It is not yet proved that the functionally homologous proteins occupy the same relative positions in the three-dimensional structure of the two types of ribosomes.

Ribosomes found inside mitochondria or chloroplasts of eukaryotic cells are placed in the 70 S class because they are sensitive to chloramphenicol but not cycloheximide. However, their physical and chemical characteristics vary greatly among different phylogenetic groups. For example, the sedimentation coefficients ($s_{20,w}$) for mitochondrial ribosomes from fungi, protozoa and plants are 70 to 80, whereas those from animals are 55 to 60. A comprehensive review of mitochondrial ribosomes from a variety of sources is available (Neupert, 1977); the discussion which follows is limited to those from animals.

In spite of the small sedimentation coefficient for animal mitochondrial ribosomes, these ribosomes are similar in mass and dimensions to *E. coli*

70 S ribosomes and dissociate into 30 S and 40 S subunits. They differ dramatically from *E. coli* ribosomes in composition, however. The 30 S subunits contain one piece of RNA and about 40 proteins; 40 S subunits consist of a single piece of RNA and about 50 proteins. Two striking differences are evident: (1) mitochondrial ribosomes are composed mostly of protein, ranging between 65 and 75% protein, compared to 33% for *E. coli* ribosomes, and (2) they lack the 5 S species of rRNA. Since it appears that mitochondrial ribosomes participate in protein synthesis in the same way as other ribosomes, their tremendous differences in composition and structure are surprising and raise interesting questons about how ribosomes function.

3. Eukaryotic Ribosomes

Ribosomes from the cytoplasm of eukaryotic cells are larger than prokaryotic ribosomes, having a mass of about 4.5×10^6 daltons and a sedimentation coefficient of about 80 S. Like their bacterial counterparts, however, they are composed of two subunits. The smaller 40 S subunit has a mass of 1.5×10^6 daltons and is composed of 1 molecule of rRNA (18 S) and about 30 proteins. The larger 60 S subunit varies in size between species, from 2.4×10^6 daltons (plants) to 3.0×10^6 daltons (mammals), and contains 3 molecules of RNA (28 S, 5.8 S, and 5 S) and 35 to 50 proteins. The greater size of 80 S compared to 70 S ribosomes is due in large part to a larger number of proteins. The mass of 80 S ribosomes consists of about 50% protein compared to 33% for 70 S ribosomes. The functional significance of the higher protein content is unclear since both classes of ribosomes participate in similar reactions. In spite of the increased size and protein content, 40 S and 60 S subunits resemble their prokaryotic counterparts when analyzed by electron microscopy (Lutsch *et al.*, 1972; Lake *et al.*, 1974): the 40 S subunit is divided into "head" and "body" regions by a constriction; the 60 S subunit appears either rounder or more asymmetric and skiff-shaped.

Ribosomes from eukaryotic cells are found in two states: those free in the cytoplasm and those attached to the endoplasmic reticulum. It is generally agreed that the two kinds of ribosomes are equivalent in composition and structure. It is postulated that the information for binding polysomes to membranes is contained in the nascent protein; a region at its N- terminus, called the "signal," is recognized by a receptor in the endoplasmic reticulum. The signal hypothesis is described in greater detail by Lusis and Swank (Chapter 8) and Redman and Banerjee (Chapter 10) of this volume.

The properties and structure of 80 S ribosomes have been described in a number of recent, comprehensive reviews (Bielka and Stahl, 1978; Wool,

1979). We shall briefly examine the composition and structure of eukaryotic ribosomes and limit the scope of this chapter to mammalian species. Some properties of 80 S ribosomes are summarized in Table I.

a. **Composition.** The 80 S ribosomes contain four different molecules of RNA. The size of 18 S rRNA (0.7×10^6 daltons) is similar from numerous species, while that of the 28 S rRNA ($\sim 1.7 \times 10^6$ daltons) varies somewhat. Both rRNA's are extensively modified, the 28 S and 18 S rRNA from HeLa cells containing 71 and 46 methyl groups, respectively (Maden and Salim, 1974). Pseudouridine and other modified bases or sugars are found as well. Little sequence information is yet available on these rRNA's, however. It is noteworthy that a sequence of 8 nucleotides at the 3'-terminus of 18 S rRNA is identical in a wide variety of species, from yeast to mammals (Baralle, 1977): G-A-U-C-A-U-U-A-3'. Fingerprint analysis of oligonucleotide fragments reveals a high degree of similarity between species, which is confirmed by DNA–rRNA hybridization experiments. *Xenopus* 18 S rRNA has about 60% common base sequences with human 18 S rRNA, but no complementarity with prokaryotic rRNA. The sequence of 5 S rRNA (120 nucleotides) has been determined (Forget and Weissman, 1969) and is identical in human, mouse, rat, and rabbits. The 5.8 S rRNA's (150–160 nucleotides) also have been sequenced (Erdmann, 1979). It appears that the 5.8 S rRNA is functionally equivalent to the 5 S rRNA from prokaryotes (Wrede and Erdmann, 1977), and that the eukaryotic 5 S rRNA is the extra molecule of RNA in 80 S ribosomes.

Most of the approximately 70 proteins of the 80 S ribosome have been isolated and identified. Many of the proteins have been purified to near-homogeneity by classic protein fractionation techniques (Wool, 1979) or, on a smaller scale, by preparative two-dimensional polyacrylamide gel electrophoresis (Bielka and Stahl, 1978). Their molecular weights (rat liver) range from 11,200 to 31,100 (number average 21,400) for the small subunit and from 11,500 to 41,800 (number average 21,200) for the large subunit. Amino acid compositions are known for more than 50 proteins, and N-terminal sequences are being determined. Most are basic proteins, rich in lysine and arginine, although a number show acidic properties. All, or nearly all, are thought to be unique. Eukaryotic ribosomal proteins are, therefore, generally similar to those from bacteria, but differ from them primarily in being larger and more numerous.

A precise determination of the number of proteins in the mammalian ribosome has not yet been made. Estimates are based upon fractionation of the proteins by two-dimensional polyacrylamide gel electrophoresis. A number of protein nomenclatures were developed (e.g., Sherton and Wool, 1972; Welfle *et al.*, 1972) which differ because of different gel systems or species of ribosomes analyzed. A common nomenclature for mammalian ribosomal proteins has been proposed recently (McConkey *et al.*, 1979). Unfortunately, identification of ribosomal proteins by gel electrophoresis and by preparative purification procedures gives conflicting results. Proteins have been purified that were not identified in gels, and others apparent in the gel systems could not be isolated preparatively. Although some of the discrepancies may be trivial, it is clear that further work and careful protein chemistry are required to resolve these problems.

Ribosomal proteins from different tissues of the same animal show essentially identical two-dimensional polyacrylamide gel electrophoretic patterns. Comparison of patterns obtained from different species indicates that the ribosomal proteins are quite similar and have evolved slowly during vertebrate evolution. However, significant differences in a few proteins are detected when a number of related two-dimensional gel systems are used (Madjar *et al.*, 1979). A detailed immunochemical comparison has been made recently between the ribosomal proteins of chicken and rat (Fisher *et al.*, 1978). Antisera specific for the proteins of either species recognize only about 20% of common determinants, suggesting considerable differences in the proteins. Immunochemical comparison of ribosomal proteins from eukaryotic and prokaryote cells do not reveal any crossreacting proteins, with the exception of the prokaryotic proteins L7/L12 and their counterparts, L40/L41 (Stöffler *et al.*, 1974). *Escherichia coli* L7/L12 and an immunologically related acidic protein from yeast are functionally interchangeable, indicating high conservation of these proteins (Wool and Stöffler, 1974; Howard *et al.*, 1976).

Ribosomal proteins are phosphorylated both *in vitro* and in intact cells. A single 40 S protein is labeled when rats or rabbit reticulocytes are administered [^{32}P]phosphate. The extent of phosphorylation in rat liver is influenced by liver regeneration, the diabetic state, and insulin. Two acidic proteins P1 and P2 from the 60 S subunit are also phosphorylated (Tsurugi *et al.*, 1978). The phosphorylation of other ribosomal proteins is observed on infection of HeLa cells with vaccinia virus (Kaerlein and Horak, 1978), and changes are detected in ribosomes obtained from cells harvested in different physiological states. These results suggest that phosphorylation may influence the activity of the ribosome (see Chapter 2, this volume). A number of ribosomal proteins in both subunits are phosphorylated *in vitro* with a variety of protein kinases. No change in ribosome activity has been found due to phosphorylation, however.

b. Structure. Little information about the structure of eukaryotic ribosomes is available at this time. The rRNA's possess considerable secondary structure, and probably interact with ribosomal proteins in a manner similar to bacterial rRNA's. The topography of ribosomal proteins has not yet been studied extensively. The immune electron microscopic approach appears feasible, since suitable images of subunits are obtained, and specific antibodies are becoming available. The application of cross-linking techniques is beginning also. Those techniques requiring reconstitution of subunits are more limited, since the total reconstitution of eukaryotic ribosomes has not been reported. Partial stripping of proteins from particles and subsequent reconstitution into active subunits has been carried out, however.

Identification of ribosomal proteins involved in specific functional regions already has begun. A number of 60 S proteins have been identified which react with acyl-aminoacyl-tRNA derivatives (P site affinity reagents which mimic peptidyl-tRNA) (Czernilofsky *et al.,* 1977). Bielka and Stahl (1978) tentatively list specific proteins involved in soluble factor, tRNA, and mRNA binding and in the peptidyltransferase center. The site of binding of initiation factor eIF-3 on native 40 S subunits has been visualized by electron microscopy (Emanuilov *et al.,* 1978). Initiation factor eIF-2 has been cross-linked to rat liver ribosomal proteins S2, S3, and S18 (Westermann *et al.,* 1979). These results demonstrate that the application of techniques developed for bacterial ribosomes are suitable for the study of eukaryotic ribosomes and suggest that much new data on the structure of mammalian ribosomes can be expected soon.

C. Messenger RNA

Messenger RNA's (mRNA's) are transcribed from DNA and contain the genetic information which is translated on the ribosome into protein. In prokaryotic cells, RNA transcripts are frequently polycistronic, possessing a number of translational start and stop signals. Translation of prokaryotic mRNA's is coupled to transcription, i.e., protein synthesis occurs while the mRNA transcript is being synthesized. In eukaryotic cells, transcription occurs in the nucleus, and translation is uncoupled, occurring subsequently in the cytoplasm. Precursors to mRNA are synthesized (hnRNA) and are spliced and processed to smaller, monocistronic mRNA's which are then transported from nucleus to cytoplasm as messenger ribonucleoprotein (mRNP) complexes. Detailed descriptions of the properties and biosynthesis of mRNA's are outside the scope of this chapter. These topics are treated in Vol. III of this series, and in Chapters 3 and 4 of this volume. The general features of mRNAs which affect translation, and especially initiation of protein synthesis, are considered here.

1. Prokaryotic mRNA's

Early studies in bacterial translation relied heavily on synthetic mixed polynucleotides, especially those containing AUG or GUG codons, and more defined polymers such as $AUG(U)_n$. So-called "natural" mRNA's were limited to the RNA genomes of the bacteriophages $Q\beta$, MS2, and R17. The latter studies indicated that the secondary structure of viral RNA's plays a key role in masking false initiation sites or suppressing proper initiation at certain cistrons. Most studies on the mechanism of protein synthesis to date have utilized only the above mRNA's. This is unfortunate, because the translation of bacterial mRNA's may differ in some details from that of phage RNA's. Because the half-lives of mRNA's are generally very short, isolation of discrete transcripts is difficult or impractical. Were they available, the addition of complete mRNA's to *in vitro* protein synthesis assay systems would nonetheless not closely mimic the *in vivo* situation. Translation is coupled to transcription; therefore, the interaction of mRNA with ribosomes during the initiation phase is limited to the initiation region near the 5'-terminus of the cistron. Little or no interaction with distant parts of the mRNA is likely to occur, since these regions are thought to be either not yet synthesized or already covered by ribosomes.

The sequences of the initiator regions of many mRNA's are known (Steitz, 1979). Nearly all cistrons begin with the initiator codon AUG, a few utilize GUG, and intracistronic reinitiation following nonsense mutations is known to involve UUG and CUG as well. In nearly all cases (the λ phage repressor cistron is an exception), the initiator codon is located some distance from the 5'-terminus of the RNA. Most mRNA's contain a purine-rich region centering about 10 base pairs on the 5'-side of the AUG codon which is thought to interact with ribosomal RNA (see Section IV,A,3); the sequence GGAG is common to many mRNA's. At the end of the cistron, termination of protein synthesis is coded by the triplets UAA, UAG, or UGA. Often two adjacent termination codons are used. A variable number of nontranslated nucleotides are found between the termination codons and the 3'-terminus of the mRNA. In polycistronic mRNA's, the initiator codon for the next cistron is near, but not necessarily directly following, the termination codons of the preceding cistron.

2. Eukaryotic mRNA's

The study of eukaryotic translation benefits from the availability of numerous purified mRNA's which code for specific proteins. An extensive list of such mRNA's has been published recently (Bielka and Stahl, 1978). Eukaryotic mRNA's differ from bacterial mRNA's in many ways. As a class, they are more stable metabolically and appear to be invariably monocistronic. Since translation occurs in the cytoplasm, uncoupled from the site of transcription and processing, an mRNA first presents itself to ribosomes as a complete RNA. *In vivo,* mRNA's arrive in the cytoplasm complexed with proteins (mRNP's) (see Chapter 4, this volume). It is

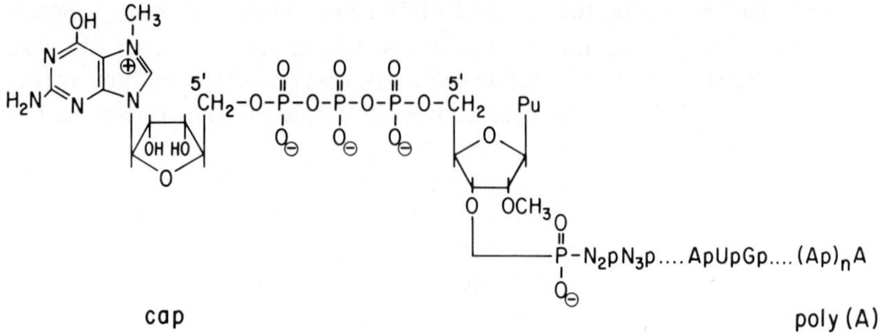

Fig. 4. Structure of eukaryotic mRNA's. The chemical structure of the "cap" located at the 5'-terminus of most eukaryotic mRNAs is shown. Pu stands for purine; N is any base. The initiator codon, AUG, is shown, separated by variable distances (dotted lines) from the cap structure of the polyadenylic acid sequence shown at the 3'-terminus of the mRNA.

should be noted that most *in vitro* studies of the mechanism of protein synthesis have utilized mRNA's free of protein.

Certain structural features are shared by most eukaryotic mRNA's. The mRNA's are "capped" at the 5'-terminus by an inverted 7-methylguanosine attached at its 5'-hydroxyl by a triphosphate bridge to the 5'-hydroxyl of the RNA (see Fig. 4). The first two bases of the RNA are usually methylated purines. Eukaryotic mRNA's are methylated internally at a frequency of about one methyl group per 500 nucleotides. Finally, the 3'-terminus may possess a poly(A) tail of 50–200 residues. Capping and polyadenylation take place posttranscriptionally in the nucleus or in the cytoplasm. The "cap" structure is functionally important during initiation (see Section IV,A,3). The cap and the poly(A) tail protect the mRNA from exonucleases, thus conferring stability to the RNA. Nearly all mRNA's, either cellular or viral, are capped; however, picornavirus mRNA's lack the cap structure. Most cellular and viral mRNA's are polyadenylated at the 3'-terminus, although histone mRNA's and a sizable proportion of other mRNA's lack the poly(A) tail (Grady *et al.*, 1978). The structures and properties of mRNAs have been reviewed recently (Shatkin, 1976; Kozak, 1978; Revel and Groner, 1978).

D. Soluble Protein Factors

Soluble factors are proteins which promote the various reactions comprising protein synthesis. Although most factors act while bound to the ribosome, they are differentiated from the structural ribosomal proteins by the fact that they are only transiently associated with ribosomes at specific times during protein synthesis and are subsequently released. Factors are categorized by the particular phase of protein synthesis which they promote. Thus there are initiation, elongation, and termination factors. In order to identify these factors, *in vitro* assay systems of protein synthesis, or partial reactions thereof, have been developed and the various factor proteins have been isolated and purified. A large number of discrete polypeptides have been purified from mammalian cells: about 20 for initiation, 4 for elongation, and 1 for termination. Bacterial factors are less numerous, there being only three polypeptides identified for each of the three phases of protein synthesis. These factors are listed in Tables II and III and are described in detail in the sections below.

1. Initiation Factors

a. **Prokaryotic Factors.** Bacterial initiation factors, especially from *Escherichia coli,* have been studied extensively and three classes of factors have been isolated: IF-1, IF-2, and IF-3. In crude cell extracts the factors appear to be localized primarily on native 30 S ribosomal subunits and cosediment with ribosomes on centrifugation. However, they may be dissociated from ribosomes in high salt buffers. Numerous assays have been employed for their detection, the most common being the stimulation of fMet-tRNA binding to 70 S ribosomes in the presence of synthetic or viral mRNA's. Their purification involves classic methods, and essentially homogenous preparations are obtained (Wahba and Miller, 1974; Dondon *et al.*, 1974; Hershey *et al.*, 1977; Voorma *et al.*, 1979). Functionally, IF-2

TABLE II

Soluble Protein Factors from *Escherichia coli*

	Molecular weight	Mole factor per mole ribosome	Function
I. Initiation			
IF-1	9,000	0.18	Promotes IF-2 and IF-3 functions
IF-2a	115,000	0.15	Binds fMet-tRNA; GTPase
IF-2b	90,000		
IF-3	21,000	0.14	mRNA binding; dissociation
II. Elongation			
EF-Tu	44,000	5–7	Binds AA-tRNA; GTPase
ET-Ts	30,000	1	GTP-GDP exchange on EF-Tu
EF-G	80,000	1	Translocation; GTPase
III. Termination			
RF-1	44,000	0.02	Recognizes UAA and UAG
RF-2	47,000	0.02	Recognizes UAA and UGA
RF-3	46,000	—	Promotes RF-1 and RF-2; GTPase

TABLE III

Soluble Protein Factors from Rabbit Reticulocytes

	Molecule weight	Number of polypeptides	Functions
I. Initiation			
eiF-1	15,000	1	Promotes mRNA binding
eIF-2	150,000	3	Forms ternary complex with Met-tRNA
eIF-3	~700,000	~9	Promotes Met-tRNA and mRNA binding; dissociation
eIF-4A	49,000	1	Promotes mRNA binding
eIF-4B	80,000	1	Promotes mRNA binding
eIF-4C	17,500	1	Promotes Met-tRNA binding
eIF-4D	16,500	1	Stimulates Met-puromycin synthesis
eIF-5	150,000	1	Required for 80 S complex formation
Co-EIF-2	22,000	1	Stimulates ternary complex formation
Cap binding	24,000	1	Binds to cap of mRNA
II. Elongation			
EF-1_α	~55,000	1	Forms ternary complex with AA-tRNA; binds to Rb; GTPase
EF-1_β	30,000	1	Promotes exchange of GTP/GDP on
EF-1_γ	55,000	1	EF-1_α
EF-2	100,000	1	Translocation; GTPase
III. Termination			
RF	56,500	1	Promotes termination; GTPase

associated with the binding of fMet-tRNA to the 30 S ribosomal subunit, and IF-3 is involved in mRNA binding and ribosome dissociation. IF-1's function is less clear; it appears to enhance the action of the other two factors.

IF-1 and IF-3 are small, basic, thermostable proteins with molecular weights of about 9,000 and 21,000, respectively. Two-dimensional gel electrophoresis indicates that there are two molecular forms of IF-3 (Suryanarayana and Subramanian, 1977). The sequences of both forms of IF-3 have been reported recently (Brauer and Wittmann-Liebold, 1977) and show that the two forms have almost identical primary structures, one lacking six amino acid residues at the N terminus. Curiously, a minor form of IF-3 methylated on the N-terminal methionine was also detected. In contrast to IF-1 and IF-3, IF-2 is a large, acidic, heat-labile protein. Two molecular weight forms have been isolated: IF-2a, $M_r = 115,000$; and IF-2b, $M_r = 90,000$. The various forms of IF-2 and IF-3 appear to have identical activities in most assays for initiation (see Section IV,A,1); however, in a highly purified DNA-coupled assay system, IF-2a, but not IF-2b, stimulated the synthesis of active β-galactosidase (Eskin *et al.*, 1978). Some of the physical and functional properties of the three initiation factors are given in Table II and are described in greater detail in numerous reviews (Hazelkorn and Rothman-Denes, 1973; Revel, 1977; Grunberg-Manago *et al.*, 1978).

Antibodies specific for the three initiation factors have been made in rabbits (Howe *et al.*, 1978). The initiation factors are antigenically distinct from each other and from ribosomal proteins and other molecules in crude cell lysates. A thermosensitive mutant with a thermolabile IF-3 has been isolated and characterized recently (Springer *et al.*, 1977). This is the only example of the identification of a mutant gene for an initiation factor.

b. Eukaryotic Factors. The process of initiation in eukaryotic cells is more complex than in prokaryotic cells, as dramatically shown by the large number of initiation factors discovered. Beginning with the pioneering studies of Staehelin and Anderson, ten different factors have been purified or identified to date, and it appears likely that still more will be discovered in the future. Initiation factors have been studied from a variety of cell types, but most intensively from rabbit reticulocytes, where eight factors have been purified to near-homogeneity by techniques similar to those used for prokaryotic factors (Safer *et al.*, 1976; Schreier *et al.*, 1977; Benne *et al.*, 1978a). The factors, named eIF-1, eIF-2, eIF-3, eIF-4A, eIF-4B, eIF-4C, eIF-4D, and eIF-5, stimulate the synthesis of globin or methionylpuromycin in cell-free systems dependent on globin mRNA. Two additional proteins are implicated in the initiation process: a 24,000 dalton protein which appears to bind to the cap portion of mRNA (Sonenberg *et al.*, 1978) and a 22,000 dalton protein, called Co-EIF-1 (Dasgupta *et al.*, 1976), which interacts with eIF-2 during ternary complex formation.

Of the ten initiation factors, eight are single polypeptide chains and two are complex factors. eIF-2 is composed of three, and eIF-3 of at least nine different polypeptides. The molecular weights, subunit structures, and functions are shown in Table III. The factor polypeptides range in size from 15,000 to 150,000 daltons, and both acidic and basic proteins are found. Together they comprise 20 nonidentical polypeptide chains, totaling about 1,200,000 daltons of proteins. None of the factor polypeptides corresponds to any of the proteins found in ribosomal subunits previously washed in high salt buffer (Benne *et al.*, 1978a).

The complex factor eIF-3 is a large, multicomponent protein (Benne and Hershey, 1976). It is thought to contain nine major polypeptides ranging in molecular weight from 28,000 to 140,000, two of which are present in two copies each. The sum of the molecular weights of the eleven subunits is 724,000 daltons, consistent with the factor's sedimentation properties. The number and stoichiometry of the components changes little during fractionation on ion exchange columns, density gradients or nondenaturing gels, and all of the major components bind to 40 S ribosomal subunits. Similar factors are found in other mammalian species and in wheat germ. Nevertheless, an adequate characterization of eIF-3 has not yet been made. It is not known whether all of the major polypeptides are necessary for eIF-3 function and whether any of the subunits is derived from another by limited proteolysis. It is uncertain whether the eIF-3 complex is unique or heterogeneous in composition; different forms of the factor could possibly discriminate between different classes of mRNA (see Section IV,A,3).

eIF-2 is less complex, containing three nonidentical polypeptides. The stoichiometry of the three proteins is probably 1 : 1 : 1, although it is possible that two copies of the smallest polypeptide are present. Like eIF-3, the eIF-2 complex retains its integrity throughout initiation. All of the subunits bind together to the 40 S subunit and release simultaneously during 80 S initiation complex formation (Benne and Hershey, 1978).

Eukaryotic initiation factors may be covalently modified. Three of the factors, eIF-2, eIF-3, and eIF-4B, are phosphorylated in intact rabbit reticulocytes (Benne *et al.*, 1978b), probably by a cyclic-AMP-independent protein kinase (Issinger *et al.*, 1976; Traugh *et al.*, 1976). Inhibition of protein synthesis in reticulocyte lysates by hemin deprivation or addition of double-stranded RNA is correlated with the phosphorylation of eIF-2 (Revel and Groner, 1978; Chapter 2, this volume). Ilan and Ilan (1976) reported that eIF-3 is a glycoprotein, and other initiation factors may be glycosylated as well.

2. Elongation Factors

a. **Prokaryotic Factors.** Two peptide chain elongation factors in *Escherichia coli* were first identified in Lipmann's laboratory (Allende *et al.*, 1964). The elongation factors, named EF-T and EF-G, promote the binding of aminoacyl-tRNA to ribosomes and the translocation of peptidyl-tRNA and mRNA on ribosomes, respectively. Each factor also stimulates a ribosome-dependent GTP hydrolysis reaction. The structures and functions of these factors have been reviewed in detail (Brot, 1977; Miller and Weissbach, 1977).

EF-T activity is often assayed in a protein synthesis system containing labeled Phe-tRNA, washed ribosomes, poly(U), and EF-G. Two homogeneous proteins have been purified from the postribosomal supernatant fraction of cell lysates: EF-Tu, which is heat labile, and EF-Ts, which is heat stable (Gordon *et al.*, 1971; Miller and Weissbach, 1974). EF-Tu interacts with a variety of molecules. It forms a binary complex with either EF-Ts or with the guanine nucleotides, GDP or GTP. The high affinity of EF-Tu for GDP provides a convenient assay for the factor: the binding of radioactive GDP. The EF-Tu · GTP complex binds aminoacyl-tRNA to form a ternary complex, which in turn binds to the ribosome. In addition to these noncovalent binding sites, EF-Tu carries the active catalytic site for GTP hydrolysis (see Section III,B,1 for details of EF-Tu function). In spite of the complexity of its interactions, EF-Tu is not a large protein; its molecular weight is about 44,000. EF-Tu contains a Zn^{2+} atom and three free cysteine residues, one of which is required for GDP or GTP binding, another for aminoacyl-tRNA binding. The factor alone is very unstable, but binary complex formation with EF-Ts, GDP, or GTP protects the protein from thermal denaturation. EF-Tu · GDP, EF-Tu · GTP, and EF-Tu · EF-Ts complexes crystallize readily in forms suitable for X-ray diffraction analyses. The primary sequence and three-dimensional structure of EF-Tu are under investigation, as is the structure of EF-Tu in complexes with aminoacyl-tRNA, GTP, and/or GDP (Morikawa *et al.*, 1978).

EF-Ts functions by catalyzing the rate of exchange of guanine nucleotides bound to EF-Tu. This forms the basis of a simple assay for the factor. The molecular weight of EF-Ts is about 30,000, and the protein contains two free cysteine residues, one being essential for its binding to EF-Tu. Temperature-sensitive mutants of EF-Tu and EF-Ts have been reported (Gordon *et al.*, 1972).

EF-G catalyzes the hydrolysis of GTP in a ribosome-dependent reaction and acts during the translocation reaction (see Section III,B,1). It is assayed either by its GTPase activity or by its ability to complement EF-T in polyphenylalanine synthesis. The factor is a single polypeptide with a molecular weight of about 80,000. It is an acidic protein and is especially sensitive to thiol reagents. The factor has been purified to homogeneity and has been crystallized (Kaziro and Inoue, 1968; Parmeggiani, 1968). Two kinds of EF-G mutants have been described: those isolated from a fusidic acid-resistant organism (Kinoshita *et al.*, 1968) and those containing a temperature-sensitive EF-G (Tocchini-Valentini and Mattoccia, 1968).

b. Eukaryotic Factors. Comparable factors have been isolated from eukaryotic cells (Miller and Weissbach, 1977) as described first for rat liver by Fessenden and Moldave (1961) and for rabbit reticulocytes by Bishop and Schweet (1961). Elongation factor EF-1, analogous to EF-Tu, has been isolated and purified from a wide variety of tissues. It exists in multiple forms ranging in molecular weight from 5×10^4 to greater than 1×10^6. The heavier forms (EF-1$_H$) are aggregates of the light form (EF-1$_L$ or EF-1$_\alpha$) and often contain lipid material. Both forms are active in promoting the binding of aminoacyl-tRNA to ribosomes, but EF-1$_L$ binds guanine nucleotides more tightly and to a greater extent. EF-1$_L$ isolated from different species is similar, all having molecular weights between 50,000 and 60,000. The protein is unusual in that it contains no free cysteine.

EF-1 forms relatively stable binary complexes with GTP or GDP. The guanine nucleotides bind better to the light forms than to the heavy forms, and can cause the dissociation of the heavy forms to EF-1$_L$. In contrast to EF-Tu, EF-1 binds GTP somewhat more tightly than GDP. Exchange of EF-1-bound GDP for GTP is catalyzed by another factor called EF-1$_{\beta\gamma}$, purified from pig liver (Motoyoshi and Iwasaki, 1977). This factor, analogous in function to EF-Ts, is composed of two subunits, EF-1$_\beta$ (MW 30,000) and EF-1$_\gamma$ (MW 55,000). A ternary complex containing one molecule each of EF-1$_\alpha$, aminoacyl-tRNA, and GTP can be isolated. The heavy forms, EF-1$_H$, are not found in such complexes. Although EF-1 is a very heat-labile protein, its stability increases dramatically in the ternary complex.

Elongation factor EF-2 is analogous to the bacterial factor, EF-G. It stimulates three reactions: polypeptide elongation in the presence of EF-1; translocation of peptidyl-tRNA from the puromycin-insensitive to the puromycin-sensitive state; and ribosome-dependent GTP hydrolysis. EF-2 has been purified to near homogeneity from rat liver (Galasinski and Moldave, 1969) and a variety of other sources. The factor is a large, acidic protein comprised of a single polypeptide with a molecular weight of about 100,000. The mammalian factor differs from its prokaryotic analogue in that it forms stable binary complexes with guanine nucleotides. The nucleotides compete with each other for a single binding site in the factor, GDP binding about ten times more tightly than GTP (Mizumoto et al., 1974; Henriksen et al., 1975). Another interesting feature of EF-2 is its covalent modification and inhibition by diphtheria toxin. The toxin catalyzes the transfer of the ADP-ribose portion of nicotinamide adenosine diphosphate (NAD) to a specific site in EF-2 (Honjo et al., 1968).

3. Release Factors

a. **Prokaryotic Factors.** The termination codon-specified hydrolysis of nascent peptidyl-tRNA and release of the peptide from the ribosome is catalyzed by protein factors found in the postribosomal supernatant fraction (Ganoza, 1966). Two different factors have been purified to near-homogeneity: RF-1 (MW 44,000) recognizes the termination codons UAA and UAG but not UGA; RF-2 (MW 47,000) is active with UAA and UGA (Milman et al., 1969; Klein and Capecchi, 1971). The factors may be assayed either (a) by peptide release with an amber mutant of bacteriophage R17 containing the nonsense termination codon UAG following the sixth codon in the coat protein cistron (Capecchi, 1967); (b) by formylmethionine release from ribosome-bound fMet-tRNA$_i$ with the triplets UAA, UAG or UGA; or (c) by stimulation of termination triplet binding to ribosomes. A third factor, RF-3 (MW 46,000), stimulates the action of RF-1 and RF-2, but has no release activity of its own (Capecchi and Klein, 1969; Milman et al., 1969). In contrast to RF-1 and RF-2, RF-3 binds GTP or GDP.

b. Eukaryotic Factors. Mammalian cells appear to contain only a single release factor which recognizes all three termination codons (Tate et al., 1973). Release factors (RF's) from a variety of mammalian sources function similarly. The RF from rabbit reticulocytes is a dimer composed of identical subunits of molecular weight 56,500 (Caskey, 1977). The factor catalyzes the hydrolysis of GTP in a ribosome-dependent reaction stimulated by tetranucleotides containing the termination codon sequences.

4. Overview

The soluble protein factors isolated and purified from prokaryotic cells (*Escherichia coli*) and mammalian cells (rabbit reticulocytes) are listed in Tables II and III. Are the identified components both necessary and sufficient for protein synthesis *in vivo?* These two important questions are discussed in light of the material presented above.

1. Have all of the factors been identified? The only approach available for studies of protein synthesis components is to identify and purify all factors which stimulate any of the available assay systems. When all of the components have been purified, it should be possible to construct a totally defined, active system for protein synthesis. Impressive advances along these lines have been made in Weissbach's laboratory (Kung et al., 1977), where a DNA-linked protein synthesis system has been constructed from highly purified bacterial components. Unfortunately the reaction rates obtained *in vitro* are much slower (tenfold or more) than the rates of protein synthesis in intact cells. This may be due either to the difficulty in optimizing so complex a system or to the absence of required components. It is, therefore, impossible at this time to conclude rigorously that all of the components for protein synthesis have been identified from either bacterial or eukaryotic cells.

2. Is each factor necessary for protein synthesis *in vivo?* Althou h each of the factors shows stimulatory activity in one or more of the assays devised, this does not alone constitute proof that the factor actually functions during protein synthesis in intact cells. Assay systems, especially

with highly purified components, are innately artificial, and stimulation by some components may be artifactual. One way to approach this problem is to obtain conditional mutant genes for the factors and show that protein synthesis is affected at the nonpermissive condition. Such mutants have been identified for two of the bacterial elongation factors (Tocchini-Valentini and Matoccia, 1968; Gordon *et al.*, 1972) and for one of the bacterial initiation factors (Springer *et al.*, 1977). Mutant genes for bacterial protein synthesis factors have been hard to obtain, and no examples are yet known from mammalian cells. A second approach is to use purified antibodies against single components and test whether protein synthesis is inhibited in crude cell lysates. Antibodies against bacterial elongation factors and an initiation factor inhibit protein synthesis, but this promising approach has not been applied to eukaryotic cells. As yet, only a few of the soluble protein synthesis factors are proved to be necessary.

III. BIOSYNTHESIS OF TRANSLATIONAL COMPONENTS

The cellular levels of ribosomes and associated macromolecules define the potential capacity of the cell for protein synthesis. Since the translational machinery is large and, therefore, relatively expensive to produce (ribosomes account for about 40% of the cell mass of rapidly growing bacteria, for example), it is reasonable to expect that levels of translational components are strictly regulated and are not present in excess of cell need. This appears to be true in bacteria growing exponentially in rich medium, where ribosome synthesis may be growth rate limiting. The level of ribosomes in mammalian cells reflects the differentiated characteristics of the cell and is high in those cells which synthesize a great deal of protein.

A fundamental understanding of the regulation and mechanism of protein synthesis, therefore, includes knowledge of the cellular levels of the translational machinery and the mechanisms of their synthesis and assembly into functional components. Of special interest is the absolute concentration of ribosomes and the ratios of other components, such as tRNA's, synthetases, and soluble factors to the ribosomes. Are these values different for cells in different physiological states? Is synthesis of the various components coordinately controlled? What processing is required for full functional activity? Finally, how are the genes for translational components organized in the genome? Considerable progress has been made toward answering these questions in bacterial systems, whereas little information has been developed for eukaryotic cells. I shall, therefore, focus attention on studies of bacteria, especially *Escherichia coli*.

A. Ribosomes

1. Prokaryotes

A rapidly growing bacterial cell contains about 30,000 ribosomes per genome. This is equivalent to an absolute concentration in the order of $10^{-5} M$. The 30 S and 50 S ribosomal subunits are assembled from newly synthesized RNA and proteins (Schlessinger, 1974). The two large rRNA's are synthesized as precursors, called p16 S and p23 S, from a single transcript, named p30 S. The p16 S is located at the 5'-end of the transcript and is cleaved during the synthesis of the spacer region separating p16 S from p23 S (for recent reviews, see Perry, 1976; Chapter 9 in Volume III). A precursor of 5 S rRNA is synthesized separately. The precursor forms are covalently modified (e.g., by methylation) and associate immediately with ribosomal proteins. Assembly of the particles is likely ordered and probably is similar to the *in vitro* assembly process discussed in Section II,B,1. Secondary trimming of p16 S and p23 S by endonucleases occurs in the so-called preribosomal particles. Free ribosomal proteins are present in cells at low levels and are synthesized from mRNA's in the normal manner. Most of the ribosomal proteins are produced in the same form as found in mature particles; large polyprotein precursors are not found, but a precursor form of S20 has been identified (Machie, 1977).

The synthesis of many ribosomal proteins is coordinated with rRNA synthesis (Kjeldgaard and Gausing, 1974; Nierlich, 1978). During steady-state growth, the rates of rRNA and ribosomal protein synthesis are comparable. When cells are placed in a richer medium (shift-up experiments), both components increase rapidly. When the cells are partially starved for an essential amino acid, the synthesis of both rRNA and ribosomal proteins decreases (stringent response mediated by ppGpp). The coordinate control of the synthesis of ribosomal proteins occurs at the transcriptional level. Whether or not ribosomal proteins function as transcriptional repressors of their own gene expression is unclear.

2. Eukaryotes

The synthesis of eukaryotic ribosomal components and their assembly into ribosomal subunits appear to follow the same general pathway found in bacterial cells (Warner, 1974). The major difference is that the site of rRNA synthesis and subunit assembly is the nucleolus. The large rRNA's are transcribed as a single precursor transcript of about 45 S and are processed by endonucleolytic cleavage and covalent modification to mature 5.8 S, 18 S, and 28 S rRNA's; 5 S rRNA is transcribed separately. The ribosomal proteins are synthesized in the cytoplasm, migrate to the nucleolus, and associate with the rRNA. Synthesis of ribosomal components occurs throughout interphase and represents a significant (up to 5%) proportion of a cell's total protein synthesis. Regulation of ribosome biogenesis is poorly understood, but the major critical events appear to be posttranscriptional (Hadjiolov, 1977).

B. Other Prokaryotic Components

Transfer RNA's as a group comprise 12–18% of total bacterial RNA in rapidly growing cells. This level corresponds to about 10 tRNA's per ribosome. In more slowly growing bacteria the ratio of tRNA to ribosomes is somewhat greater, and up to 25% of the total RNA is tRNA. This implies that tRNA and rRNA synthesis are regulated by somewhat different mechanisms. However, coordinate synthesis of tRNA's and rRNA is observed in shift-up experiments and during the stringent response (Morgan and Söll, 1978; Nierlich, 1978).

Specific protein levels in crude cell extracts may be readily measured by two-dimensional polyacrylamide gel isoelectric focusing techniques or by immunochemical approaches. These methods have been used to study aminoacyl-tRNA synthetase levels (Neidhardt et al., 1975). Values of about 1 specific aminoacyl-tRNA synthetase enzyme per 50 ribosomes are obtained for rapidly growing cells. Synthetase levels decrease with decreasing steady-state growth rate and increase dramatically in shift-up experiments, generally following ribosome levels. Most synthetases appear not to be under stringent control, however, but may be regulated by specific metabolites. When the concentration of the cognate amino acid is limited, the activity of some synthetases rises either transiently or more long lastingly. The amino acid seems to derepress the synthetase gene. No uniform pattern of regulation appears to apply to all of the synthetases of a bacterial cell; a variety of complex control mechanisms may be involved. The recent isolation of mutants altered in synthetase regulation should be of great assistance in these studies.

The cellular levels of initiation, elongation, and termination factors have been determined by immunochemical and gel electrophoretic methods. EF-G and EF-Ts are present at about 1 molecule per ribosome, while EF-Tu comprises about 5% of the cell's protein and is present in 5 to 7 copies per ribosome (Furano and Wittel, 1976). Thus EF-Tu and tRNA are present in approximately stoichiometric amounts. Elongation factor synthesis is under stringent control and is coordinately regulated with ribosomal synthesis at different growth rates (Miyajima and Kaziro, 1978). Initiation factor levels have been determined by a radioimmunoassay in lysates of rapidly growing cells. The levels of IF-1, IF-2, and IF-3 are equimolar, each present is about 1 copy per 7 ribosomes (Howe et al., 1978). Thus the levels of initiation factors are substantially below the levels found for elongation factors, but appear to be in slight excess over the number of native ribosomal subunits found in cells. The abundance of termination factors has been estimated to be 1 each per 50 ribosomes (Klein and Capecchi, 1971), thereby approximating the levels found for the synthetases.

C. The Organization of Genes for Translational Components

The coordinate expression of the genes for ribosomal proteins, tRNA's, and some soluble protein factors suggests that many of these elements may be present on the same operon. Mutant genes for a number of the ribosomal proteins, many synthetases, the elongation factors and initiation factor IF-3 have been identified and mapped (Nomura and Morgan, 1977). These genes are located throughout the genome, although many are clustered together in specific regions. With the advent of recombinant DNA techniques, tremendous strides have been made in defining the order of the translational genes and their organization into transcriptional units or operons (Nomura, 1976). These studies, coupled with rapid DNA sequencing techniques, will soon lead to the elucidation of the sequences of the genes and their products, and of the controlling elements for each operon.

The map positions for many E. coli translational components are shown in Fig. 5, and the results are generalized as follows. The genes for rRNA occur in about 6 or 7 copies distributed around the genome. In each operon, the 16 S region precedes the 23 S region and is separated by a different tRNA gene. tRNA genes are frequently clustered in small groups. Ribosomal proteins are grouped into rather large polycistronic operons, some of which contain genes for the elongation factors and RNA polymerase subunits. The exceptionally high levels of EF-Tu are explained in part by there being two genes for this protein. The aminoacyl-tRNA synthetase genes are scattered throughout the genome; their organization into transcriptional units is not known. The IF-3 gene has been obtained on a transducing phage which also carries the genes for the phenylalanine synthetase subunits (Springer et al., 1977), but it is not known whether the factor and synthetase proteins are linked transcrip-

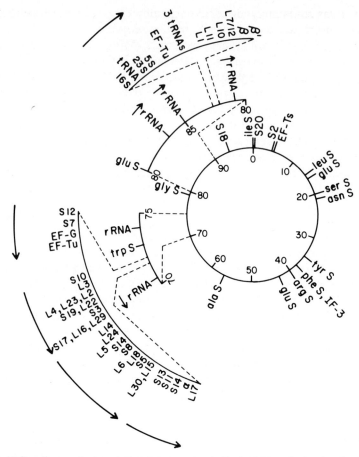

Fig. 5. Genetic map for translational components in *Escherichia coli*. On the circular map and its expanded portions are shown the locations of the genes for ribosomal RNA (rRNA); ribosomal proteins (e.g., S20, L11); aninoacyl-tRNA synthetases (e.g., tyr S); only a few of the tRNA's known; soluble factors for protein synthesis; and the α, β, and β' subunits of DNA-dependent RNA polymerase. The arrows indicate the direction of transcription of identified operons. Most of the information was obtained from Nomura and Morgan (1977) and Morgan and Söll (1978).

tionally. The other initiation factor and termination factor genes have not yet been identified.

Since tools are available to identify the genes for essentially all of the translational components, whether or not mutant genes are found, we can anticipate new precise informaton about the mapping and organization of all of these genes. Recombinant DNA techniques can also provide pure

DNA fractions containing one or more specific operons which can be used to study transcriptional activity and regulation.

IV. PATHWAY AND MECHANISM

A. Initiation

Initiation of protein synthesis is the process whereby an initiation complex containing the unique initiator tRNA, mRNA, and the ribosome forms and enters the elongation phase of translation. The general pathway of initiation is quite similar in both prokaryotic and eukaryotic cells. Binding of the initiator methionyl-tRNA and mRNA to the small ribosomal subunit is promoted by initiation factors. This preinitiation complex is joined by the large ribosomal subunit; initiation factors are released at various stages; and the initiation complex is formed. Of critical importance are the interactions which assure the proper recognition and positioning of the mRNA so that protein synthesis begins at the correct codon and that translation of the message is in phase. Elucidation of the molecular mechanism of initiation is also required in order to understand how translation is controlled. The pathways and some aspects of mechanism are described below for bacterial and mammalian cells. Detailed reviews of recent advances in this area contain references to much of the work discussed below (Revel, 1977; Grunberg-Manago *et al.,* 1978; Revel and Groner, 1978; Safer and Anderson, 1978).

1. The Prokaryotic Pathway

The mechanism of initiation in *Escherichia coli* has been studied extensively and the components have been identified (see Section II,D,1). A tentative pathway for initiation is given in Fig. 6A. In the first step, IF-1 increases the rate of dissociation of 70 S ribosomes, while IF-3 binding to 30 S subunits prevents the reassociation of the subunits. IF-2 also joins the 30 S ribosomal subunit and a trifactor complex is formed. In the next step, mRNA, fMet-tRNA$_i$, and GTP bind to the trifactor complex in an unknown order (see below). IF-3 is released following fMet-tRNA$_i$ binding. The presence of the other components in the 30 S preinitiation complex has been shown directly with radioactively labeled molecules. In the next step, the 50 S subunit joins the 30 S complex with the concomitant release of IF-1. In the final step GTP hydrolysis occurs and IF-2 is ejected along with GDP and P$_i$. The role of GTP hydrolysis is unclear, but may involve stimulation of the rate of release of IF-2. The GTPase reaction is more similar to that of EF-Tu than of EF-G (see Section IV,B,1, p. 47), since

(A) PROKARYOTES

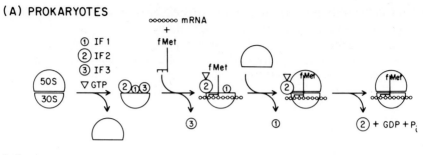

(B) EUKARYOTES

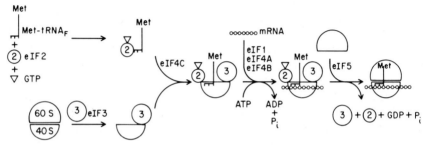

Fig. 6. Pathways for initiation of protein synthesis. (A) Prokaryotes; (B) eukaryotes.

the fMet-tRNA$_i$ and mRNA are not translocated on the ribosome. After the release of IF-2, the initiation complex is fully competent to enter the elongation phase of protein synthesis.

A major problem in defining the pathway is to determine the order of binding of fMet-tRNA$_i$ and mRNA to the 30 S ribosomal subunit. The usual approach is to use radioactive components, isolate various stable complexes, and arrange the complexes in a logical order. A complex of mRNA and 30 S subunit can be isolated, while stable fMet-tRNA$_i$ binding to ribosomes requires the presence of mRNA. This suggests that mRNA binding precedes fMet-tRNA$_i$ binding. Alternatively, formation of a weak ribosomal complex with fMet-tRNA$_i$ may be obligatory for proper mRNA binding. A kinetic analysis of these reactions indicates that either order of binding is possible (Gualerzi et al., 1977). When the binding of initiation factors is considered, the problem becomes even more confusing. There is evidence for an interaction between IF-2 and fMet-tRNA$_i$ and for the formation of a ternary complex with these components and GTP (Van der Hofstad et al., 1977). Similarly, IF-3 is implicated in mRNA binding and binds to mRNA's in the absence of ribosomes. Alternatively, the three factors, IF-1, IF-2, and IF-3, bind cooperatively to 30 S subunits in the absence of mRNA and fMet-tRNA$_i$ (Langberg et al., 1977) and are found on native 30 S subunits in cell extracts. Since all of the isolated intermediates cannot be accommodated in a single pathway, either there is more than one route for the formation of the 30 S preinitiation complex or some of the intermediate complexes are not actually formed *in vivo* and are artifacts of the experimental conditions. More sophisticated kinetic analyses are required to elucidate the precise order of addition of initiation components to ribosomes (see Weiel *et al.*, 1978).

2. The Eukaryotic Pathway

The pathway of assembly of the 80 S initiation complex has been studied most extensively with components from rabbit reticulocytes. Two major approaches to the problem have been pursued: intermediates are isolated and identified from crude cell lysates, often following the addition of an antibiotic inhibitor of initiation, and intermediate complexes are assembled *in vitro* from purified components. A tentative scheme for mammalian initiation is shown in Fig. 6B. Because of the enormous complexity of the initiation factors, their precise functional role in the pathway is not yet clear.

As with prokaryotes, initiation begins with the dissociation of 80 S ribosomes into subunits. Mammalian dissociation factor activity is poorly characterized, but it is proposed that the multicomponent factor eIF-3 is responsible (Thompson *et al.*, 1977). All of the protein components of eIF-3 bind to 40 S subunits in the absence of other initiation components (Benne and Hershey, 1976), and a native 40 S subunit which contains the appropriate amount of protein but no Met-tRNA$_i$ or mRNA has been identified (Ayuso-Parilla *et al.*, 1973). None of the other initiation factors binds to 40 S subunits in the absence of other initiation components. Thus the eIF-3 · 40 S complex may be the first stable intermediate in the pathway.

In a parallel set of reactions, the initiator tRNA (Met-tRNA$_i$) forms a stable ternary complex with GTP and eIF-2 (Levin and Kyner, 1971; Chen *et al.*, 1972; Dettman and Stanley, 1972). GTP is not hydrolyzed, and ternary complexes may be formed with nonhydrolyzable analogues of GTP. GDP is a potent inhibitor of the reaction and may play a role in regulating initiation through the "energy charge" of the cell (Walton and Gill, 1976). When an energy generating system is present to assure that GDP is fully converted to GTP, all of the eIF-2 is present as ternary complex (Benne *et al.*, 1979). The results indicate that at physiological concentrations the three components alone associate tightly. A number of reports suggest that additional factors stimulate ternary complex formation, namely, Co-EIF-1 (Dasgupta *et al.*, 1976) and ESP (DeHaro *et al.*, 1978). Since the assay conditions were not adequately controlled, it is not yet clear whether these and/or other factors are truly involved in the reactions. It is possible that one or more factors influence the extent of GDP inhibition of ternary complex formation or help to recycle eIF-2 by a mechanism similar to EF-Ts action on EF-Tu · GDP.

The Met-tRNA$_i$ · eIF-2 · GTP complex binds to the 40 S subunit in the absence of other components for initiation (Schreier and Staehelin, 1973; Adams *et al.*, 1975); eIF-3 and eIF-4C enhance the 40 S binding while mRNA exerts no effect (Benne *et al.*, 1976; Schreier *et al.*, 1977). Since mRNA · 40 S complexes are not detected in the absence of bound Met-tRNA$_i$, the results suggest that Met-tRNA$_i$ binding precedes mRNA binding. Further evidence is obtained from analysis of initiation complexes in cell lysates: 40 S initiation complexes were observed with bound Met-tRNA$_i$ and protein, but no evidence for bound mRNA was obtained (Hirsch *et al.*, 1973; Sundkvist and Staehelin, 1975).

The mRNA binding reaction is complicated and thus far poorly characterized. It is especially interesting since it is the step during which a particular mRNA is selected for translation and hence may involve translational controls. mRNA binding is promoted by eIF-4A and eIF-4B, and requires the presence of eIF-2, eIF-3, and Met-tRNA$_i$ on the 40 S subunit (Schreier *et al.*, 1977; Benne and Hershey, 1978). eIF-1 and eIF-4C stimulate marginally. In addition, mRNA binding involves the hydrolysis of ATP (Marcus, 1970; Schreier and Staehelin, 1973); nonhydrolyzable analogues inhibit the reaction. The components responsible for ATP hydrolysis have not been identified, nor is the functional role of ATP known. It is possible that some of the initiation factors bind to the mRNA before it joins the 40 S complex. Evidence for such interactions has been obtained for eIF-3 and eIF-4B (Shafritz *et al.*, 1976; Brown-Luedi *et al.*, 1978), but the significance of the results is unknown. A protein of 24,000 daltons found in preparations of eIF-3 and eIF-4B can be cross-linked to the capped portion of mRNA's and may also be involved in initiation (Sonenberg *et al.*, 1978).

The 40 S preinitiation complex, like its prokaryotic counterpart, therefore is composed of the small ribosomal subunit, met-tRNA$_i$ and mRNA. The presence of eIF-2 and eIF-3 in the complex has been shown with radioactive factors by *in vitro* construction of the complexes (Benne and Hershey, 1978; Trachsel and Staehelin, 1978) and by analysis of lysates supplemented with labeled factors (Safer *et al.*, 1978). There is as yet no firm evidence for the stable binding of the other initiation factors implicated in the pathway. The complex also contains intact GTP (Trachsel *et al.*, 1977).

Junction of the 40 S preinitiation complex with the 60 S ribosomal subunit to form an 80 S initiation complex requires the presence of eIF-5 in catalytic quantities and the hydrolysis of GTP to GDP and P_i. eIF-2 and eIF-3 are released at this stage (Benne and Hershey, 1978; Trachsel and Staehelin, 1978). The 80 S complex may not yet be fully competent for the subsequent steps of elongation, however; eIF-4D stimulates the activity of the complex in the model reaction, methionylpuromycin synthesis (Benne and Hershey, 1978). The junction reaction appears to be very rapid, since complete 40 S preinitiation complexes are not observed in cell lysates. Edeine inhibits the junction reaction and allows the isolation of 48 S complexes containing 40 S subunits, Met-tRNA and mRNA (Safer *et al.*, 1978).

3. Molecular Interactions in mRNA Binding

The step at which mRNA binds to ribosomes is probably the rate-limiting reaction in protein synthesis. It is also the step during which the ribosome selects the correct initiator site in the mRNA and discriminates between different mRNA's. Translational control mechanisms can be expected to operate during this critical reaction. For these reasons it is desirable to elucidate the detailed molecular mechanism of mRNA binding to ribosomes. The events are as yet poorly understood, however. Examples are known of the preferred binding of one mRNA over another, but it is not clear whether the more efficiently translated mRNA binds more rapidly, forms more stable complexes (i.e., dissociates from the ribosome more slowly), or influences the rate of a subsequent step by some ''allosteric'' mechanism. Some progress has been made in determining how the structure of the mRNA and ribosome contribute to mRNA binding and how protein factors influence these interactions.

a. **Shine and Dalgarno Hypothesis for Prokaryotes.** Ribosomes from different bacterial species initiate protein synthesis on various mRNA's with different efficiencies. The discriminating property was ascribed to the 30 S subunit (Lodish, 1970), and more precisely to the 16 S RNA and ribosomal protein S12 by mixed reconstitution experiments (Held *et al.*, 1974a). The base sequences of initiator regions in mRNA's were determined by analyzing mRNA fragments protected from nuclease digestion by ribosomal initiation complexes (Steitz, 1969). Based on these sequences and the sequence of the 3'-terminus of 16 S RNA, Shine and Dalgarno (1974) proposed an ingenious hypothesis which states that nucleotides near the 3'-terminus of 16 S rRNA form base pairs with a complementary sequence of nucleotides in the mRNA near the 5'-side of the AUG initiator site. Examination of the known initiator regions of prokaryotic cistrons indicates that most have a purine-rich sequence centered about 10 nucleotides from the AUG (see review by Steitz, 1979). The common sequence GGAG in the mRNA's is complementary to the se-

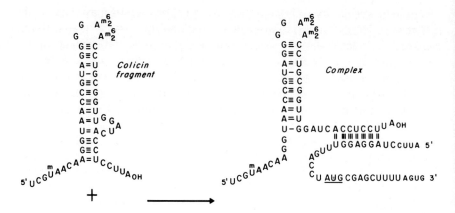

5' AUUCCUAGGAGGUUUGACCU<u>AUG</u>CGAGCUUUUAGUG 3'

R17 A protein initiator region

Fig. 7. Proposed interaction between ribosomal and messenger RNA's. The figure shows the hydrogen binding interaction between the colicin fragment of 16 S rRNA and the initiator region of the bacteriophage R17 A protein cistron, as postulated by Shine and Dalgarno (1974). From Steitz (1977a), reproduced with permission.

quence CUCC in the 16 S rRNA and the two regions are thought to interact as shown in Fig. 7.

Evidence in support of the Shine and Dalgarno hypothesis was obtained by Steitz and Jakes (1975), who treated initiation complexes (containing the mRNA initiation fragment from the maturation A protein cistron of phage R17) with Col E3 and isolated a mRNA–rRNA hybrid. This would be considered strong evidence for the existence of the interaction in the initiation complex if the following possibility could be ruled out: base pairing might have occurred during isolation because of a high local concentration of the two RNA's. A mutation in the bacteriophage T7 gene 0.3 initiator region, which reduces complementarity in the Shine–Dalgarno sequence, causes a tenfold decrease in the rate of protein synthesis (Dunn *et al.*, 1978). The phage λ repressor mRNA, which begins with the initiator codon AUG and therefore lacks the Shine–Dalgarno sequence, is inefficiently translated.

Two major RNA–RNA interactions may, therefore, be involved in mRNA binding: the Shine–Dalgarno base pairing to 16 S RNA as discussed above and the codon interaction with the anticodon of fMet-$tRNA_i$. The usual codon for initiation is AUG, although GUG sometimes is used. These interactions may be modulated by proteins such as IF-3 and S1, which promote mRNA binding (Steitz *et al.*, 1977b) and IF-2, which is required for fMet-$tRNA_i$ binding. The relative contribution of each to the specificity of mRNA binding is unknown and may differ with different mRNA's. It is possible that additional portions of the mRNA sequence interact with other domains in the 30 S ribosomal subunit. More structural and kinetic information is required in order to elucidate the mechanism of mRNA binding.

b. Eukaryotic mRNA Interactions. The structures of eukaryotic mRNA's differ in a number of ways from prokaryotic mRNA's; they are monocistronic and are usually capped at the 5'-terminus and polyadenylated at the 3'-terminus (see Section II,C,2). Eukaryotic mRNA's generally are not translated accurately or efficiently by prokaryotic ribosomes. Eukaryotic ribosomes, in contrast, translate a wide variety of eukaryotic mRNA's and also prokaryotic mRNA's. Three features of mRNA translation are characteristic of eukaryotic cells. First, protein synthesis almost always begins with the AUG codon located closest to the 5'-terminus of the mRNA. Second, capped mRNA's are translated greater than threefold more efficiently than uncapped mRNA's. However, some mRNA's, e.g., EMC viral RNA, are very efficiently translated and yet contain no cap structure. Third, eukaryotic systems apparently are unable to initiate within the interior of mRNA's. The distance separating the cap structure and the initiator AUG codon varies considerably, from about 10 to greater than 200 nucleotides in different mRNA's; however, the efficiency of mRNA translation does not correlate with the variable distances. There is little evidence for complementarity between the 3'-terminus of 18 S rRNA and mRNA sequences proximal to the AUG initiator codon. The 3'-terminal sequence of 18 S rRNA is highly conserved in eukaryotic species, but only a few mRNA's possess sequences complementary to it. Therefore, an interaction comparable to that postulated by Shine and Dalgarno for prokaryotes appears unlikely in eukaryotes. In summary, little is yet known about the molecular interactions involved in mRNA binding to eukaryotic ribosomes and the roles played by the initiation factors. Two reviews describe recent progress in relating the structure of mRNA's to their function in initiation of protein synthesis (Kozak, 1978; Revel and Groner, 1978).

c. Are There mRNA-Specific Initiation Factors? It has been proposed that subspecies of initiation factors which are specific for different classes of mRNA's exist. In *Escherichia coli*, IF-3 was fractionated into two or more subspecies which differentially stimulated the translation of bacterial mRNA and late phage T4 mRNA (Revel *et al.*, 1970; Lee-Huang and Ochoa, 1971). In contradiction to such reports, other groups have been unable to isolate subspecies of IF-3 with different mRNA specificities (Schiff *et al.*, 1974; Spremulli *et al.*, 1974). The problem is unresolved. In addition to multiple species of IF-3, the Revel and the Ochoa group also characterized other proteins which selectively inhibited the translation of phage MS2 RNA or late T4 mRNA. One of the proteins, called interference factor i, was purified and identified as ribosomal protein S1. The presence of S1 is obligatory for mRNA binding to 30 S subunits and for translation (Van Duin and Van Knippenberg, 1974). When added in great excess, it is inhibitory and shows interference factor activity. Similar observations have been made with ribosomal protein S21 (Held *et al.*, 1974b). Since S1, and ribosomal proteins, in general, are not present in cells in excess over ribosomal particles (Miyajima and Kaziro, 1978), the interference phenomenon observed is likely to be an artifact of the *in vitro* experiments where nonphysiological levels or ratios of components were used.

In eukaryotic systems, a factor thought to be specific for EMC RNA translation was purified from Krebs ascites cells (Wigle and Smith, 1973), but was shown later to be identical to eIF-4A (Staehelin *et al.*, 1975). Apparent discrimination was due to a quantitative difference in the amounts of initiation factors required for optimal *in vitro* synthesis. There are reports that initiation factor eIF-3 is heterogeneous and that subspecies specifically stimulate the translation of certain classes of mRNA. Heywood *et al.* (1974) partially purified eIF-3 from chick muscle which stimulates translation of myosin mRNA but not globin mRNA, while the converse was true of reticulocyte eIF-3. These studies are controversial and difficult to analyze, partly because the components studied were not pure. Because a great variety of mRNA's are translated efficiently in wheat germ, reticulocyte, and other cell lysates, it remains conjectural whether protein factors exist which specifically stimulate the utilization of one class of mRNA's but do not function for other classes. There are as yet few studies with highly purified components of initiation which might elucidate the problem. Even with such systems, however, great care will be required to avoid artifacts similar to those experienced with the simpler bacterial system.

4. Role of the Ribosome

A complete description of the molecular events of initiation must include the contributions made by the structural components of the ribosome. Considerable progress has been made in correlating bacterial ribosome structure with function, and recent reviews describe and document these advances in detail (Brimacombe *et al.*, 1978; Grunberg-Manago *et al.*, 1978). Studies on eukaryotic ribosomes are only just beginning (Bielka and Stahl, 1978), so I shall focus attention exclusively on the results obtained from bacteria.

The involvement of rRNA in initiation has already been discussed above (Section IV,A,3). Efforts have been made to identify the 30 S ribosomal proteins which comprise the region where initiation occurs. The following experimental approaches have been used. (1) Initiation factors bound to the 30 S subunit are cross-linked to ribosomal proteins with bifunctional reagents, and the cross-linked proteins are identified. All three initiation factors are cross-linked to a number of common proteins (Langberg *et al.*, 1977), suggesting that the factors bind contiguously on the ribosomal surface. The major cross-linked proteins are S1, S11, S12, S13, and S19. The 3'-terminus of 16 S rRNA may be chemically cross-linked to some of the same proteins and to all three bound initiation factors as well. Thus the factors bind near the sequence of rRNA implicated in base-pair interactions with mRNA. (2) mRNA analogues are linked to ribosomal proteins by utilizing electrophilic derivatives of oligonucleotides or by photolysis (Fiser *et al.*, 1975; Pongs *et al.*, 1975). S1, S4, S12, S18,and S21 are implicated. (3) Affinity analogues of fMet-tRNA$_f$ react with S3, S7, S13, and S14 (Girshovich *et al.*, 1974). (4) Single omission reconstitution experiments of 30 S subunits implicate ribosomal proteins specifically required for initiation functions. S12 and S21 are required for AUG-directed fMet-tRNA binding (Nomura and Held, 1974).

The various proteins implicated above form a neighborhood on the 30 S particle in the head region of the model shown in Fig. 3A. Considerable refinement of the three-dimensional structure is needed in order to more clearly understand the individual functions of these proteins.

B. The Elongation Cycle

The elongation phase of protein synthesis is conveniently divided into three steps (see Fig. 8): the binding of aminoacyl-tRNA, peptide bond formation, and translocation. The reactions occur on the surface of the ribosome and involve two sites for tRNA binding: the A site, where the incoming aminoacyl-tRNA binds, and the P site, where peptidyl-tRNA binds just prior to peptide bond formation. The elongation steps are preceded by an extra cycle reaction, the formation of a ternary complex containing the aminoacyl-tRNA, EF-Tu (or EF-1), and GTP. During the aminoacyl-tRNA binding reaction, a particular ternary complex is selected from the various species available on the basis of the codon–anticodon interaction between the ribosome-bound mRNA and the tRNA. Peptide bond formation occurs by transfer of the nascent peptide from its tRNA in the P site to the α-amino group of the newly bound aminoacyl-tRNA in the A site. Finally, the translocation reaction involves the EF-G

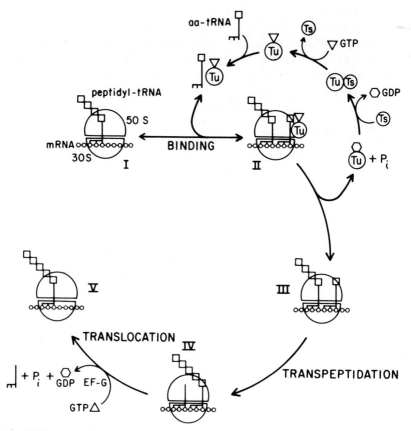

Fig. 8. The alongation cycle. The reactions in the cycle, and those in the elongation factor EF-Tu cycle, are explained in the text. Intermediates I and V are identical except that the peptidyl-tRNA in V is one amino acid longer.

(or EF-2) promoted movement of the peptidyl-tRNA from the A site to the P site, with the concomitant movement of the mRNA by three nucleotides. The result is a ribosomal complex with a peptidyl-tRNA one amino acid longer and with the next mRNA codon available for interaction with its cognate ternary complex. The pathways and mechanisms of elongation in prokaryotic and eukaryotic cells are quite similar and have been reviewed extensively (Weissbach and Ochoa, 1976; Weissbach and Pestka, 1977; Grunberg-Manago *et al.*, 1978). Besides the question of the overall mechanism, two aspects of elongation are especially interesting. (1) How is the correct aminoacyl-tRNA selected with high fidelity? and (2) Does translocation share a common fundamental mechanism with other biological reactions involving unidirectional movement?

1. The Prokaryotic Mechanism

a. **Ternary Complex Formation.** EF-Tu forms a binary complex with GTP, which in turn reacts rapidly with aminoacyl-tRNA to form a stable ternary complex with a dissociation constant of about $10^{-8}M$ (Miller *et al.*, 1973). The ternary complex may be isolated by gel filtration chromatography or is readily assayed by nitrocellulose filtration that measures a decrease in bound radioactive GTP as EF-Tu · GTP is converted to ternary complex. The tRNA must be charged, and the aminoacyl group may not be acylated. The initiator tRNA, Met-tRNA$_i$, whether formylated or not, does not bind to EF-Tu · GTP. There appears to be no preference for the positional isomers of aminoacyl-tRNA; both 2'- and 3'-deoxyadenosine analogues of aminoacyl-tRNA bind equally well (Hecht *et al.*, 1977). The features of tRNA structure recognized by EF-Tu are not well known, however. Results of X-ray diffraction studies of ternary complexes are eagerly awaited. Nuclear magnetic resonance spectra of aminoacyl-tRNA as a free molecule or bound in ternary complex are essentially identical, which indicates that extensive changes in the secondary and tertiary structures of tRNA do not occur following complex formation (Schulman *et al.*, 1974).

EF-Tu binds to GDP more tightly than to GTP [K_d = 4.9 × 10^{-9} and 3.6 × 10^{-7} M, respectively (Arai *et al.*, 1974)]. However, the EF-Tu · GDP complex binds to aminoacyl-tRNA with an affinity 10^5-fold less than EF-Tu · GTP. EF-Ts plays a role in converting the EF-Tu · GDP complex, which is formed following ribosome binding, to the EF-Tu · GTP complex. The reactions involved are shown in the EF-Tu cycle in Fig. 8. Thus, EF-Tu participates in a cyclic reaction sequence which is driven from equilibrium by GTP hydrolysis on the ribosome. EF-Ts catalyzes the rate of exchange of EF-Tu-bound GDP for GTP. Since the level of EF-Tu in cells is high, approximately equal to that for the sum of tRNA's, most charged tRNA's are present as ternary complexes provided that the energy charge of the cell is sufficiently high. The importance of this fact for proofreading is discussed below in Section IV,B,4.

b. **Aminoacyl-tRNA Binding to Ribosomes.** A ternary complex binds rapidly to ribosomes carrying the appropriate mRNA codon in the A site. Correct binding is followed by GTP hydrolysis to GDP and P$_i$. EF-Tu · GDP is released from the ribosome and peptide bond formation occurs (see Fig. 8). The EF-Tu · GDP is then free to be recycled into another ternary complex. When the codon–anticodon sequences are not complementary, binding and GTP hydrolysis do not readily occur. If the nonhydrolyzable analogue of GTP guanylyl-5'-methylenediphosphonate (GMP-PCP) is substituted, the ternary complex binds, but GTP hydrolysis is blocked and EF-Tu release and peptide bond formation are prevented (Shorey *et al.*, 1971).

The catalytic center for GTP hydrolysis is located in the EF-Tu molecule, not in the ribosome. The ternary complex is quite stable, but in the presence of the antibiotic, kirromycin, a ribosome-independent GTPase activity is induced (Chinali *et al.*, 1977). Ribosomes also stimulate GTP hydrolysis with the binary complex EF-Tu · GTP in the absence of aminoacyl-tRNA. Comparison of the kinetics of GTP hydrolysis with binary and ternary complexes indicates that EF-Tu · GTP ($K_m = 4 \times 10^{-6}$ *M*) binds to ribosomes nearly as tightly as the ternary complex ($K_m = 1 \times 10^{-6}$ *M*). Thus the binding of ternary complexes is stabilized by at least two interactions: codon–anticodon base pair formation and EF-Tu–ribosome binding.

The precise role of EF-Tu and GTP hydrolysis in aminoacyl-tRNA binding is not yet understood. Aminoacyl-tRNA alone binds to ribosomes in a mRNA-dependent reaction, but not nearly so rapidly as the ternary complex. Higher concentrations of Mg^{2+} are required, and factor-independent binding occurs under these conditions with uncharged tRNA's as well. Thus EF-Tu discriminates against uncharged tRNA's and stimulates the overall rate of aminoacyl-tRNA binding. A further role may be to prevent entry of the aminoacyl group into the peptidyltransferase site until the correct interaction of the anticodon with the mRNA can be monitored by the GTPase step. The presence of EF-Tu and the hydrolysis of GTP, therefore, may serve the mechanism whereby the correct ternary complex is selected with high fidelity. This important problem is discussed in greater detail in Section IV,B,4 below. GTP hydrolysis also serves to eject EF-Tu rapidly so that peptide bond formation can occur.

c. Peptide Bond Formation. Peptide bond formation proceeds by transfer of the peptide moiety of peptidyl-tRNA (bound in the P site) to the α-amino group of aminoacyl-tRNA (bound in the A site). The chemistry is relatively simple, involving an O to N acyl shift. The products are a new peptidyl-tRNA, which is one amino acid residue longer, in the A site, and a stripped tRNA in the P site. The enzyme peptidyltransferase catalyzes the reaction; it is a structural part of the 50 S ribosomal subunit. The sites in the peptidyltransferase center which bind the 3′-termini of the tRNA substrates in the A and P sites are named the A′ and P′ sites (Pestka, 1972). The specificity of these sites has been studied with aminoacyl-tRNA fragments and with the antibiotic puromycin, which mimics aminoacyl-tRNA. In the presence of methanol, the substrates react in the absence of 30 S subunits and mRNA (Monro *et al.*, 1968). The smallest fragment with donor activity (P′ site binding) is acyl-aminoacyl-(3′)ACC(5′); its reduced activity relative to acyl-aminoacyl-tRNA suggests that other parts of the tRNA molecule are also involved in binding to the 50 S subunit. At the A' site, 3'-O-aminoacyl isomers are the preferred acceptors (Ringer *et al.*, 1975), the smallest exemplified by puromycin, a simple nucleoside derivative. Peptidyltransferase can catalyze a transesterification reaction, using phenyllactyl-tRNA^{Phe} as substrate during *in vitro* protein synthesis (Fahnestock and Rich, 1971). Attempts to identify a peptidyl–ribosome intermediate have been unsuccessful, and the detailed mechanism of peptide bond formation remains unclear. Reaction details have been reviewed recently (Harris and Pestka, 1977).

d. Translocation. Translocation involves the movement of the newly formed peptidyl-tRNA from the A site to the P site, the concomitant movement of the associated mRNA by three nucleotides, and the ejection of stripped tRNA from the P site (see Fig. 8). The reaction is promoted by EF-G and the hydrolysis of GTP to GDP and P$_i$. EF-G binds GTP, although with much less affinity ($K_a = 2.7 \times 10^4 M^{-1}$) than EF-Tu. Kinetic studies of the GTPase reaction indicate an ordered sequence of steps: EF-G forms a binary complex with GTP, the complex binds to ribosomes, GTP hydrolysis occurs, an EF-G · GDP complex is released, and the EF-G · GDP complex dissociates (Rohrbach and Bodley, 1976). A stable EF-G · ribosome complex is formed when GMP-PCP is substituted for GTP or when the antibiotic fusidic acid is added. In the latter case, GTP hydrolysis occurs but release of EF-G · GDP is prevented. EF-G · GTP complexes can bind to ribosomes in the absence of any tRNA's, or if peptidyl-tRNA occupies the A site and stripped tRNA occupies the P site. The presence of aminoacyl-tRNA in the A site or peptidyl-tRNA in the P site blocks EF-G binding. Prebound EF-Tu also blocks EF-G binding, and the converse is true as well. Thus, the two elongation factors appear to have overlapping binding sites and cannot bind simultaneously to the same ribosome (Modolell and Vasquez, 1973).

Catalytic amounts of EF-G with GTP (but not GMP-PCP) cause the rapid translocation of peptidyl-tRNA from the A site to the P site. Occupancy in the A or P sites is measured operationally by the reactivity of the peptidyl-tRNA with puromycin; peptidyl-tRNA in the A site is unreactive, but when in the P site the peptide moiety is transferred to puromycin. (Attempts to define ribosomal A and P sites structurally are discussed in Section IV,B,3 below.) Alternatively, translocation can be measured by analyzing the nucleotides in mRNA which are protected by the ribosome from nuclease digestion (Thach and Thach, 1971). No movement of mRNA occurs during initiation, aminoacyl-tRNA binding, or peptide bond formation; only following translocation is there a change in the protected nucleotides.

When stoichiometric amounts of EF-G are used with GMP-PCP (or with GTP and fusidic acid), peptidyl-tRNA prebound in the A site be-comes puromycin reactive. Thus GTP hydrolysis is not required for trans-location per se, but is needed for the efficient recycling of EF-G. The presence of bound EF-G does not prevent peptidyltransferase activity but does inhibit the binding of aminoacyl-tRNA with EF-Tu. The role of GTP hydrolysis appears to be one of accelerating the release of EF-G · GDP. One molecule of GTP is thought to be hydrolyzed per translocation event (Cabrer et al., 1976). Thus EF-G is only transiently attached to ribosomes during the translocation step and is mostly free in solution, a fact consis-tent with the observation that EF-G is found primarily in the post-ribosomal supernatant.

The molecular events of translocation must provide for the movement of peptidyl-tRNA and mRNA relative to the ribosome without losing either from the surface of the particle. A number of models for this have been proposed (see Brot, 1977), but a compelling hypothesis must await more precise knowledge of the structure and function of the translational machinery.

2. The Eukaryotic Mechanism

The pathway and mechanism of elongation in eukaryotic cells appear to be very similar to those in bacterial cells. The same sequence of reactions occurs during the elongation cycle, and the eukaryotic elongation factors EF-1 and EF-2 (reviewed by Weissbach and Ochoa, 1977b) function like EF-Tu and EF-G. This is strikingly shown by the interchangeability of EF-Tu and EF-1 on both E. coli and Krebs ascites ribosomes (Grasmuk et al., 1977a). There are some differences in details, however, and these are described below.

EF-1 binds to GTP and forms a ternary complex with aminoacyl-tRNA's (Nagata et al., 1976), but not with initiator methionyl-tRNA$_i$. The complexes appear to be less stable than their prokaryotic counterparts, and are difficult to isolate. Ternary complexes bind to ribosomes, GTP hydrolyzes and EF-1 · GDP is formed (Weissbach et al., 1973). It is not proved, however, that the ternary complex is an obligatory intermediate and that EF-1 · GDP is released from the ribosome. Proteins which stimu-late the exchange of GTP for GDP in EF-1 complexes have been de-scribed (see Section II,D,2). Studies with radioactive EF-1 indicate that the factor remains bound to the ribosome during successive cycles of elongation (Grasmuk et al., 1977b) and that bound EF-1 does not prevent the subsequent binding of EF-2. On the other hand, low levels of EF-1 appears to act catalytically during protein synthesis. The determination of whether a factor is released from the ribosome during cycles of elongation is complicated by heterogeneous populations of active and inactive ribosomes and the lack of precise kinetic measurements. Further work is required to resolve whether the prokaryotic and eukaryotic mechanisms differ in this fundamental way, or which mechanism is correct.

EF-2, like its counterpart EF-G, binds GTP, but with much greater affinity ($K_d = 2 \times 10^{-6}\ M$; Henriksen et al., 1975). It binds GDP even more tightly ($K_d = 4 \times 10^{-7}\ M$) and contains one nucleotide binding site per molecule (Mizumoto et al., 1974). The binary complexes interact with ribosomes, but only that prepared with GMP-PCP forms a stable complex. It is presumed that EF-2 leaves the ribosome following the translocation step during each turn of the elongation cycle.

EF-2 activity is inhibited by diphtheria toxin in a unique reaction involving nicotinamide adenosine diphosphate (NAD) (Collier, 1967; Honjo et al., 1968). The toxin catalyzes the transfer of the adenosine diphosphate-ribose (ADPR) moiety of NAD to a specific site in the factor to form a covalently modified EF-2. The ADPR–EF-2 can form a binary complex with GTP which presumably binds to ribosomes, but translocation activity is inhibited.

3. Structure–Function Correlates of the Ribosome

The identification of ribosomal components involved in the reactions of elongation is an important step toward development of a detailed molecular mechanism for protein synthesis. Of special interest are the structures of the A and P sites for tRNA binding and the peptidyltransferase center. Experiments for identifying the ribosomal proteins in functional centers involve affinity labeling with derivatives of tRNA's, antibiotics, mRNA's, and GTP; cross-linking of soluble factors; and the study of protein-deficient particles. The results until now have been obtained almost exclusively with the prokaryotic system, while studies on eukaryotic ribosomes are just beginning. They indicate that the 50 S subunit plays a dominant role in elongation, in contrast to initiation, in which the 30 S subunit is more critical. The structural correlates of bacterial ribosome function have been reviewed recently and provide detailed accounts of the experiments (Pellegrini and Cantor, 1977; Grunberg-Manago et al., 1978).

Transfer RNA derivatives bound at the P site label proteins L2, L11, L18, and L27. When the reactive group is placed farther from the 3′-terminus of the tRNA, L24 and L32/L33 are labeled. Transfer RNA derivatives occupying the A site label L16 and also L2 and L27. The involvement of 5 S rRNA in tRNA binding is controversial. Some of the above proteins may comprise the peptidyltransferase center, since tRNA derivatives would be expected to bind and react there. Partial reconstitution experiments implicate L16 in peptide bond formation and other evidence indicates that L11 and L2 are involved as well. Chloramphenicol derivatives, which inhibit peptidyltransferase activity, react with L16, while puromycin analogues bind to L23 and S14. The aminoglycoside antibiotics, which affect A site functions and aminoacyl-tRNA binding fidelity, interact with the 30 S proteins S4, S9, and S12.

Cross-linking studies implicate ten or more proteins in the binding site of EF-Tu: L1, L5, L7/L12, L15, L20, L30, and L33 with the reagent p-nitrophenylchloroformate and L23 and L28 >L1, L3, and L24 with 2-iminothiolane. EF-G is linked to L7/L12 and possibly to L2 and L6. The proteins L7/L12 can also be cross-linked to IF-2 and RF-2 and are required for the GTPase activities of all four factors.

Most of the 50 S ribosomal proteins implicated above form a compact neighborhood in the model of the 50 S subunit shown in Fig. 3B. The region is located in the "seat" of the chair form, which is also part of the region involved in the interface with 30 S subunits. Stöffler and Wittmann (1977) and others suggest that the two subunits form a tunnel through which the mRNA travels and in which the tRNA's bind. It seems that numerous proteins participate in each of the specific functions of the ribosome. The high degree of cooperation between the components appears to be a general feature of the ribosome. The results cited above still are insufficient to explain the molecular events of translocation, although numerous hypotheses have been made to account for this reaction (see review by Brot, 1977). As more precise structural information on ribosomes becomes available, we can expect further refinements of such speculative models.

4. tRNA Selection and Proofreading

During protein synthesis *in vivo,* it is estimated that very few translational errors occur. The frequency of mistranslation of isoleucine as valine (Loftfield and Vanderjagt, 1972) or arginine as cysteine (Edelmann and Gallant, 1977) is estimated to be of the order of 1 in 10^4. The discriminating interaction is presumably between the three nucleotides in the mRNA codon and those in the anticodon of the tRNA. The amino acid itself plays no role, as proved by the following elegant experiment (Chapeville and Rouget, 1972): incorporation into protein of alanine from alanyl-tRNA^{Cys} (obtained from cysteinyl-tRNA^{Cys} by reduction) was dependent on mRNA containing codons for cysteine but not alanine. How does the translational machinery achieve the observed precision?

The energy of the codon–anticodon interaction presumably is not sufficient to explain the observed discrimination for a single-step reaction. Such energies are virtually impossible to calculate precisely, however. Binding studies of the anticodon regions of tRNA's to oligonucleotides or to other tRNA's show that the binding energies do not follow the quantitative rules for base pairing between complimentary RNA strands. Furthermore, some tRNA species appear to use only the first two letters of the codon; that is, they read a set of four codons which differ only in the third letter. In others, the hypermodified purine base on the 5'-side of the anticodon contributes to the strength of the codon–anticodon interaction. Codon specificity may be influenced by other structural features of the tRNA as well. The ribosome presumably creates a special environment, as yet undefined, which enhances the ability to discriminate. Finally, nondiscriminatory interactions common to all tRNA's, e.g., through the EF-Tu molecule, influence ternary complex binding.

The rate of ternary complex binding to ribosomes is very rapid and may be essentially diffusion controlled. Discrimination more likely occurs by differing rates for the reverse reaction, improper ternary complexes being released from the ribosome at a faster rate than proper ternary complexes. A comparison of cognate and noncognate complexes indicates that the latter have lifetimes about 1% of those for proper complexes. It is clear that insufficient discrimination occurs during a single binding step alone (Fig. 8). A number of hypotheses suggesting how higher levels of discrimination may be achieved have been proposed (Hopfield, 1974; Ninio, 1975). It is postulated that there are at least two points during which aminoacyl-tRNA may dissociate from the ribosome, thus allowing the codon–anticodon discrimination to function twice. In relation to the reaction scheme presented in Fig. 8, the ternary complex may dissociate from intermediate II, and the aminoacyl-tRNA may dissociate from intermediate III. The two intermediates are driven from equilibrium by an energy-consuming step, the EF-Tu-catalyzed hydrolysis of GTP. It is also necessary that the rate of aminoacyl-tRNA binding to the ribosome to form intermediate III be very slow. This is accomplished either by assigning a high energy state to intermediate III or by sequestering the aminoacyl-tRNA into ternary complex. A consequence of the proofreading mechanism is that the rate of protein synthesis is limited; in effect, accuracy costs time. Evidence consistent with this scheme shows that more GTP is hydrolyzed per peptide bond formed with near-cognate ternary complexes than with cognate complexes (Thompson and Stone, 1977). Another way to achieve further discrimination is to make the reaction velocity of the GTP hydrolysis step sensitive to the codon–anticodon interaction.

The fidelity of translation is strongly influenced by antibiotics and ribosomal mutations. The aminoglycosides stimulate misreading of the first two letters of codons (e.g., streptomycin) or of the second and third letters (e.g., neomycin). *Str* A mutants translate with higher fidelity than wild-type strains, while *ram* mutants show a higher rate of errors. These mutations are in the genes for ribosomal proteins S4, S5, and S12, but the precise effects in the molecular interactions discussed above are not known. A kinetic analysis of missense and nonsense suppression with these mutants has been made (Ninio, 1974). It is clear that great accuracy is required for the synthesis of active enzymes. The cell must, therefore, balance its need for such accuracy against its need for rapid protein synthesis.

C. The Termination Pathway

Termination of protein synthesis results in hydrolysis of the completed peptide from its tRNA, followed by release from the ribosome of the peptide, its tRNA and the mRNA. The process begins with the peptidyl-tRNA in the P site [intermediate I or V (Fig. 8)] and one of the three nonsense codons, UAA, UAG, or UGA, in the A site. This configuration is recognized by the appropriate release factor, which binds to the ribosome. In eukaryotic cells, RF binding requires GTP; in bacterial cells it does not. In cells containing a suppressor tRNA which can recognize the termination codon, the suppressor tRNA and the release factor compete for binding to the ribosome. The presence of RF on the ribosome activates the peptidyltransferase center, which transfers the peptidyl moiety to water. The various macromolecules then dissociate from the ribosome. The release of RF is stimulated by GTP hydrolysis in eukaryotic cells.

The termination process in bacterial cells differs little from that in mammalian cells. Prokaryotic cells contain two codon-specific RF's, whereas mammalian cells contain one factor which recognizes all three nonsense codons. GTP is involved in eukaryotic cells, but may also be involved in the function of RF-3 in prokaryotes. The interaction of bacterial RF's with ribosomes has been studied in some detail. The RF's interact with ribosomal proteins L7/L12 and S9 and are prevented from binding by prebound EF-G or EF-Tu. The nature of the molecular interactions of RF with the nonsense codons or with the peptidyltransferase center is unclear. A recent review of the details of the termination process in prokaryotic and eukaryotic cells is available (Caskey, 1977).

V. PERSPECTIVES

This review describes the impressive amounts of new knowledge generated over the past 20 years by studies of protein synthesis in both prokaryotic and eukaryotic cells. Progress has been somewhat more rapid with bacterial systems, especially in determining the structures of the components for translation, but protein synthesis in eukaryotic cells is now understood nearly as well. The similarities shared by the two systems are striking and indicate that many of the components and the overall mechanism have been highly conserved during evolution.

Essentially all of the components of the translational machinery from *Escherichia coli* are thought to be defined; the macromolecules have been purified, and all their primary sequences will soon be available. Furthermore, much information has been obtained about their levels in cells and how such levels are controlled by regulating the expression of their genes. In eukaryotic cells, nearly all of the components have also been identified, but much less is known about their structures, levels, and gene organization. Eukaryotic components differ from bacterial components primarily by being larger and more numerous, as observed especially for ribosomes and the soluble protein factors. Since the overall functions of the two translational systems are identical, the greater complexity in eukaryotes suggests that translational control mechanisms may operate at the level of the 80 S ribosome. Although a wealth of information exists about these components from both cell types, it must be emphasized that proof is lacking for the requirement *in vivo* of many of the identified macromolecules, and that still additional components may yet be discovered.

The broad outlines of when and how the translational components interact are known for both prokaryotic and eukaryotic cells, and it is likely that no dramatic changes in our general understanding of protein synthesis will occur. Nevertheless, a rigorous determination of the pathways and the elucidation of the molecular mechanisms involved are yet to be accomplished. A great deal of future experimentation will doubtlessly be directed toward problems of mechanism.

The process of protein synthesis involves rather few covalent bond forming or breaking reactions, but is characterized by a large number of noncovalent interactions. The covalent reactions are the aminoacylation of tRNA, peptide bond formation, and the hydrolysis of ATP and GTP; these appear to be enzymologically straightforward. The noncovalent interactions contribute to the structure of the ribosome and occur during the binding of aminoacyl-tRNA's and mRNA to ribosomes and during the function of the protein synthesis factors. The fundamental nature of these molecular interactions is still poorly understood, but must be known if we are to understand the mechanism of protein synthesis. Perhaps best characterized are RNA–RNA interactions, which primarily involve Watson–Crick base pairing. They are seen in tRNA structures and are known to occur between the codon and anticodon regions of mRNA and tRNA's. RNA–RNA interactions are also found in the structure of the ribosome and possibly between rRNA and mRNA's or tRNA's. Much less is known about the binding of proteins to RNA or to other proteins. In order to elucidate the nature of the RNA–RNA, RNA–protein, and protein–protein interactions which occur during protein synthesis, it is imperative to determine the three-dimensional structures of the translational components. X-Ray crystallographic analyses of relatively simple systems, such as complexes of aminoacyl-tRNA's with their synthetases or with elongation factor EF-Tu, are in progress and should shed light on how proteins recognize and bind to nucleic acids. Resolution of the structure of the ribosome and its complexes at a 2–3 Å level will also be required. Most current structural studies of the bacterial ribosome have a resolving power of only 30–40 Å, and, therefore, are not able to provide the kind of information necessary for answering detailed questions of molecular mechanism. Sufficiently precise determinations of the structure of ribosomes are a formidable task. Further studies on ribosome structure and the molecular mechanism of protein synthesis will involve increasingly more sophisticated biophysical techniques.

A second feature of translation which needs further development and experimentation is the kinetics of the reaction steps. Some kinetic parameters of a few reactions are known, e.g., the K_m's for the binding of nucleotides to factors. Rate constants for the various reactions in the proposed pathway are required. Ultimately, the fidelity of protein synthesis, the selection of mRNA's for translation, and the regulation of protein synthesis must be quantitatively characterized and explained by kinetic data.

A combination of precise structural and kinetic information will bring us closer to our goal of explaining protein synthesis in terms of well-understood chemical forces. Knowledge of molecular mechanisms, coupled with results from genetic and physiological studies, are essential for a comprehensive understanding of the translational process and its control.

REFERENCES

Adams, S. L., Safer, B., Anderson, W. F., and Merrick, W. C. (1975). Eukaryotic initiation complex formation. Evidence for two distinct pathways. *J. Biol. Chem.* **250,** 9083–9089.
Allende, J. E., Monro, R., and Lipmann, F. (1964). Resolution of the *E. coli* amino acyl sRNA transfer factor into two complementary fractions. *Proc. Natl. Acad. Sci. U.S.A.* **51,** 1211–1216.
Amils, R., Matthews, E. A., and Cantor, C. R. (1978). An efficient *in vitro* total reconstitution of the *Escherichia coli* 50 S ribosomal subunit. *Nucleic Acids Res.* **5,** 2455–2470.

Arai, K., Kawakita, M., and Kaziro, Y. (1974). Studies on the polypeptide elongation factors from *E. coli*. V. Properties of various complexes containing EF-Tu and EF-Ts. *J. Biochem. (Tokyo)* **76**, 293–306.

Ayuso-Parilla, M., Henshaw, E. C., and Hirsch, C. A. (1973). The ribosome cycle in mammalian protein synthesis. III. Evidence that the nonribosomal proteins bound to the native smaller subunit are initiation factors. *J. Biol. Chem.* **248**, 4386–4393.

Bandyopadhyay, A. K., and Deutscher, M. P. (1973). Lipids associated with the aminoacyl-transfer RNA synthetase complex. *J. Mol. Biol.* **74**, 257–261.

Baralle, F. E. (1977). Structure-function relationship of 5′ non-coding sequence of rabbit α- and β-globin mRNA. *Nature (London)* **267**, 279–281.

Benne, R., and Hershey, J. W. B. (1976). Purification and characterization of initiation factor IF-E3 from rabbit reticulocytes. *Proc. Natl. Acad. Sci. U.S.A.* **73**, 3005–3009.

Benne, R., and Hershey, J. W. B. (1978). The mechanism of action of protein synthesis initiation factors from rabbit reticulocytes. *J. Biol. Chem.* **253**, 3078–3087.

Benne, R., Wong, C., Luedi, M., and Hershey, J. W. B. (1976). Purification and characterization of initiation factor IF-E2 from rabbit reticulocytes. *J. Biol. Chem.* **251**, 7675–7681.

Benne, R., Brown-Luedi, M., and Hershey, J. W. B. (1978a). Purification and characterization of protein synthesis initiation factors eIF-1, eIF-4C, eIF-4D and eIF-5 from rabbit reticulocytes. *J. Biol. Chem.* **253**, 3070–3077.

Benne, R., Edman, J., Traut, R. R., and Hershey, J. W. B. (1978b). Phosphorylation of eukaryotic protein synthesis initiation factors. *Proc. Natl. Acad. Sci. U.S.A.* **75**, 108–112.

Benne, R., Amesz, H., Hershey, J. W. B., and Voorma, H. O. (1979). The activity of eukaryotic initiation factor eIF-2 in ternary complex formation with GTP and met-tRNA$_i$. *J. Biol. Chem.* **254**, 3201–3205.

Bielka, H., and Stahl, J. (1978). Structure and function of eukaryotic ribosomes. *Int. Rev. Biochem.* **18**, 79–168.

Bishop, J. O., and Schweet, R. S. (1961). Role of glutathione in transfer of amino acids from amino acyl-ribonucleic acid to ribosomes. *Biochim. Biophys. Acta* **49**, 235–236.

Branlant, C., SriWidada, J., Krol, A., and Ebel, J.-P. (1977). RNA sequences in ribonucleoprotein fragments of the complex formed from ribosomal 23-S RNA and ribosomal protein L24 of *Escherichia coli*. *Eur. J. Biochem.* **74**, 155–170.

Brauer, D., and Wittmann-Liebold, B. (1977). The primary structure of the initiation factor IF-3 from *Escherichia coli*. *FEBS Lett.* **79**, 269–275.

Brimacombe, R., Stöffler, G., and Wittmann, H. G. (1978). Ribosome structure. *Annu. Rev. Biochem.* **47**, 217–249.

Brosius, J., Palmer, M. L., Kennedy, P. J., and Noller, H. F. (1978). Complete nucleotide sequence of a 16 S ribosomal RNA gene from *Escherichia coli*. *Proc. Natl. Acad. Sci. U.S.A.* **75**, 4801–4805.

Brot, N. (1977). Translocation. *In* "Molecular Mechanisms of Protein Biosynthesis" (H. Weissbach and S. Pestka, eds.), pp. 375–411. Academic Press, New York.

Brownlee, G. G., Sanger, F., and Barrell, B. G. (1968). The sequence of 5S ribosomal ribonucleic acid. *J. Mol. Biol.* **34**, 379–412.

Brown-Luedi, M. L., Benne, R., Yau, P., and Hershey, J. W. B. (1978). Interaction of ribosomes and RNA's with purified initiation factors from reticulocytes. *Fed. Proc., Fed. Am. Soc. Exp. Biol.* **37**, 1307.

Cabrer, B., San-Millan, M. J., Vazquez, D., and Modolell, J. (1976). Stoichiometry of polypeptide chain elongation. *J. Biol. Chem.* **251**, 1718–1722.

Cantoni, G. L., and Davies, D. R. (1971). "Procedures in Nucleic Acid Research." Harper, New York.

Cantor, C. R., Huang, K., and Fairclough, R. (1974). Fluorescence spectroscopic approaches to study of three-dimensional structure of ribosomes. *In* "Ribosomes" (M. Nomura, A. Tissières, and P. Lengyel, eds.), pp. 587–599. Cold Spring Harbor Lab., Cold Spring Harbor, New York.

Capecchi, M. R. (1967). Polypeptide chain termination *in vitro:* Isolation of a release factor. *Proc. Natl. Acad. Sci. U.S.A.* **58**, 1144–1151.

Capecchi, M. R., and Klein, H. A. (1969). Characterization of three proteins involved in polypeptide chain termination. *Cold Spring Harbor Symp. Quant. Biol.* **34**, 469–477.

Carbon, P., Ehresmann, C., Ehresmann, B., and Ebel, J. P. (1978). The sequence of *Escherichia coli* ribosomal 16 S RNA determined by rapid gel methods. *FEBS Lett.* **94**, 152–156.

Caskey, C. T. (1977). Peptide chain termination. *In* "Molecular Mechanisms of Protein Biosynthesis" (H. Weissbach, and S. Pestka, eds.), pp. 443–465. Academic Press, New York.

Chapeville, F., and Rouget, P. (1972). Aminoacyl-tRNA synthetases. *In* "The Mechanism of Protein Synthesis and Its Regulation" (L. Bosch, ed.), pp. 5–32. North-Holland Publ., Amsterdam.

Chen, Y. C., Woodley, C. L., Bose, K. K., and Gupta, N. K. (1972). Protein synthesis in rabbit reticulocytes: Characteristics of a Met-tRNA$_f^{Met}$ binding factor. *Biochem. Biophys. Res. Commun.* **48**, 1–9.

Chinali, G., Wolf, H., and Parmeggiani, A. (1977). Effect of kirromycin on elongation factor Tu: Location of the catalytic center for ribosome elongation-factor-Tu GTPase activity on the elongation factor. *Eur. J. Biochem.* **75**, 55–65.

Clark, B. F. C. (1977). Correlation of biological activities with structural features of transfer RNA. *Prog. Nucleic Acid Res. Mol. Biol.* **20**, 1–19.

Collier, R. J. (1967). Effect of diphtheria toxin on protein synthesis: Inactivation of one of the transfer factors. *J. Mol. Biol.* **25**, 83–98.

Crick, F. H. C. (1966). Codon-anticodon pairing: The wobble hypothesis. *J. Mol. Biol.* **19**, 548–555.

Czernilofsky, A. P., Collatz, E., Gressner, A. M., and Wool, I. G. (1977). Identification of the tRNA-binding sites on rat liver ribosomes by affinity labeling. *Mol. Gen. Genet.* **153**, 231–235.

Dasgupta, A., Majumdar, A., George, A. D., and Gupta, N. K. (1976). Protein synthesis in rabbit reticulocytes. XV. Isolation of a ribosomal protein factor (CO-EIF-1) which stimulates met-tRNA$_f^{Met}$ binding to EIF-1. *Biochem. Biophys. Res. Commun.* **71**, 1234–1241.

DeHaro, C., Datta, A., and Ochoa, S. (1978). Mode of action of the hemin-controlled inhibitor of protein synthesis. *Proc. Natl. Acad. Sci. U.S.A.* **75**, 243–247.

Dettman, G. L., and Stanley, W. M. (1972). Recognition of eukaryotic initiator tRNA by an initiation factor and the transfer of the methionine moiety into peptide linkage. *Biochim. Biophys. Acta* **287**, 124–133.

Deutscher, M. P. (1974). Rabbit liver tRNA nucleotidyltransferase. *In* "Methods in Enzymology" (L. Grossman and K. Moldave, eds.), Vol. 29, Part E, pp. 706–716. Academic Press, New York.

Dondon, J., Godefroy-Colburn, T., Graffe, M., and Grunberg-Manago, M. (1974). IF-3 requirements for initiation complex formation with synthetic messengers in *E. coli* system. *FEBS Lett.* **45**, 82–87.

Dunn, J. J., Buzash-Pollert, E., and Studier, F. W. (1978). Mutations of bacteriophage T7 that affect initiation of synthesis of the gene 0.3 protein. *Proc. Natl. Acad. Sci. U.S.A.* **75**, 2741–2745.

Dzionara, M., Robinson, S. M. L., and Wittmann-Liebold, B. (1977). Secondary structures of proteins from the 30 S subunit of the *Escherichia coli* ribosome. *Hoppe-Seyler's Z. Physiol. Chem.* **358**, 1003–1019.

Edelmann, P., and Gallant, J. (1977). Mistranslation in *E. coli. Cell* **10**, 131–137.

Ehresmann, C., Stiegler, P., Mackie, G. A., Zimmermann, R. A., Ebel, J. P., and Fellner, P. (1975). Primary sequence of the 16 S ribosomal RNA of *Escherichia coli. Nucleic Acids Res.* **2**, 265–278.

Eigner, E. A., and Loftfield, R. B. (1974). Kinetic techniques for the investigation of amino acid: tRNA ligases (aminoacyl-tRNA synthetases, amino acid activating enzymes). *In* "Methods in Enzymology" (L. Grossman and K. Moldave, eds.), Vol. 29, Part E, pp. 601–619. Academic Press, New York.

Emanuilov, I., Sabatini, D. D., Lake, J. A., and Freienstein, C. (1978). Localization of eukaryotic initiation factor 3 on native small ribosomal subunits. *Proc. Natl. Acad. Sci. U.S.A.* **75**, 1389–1393.

Erdmann, V. A. (1979). Collection of published 5 S and 5.8 S RNA sequences and their precursors. *Nucleic Acids Res.* **6**, r29–r44.

Eskin, B., Treadwell, B., Redfield, B., Spears, C., Kung, H., and Weissbach, H. (1978). Activity of different forms of initiation factor 2 in the *in vitro* synthesis of β-galactosidase. *Arch. Biochem. Biophys.* **189**, 531–534.

Fahnestock, S., and Rich, A. (1971). Ribosome-catalyzed polyester formation. *Science* **173**, 340–343.

Fessenden, J. M., and Moldave, K. (1961). Evidence for two protein factors in the transfer of amino acids from soluble-RNA to ribonucleoprotein particles. *Biochem. Biophys. Res. Commun.* **6**, 232–235.

Fiser, I., Scheit, K. H., Stöffler, G., and Küchler, E. (1975). Proteins at the mRNA binding site of the *Escherichia coli* ribosome. *FEBS Lett.* **56**, 226–229.

Fisher, N., Stöffler, G., and Wool, I. G. (1978). Immunological comparison of the proteins of chicken and rat liver ribosomes. *J. Biol. Chem.* **253**, 7355–7360.

Forget, B. G., and Weissman, S. M. (1969). The nucleotide sequence of ribosomal 5 S ribonucleic acid from KB cells. *J. Biol. Chem.* **244**, 3148–3165.

Furano, A. V., and Wittel, F. P. (1976). Effect of the *RelA* gene on the synthesis of individual proteins *in vivo. Cell* **8**, 115–122.

Galasinski, W., and Moldave, K. (1969). Purification of aminoacyltransferase II (translocation factor) from rat liver. *J. Biol. Chem.* **244**, 6527–6532.

Ganoza, M. C. (1966). Polypeptide chain termination in cell-free extracts of *E. coli. Cold Spring Harbor Symp. Quant. Biol.* **31**, 273–278.

Gauss, D. H., Grüter, F., and Sprinzl, M. (1979). Compilation of tRNA sequences. *Nucleic Acids Res.* **6**, r1–r19.

Geisser, M., Tischendorf, G. W., and Stöffler, G. (1973). Comparative immunological and electrophoretic studies of ribosomal proteins of *Bacillaceae. Mol. Gen. Genet.* **127**, 129–145.

Girschovich, A. S., Bochkareva, E. S., and Pozdnyakov, V. A. (1974). Affinity labelling of functional centers of *Escherichia coli* ribosomes. *Acta Biol. Med. Ger.* **33**, 639–648.

Gordon, J., Lucas-Lenard, J., and Lipmann, F. (1971). Isolation of bacterial chain elongation factors. *In* "Methods in Enzymology" (K. Moldave and L. Grossman, eds.), Vol. 20, Part C, pp. 281–291. Academic Press, New York.

Gordon, J., Baron, L. S., and Schweiger, M. (1972). Chromosomal localization of the structural genes of the polypeptide chain elongation factors. *J. Bacteriol.* **110**, 306–312.

Grady, L. J., North, A. B., and Campbell, W. P. (1978). Complexity of poly (A$^+$) and poly

(A⁻) polysomal RNA in mouse liver and cultured mouse fibroblasts. *Nucleic Acids Res.* **5**, 697–712.

Grasmuk, H., Nolan, R. D., and Drews, J. (1977a). Interchangeability of elongation factor-Tu and elongation factor-1 in aminoacyl-tRNA binding to 70 S and 80 S ribosomes. *FEBS Lett.* **82**, 237–242.

Grasmuck, H., Nolan, R. D., and Drews, J. (1977b). Further evidence that elongation factor 1 remains bound to ribosomes during peptide chain elongation. *Eur. J. Biochem.* **79**, 93–102.

Grunberg-Manago, M., Buckingham, R. H., Cooperman, B. S., and Hershey, J. W. B. (1978). Structure and function of the translation machinery. *Symp. Soc. Gen. Microbiol.* **28**, 27–110.

Gualerzi, C., Risuleo, G., and Pon, C. L. (1977). Initial rate kinetic analysis of the mechanism of initiation complex formation and the role of initiation factor IF-3. *Biochemistry* **16**, 1684–1689.

Hadjiolov, A. A. (1977). Patterns of ribosome biogenesis in eukaryotes. *Trends Biochem. Sci.* **2**, 84–86.

Hardy, S. J. S. (1975). The stoichiometry of the ribosomal proteins of *Escherichia coli. Mol. Gen. Genet.* **140**, 253–274.

Harris, R. J., and Pestka, S. (1977). Peptide bond formation. *In* "Molecular Mechanisms of Protein Biosynthesis" (H. Weissbach and S. Pestka, eds.), pp. 413–442. Academic Press, New York.

Hazelkorn, R., and Rothman-Denes, L. B. (1973). Protein synthesis. *Annu. Rev. Biochem.* **42**, 397–438.

Hecht, S. M., Tan, K. H., Chinault, A. C., and Arcari, P. (1977). Isomeric aminoacyl-tRNAs are both bound by elongation factor Tu. *Proc. Natl. Acad. Sci. U.S.A.* **74**, 437–441.

Held, W. A., Gette, W. R., and Nomura, M. (1974a). Role of 16 S ribosomal ribonucleic acid and the 30 S ribosomal protein S12 in the initiation of natural messenger ribonucleic acid translation. *Biochemistry* **13**, 2115–2122.

Held, W. A., Nomura, M., and Hershey, J. W. B. (1974b). Ribosomal protein S21 is required for full activity in the initiation of protein synthesis. *Mol. Gen. Genet.* **128**, 11–22.

Henriksen, O., Robinson, E. A., and Maxwell, E. S. (1975). Interactions of guanosine nucleotides with elongation factor 2. I. Equilibrium dialysis studies. *J. Biol. Chem.* **250**, 720–724.

Hershey, J. W. B., Yanov, J., Johnston, K., and Fakunding, J. L. (1977). Purification and characterization of protein synthesis initiation factors IF-1, IF-2 and IF-3 from *Escherichia coli. Arch. Biochem. Biophys.* **182**, 626–638.

Heywood, S. M., Kennedy, D. S., and Bester, A. J. (1974). Separation of specific initiation factors involved in the translation of myosin and myoglobin messenger RNAs and the isolation of a new RNA involved in translation. *Proc. Natl. Acad. Sci. U.S.A.* **71**, 2428–2431.

Higo, K., Held, W., Kahan, L., and Nomura, M. (1973). Functional correspondence between 30 S ribosomal proteins of *Escherichia coli* and *Bacillus stearothermophilus. Proc. Natl. Acad. Sci. U.S.A.* **70**, 944–948.

Higo, K.-I., and Loertscher, K. (1974). Amino-terminal sequences of some *Escherichia coli* 30 S ribosomal proteins and functionally corresponding *Bacillus stearothermophilus* ribosomal proteins. *J. Bacteriol.* **118**, 180–186.

Hirsch, C. A., Cox, M. A., van Venrooij, W. J. W., and Henshaw, E. C. (1973). The ribosome cycle in mammalian protein synthesis. II. Association of the native smaller ribosomal subunit with protein factors. *J. Biol. Chem.* **248**, 4377–4385.

Holley, R. W., Apgar, J., Everett, G. A., Madison, J. T., Marquisee, M., Merrill, S. H., Penswick, J. R., and Zamir, A. (1965). Structure of a ribonucleic acid. *Science* 147, 1462–1465.

Honjo, T., Nishizuka, Y., and Hayaishi, O. (1968). Diphtheria toxin-dependent adenosine diphosphate ribosylation of aminoacyl transferase II and inhibition of protein synthesis. *J. Biol. Chem.* 243, 3553–3555.

Hopfield, J. J. (1974). Kinetic proofreading: A new mechanism for reducing errors in biosynthetic processes requiring high specificity. *Proc. Natl. Acad. Sci. U.S.A.* 71, 4135–4139.

Howard, G. A., Smith, R. L., and Gordon, J. (1976). Chicken liver ribosomes: Characterization of cross-reaction and inhibition of some functions by antibodies prepared against *Escherichia coli* ribosomal proteins L7 and L12. *J. Mol. Biol.* 106, 623–637.

Howe, J. G., Yanov, J., Meyer, L., Johnston, K., and Hershey, J. W. B. (1978). Determination of protein synthesis initiation factor levels in crude lysates of *Escherichia coli* by a sensitive radioimmune assay. *Arch. Biochem. Biophys.* 191, 813–820.

Ilan, J., and Ilan, J. (1976). Requirement for homologous rabbit reticulocyte initiation factor 3 for initiation of α- and β-globin mRNA translation in a crude protozoal cell-free system. *J. Biol. Chem.* 251, 5718–5725.

Issinger, O.-G., Benne, R., Hershey, J. W. B., and Traut, R. R. (1976). Phosphorylation *in vitro* of eukaryotic initiation factors IF-E2 and IF-E3 by protein kinases. *J. Biol. Chem.* 251, 6471–6474.

Kaerlein, M., and Horak, I. (1978). Identification and characterization of ribosomal proteins phosphorylated in vaccinia-virus-infected HeLa cells. *Eur. J. Biochem.* 90, 463–469.

Kaziro, Y., and Inoue, N. (1968). Crystalline G factor from *Escherichia coli*. *J. Biochem. (Tokyo)* 64, 423–425.

Kenny, J. W., and Traut, R. R. (1979). Identification of fifteen neighboring protein pairs in the *Escherichia coli* 50 S ribosomal subunit crosslinked with 2-iminothiolane. *J. Mol. Biol.* 127, 243–263.

Kim, S. H., Suddath, F. L., Quigley, G. J., McPherson, A., Sussman, J. L., Wang, A. H. J., Seeman, N. C., and Rich, A. (1974). Three-dimensional tertiary structure of yeast phenylalanine transfer RNA. *Science* 185, 435–440.

Kinoshita, T., Kawano, G., and Tanaka, N. (1968). Association of fusidic acid sensitivity with G factor in a protein-synthesizing system. *Biochem. Biophys. Res. Commun.* 33, 769–773.

Kisselev, L. L., and Favorova, O. O. (1974). Aminoacyl-tRNA synthetases: Some recent results and achievements. *Adv. Enzymol. Relat. Subj. Biochem.* 40, 141–238.

Kjeldgaard, N. O., and Gausing, K. (1974). Regulation of biosynthesis of ribosomes. In "Ribosomes" (M. Nomura, A. Tissières, and P. Lengyel, eds.), pp. 369–392. Cold Spring Harbor Lab., Cold Spring Harbor, New York.

Klein, H. A., and Capecchi, M. R. (1971). Polypeptide chain termination: Purification of the release factors, R_1 and R_2, from *Escherichia coli*. *J. Biol. Chem.* 246, 1055–1061.

Kozak, M. (1978). How do eucaryotic ribosomes select initiation regions in messenger RNA? *Cell* 15, 1109–1123.

Kung, H., Redfield, B., Treadwell, B. V., Eskin, B., Spears, C., and Weissbach, H. (1977). DNA-directed *in vitro* synthesis of β-galactosidase. *J. Biol. Chem.* 259, 6899–6894.

Kurland, C. G. (1977). Structure and function of the bacterial ribosome. *Annu. Rev. Biochem.* 46, 173–200.

Lake, J. A., Sabatini, D. D., and Nonomura, Y. (1974). Ribosome structure as studied by electron microscopy. In "Ribosomes" (M. Nomura, A. Tissières, and P. Lengyel, eds.), pp. 543–557, Cold Spring Harbor, Lab., Cold Spring Harbor, New York.

Langberg, S., Kahan, L., Traut, R. R., and Hershey, J. W. B. (1977). Binding of protein synthesis initiation factor IF-1 to 30 S ribosomal subunits: Effects of other initiation factors and identification of proteins near the binding site. *J. Mol. Biol.* **117**, 307–319.

Lee-Huang, S., and Ochoa, S. (1971). Messenger discriminating species of initiation factor F_3. *Nature (London), New Biol.* **234**, 236–239.

Levin, D. H., and Kyner, D. (1971). Specific formation *in vitro* of a ribosomal protein-GTP-Met-tRNA$_f$ ternary complex from eukaryotic sources. *Fed. Proc., Fed. Am. Soc. Exp. Biol.* **30**, 1289.

Littlefield, J. W., Keller, E. B., Gross, J., and Zamecnik, P. C. (1955). Studies on cytoplasmic ribonucleoprotein particles from the liver of the rat. *J. Biol. Chem.* **217**, 111–123.

Lodish, H. F. (1970). Specificity in bacterial protein synthesis: Role of initiation factors and ribosomal subunits. *Nature (London)* **226**, 705–707.

Loftfield, R. B. (1972). The mechanism of aminoacylation of transfer RNA. *Prog. Nucleic Acid Res. Mol. Biol.* **12**, 87–128.

Loftfield, R. B., and Eigner, E. A. (1969). Mechanism of action of amino acid transfer ribonucleic acid ligases. *J. Biol. Chem.* **244**, 1746–1754.

Loftfield, R. B., and Vanderjagt, D. (1972). The frequency of errors in protein biosynthesis. *Biochem. J.* **128**, 1353–1356.

Lutsch, G., Bielka, H., Wahn, K., and Stahl, J. (1972). Studies on the structure of animal ribosomes. *Acta Biol. Med. Ger.* **29**, 851–876.

Lutter, L. C., and Kurland, C. G. (1973). Reconstitution of active ribosomes with cross-linked proteins. *Nature (London), New Biol.* **243**, 15–17.

McConkey, E. H., Bielka, H., Gordon, J., Lastick, S. M., Lin, A., Ogata, K., Reboud, J.-P., Traugh, J. A., Traut, R. R., Warner, J. R., Welfle, H., and Wool, I. G. (1979). Proposed uniform nomenclature for mammalian ribosomal proteins. *Mol. Gen. Genet.* **169**, 1–6.

Machie, G. A. (1977). Evidence for a precursor-product relationship in the biosynthesis of ribosomal protein S20. *Biochemistry* **16**, 1391–1398.

Maden, B. E. H., and Salim, M. (1974). The methylated nucleotide sequences in HeLa cell ribosomal RNA and its precursors. *J. Mol. Biol.* **88**, 133–164.

Madjar, J.-J., Arpin, M., Buisson, M., and Reboud, J.-P. (1979). Spot positions of rat liver ribosomal proteins by four different two-dimensional electrophoreses in polyacrylamide gels. *Mol. Gen. Genet.* **171**, 121–134.

Marcus, A. (1970). Tobacco mosaic virus ribonucleic acid-dependent amino acid incorporation in a wheat embryo system *in vitro*. *J. Biol. Chem.* **245**, 955–961.

Miller, D. L., and Weissbach, H. (1974). Elongation factor Tu and the aminoacyl-tRNA · EFTu · GTP complex. *In* "Methods in Enzymology" (L. Grossman and K. Moldave, eds.), Vol. 30, Part F, pp. 219–232.

Miller, D. L., and Weissbach, H. (1977). Factors involved in the transfer of aminoacyl-tRNA to the ribosome. *In* "Molecular Mechanisms of Protein Biosynthesis" (H. Weissbach and S. Pestka, eds.), pp. 323–373. Academic Press, New York.

Miller, D. L., Cashel, M., and Weissbach, H. (1973). The interaction of guanosine 5'-diphosphate, 2'(3')-diphosphate with the bacterial elongation factor Tu. *Arch. Biochem. Biophys.* **154**, 675–682.

Milman, G., Goldstein, J., Scolnick, E., and Caskey, T. (1969). Peptide chain termination. III. Stimulation of *in vitro* termination. *Proc. Natl. Acad. Sci. U.S.A.* **63**, 183–190.

Miyajima, A., and Kaziro, Y. (1978). Coordination of levels of elongation factors Tu, Ts and G, and ribosomal protein S1 in *Escherichia coli*. *J. Biochem. (Tokyo)* **83**, 453–462.

Mizumoto, K., Iwasaki, K., and Kaziro, Y. (1974). Studies of polypeptide elongation factor

2 from pig liver. III. Interaction with guanine nucleotides in the presence and absence of ribosomes. *J. Biochem. (Tokyo)* **76**, 1269–1280.

Mizushima, S., and Nomura, M. (1970). Assembly mapping of 30 S ribosomal proteins from *E. coli. Nature (London)* **226**, 1214–1218.

Modolell, J., and Vazquez, D. (1973). Inhibition by aminoacyl transfer ribonucleic acid of elongation factor G-dependent binding of guanosine nucleotide ribosomes. *J. Biol. Chem.* **248**, 488–493.

Monro, R. E., Cerná, J., and Marcker, K. (1968). Ribosome-catalyzed peptidyl transfer: Substrate specificity at the p-site. *Proc. Natl. Acad. Sci. U.S.A.* **61**, 1042–1049.

Moore, P. B., Langer, J. A., Schoenborn, B. P., and Engelman, D. M. (1977). Triangulation of proteins in the 30 S ribosomal subunit of *Escherichia coli. J. Mol. Biol.* **112**, 199–234.

Morgan, S. D., and Söll, D. (1978). Regulation of the biosynthesis of amino acid: tRNA ligases and of tRNA. *Annu. Rev. Microbiol.* **32**, 181–207.

Morikawa, K., La Cour, T. F. M., Nyborg, J., Rasmussen, K. M., Miller, D. L., and Clark, B. F. C. (1978). High resolution X-ray crystallographic analysis of a modified form of elongation factor Tu: Guanosine diphosphate complex. *J. Mol. Biol.* **125**, 325–338.

Motoyoshi, K., and Iwasaki, K. (1977). Resolution of the polypeptide chain elongation factor-$1_{\beta\gamma}$ into subunits and some properties of the subunits. *J. Biochem. (Tokyo)* **82**, 703–708.

Nagata, S., Iwasaki, K., and Kaziro, Y. (1976). Interaction of the low molecular weight form of elongation factor 1 with guanine nucleotides and aminoacyl-tRNA. *Arch. Biochem. Biophys.* **172**, 168–177.

Neidhardt, F. C., Parker, J., and McKeever, W. G. (1975). Function and regulation of aminoacyl-tRNA synthetases in prokaryotic and eukaryotic cells. *Annu. Rev. Microbiol.* **29**, 215–250.

Neupert, W. (1977). Mitochondrial ribosomes. *Horiz. Biochem. Biophys.* **3**, 257–296.

Nierhaus, K. H., and Dohme, F. (1974). Total reconstitution of functionally active 50 S ribosomal subunits from *Escherichia coli. Proc. Natl. Acad. Sci. U.S.A.* **71**, 4713–4717.

Nierlich, D. P. (1978). Regulation of bacterial growth, RNA and protein synthesis. *Annu. Rev. Microbiol.* **32**, 393–432.

Ninio, J. (1974). A semi-quantitative treatment of missense and nonsense suppression in the *str* A and *ram* ribosomal mutants of *Escherichia coli:* Evaluation of some molecular parameters of translation *in vivo. J. Mol. Biol.* **84**, 297–313.

Ninio, J. (1975). Kinetic amplification of enzyme discrimination. *Biochimie* **57**, 587–595.

Noller, H. F. (1974). Topography of 16 S RNA in 30 S ribosomal subunits. Nucleotide sequences and location of sites of reaction with kethoxal. *Biochemistry* **13**, 4694–4703.

Nomura, M. (1976). Organization of bacterial genes for ribosomal components: Studies using novel approaches. *Cell* **9**, 633–644.

Nomura, M., and Erdmann, V. A. (1970). Reconstitution of 50 S ribosomal subunits from dissociated molecular components. *Nature (London)* **228**, 744–748.

Nomura, M., and Held, W. A. (1974). Reconstitution of ribosomes: Studies of ribosome structure, function and assembly. *In* "Ribosomes" (M. Nomura, A. Tissières, and P. Lengyel, eds.), pp. 193–223. Cold Spring Harbor Lab., Cold Spring Harbor, New York.

Nomura, M., and Morgan, E. A. (1977). Genetics of bacterial ribosomes. *Annu. Rev. Genet.* **11**, 297–347.

Nomura, M., Tissières, A., and Lengyel, P., eds. (1974). "Ribosomes." Cold Spring Harbor Lab., Cold Spring Harbor, New York.

Ofengand, J. (1977). tRNA and aminoacyl-tRNA synthetases. In "Molecular Mechanisms of Protein Biosynthesis" (H. Weissbach and S. Pestka, eds.), pp. 7–79. Academic Press, New York.

Parmeggiani, A. (1968). Crystalline transfer factors from Escherichia coli. Biochem. Biophys. Res. Commun. 30, 613–619.

Pearson, R. L., Weiss, J. F., and Kelmers, A. D. (1971). Improved separation of transfer RNA's on polychlorotrifluoroethylene-supported reversed-phase chromotography columns. Biochim. Biophys. Acta 228, 770–774.

Pellegrini, M., and Cantor, C. R. (1977). Affinity labeling of ribosomes. In "Molecular Mechanisms of Protein Biosynthesis" (H. Weissbach and S. Pestka, eds.), pp. 203–244. Academic Press, New York.

Perry, R. P. (1976). Processing of RNA. Annu. Rev. Biochem. 45, 605–629.

Pestka, S. (1972). Studies on transfer ribonucleic acid-ribosome complexes. XIX. Effect of antibiotics on peptidyl puromycin synthesis on polyribosomes from Escherichia coli. J. Biol. Chem. 247, 4669–4678.

Pongs, O., Stöffler, G., and Lanka, E. (1975). The codon binding site of the Escherichia coli ribosome as studied with a chemically reactive A-U-G analog. J. Mol. Biol. 99, 301–315.

Reid, B. R. (1977). Synthetase-tRNA recognition. In "Nucleic Acid-Protein Recognition" (H. J. Vogel, ed.), pp. 375–390. Academic Press, New York.

Revel, M. (1977). Initiation of messenger RNA translation into protein and some aspects of its regulation. In "Molecular Mechanisms of Protein Biosynthesis" (H. Weissbach and S. Pestka, eds.), pp. 245–321. Academic Press, New York.

Revel, M., and Groner, Y. (1978). Post-transcriptional and translational controls of gene expression in eukaryotes. Annu. Rev. Biochem. 47, 1079–1126.

Revel, M., Aviv (Greenshpan), H., Groner, Y., and Pollack, Y. (1970). Fractionation of translation initiation factor B(F3) into cistron-specific species. FEBS Lett. 9, 213–217.

Rich, A. (1977). The molecular structure of transfer RNA and its interaction with synthetases. In "Nucleic Acid-Protein Recognition" (H. J. Vogel, ed.), pp. 281–291. Academic Press, New York.

Rich, A. (1978). Transfer RNA: Three-dimensional structure and biological function. Trends Biochem. Sci. 3, 34–37.

Rich, A., and RajBhandary, U. L. (1976). Transfer RNA: Molecular structure, sequence, and properties. Annu. Rev. Biochem. 45, 805–860.

Ringer, D., Quiggle, K., and Chládek, S. (1975). Recognition of the 3' terminus of 2'-O-aminoacyl transfer ribonucleic acid by the acceptor site of ribosomal peptidyltransferase. Biochemistry 14, 514–520.

Robertus, J. D., Ladner, J. E., Finch, J. T., Rhodes, D., Brown, R. S., Clark, B. F. C., and Klug, A. (1974). Structure of yeast phenylalanine tRNA at 3 Å resolution. Nature (London) 250, 546–551.

Rohrbach, M. S., and Bodley, J. W. (1976). Steady state kinetic analysis of the mechanism of guanosine triphosphate hydrolysis catalyzed by Escherichia coli elongation factor G and the ribosome. Biochemistry 15, 4565–4569.

Safer, B., and Anderson, W. F. (1978). The molecular mechanism of hemoglobin synthesis and its regulation in the reticulocyte. Crit. Rev. Biochem. 5, 261–290.

Safer, B., Adams, S. L. Kemper, W. M., Berry, K. W., Lloyd, M., and Merrick, W. C. (1976). Purification and characterization of two initiation factors required for maximal activity of a highly fractionated globin mRNA translation system. Proc. Natl. Acad. Sci. U.S.A. 73, 2584–2588.

Safer, B., Kemper, W., and Jagus, R. (1978). Identification of a 48 S preinitiation complex in reticulocyte lysate. *J. Biol. Chem.* **253**, 3384–3386.
Schiff, N., Miller, M. J., and Wahba, A. J. (1974). Purification and properties of chain initiation factor 3 from T4-infected and uninfected *Escherichia coli* MRE600. *J. Biol. Chem.* **249**, 3797–3802.
Schlessinger, D. (1974). Ribosome formation in *Escherichia coli*. In "Ribosomes" (M. Nomura, A. Tissières, and P. Lengyel, eds.), pp. 393–416. Cold Spring Harbor Lab., Cold Spring Harbor, New York.
Schreier, M. H., and Staehelin, T. (1973). Functional characterization of five initiation factors for mammalian protein synthesis. In "Regulation of Translation in Eukaryotes" (E. Bautz, ed.) 24th Mosbacher Colloq., pp. 335–349. Springer-Verlag, Berlin and New York.
Schreier, M. H., Erni, B., and Staehelin, T. (1977). Initiation of mammalian protein synthesis. I. Purification and characterization of seven initiation factors. *J. Mol. Biol.* **116**, 727–753.
Schulman, L. H., Pelka, H., and Sundari, R. M. (1974). Structural requirements for recognition of *Escherichia coli* initiator and non-initiator transfer ribonucleic acids by bacterial T factor. *J. Biol. Chem.* **249**, 7102–7110.
Shafritz, D. A., Weinstein, J. A., Safer, B., Merrick, W. C., Weber, L. A., Hickey, E. D., and Baglioni, C. (1976). Evidence for role of m⁷G⁵'-phosphate group in recognition of eukaryotic mRNA by initiation factor IF-M₃. *Nature (London)* **261**, 291–294.
Shatkin, A. J. (1976). Capping of eucaryotic mRNAs. *Cell* **9**, 645–653.
Sherton, C. C., and Wool, I. G. (1972). Determination of the number of proteins in liver ribosomes and ribosomal subunits by two-dimensional polyacrylamide gel electrophoresis. *J. Biol. Chem.* **247**, 4460–4467.
Shine, J., and Dalgarno, L. (1974). The 3'-terminal sequence of *Escherichia coli* 16 S ribosomal RNA: Complementarity of nonsense triplets and ribosome binding sites. *Proc. Natl. Acad. Sci. U.S.A.* **71**, 1342–1346.
Shorey, R. L., Ravel, J. M., and Shive, W. (1971). The effect of guanylyl-5'-methylene diphosphate on binding of aminoacyl-transfer ribonucleic acid to ribosomes. *Arch. Biochem. Biophys.* **146**, 110–117.
Söll, D., and Schimmel, P. R. (1974). Aminoacyl-tRNA synthetases. In "The Enzymes" (P. D. Boyer, ed.), 3rd ed., Vol. 10, pp. 489–538. Academic Press, New York.
Sommer, A., and Traut, R. R. (1976). Identification of neighboring protein pairs in the *Escherichia coli* 30 S ribosomal subunit by crosslinking with methyl 4-mercaptobutyrimidate. *J. Mol. Biol.* **106**, 995–1015.
Sonenberg, N., Morgan, M. A., Merrick, W. C., and Shatkin, A. J. (1978). A polypeptide in eukaryotic initiation factors that crosslinks specifically to the 5'-terminal cap in mRNA. *Proc. Natl. Acad. Sci. U.S.A.* **75**, 4843–4847.
Spremulli, L. L., Haralson, M. A., and Ravel, J. M. (1974). Effect of T4 infection on initiation of protein synthesis and messenger specificity of initiation factor 3. *Arch. Biochem. Biophys.* **165**, 581–587.
Springer, M., Graffe, M., and Grunberg-Manago, M. (1977). Characterization of an *E. coli* mutant with a thermolabile initiation factor IF-3 activity. *Mol. Gen. Genet.* **151**, 17–26.
Staehelin, T., Trachsel, H., Erni, B., Boschetti, A., and Schreier, M. H. (1975). The mechanism of initiation of mammalian protein synthesis. *Mol. Interactions Genet. Transl., Fed. Eur. Biochem. Soc. Meet., 10th, 1975*, pp. 309–323.
Steitz, J. A. (1969). Polypeptide chain initiation: Nucleotide sequences of the three ribosomal binding sites in bacteriophage R17 RNA. *Nature (London)* **224**, 957–964.

Steitz, J. A. (1979). Genetic signals and nucleotide sequences in messenger RNA. *In* "Biological Regulations and Development" (R. F. Goldberger, ed.), Vol. 1, pp. 349–399. Plenum, New York.

Steitz, J. A., and Jakes, K. (1975). How ribosomes select initiator regions in mRNA: Base pair formation between the 3'-terminus of 16 S rRNA and the mRNA during initiation of protein synthesis in *Escherichia coli. Proc. Natl. Acad. Sci. U.S.A.* **72**, 4734–4738.

Steitz, J. A., Sprague, K. U., Steege, D. A., Yuan, R. C., Laughrea, M., Moore, P. B., and Wahba, A. J. (1977a). RNA–RNA and protein–RNA interactions during the initiation of protein synthesis. *In* "Nucleic Acid-Protein Recognition" (H. J. Vogel, ed.), pp. 491–508. Academic Press, New York.

Steitz, J. A., Wahba, A. J., Laughrea, M., and Moore, P. B. (1977b). Differential requirements for polypeptide chain initiation complex formation at the three bacteriophage R17 initiator regions. *Nucleic Acids Res.* **4**, 1–15.

Stöffler, G. (1974). Structure and function of the *Escherichia coli* ribosome: Immunochemical analysis. *In* "Ribosomes" (M. Nomura, A. Tissières, and P. Lengyel, eds.), pp. 615–667. Cold Spring Harbor Lab., Cold Spring Harbor, New York.

Stöffler, G., and Wittmann, H. G. (1977). Primary structure and three-dimensional arrangement of proteins within the *Escherichia coli* ribosome. *In* "Molecular Mechanisms of Protein Biosynthesis" (H. Weissbach and S. Pestka, eds.), pp. 117–202. Academic Press, New York.

Stöffler, G., Wool, I. G., Lin, A., and Rak, K. H. (1974). The identification of the eukaryotic ribosomal proteins homologous with *Escherichia coli* proteins L7 and L12. *Proc. Natl. Acad. Sci. U.S.A.* **71**, 4723–4726.

Stuhrmann, H. B., Koch, M. H. J., Parfait, R., Haas, J., Ibel, K., and Crichton, R. R. (1977). Shape of the 50 S subunit of *Escherichia coli* ribosomes. *Proc. Natl. Acad. Sci. U.S.A.* **74**, 2316–2320.

Sun, T.-T., Bickle, T. A., and Traut, R. R. (1972). Similarity in the size and number of ribosomal proteins from different prokaryotes. *J. Bacteriol.* **111**, 474–480.

Sundkvist, I. C., and Staehelin, T. (1975). Structure and function of free 40 S ribosome subunits: Characterization of initiation factors. *J. Mol. Bioi.* **99**, 401–418.

Suryanarayana, T., and Subramanian, A. R. (1977). Separation of two forms of IF-3 in *Escherichia coli* by two-dimensional gel electrophoresis. *FEBS Lett.* **79**, 264–268.

Tate, W. P., Beaudet, A. L., and Caskey, C. T. (1973). Influence of guanine nucleotides and elongation factors on interaction of release factors with the ribosome. *Proc. Natl. Acad. Sci. U.S.A.* **70**, 2350–2352.

Thach, S. S., and Thach, R. E. (1971). Translocation of messenger RNA and "accommodation" of fMet-tRNA. *Proc. Natl. Acad. Sci. U.S.A.* **68**, 1791–1795.

Thompson, H. A., Sadnik, I., Scheinbuks, J., and Moldave, K. (1977). Studies on native ribosomal subunits from rat liver. Purification and characterization of a ribosome dissociation factor. *Biochemistry* **16**, 2221–2230.

Thompson, R. C., and Stone, P. J. (1977). Proofreading of the codon–anticodon interaction on ribosomes. *Proc. Natl. Acad. Sci. U.S.A.* **74**, 198–202.

Tocchini-Valentini, G. P., and Mattoccia, E. (1968). A mutant of *E. coli* with an altered supernatant factor. *Proc. Natl. Acad. Sci. U.S.A.* **61**, 146–151.

Trachsel, H., and Staehelin, T. (1978). Binding and release of eukaryotic initiation factor eIF-2 and GTP during protein synthesis initiation. *Proc. Natl. Acad. Sci. U.S.A.* **75**, 204–208.

Trachsel, H., Erni, B., Schreier, M. H., and Staehelin, T. (1977). Initiation of mammalian protein synthesis. II. The assembly of the initiation complex with purified initiation factors. *J. Mol. Biol.* **116**, 755–767.

Traub, P., and Nomura, M. (1968). Structure and function of *E. coli* ribosomes, V. Reconstitution of functionally active 30 S ribosomal particles from RNA and proteins. *Proc. Natl. Acad. Sci. U.S.A.* **59,** 777–784.

Traugh, J. A., Tahara, S. M., Sharp, S. B., Safer, B., and Merrick, W. C. (1976). Factors involved in initiation of haemoglobin synthesis can be phosphorylated *in vitro*. *Nature (London)* **263,** 163–165.

Traut, R. R., Heimark, R. L., Sun, T.-T., Hershey, J. W. B., and Bollen, A. (1974). *In* "Ribosomes" (M. Nomura, A. Tissières, and P. Lengyel, eds.), pp. 271–308. Cold Spring Harbor Lab., Cold Spring Harbor, New York.

Tsurugi, K., Collatz, E., Todokow, K., Ulbrich, N., Lightfoot, H. N., and Wool, I. G. (1978). Isolation of eukaryotic ribosomal proteins. *J. Biol. Chem.* **253,** 946–955.

Van der Hofstad, G. A. J. M., Foekens, J. A., Bosch, L., and Voorma, H. O. (1977). The involvement of a complex between formylmethionyl-tRNA and initiation factor IF-2 in prokaryotic initiation. *Eur. J. Biochem.* **77,** 69–75.

Van Duin, J., and Van Knippenberg, P. H. (1974). Functional heterogeneity of the 30 S ribosomal subunit of *Escherichia coli*. *J. Mol. Biol.* **84,** 185–195.

Van Holde, K. E., and Hill, W. E. (1974). General physical properties of ribosomes. *In* "Ribosomes" (M. Nomura, A. Tissières, and P. Lengyel, eds.), pp. 53–91. Cold Spring Harbor, Lab., Cold Spring Harbor, New York.

Voorma, H. O., Benne, R., Naaktgeboren, N., and Van der Hofstad, G. (1979). *In* "Methods in Enzymology" (K. Moldave and L. Grossman, eds.), Vol. 60, pp. 124–135. Academic Press, New York.

Wahba, A. J., and Miller, M. J. (1974). Chain initiation factors from *Escherichia coli*. *In* "Methods in Enzymology" (L. Grossman and K. Moldave, eds.), Vol. 30, Part F, pp. 3–18. Academic Press, New York.

Walton, G. M., and Gill, G. N. (1976). Regulation of ternary [Met-tRNA$_f$ · GTP · eukaryotic initiation factor[2]] protein synthesis initiation complex formation by the adenylate energy charge. *Biochim. Biophys. Acta* **418,** 195–203.

Warner, J. R. (1974). The assembly of ribosomes in eukaryotes. *In* "Ribosomes" (M. Nomura, A. Tissières, and P. Lengyel, eds.), pp. 461–488. Cold Spring Harbor Lab., Cold Spring Harbor, New York.

Weiel, J., Hershey, J. W. B., and Levison, S. A. (1978). Fluorescence polarization studies of the binding of fluorescein-labeled initiation factor IF-3 to 30 S ribosomal subunits from *Escherichia coli*. *FEBS Lett.* **87,** 103–106.

Weissbach, H., and Ochoa, S. (1976). Soluble factors required for eukaryotic protein synthesis. *Annu. Rev. Biochem.* **45,** 191–216.

Weissbach, H., and Pestka, S., eds. (1977). "Molecular Mechanisms of Protein Biosynthesis." Academic Press, New York.

Weissbach, H., Redfield, B., and Moon, H.-M. (1973). Further studies on the interactions of elongation factor 1 from animal tissues. *Arch. Biochem. Biophys.* **156,** 267–275.

Welfle, H., Stahl, J., and Bielka, H. (1972). Studies on proteins of animal ribosomes. XIII. Enumeration of ribosomal proteins of rat liver. *FEBS Lett.* **26,** 228–232.

Westermann, P., Neumann, W., Bommer, U.-A., Bielka, H., Nygard, O., and Hultin, T. (1979). Crosslinking of initiation factor eIF-2 to proteins of the small subunit of rat liver ribosomes. *FEBS Lett.* **97,** 101–104.

Wigle, D. T., and Smith, A. E. (1973). Specificity in initiation of protein synthesis in a fractionated mammalian cell-free system. *Nature (London), New Biol.* **242,** 136–140.

Winkelmann, D., and Kahan, L. (1978). Accessibility of the antigenic determinants of ribosomal protein S4 on the 30 S ribosomal subunit and assembly intermediates. *Fed. Proc., Fed. Am. Soc. Exp. Biol.* **37,** 1739.

68   John W. B. Hershey

Wool, I. G. (1979). The structure and function of eukaryotic ribosomes. *Annu. Rev. Biochem.* **48,** 719–754.

Wool, I. G., and Stöffler, G. (1974). Structure and function of eukaryotic ribosomes. *In* "Ribosomes" (M. Nomura, A. Tissières, and P. Lengyel, eds.), pp. 417–460. Cold Spring Harbor, Lab., Cold Spring Harbor, New York.

Wrede, P., and Erdmann, V. A. (1977). *Escherichia coli* 5 S RNA binding proteins L18 and L25 interact with 5.8 S RNA but not with 5 S RNA from yeast ribosomes. *Proc. Natl. Acad. Sci. U.S.A.* **74,** 2706–2709.

Yaguchi, M., Matheson, A. T., and Visentin, L. P. (1974). Procaryotic ribosomal proteins: N-terminal sequence homologies and structural correspondence of 30 S ribosomal proteins from *Escherichia coli* and *Bacillus stearothermophilus*. *FEBS Lett.* **46,** 296–300.

2

Regulation of Eukaryotic Protein Synthesis

Gisela Kramer and Boyd Hardesty

I. INTRODUCTION

In general, the elements of peptide initiation and translational regulation in eukaryotic organisms are different in structure, more numerous, more complex, and not as well characterized as their prokaryotic counterparts.

69

The material presented here should be considered to be a progress report on an emerging and rapidly changing area. Examples have been selected to provide an overview of the primary elements of translational control as they are currently recognized and understood. However, many of the phenomena associated with translational control in eukaryotes have proved to be experimentally difficult to elucidate, and the underlying biochemical mechanisms have not been resolved. Conflicting results and divergent interpretations are common. We have made a special effort to indicate uncertainties and unresolved controversies.

We will consider only those biochemical events that affect the steps of protein synthesis in which ribosomes are involved. Regulation also may occur at a number of other posttranscriptional sites, such as processing, transport, storage and stability of mRNA, and processing of newly synthesized peptides or proenzymes into an active form.

Steps of protein synthesis are conveniently subdivided into peptide initiation, elongation, and termination. Polypeptide initiation appears to be the rate-limiting step of protein synthesis in most systems and the point at which most translational control mechanisms operate. The mechanism of eukaryotic peptide initiation has been described and discussed in the preceding chapter of this volume. A summary diagram is presented in Fig. 1.

Eukaryotic peptide initiation involves at least seven factors, probably more, and requires the hydrolysis of GTP and ATP. The initiation factor,

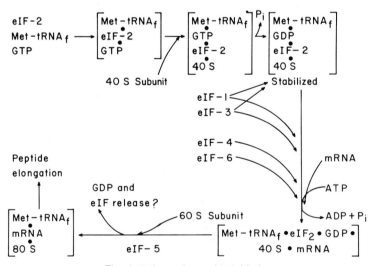

Fig. 1. Eukaryotic peptide initiation.

eIF-2, forms a ternary complex with Met-tRNA$_f$ and GTP. This ternary complex can bind to 40 S ribosomal subunits. There is good evidence that binding of the initiator tRNA, Met-tRNA$_f$, to 40 S ribosomal subunits precedes binding of mRNA. It is likely that other factors influence this binding reaction and the activity of eIF-2 (Majumdar et al., 1977; Dasgupta et al., 1978). Evidence has been presented (Odom et al., 1978) and recently confirmed (Merrick, 1979) that GTP hydrolysis occurs concomitantly with or immediately following proper binding of the ternary complex to 40 S ribosomal subunits. Two of the three subunits of eIF-2 may be phosphorylated by different protein kinases. Phosphorylation of at least the smallest subunit alters the biological activity of eIF-2 and is a primary site of translational control as discussed below. A number of translational control systems appear to involve binding of mRNA to 40 S ribosomal subunits. The mechanism and role of initiation factors in the latter reaction are less clear than for the binding of Met-tRNA$_f$. A number of protein factors and ATP are required. Very little is known of the specific chemical and physical roles these factors play in binding of mRNA to the ribosomes.

Regulatory mechanisms involving binding of Met-tRNA$_f$ or mRNA will be treated separately in the primary sections of this chapter. The mechanisms by which interferon and peptide hormones effect translational control in target cells appears to be complex and will be considered separately.

II. REGULATION AT THE STEP OF MET-tRNA$_f$ BINDING TO 40 S RIBOSOMAL SUBUNITS

A. Control of Protein Synthesis in Reticulocytes by Heme

Early observations demonstrated that protein synthesis in reticulocytes (Kruh and Borsook, 1956; Bruns and London, 1965) and their cell-free lysates (Adamson et al., 1968; Zucker and Schulman, 1968) is regulated by heme. When a reticulocyte lysate is incubated at 37°C in the absence of added hemin, the initial rate of leucine incorporation is normal but decreases abruptly to a low rate, typically 2 to 10% of the initial rate, after about 5 min of incubation. The exact time of shutoff varies somewhat with individual lysates, apparently reflecting variation in the amount of endogenous heme, the initial rate of protein synthesis, and other factors. This striking inhibition in the absence of added hemin has been shown in many studies to involve a block in the initiation of new peptides with no detectable effect on peptide elongation (Adamson et al., 1968; Zucker and

Schulman, 1968; Grayzel *et al.*, 1966; Waxman and Rabinovitz, 1966; Howard *et al.*, 1970). Polysomes are converted to 80 S ribosomes and ribosomal subunits as nascent peptides are completed and released. Inhibition is associated with the disappearance of Met-tRNA$_f$ on 40 S ribosomal subunits and appears to involve inhibition of the reaction by which the ternary complex formed between Met-tRNA$_f$, eIF-2, and GTP is bound to small ribosomal subunits (Balkow *et al.*, 1973; Legon *et al.*, 1973). The inhibition is potentiated by millimolar concentrations of ATP, but is reversed by 2 mM GTP, or high concentrations of cAMP, in the range of 2–10 mM (Balkow *et al.*, 1975). It should be noted that the concentration of cAMP necessary to activate preparations of cAMP-dependent protein kinases half-maximally is in the order of $2 \times 10^{-7} M$ (Beavo *et al.*, 1974). cAMP is inactive in reversing inhibition of protein synthesis in heme-deficient lysates in this concentration range. It is assumed that the effect noted above does not involve a cAMP-dependent protein kinase as discussed in detail below.

Gross and Rabinovitz (1973) partially purified what appears to be a high molecular weight complex from reticulocyte postribosomal supernatant incubated in the absence of hemin. When added exogenously to a hemin-supplemented reticulocyte lysate, this material blocked peptide initiation with the same kinetics and characteristics as observed under heme deficiency. The agent responsible for the inhibition was named the "heme-controlled repressor" or HCR. It is not clear whether the inhibitory activity activated in the absence of heme is a single enzyme or several functionally related enzymes that may be physically associated in the form of a multienzyme complex under physiological conditions. The inhibitory activity observed by Gross and Rabinovitz (1972a) eluted from gel filtration columns as a protein of about 300,000 daltons. However, the inhibitor was estimated to have a molecular weight of about 95,000 in purified preparations when analyzed under denaturing as well as nondenaturing conditions (Trachsel *et al.*, 1978).

The mechanism by which HCR blocks peptide initiation appears to involve phosphorylation of the smallest subunit (38,000 daltons) of eIF-2. Highly purified preparations of HCR contain a cAMP-independent protein kinase for this peptide (Farrell *et al.*, 1977; Kramer *et al.*, 1976; Levin *et al.*, 1976), and it has been shown that HCR inhibition of protein synthesis in lysates may be overcome by the addition of exogenous eIF-2 (Ranu *et al.*, 1976). Farrell *et al.* (1978) analyzed eIF-2 taken directly from reticulocyte lysates incubated under conditions in which HCR was activated. They found a direct correlation between phosphorylation of the smallest subunit and inhibition of protein synthesis by HCR. The eIF-2- and GTP-dependent binding of Met-tRNA$_f$ to 40 S ribosomal subunits may be measured independently as a partial reaction of peptide initiation. This system was used to demonstrate directly that phosphorylated eIF-2 was inactivated for the binding reaction (Pinphanichakarn *et al.*, 1976; Kramer *et al.*, 1977; Ranu *et al.*, 1978). Highly purified eIF-2 was phosphorylated from ATP by HCR. Excess ATP remaining in the reaction mixture was destroyed by added hexokinase and glucose, and then the activity of the phosphorylated eIF-2 for binding of Met-tRNA$_f$ to 40 S subunits was measured with a nonhydrolyzable analogue of GTP such as GMP-P(CH$_2$)P (Kramer *et al.*, 1977). HCR-dependent inhibition of the binding reaction was observed in this simplified assay system with no detectable reduction in the ability of eIF-2 to form the ternary complex with Met-tRNA$_f$ and GTP, Met-tRNA$_f$ · eIF-2 · GTP. However, Trachsel and Staehelin (1978) were unable to demonstrate reduced activity of eIF-2 that was phosphorylated by HCR and then reisolated by chromatography. The basis for this apparent discrepancy in reaction systems containing purified components is not clear. Phosphatase activity for phosphorylated eIF-2 may be involved as well as the relatively low activity that is observed for many preparations of highly purified eIF-2.

There is little doubt that phosphorylation of the small subunit of eIF-2 is the causal event in inhibition of Met-tRNA$_f$ binding to 40 S ribosomal subunits. Grankowski *et al.* (1980a) have demonstrated that phosphorylated eIF-2 is reactivated by dephosphorylation with the protein phosphatase described below. Thus the eIF-2 kinase systems with their counteracting phosphatase appear to provide a physiologically important mechanism by which binding of Met-tRNA$_f$ to 40 S ribosomal subunits can be reversibly inhibited and thereby the rate of protein synthesis may be modulated.

The details of the mechanism by which HCR is activated under physiological conditions are unknown. Gross and Rabinovitz (1972b) demonstrated that in reticulocyte postribosomal supernatants an "inactive prorepressor" was converted to what they called the "reversible inhibitor" in the absence of heme. This activated inhibitor could be converted back to an inactive form by heme. However, in the prolonged absence of heme, the reversible inhibitor was converted into a form, called the "irreversible inhibitor," that was no longer sensitive to heme. The inactive proinhibitor could be converted directly to the irreversible inhibitor by treatment with the sulfhydryl reactive reagent, *N*-ethylmaleimide. The relationship of these forms of the inhibitor may be as indicated in Fig. 2.

Trachsel *et al.* (1978) and Gross and Mendelewski (1978) have isolated the reversible inhibitor that can be activated or inactived in the presence or absence of heme and that is converted to the irreversible form by treatment with *N*-ethylmaleimide. The existence of a proinhibitor, distinct

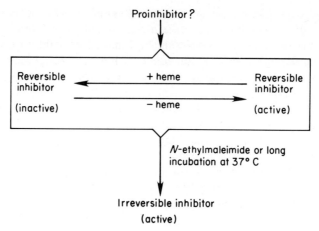

Fig. 2. Activation of the hemin-controlled translational inhibitor.

from the inactive form of the reversible inhibitor, has not been established.

The active form of the reversible inhibitor appears always to be associated with a phosphorylated peptide of about 95,000 daltons as detected by polyacrylamide gel electrophoresis with sodium dodecyl sulfate. This band corresponds in molecular weight to the active inhibitor and is concluded to be the eIF-2 kinase (Trachsel *et al.*, 1978). Phosphorylation of this peptide has been assumed to occur by autophosphorylation of the active reversible inhibitor (Trachsel *et al.*, 1978; Gross and Mendelewski, 1978). Heme is thought to interact directly with the eIF-2 kinase to prevent this phosphorylation. However, Wallis *et al.* (1980) succeeded in separating the eIF-2 kinase from an enzyme fraction containing a peptide of 90,000 daltons and demonstrated that it is involved in activation of the eIF-2 kinase. This peptide appears to be either a protein kinase for the eIF-2 kinase or to be involved in the activation of the kinase kinase. The relation of this component to the HS-HL system described below is not clear. However, considered together they indicate that the eIF-2 kinase of the HCR may be activated by a sequence of reactions that appears to function in a cascade type series.

Ochoa and co-workers (Datta *et al.*, 1977) are convinced that a cAMP-dependent protein kinase is involved in the activation of the cAMP-independent protein kinase for the 38,000 dalton subunit of eIF-2. They envision a cascade system similar to that known for the regulation of glycogen synthesis or breakdown. They have reported evidence interpreted to indicate that heme regulation involves binding of heme directly to the regulatory subunit of this hypothetical cAMP-dependent protein kinase, thereby rendering it inactive (Datta *et al.*, 1978). Several lines of evidence weigh heavily against this model. Purified preparations of the catalytic subunit of the cAMP-dependent kinase from rabbit reticulocytes have no detectable effect on protein synthesis in reticulocyte lysates (Grankowski *et al.*, 1979; Levin *et al.*, 1979). Also, the activity of cAMP-dependent protein kinases may be blocked by a naturally occurring peptide inhibitor that binds tightly to the catalytic subunit of these kinases (Walsh *et al.*, 1971). This inhibitor is active with the reticulocyte cAMP-dependent kinase in both crude and purified preparations. It would be anticipated from the cAMP-dependent system suggested by Ochoa and co-workers that addition of the inhibitor to the lysate system in the absence of heme would prevent HCR-dependent inhibition of protein synthesis. Such an effect could not be detected (Grankowski *et al.*, 1979).

Whether or not an HCR system regulated by heme occurs in all types of eukaryotic cells or is limited to cells late in the erythroid series is not known. Heme does increase protein synthesis in lysates of cells other than reticulocytes, but the effect is very small compared to that observed in reticulocyte lysates. Qualitatively similar results have been interpreted to indicate heme regulation (Beuzard *et al.*, 1973) or an effect of heme on reactions that do not involve an HCR-like activity (Cimadevilla and Hardesty, 1975). We have found that heme in the concentration range of 10 to 50 μM has small stimulatory or inhibitory effects on the activity of a number of enzymes including several aminoacyl-tRNA synthetases. The basis for these effects are not known, but we presume that they reflect nonspecific interaction of heme with some proteins.

B. eIF-2—Phosphorylation That Is Not Regulated by Heme

The initiation of protein synthesis in reticulocyte lysates is reversibly inhibited under at least six conditions that do not appear to involve heme: when double-stranded RNA (dsRNA) is added (Ehrenfeld and Hunt, 1971), when oxidized glutathione is added (Kosower *et al.*, 1971), when the lysate has been prepared from cells incubated anaerobically or with inhibitors of oxidative phosphorylation (Giloh and Mager, 1975), when the lysate is depleted of low molecular weight compounds by gel filtration (Hunt, 1976), and when the lysate is subjected to high pressure (Henderson and Hardesty, 1978) or heat (Henderson *et al.*, 1979). Each of these conditions causes inhibition of peptide initiation. It appears likely but has not been definitively proved that each condition leads to phosphorylation of eIF-2 and that the final mechanism of inhibition may be similar to that of HCR.

Addition of very small amounts (1–100 ng/ml) of dsRNA to the reticulocyte lysate system causes inhibition with kinetics that are strikingly similar to those observed with heme deficiency (Hunter *et al.*, 1975). Surprisingly, inhibition is not observed with higher concentrations of dsRNA in the order of 10 μg/ml of the incubation mixture. Farrell *et al.* (1977) presented evidence that dsRNA activates a specific protein kinase for the smallest subunit of eIF-2 and that this kinase is distinct from HCR. The kinase activated by dsRNA was found to be insensitive to antibodies for the eIF-2 kinase of the HCR system (Petryshyn *et al.*, 1979). Phosphorylation of a protein with a molecular weight of 67,000 was observed when crude ribosomes were separated from a reticulocyte lysate by gel filtration and then incubated with dsRNA and [γ-^{32}P]ATP. This protein is thought to be related to the active form of the interferon-induced eIF-2 kinase (Farrell *et al.*, 1977). These observations are particularly interesting in light of the effects of interferon on susceptible cells, as considered below (Section IV,A). Interferon is induced by dsRNA. Treatment of susceptible animal cells with interferon causes their lysates to exhibit sensitivity to dsRNA similar to that observed in reticulocyte lysates. In the absence of interferon treatment, the lysates of most cells are relatively insensitive to dsRNA.

Both the interferon-induced eIF-2 kinase and the eIF-2 kinase of the HCR system appear to phosphorylate serine at the same sites on the small subunit of eIF-2. Samuel (1979) observed an identical set of three phosphorylated tryptic peptides from eIF-2 that had been phosphorylated by the HCR kinase or a ribosome-associated eIF-2 kinase from interferon-treated human cells.

Oxidized glutathione in concentrations of about 0.5 mM or less also causes inhibition of protein synthesis in reticulocyte lysates. The kinetics of inhibition are similar to those observed with heme deficiency and dsRNA (Kosower *et al.*, 1971; Legon *et al.*, 1974; Ernst *et al.*, 1978a). Other characteristics of inhibition that are similar are disaggregation of polysomes and depletion of the 40 S · Met-tRNA$_f$ initiation complex. Inhibition is potentiated by ATP and reversed by GTP, high concentrations of cAMP, and eIF-2. Oxidized glutathione added to reticulocyte post-ribosomal supernatant also appears to cause the activation of a cAMP-independent protein kinase for the smallest subunit of eIF-2 (Ernst *et al.*, 1978a). The relationship between this enzyme and the protein kinase activated by dsRNA or under heme deficiency remains to be established.

Protein synthesis in lysates that are inhibited by heme deficiency or by oxidized glutathione may be restored by the addition of physiological concentrations of glucose 6-phosphate or other sugar phosphates that may be converted to glucose 6-phosphate by enzymes present in the lysate (Ernst *et al.*, 1978b). Removal of these and related low molecular weight compounds by gel filtration chromatography may cause the inhibition observed under those conditions. The sugar derivatives had a lower but significant effect in reversing inhibition when exogenous HCR was added to the lysate, but little or no effect on inhibition that was caused by dsRNA (Ernst et al., 1978b). It has been suggested that the effect of these compounds may reflect their ability to function in the generation of NADPH (Giloh and Mager, 1975). However, there is some doubt about the validity of this hypothesis. Addition of high concentrations of NADPH or an NADPH regenerating system does not prevent inhibition in a heme-deficient lysate (Ernst et al., 1978b). In fact, direct measurements of NADPH:NADP ratios in reticulocyte lysates show a sharp increase during heme-deficient inhibition compared with lysates actively synthesizing protein. NAD^+ at 0.16 mM has been reported to stimulate protein synthesis in the reticulocyte lysate but to cause inhibition at higher concentrations (Wu et al., 1978).

Pressure in the order of 15,000 psi (Henderson and Hardesty, 1978) or temperature (Henderson et al., 1979) may also activate a regulatory system for an eIF-2 kinase in reticulocyte lysates. Inhibition caused by this treatment has the characteristic kinetics seen under HCR activation and is reversed by GTP, high concentrations of cAMP, or eIF-2 (Henderson et al., 1979). Two inhibitors have been separated and partially purified from the postribosomal supernatant of rabbit reticulocytes. One is a heat-stable component (HS) of about 30,000 daltons that can be activated by either heat or pressure. HS has the unusual property of reverting to an inactive form if it is held at 0°C for 24–48 hr. The second inhibitor is a heat-labile component (HL) of about 45,000 daltons that is activated, apparently irreversibly, by HS. Neither HL nor HS have eIF-2 kinase activity, but HL activates an eIF-2 kinase present in the reticulocyte lysate.

Activation of the eIF-2 kinase by HL appears to involve proteolysis in that it can be blocked by soybean trypsin inhibitor. The number of reaction steps or components between HL and the eIF-2 kinase is not definitely established. However, it appears that HL may not activate the eIF-2 kinase directly but rather activate another protein kinase that phosphorylates and thereby activates the eIF-2 kinase.

Evidence has been presented for yet another type of protein kinase for the smallest subunit of eIF-2. The enzyme was purified from Friend leukemia cells (Pinphanichakarn et al., 1977) grown in cell culture. These cells are murine proerythroblasts that have been transformed by the Friend leukemia virus complex. They produce infectious virus particles and have the unusual property that they can be induced to synthesize hemoglobin and undergo differentiation by the addition of dimethyl sulfoxide and certain other aprotic solvents to the culture media. Uninduced

Friend leukemia cells synthesize little or no globin and their protein synthesis is relatively insensitive to heme (Kramer *et al.*, 1979a) as considered above.

An inhibitor of peptide initiation with a molecular weight of about 214,000 has been partially purified from unstimulated Friend leukemia cells (Pinphanichakarn *et al.*, 1977). The inhibitor is a protein kinase for the smallest subunit of eIF-2 and causes inhibition with the characteristics described earlier for HCR and other eIF-2 kinase systems. It is functionally different from HCR in that it is not regulated by heme. Whether the gene that codes for this eIF-2 kinase is part of the cellular or viral genome is not known. It may be relevant to note that the *sarc* gene of avian sarcoma virus appears to code for a cAMP-independent protein kinase (Collett and Erikson, 1978) that is associated with oncogenic transformation. However, the physiological substrate for this viral kinase is not known. It may affect polymerization of the cytoskeletal elements (Ash *et al.*, 1976) or elements of a translational control system.

C. Phosphorylation of 40 S Ribosomal Proteins

Several proteins from eukaryotic ribosomes may be phosphorylated under various growth or reaction conditions as reviewed earlier (Wool and Stöffler, 1974; Krystosek *et al.*, 1974). This phenomenon has been studied extensively in whole animals and in cell-free extracts, especially with ribosomes from rat liver and rabbit reticulocytes. No comparable phosphorylation has been observed in prokaryotes. The protein kinases catalyzing these reactions *in vitro* are not an integral part of the ribosomes; they may be washed off with high concentrations of monovalent cations. Both cAMP-dependent as well as cAMP-independent protein kinases have been implicated in the phosphorylation of different specific ribosomal proteins. A common feature of all of these studies has been the absence of a correlation between phosphorylation or dephosphorylation and a change in an activity or function when the ribosomes were isolated and assayed *in vitro*.

More recently a number of studies were carried out with mammalian cells in culture. Phosphorylation of 40 S ribosomal proteins was analyzed after the cells were kept under different physiological conditions or after virus infection. Phosphorylation of the 40 S ribosomal protein S6 was reported to be greater in preconfluent cultured baby hamster kidney fibroblasts than in postconfluent cells (Leader *et al.*, 1976). Protein S6, in the nomenclature of Lin and Wool (1974), is one of the larger proteins of the 40 S ribosomal subunit. Its molecular weight was determined by these authors to be 38,500. Lastick *et al.* (1977) demonstrated maximum phosphorylation of ribosomal protein S6 from HeLa cells 2 hr after the addition of fresh medium plus serum to cells that had been allowed to grow to high density. Phosphorylation of S6 appears to be associated with the increased growth rate that occurred after the addition of fresh medium. An interesting observation involving changes in the phosphorylation pattern of ribosomal proteins during the onset of virus infection was reported by Kaerlein and Horak (1978). Phosphorylation of the ribosomal proteins S2 and S16 occurred only when HeLa cells were infected with vaccinia virus. S6 was phosphorylated in infected as well as in uninfected cells.

These studies were performed by labeling whole cells with radioactive phosphate. A correlation between phosphorylation and differences in activity of the ribosomes involved is indirect at best. Thus far the only direct evidence for a change in activity of 40 S ribosomal subunits after their phosphorylation stems from studies with reticulocyte subunits (Kramer *et al.,* 1977). Partially purified preparations of the reticulocyte translational inhibitor HCR (described above) contain a cAMP-independent protein kinase that will phosphorylate at least one 40 S ribosomal protein. This protein has a molecular weight of about 33,000 as determined by one-dimensional SDS–polyacrylamide gel electrophoresis. After being phosphorylated by these protein kinase preparations, the 40 S ribosomal subunits are less active in a partial reaction of peptide initiation in which binding of [^{35}S]Met-tRNA$_f$ to these particles is measured. This cAMP-independent kinase activity for a 40 S ribosomal protein seems to be different from the HCR-associated eIF-2 kinase that is described above (Kramer *et al.,* 1977). Thus it appears that phosphorylation of 40 S ribosomal proteins may constitute a separate control system, distinct from the eIF-2 kinase system, that may provide translational regulation at the level of Met-tRNA$_f$ binding to 40 S ribosomal subunits during peptide initiation.

D. Phosphoprotein Phosphatase

A phosphoprotein phosphatase that functions with a protein kinase to provide a phosphorylation-dephosphorylation cycle is required to form a reversible regulatory system. Potentially, the effect of such a system can be modulated by changing the activity of either the phosphatase or the kinase. Relatively little is known of how the activity of protein phosphatases is controlled *in vivo*. In recent years phosphatase(s) involved in regulatory reactions of glycogen metabolism have been studied extensively, as reviewed by Krebs and Beavo (1979). Protein phosphatase activity from muscle or liver has been found in different molecular weight forms. The higher molecular weight components may be reduced in size by a number of techniques including ethanol precipitation. There is a disagreement about the substrate specificity of protein phosphatases. Some authors find that the enzyme will utilize a broad range of substrates. Others demonstrate specificity apparently restricted to phosphorylase *a*. A number of reports describe inhibitors of phosphoprotein phosphatase (Krebs and Beavo, 1979). Characteristically, these effectors are low molecular weight, heat-stable proteins that appear to have an extended structure in solution. Krebs and Beavo (1979) developed the concept and reviewed the evidence for a multifunctional phosphoprotein phosphatase that may be regulated by the low molecular weight, heat-stable proteins mentioned above.

Indirect evidence for phosphatase activity for phosphorylated eIF-2 in reticulocyte lysates has been presented (Safer and Jagus, 1979). Recently this enzyme was partially purified and characterized (Grankowski *et al.*, 1980a). The purified reticulocyte phosphoprotein phosphatase was found to have a rather broad substrate specificity. However, it catalyzed the efficient release of phosphate from eIF-2 that had been phosphorylated with the eIF-2 kinase of the HCR system and was shown to reverse the inhibition of eIF-2-dependent binding of Met-tRNA$_f$ to 40 S ribosomal subunits caused by this kinase. Two different low molecular weight, heat-stable, acidic proteins that differentially activate the reticulocyte phosphatase for different phosphoprotein substrates were isolated to homogeneity (Grankowski *et al.*, 1980b). These effectors appear to provide a degree of substrate specificity to the enzyme. The results with the reticulocyte phosphoprotein phosphatase appear to be consistent with the concept of a multifunctional phosphatase.

E. Polyamines

Polyamines are constituents of most, if not all living cells and have been implicated repeatedly as important elements in the control of a variety of cellular functions (Cohen, 1971). Prokaryotic cells contain putrescine and spermidine but lack the synthetic capacity to form spermine. All three of these polyamines are synthesized by eukaryotic cells. The total amount of polyamines differs dramatically in various cell types, with growth conditions, during differentiation, and following oncogenic transformation of certain cells. Putrescine is formed from ornithine by the action of ornithine decarboxylase which is precisely controlled by a complex regulatory system in eukaryotic cells (Campbell *et al.*, 1978).

The molecular mechanism by which polyamines exert their regulatory role is not clear. They appear to function as oligocations in nonspecific interactions, both with proteins and nucleic acids. However, they also have specific sites of interaction with nucleic acids and, perhaps, proteins. One of the clearest examples of a specific interaction that may be involved in regulation comes from examination of the crystal structure of tRNA^{Phe} by X-ray diffraction. Crystals of tRNA grown in the absence of polyamines are disordered and of little value for analysis by X-ray diffraction; however, highly ordered orthorhombic crystals may be grown in the presence of spermine (Kim et al., 1971). The high resolution diffraction pattern of these crystals shows that two molecules of spermine are bound in very precise positions on the tRNA^{Phe} molecule (Quigley et al., 1978). These spermine residues appear to neutralize the negative charges on closely positioned phosphates and to hold the tRNA^{Phe} molecule in a stable configuration.

The intracellular concentration of polyamines appears to affect translation. Certain auxotrophic mutants of E. coli require putrescine for growth. The 30 S ribosomal subunits isolated from these mutants grown in the absence of putrescine have a lower capacity to bind fMet-tRNA$_f$ in peptide initiation, compared with 30 S subunits isolated from cells grown with putrescine (Algranati and Goldemberg, 1977). A stimulatory effect of polyamines on in vitro translation of exogenous mRNA has been shown in several eukaryotic systems (Konecki et al., 1975; Atkins et al., 1975; Salden and Bloemendal, 1976; Hunter et al., 1977). A remarkably sharp concentration optimum at 75 μM for added spermine was observed when globin mRNA was translated on Artemia ribosomes (Konecki et al., 1975). Exogenous spermine has a marked stimulatory effect on binding of ^{125}I-labeled globin mRNA to 40 S ribosomal subunits in vitro (Kramer et al., 1979b). We have not been able to prepare reticulocyte or Artemia ribosomes with activity for peptide initiation that do not contain polyamines. A specific polyamine requirement may be a factor in the failure thus far to reconstitute eukaryotic ribosomes from their component parts as has been done for prokaryotic ribosomes.

Evidence that polyamines play a specific role in peptide initiation has come from work with edeine, a peptide antibiotic that contains spermidine (edeine A) or N-guanidylspermidine (edeine B) at its carboxyl terminal end (Hettinger and Craig, 1970). At concentrations that block de novo synthesis of peptides in vitro, edeine has a specific effect on initiation with little or no effect on peptide elongation or release (Obrig et al., 1971). Its unusually high specificity for peptide initiation has been used to advantage in distinguishing the site at which other inhibitors block protein synthesis.

Edeine is bound in a 1 : 1 molar ratio to 40 S ribosomal subunits, but does not bind to 60 S ribosomal subunits (Odom et al., 1978). The dissociation constant for the edeine · 40 S complex is less than $10^{-10} M$, which is appreciably lower than the comparable value for most antibiotics. Edeine inhibits eIF-2-dependent binding of Met-tRNA$_f$ to 40 S ribosomal subunits with GTP but not with the nonhydrolyzable GTP analogues, GMP-P(CH$_2$)P or GMP-P(NH)P (Odom *et al.*, 1978). Edeine probably inhibits peptide initiation by blocking a specific polyamine binding site that may be involved in stabilizing the Met-tRNA$_f$ · 40 S subunit complex following hydrolysis of the GTP bound to the 40 S subunit as part of the Met-tRNA$_f$ · eIF-2 · GTP ternary complex.

III. REGULATION AT THE STEP OF mRNA BINDING TO THE RIBOSOMAL INITIATION COMPLEX

A. Selection of Specific mRNA's in Prokaryotes

A number of well-documented examples led to the conclusion that all mRNA's are not necessarily translated with the same efficiency, that is, in proportion to their molar ratio. A number of mechanisms involving the tertiary structure of the mRNA, the ribosome, or RNA-specific factors appear to be responsible for this apparent messenger selection. Perhaps the clearest example of a system in which the mRNA structure appears to influence the efficiency of translation involves the RNA bacteriophages Qβ, R17, f2, and MS2 of *E. coli*. Each of these phages contains a 3.3 to 4.0 kilobase RNA that functions as a polycistronic message for three proteins: a coat protein that is normally the primary synthetic product in infected cells or cell-free systems, a peptide that functions as a subunit of RNA replicase, and a maturation factor called the A protein that is a minor but essential component of the phage virion. The three genes are in the order of A, coat, and synthetase from the 5'- to the 3'-end of the RNA. Unlike eukaryotic mRNA, which appears to be monocistronic, each coding section of the phage RNA has a separate initiation codon and ribosome attachment site. When intact phage RNA is used as mRNA in a cell-free translation system, ribosomes bind primarily to the initiation site for the coat protein, apparently because the initiation sites for the synthetase and A protein are masked in double-stranded hairpin loops. Lodish (1970a, 1971a) observed that these initiation sites for the synthetase and A protein become available for ribosome attachment and peptide initiation by disrupting the double-stranded regions of the RNA, either by heat or treatment with formaldehyde. The conclusion that the secondary structure of the mRNA is involved in the differential translation and apparent cistron selection with RNA phages was greatly strengthened by the results of Steitz (1973). Bacterial 70 S ribosomes attached to mRNA cover a 25–30 nucleotide section of the mRNA and thereby protect it from nuclease degradation. Steitz formed 70 S initiation complexes on phage mRNA then used nuclease to degrade the exposed sections of the RNA. RNA fragments containing 12–15 nucleotides on each side of the initiation codon for each of the three phage proteins were recovered from the ribosomes. When these fragments were rebound to ribosomes in a cell-free system, it was found that the segment from the A protein cistron, which had the lowest efficiency of binding to ribosomes or translation in the native RNA, now bound most efficiently. These results suggest that the efficiency of ribosome attachment and peptide initiation with intact mRNA is determined by the tertiary structure of the RNA as well as by the primary sequence of nucleotides. Initiation sites that have a base sequence potentially capable of giving efficient peptide initiation may be inefficiently utilized because they are buried in hairpin loops or are restricted by features of the tertiary structure.

These results of Steitz and others demonstrate that initiation site selection is not random even in mRNA fragments in which little or no double-stranded structure exists. Evidence for mRNA discrimination by ribosomes came from the comparison of ribosomes from different bacterial species. In contrast to ribosomes from *E. coli, B. stearothermophilus* ribosomes initiate poorly at the coat cistron. Initiation at the A protein cistron is comparable to that of *E. coli* ribosomes (Lodish, 1970b). The difference was attributed to the 30 S ribosomal subunit. Held *et al.* (1974) prepared hybrid 30 S ribosomal subunits using 16 S RNA and the corresponding ribosomal proteins from the two organisms. Discrimination against the coat protein cistron of R17 RNA was found to be related to the presence of both the 16 S RNA and the protein S21 from *B. stearothermophilus*. Ribosomal protein S1 also has been reported to contribute to differential activity of *E. coli* and *B. stearothermophilus* 30 S ribosomal subunits (Isono and Isono, 1975).

A possible basis for the role of 16 S ribosomal RNA in the selective translation of different mRNA's comes from the hypothesis proposed by Shine and Dalgarno (1974, 1975) that a nucleotide sequence near the 3'-end of the 16 S RNA base pairs directly with a common oligopurine sequence found at the 5'-side of the initiation codon of most mRNA's. This interaction between 16 S RNA and the initiator region of mRNA was proposed as the mechanism by which the proper, in phase, initiation codon is selected from the numerous in phase or out of phase AUG or GUG codons that exist in most RNA species. Strong evidence has been presented (Steitz and Jakes, 1975) in support of the Shine-Dalgarno hypothesis, and it appears most likely that initiation site selection is accomplished by this mechanism in prokaryotic systems. This attractive hypothesis also provides a basis for understanding how ribosomes may discriminate between different mRNA's or between different coding sites on polycistronic mRNA. The nucleotide sequence at the 5'-side of the initiator codon varies considerably in composition for different mRNA's. Nucleotides that could base pair with nucleotides near the 3'-end of 16 S RNA differ in number and position (Steitz and Jakes, 1975). The relative rates of formation and stability of mRNA · 30 S subunit complexes formed with various mRNA's would be anticipated to reflect these differences.

The roles played by the ribosomal proteins S12 and S1 in mRNA selection are less clear. S12 appears to form part of the binding site for IF-2 and IF-3, as indicated by cross-linking experiments (Traut *et al.*, 1974). S1 was included as a ribosomal protein in the Wittmann catalogue (Wittmann *et al.*, 1971), but can also be considered to be a ribosome-associated factor since it is easily removed from ribosomes by very low or high salt. In preparations of unwashed ribosomes, 0.6 to 0.9 copies of S1 per ribosome are found (van Knippenberg *et al.*, 1974), and it does not appear to recycle or to be released at any stage of mRNA translation (van Duin and van Knippenberg, 1974). S1 appears to be identical to a protein called interference factor i-α (Inouye *et al.*, 1974; Wahba *et al.*, 1974) and is subunit I of the Qβ RNA replicase (Kamen *et al.*, 1972; Groner *et al.*, 1972a).

When added to cell-free reaction mixtures S1 causes increased initiation at some cistrons and decreased initiation at others. Revel and co-workers (Groner *et al.*, 1972b) found that S1 (factor i-α) added to depleted ribosomes reduced initiation at the coat protein cistron but increased initiation with the replicase cistron of MS2 RNA. Similar differential effects of S1 on initiation with different mRNA's have been observed with phage T7 mRNA (Revel *et al.*, 1973) and *lac* operon mRNA from *E. coli* (Kung *et al.*, 1975).

The mechanism by which ribosomal protein S1 causes differential translation of mRNA is not clear. It appears to be located in or near the site on 30 S ribosomal subunits to which the initiation sequence of mRNA is bound. S1 can be cross-linked to the 3'-end of 16 S rRNA (Kenner, 1973) and to both IF-2 and IF-3 (Bollen *et al.*, 1975). It may function by causing a conformational change in the 30 S subunit (Szer *et al.*, 1975) or by binding to the mRNA, as suggested by Jay and Kaempfer (1974).

These are but a few of the examples demonstrating that the interactions between the initiation region of prokaryotic mRNA with cognate ribosomes and initiation factors are complex. The interaction may be affected in a variety of ways by a large number of factors and variations in reaction conditions. Apparent differential translation of mRNA species may result from differences in the primary or secondary structure of the mRNA, the structure or conformation of the ribosomes, or from factors that interact with either. It is important to appreciate that apparent specificity for synthesis of different peptides may reflect changes in binding or rate constants for the interacting components and need not necessarily involve interference or initiation factors that are specific for individual mRNA species.

B. Structural Features Specific to Eukaryotic mRNA

Almost certainly differences will be found in the mechanisms that control mRNA selection in eukaryotic and prokaryotic systems. More protein factors and both GTP and ATP hydrolysis are required for eukaryotic peptide initiation. One of the more perplexing problems of eukaryotic peptide initiation is the mechanism by which the proper in phase initiation codon of mRNA is selected. There appears to be no common nucleotide sequence at the 5'-side of the initiation codon of eukaryotic mRNA (Baralle and Brownlee, 1978). Thus, in contrast to prokaryotes, the initiation codon is unlikely to be selected by a specific base pairing between nucleotides of this region of mRNA and bases near the 3'-end of 18 S rRNA. Eukaryotic mRNA *in vivo* is associated with proteins that are likely to play an important role in initiation and regulation. It should be emphasized that most *in vitro* studies have been carried out with phenol-extracted mRNA that is free of protein. Some of the observations from these studies may not reflect physiologically significant mechanisms.

Unique structural features found only in eukaryotic mRNA also dictate caution in extrapolating mechanisms between the two types of organisms. Those special structures of eukaryotic mRNA, cap, and poly(A), may provide points for regulation that do not exist in prokaryotes.

1. Poly(A) and Cap at the Ends of mRNA

Most eukaryotic cellular and viral mRNA's contain a segment of 50 to 200 adenylic acid residues at their 3'-end (Brawerman, 1974; Revel and Groner, 1978) and 7-methylguanosine attached by an unusual 5'-to 5'-triphosphate linkage to the nucleoside at the 5'-terminus (Shatkin, 1976), to form a "cap" (Fig. 3). Both poly(A) tails and caps on mRNA's have been extensively investigated, but a firm conclusion regarding their physiological role has not been reached. Poly(A) is absent from some functionally competent mRNA species, such as histone mRNA (Adesnik and Darnell, 1972) and some viral mRNA's (Stoltzfus *et al.*, 1973). The 3'-poly(A) segment apparently contributes to the stability of the mRNA molecule by protecting it from degradation by 3'-exonucleases. Evidence supporting this hypothesis came from experiments in which both deadenylated and poly(A)-containing globin mRNA were injected into *Xenopus* oocytes. The disappearance of the two types of mRNA was followed over a period of time (Huez *et al.*, 1977). The deadenylated mRNA was degraded faster

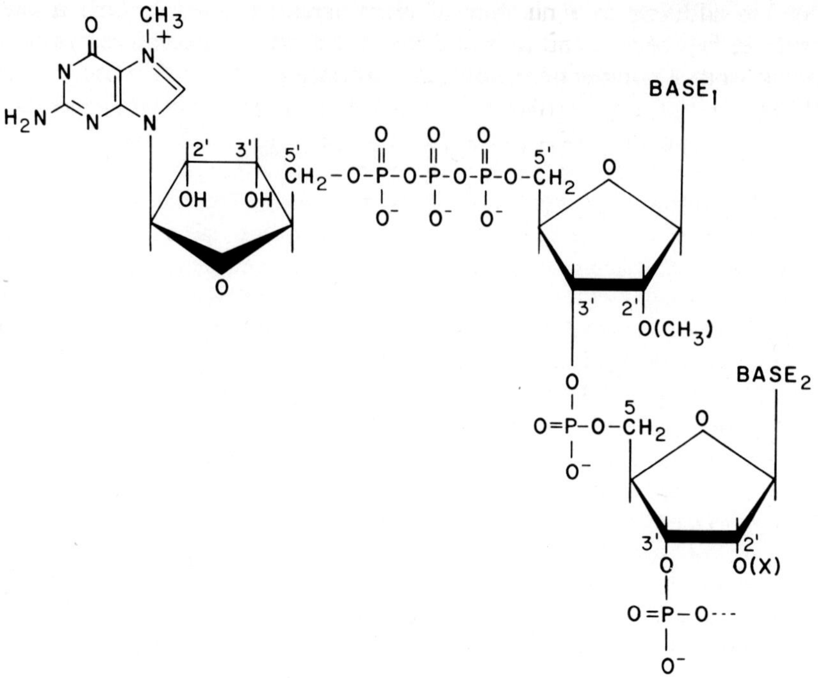

Fig. 3. Cap structure at the 5′-end of mRNA. Cap I: (X) = H; Cap II: (X) = CH_3.

than the mRNA containing poly(A). Degradation of both types of mRNA appeared to be dependent upon protein synthesis. Thus poly(A) might protect the mRNA from a 3′-exonuclease and provide something like a ticket to assure that a certain number of peptides would be formed from each mRNA molecule before it was destroyed. Although intriguing, there appears to be little evidence for such a nuclease and for this "ticket" hypothesis.

The widespread presence of cap structures on naturally occurring eukaryotic mRNA's has led to speculations that they may play an important role in translation, probably at the level of mRNA binding to 40 S ribosomal subunits. Shatkin and his co-workers (Both *et al.*, 1975b) demonstrated that a 5′-terminal m⁷G in cap was required for efficient translation of either reovirus or vesicular stomatitis virus mRNA in a cell-free, protein-synthesizing system from wheat germ. However, the dependence of eukaryotic mRNA translation on the cap structure is neither absolute nor universal. Poliovirus RNA isolated from polyribosomes of productively infected HeLa cells does not end in a cap structure but rather has pU at its 5′-terminus (Hewlett *et al.*, 1976; Nomoto *et al.*, 1976). This

RNA, in addition to a number of plant virus mRNA's lacking a cap structure, has been found to be faithfully translated in cell-free systems derived from a number of eukaryotic organisms.

The results of several studies indicate that a cap may have an important function in binding mRNA to 40 S subunits during peptide initiation by a specific interaction between the cap structure and one or more initiation factors (Filipowicz et al., 1976; Shafritz et al., 1976; Kaempfer et al., 1978). In some of these studies, binding of putative initiation factors to radioactive mRNA was measured by retention of the mRNA · protein complex on Millipore filters. Interaction of the initiation factor with the cap structure was concluded to be specific if it was competitively reduced by addition of m⁷GMP, which is considered to be a cap analogue. However, Sonenberg and Shatkin (1978) have raised serious doubts about the conclusions reached using this method. Their results demonstrate that m⁷GMP will interfere with interactions between RNA and protein that do not involve the cap structure. Inhibition of mRNA binding by m⁷GMP does not necessarily reflect a direct involvement of cap in the binding reaction. They developed a new method of determining which protein is bound to the cap structure (Sonenberg and Shatkin, 1977). The terminal ribose of a radioactively labeled cap on mRNA was oxidized by periodate and then allowed to react with the amino groups of bound protein. The resulting Schiff's base was reduced with cyanoborohydride, thus forming a relatively stable covalent bond between the cap structure and the protein. The bound protein was detected by radioactivity from the labeled cap following nuclease digestion of the mRNA. Only one peptide of 24,000 daltons was found to be significantly labeled (Sonenberg et al., 1978). It does not correspond to any of the known peptide initiation factors, but is present in preparations of eIF-3, and to a lesser extent with eIF-4B. However, it is present in less than stoichiometric amounts and does not appear to be an integral part of either factor.

2. Messenger Ribonucleoprotein Particles

Eukaryotic mRNA in polysomes or free in the cytoplasm is associated with proteins to form mRNA · protein complexes called mRNP. The mRNP not associated with polysomes are also called informosomes (Spirin, 1969). Many studies have been directed at the analysis of both the RNA and protein components of these particles. Phenol-extracted mRNA from polysomal and free particles isolated from duck erythroblasts was translated equally well in vitro. 15 S mRNP containing globin mRNA from polysomes also was translated efficiently; however, 20 S globin mRNP not associated with ribosomes was not translated in the cell-free system (Civelli et al., 1976). Analyses of mRNP isolated from a number of eukaryotic cell types demonstrate a wide variation in the number and size of the mRNP proteins.

The two major polypeptides found in nearly all mRNP preparations that have been analyzed exhibit molecular weights of approximately 52,000 and 78,000. Blobel (1973) found these two proteins in polysomal mRNP from rat liver or L cells. He labeled mouse L cells with either radioactive uridine or radioactive adenosine and isolated the mRNP from the polysome fraction. After digestion of the accessible mRNA by pancreatic ribonuclease, the 78,000 dalton protein was found to be associated with an RNA fragment, the base composition of which was about 80% adenine. Thus it seems likely that this protein covers the 3'-end of mRNA.

Recent studies on free cytoplasmic mRNP from Ehrlich ascites cells (Jeffery, 1977) and duck erythroblasts (Vincent et al., 1977) revealed a spectrum of proteins that vary in number and size. The peptide pattern seen after SDS polyacrylamide gel electrophoresis of free 20 S globin mRNP from duck erythroblasts differs from that of the polysomal globin mRNP. These authors postulate a selective recognition of specific mRNA by accompanying proteins that negatively control gene expression in the cytoplasm as proposed earlier by Spirin for informosomes (Spirin, 1969). Kinetic studies of Infante and co-workers (Dworkin et al., 1977) on mRNA synthesis is sea urchin embryos do not support a simple precursor-product relationship between mRNA in free mRNP and mRNA bound in polysomes.

Heywood and co-workers (Bester et al., 1975) propose a different model to explain why cytoplasmic mRNP will not bind to ribosomes. Small RNA molecules, called translational control RNA (tcRNA), were isolated from embryonic chicken muscle, and two classes were found. One class, associated with free mRNP, inhibited the translation of mRNA from free mRNP in a cell-free system, but had no effect on the translation of mRNA from polysomes. The other class of tcRNA, isolated from polysomes, had little inhibitory activity. The two classes differed in that the second contained fewer uridylate residues, which comprise about 50% of the nucleotides in the first class of tcRNA. Bester et al. (1975) suggested that tcRNA from free mRNP would form a stable double-stranded structure with the poly(A) at the 3'-end of the mRNA and thereby prevent the mRNA from being translated. There is very little additional published data to support this hypothesis.

C. Competition for Different mRNA's

Evidence from a number of sources demonstrate convincingly that the amount of protein synthesized in eukaryotic systems is not necessarily directly related to the amount of mRNA present. The various mRNA's of reovirus are made in equal amounts in infected cells; however, up to a tenfold difference was observed in the amounts of the different proteins synthesized (Both *et al.*, 1975a). Several studies have shown differential translation when two or more mRNA's were present in or added to cell-free translational systems (Lebleu *et al.*, 1972; Lawrence and Thach, 1974; Blair *et al.*, 1977, and as discussed below).

Perhaps the most thoroughly documented example is the synthesis of the α- and β-chains of hemoglobin. These two peptides are synthesized on different mRNA's in nearly a 1 : 1 ratio in intact cells under normal conditions. The rates of peptide elongation and release are the same for the two mRNA's (Lodish and Jacobsen, 1972); however, Lodish (1971b) demonstrated that the polysomes carrying β-globin mRNA contained about 1.5 times as many ribosomes as the polysomes containing α-globin mRNA. He concluded that the ratio of α-globin mRNA to β-globin mRNA must be 1.5 and that a rate-limiting step of peptide initiation must be 1.5 times more efficient with β-globin mRNA than with α-globin mRNA. These conclusions were tested by using the antibiotics cycloheximide or sparsomycin to slow down peptide elongation so that this step, rather than peptide initiation, became the rate-limiting step of the reaction sequence. Under these conditions the ratio of α-chain to β-chain synthesis increased to about 1.4. These data support the hypothesis that equimolar synthesis of the α- and β-globin peptides results from differences of the respective mRNA in the rate-limiting reaction of peptide initiation. However, it has not been possible by direct measurement to confirm the conclusion that the α- to β-chain mRNA ratio is 1.5 : 1. Lingrel and co-workers (Morrison *et al.*, 1974) separated mRNAs for the α- and β-chains by polyacrylamide gel electrophoresis and found that globin mRNA from both intact reticulocytes and isolated polysomes contained more β-chain than α-chain mRNA. Kramer *et al.* (1975) found the α : β chain ratio synthesized from mRNA in the ribosomal salt wash fraction to be 4 : 1; however, the corresponding value for 9 S RNA isolated from salt-washed ribosomes was 1 : 4. The corresponding ratio of globin mRNA isolated from dimethyl sulfoxide-stimulated Friend leukemia cells was 1 : 1. These ratios remained constant under all of the conditions tested and appeared to reflect the mRNA composition. Thus, it appears that the kinetic model proposed by Lodish (1976) is not adequate to explain all of the phenomena observed.

D. Virus-Mediated Inhibition of Host Cell mRNA Translation

A number of reports describe virus-induced shut-off of the synthesis of host cell proteins and preferential synthesis of viral proteins. In HeLa cells infected by poliovirus, the synthesis of host cell proteins is reduced by 80 to 90% 2 to 4 hr following the initial infection, at which time the synthesis of viral protein is maximal. Preexisting host cell mRNA is not degraded in infected cells (Baltimore, 1969). Three different mechanisms have been proposed to explain the effect of virus infection.

1. Virus-induced modification of initiation factors or an altered requirement for the factors
2. Virus-induced change to an intracellular ionic composition or strength that favors preferential synthesis of viral proteins
3. Ribosome modifications either by virus-induced phosphorylation or binding of a virus-induced protein to inhibit synthesis of host cell proteins

Experimental evidence has been presented for each of these mechanisms. Golini *et al.* (1976) reported that encephalomyocarditis (EMC) virus RNA is translated more efficiently than either cellular mRNA or globin mRNA in a cell-free system derived from MOPC ascites cells. However, addition of the initiation factor eIF-4B from rabbit reticulocytes restored the high efficiency of translation of globin mRNA or MOPC 10 S mRNA in the presence of the EMC RNA. Recently, Rose *et al.* (1978) found that vesicular stomatitis virus (VSV) mRNA is translated in polio-infected HeLa cell lysates only when the lysates were prepared within 2 to 3 hr of infection but not thereafter as is the case with the uninfected lysates. Addition of the initiation factor, eIF-4B, to lysates derived from HeLa cells 3 hr or later after infection allow translation of VSV mRNA. The authors suggest that initiation factor eIF-4B may be inactivated in infected HeLa cells and that it may bind to the cap structure of cellular mRNA. Polio RNA does not contain a 5'-cap, and thus might be translatable without eIF-4B late after infection. It has not been possible to substantiate this hypothesis. Shatkin and co-workers did not detect cross-linking of any of the known initiation factors to the cap structure as described above. Also, it appears that highly purified eIF-4B did not promote translation of capped mRNA late after infection (Rose *et al.*, 1978). The original observation may have been due to a contaminating factor in the partially purified preparation of eIF-4B.

An interesting model for shut-off of host protein synthesis has been proposed by Carrasco and Smith (1976) based on their observation that viral mRNA's are translated best *in vitro* at relatively high monovalent cation concentrations, apparently through an effect on peptide initiation. In contrast to viral mRNA, host cell mRNA is translated best at lower salt concentrations. Thus, these authors propose that viruses may cause changes in the outer membrane of the host cell that result in increased intracellular salt concentrations.

Observation made by Koch and co-workers (reviewed in Koch *et al.*, 1976) appear to support this hypothesis. The synthesis of viral and host cell proteins was examined in intact cells that had been transferred to hypertonic medium shortly after viral infection. The pattern of mRNA translation was shifted under these conditions so that proteins normally synthesized only late after virus infection were synthesized immediately. These authors propose that different eukaryotic and viral mRNA species have a characteristic efficiency for translation and that this may be altered by different concentrations of salt.

As mentioned above, Kaerlein and Horak (1978) reported that ribosomal proteins, S2 and S16, of the small subunit are phosphorylated during vaccinia virus infection of HeLa cells. These proteins were not detectably phosphorylated in uninfected cells. Phosphorylation was associated in time with shut-off of host cell protein synthesis; however, a cause and effect relationship between the two phenomena remains to be established. Ben-Hamida and Beaud (1978) made the interesting observation that vaccinia virus cores inhibit translation of natural mRNA in the reticulocyte lysate. The virus cores had no effect on poly(U)-directed synthesis of polyphenylalanine.

IV. FACTORS WITH RECEPTOR SITES ON THE CELL MEMBRANE

Interferon, peptide hormones, and protein growth factors have a common feature of being active in low concentrations with target cells by virtue of interaction with specific receptor sites on the outer surface of the cell membrane. Some evidence indicates that the mechanism by which these factors elicit physiological changes within target cells may have common elements (Grollman *et al.*, 1978). Most of these factors affect a number of different biochemical systems, frequently including protein synthesis. Most of the factors cause changes at the transcriptional level; however, there is evidence that some of the factors also affect translational events. The effects of interferon at the translational level are particularly well documented.

Little is known of the biochemical events by which most of these factors trigger internal responses within the target cell, and whether they function by virtue of mechanistically comparable systems is not clear. Also, it is not known whether or not interaction of the factor with its cell surface receptor is sufficient to trigger subsequent biochemical and physiological changes within the interior of the cell (cf. review by Insel, 1978). Alternatively, cellular response to the factor or hormone may require that it be taken into the interior of the cell to become biochemically active. It appears that at least insulin and epidermal growth factor are internalized with their receptors and then degraded (Schlessinger *et al.*, 1978; Haigler *et al.*, 1978). Internalized factors or their degradation products may function in the regulation of cAMP-independent protein kinases. As yet there is little direct evidence to support this hypothesis, however, it may be of special significance in this regard that nerve growth factor itself appears to be a specific protease (Orenstein *et al.*, 1978).

A. Interferons

Interferons are species-specific glycoproteins that vary in size and structure depending on their origin. They are formed in and released from most animal cells after virus infection. Interferons will induce an "antiviral state" only in cells from the same species; however, they are not virus specific. Productive cell types may be triggered to produce interferon by viral double-stranded RNA (dsRNA) and some synthetic double-stranded polymers such as poly(I) · poly(C). The synthesis of interferon itself appears to be under a form of posttranscriptional control, apparently involving inactivation of interferon mRNA (Sehgal *et al.*, 1977). Interferon-treated cells that are in the "antiviral state" are insensitive to replication of a broad range of viruses. Many questions concerning the mechanism of interferon action are unanswered, but viral protein synthesis clearly is a major target of interferon activity.

Protein synthesis by extracts from interferon-treated cells show an enhanced sensitivity to double-stranded RNA (Kerr *et al.*, 1974, 1976), and the inhibitory effect of dsRNA on protein synthesis in reticulocyte lysates has already been noted. Three different aspects of the molecular events triggered by interferon have received particular attention in several laboratories. These are (1) the formation of low molecular weight substances that inhibit protein synthesis, (2) the synthesis or activation of an endoribonuclease that may have hydrolytic specificity for certain tRNA and/or mRNA species, and (3) activation of a protein kinase for eIF-2 and an uncharacterized protein.

1. An Oligonucleotide Inhibitor of Unusual Structure

Kerr and co-workers detected and characterized a low molecular weight inhibitor of protein synthesis that is formed when extracts from interferon-treated mouse L cells are incubated with dsRNA. This inhibitor is synthesized from ATP by an enzyme that binds dsRNA. Its structure was found to be pppA2′p5′A2′p5′A (Kerr and Brown, 1978). The oligonucleotide is active at subnanomolar concentrations and is thought to affect peptide elongation rather than initiation. The inhibitor does not appear to interact directly with a component required for peptide elongation. It has been suggested that it may activate a ribonuclease that has been observed in extracts of interferon-treated cells as described below (Kerr *et al.*, 1979).

The oligonucleotide with the unusual 2′–5′ linkage is not restricted to interferon-treated mouse cells. What appears to be an identical inhibitor has been described by Ball and White (1978) for interferon-treated primary chicken embryo cells. Furthermore, Hovanessian and Kerr (1978) synthesized the same type of oligonucleotide with enzyme(s) from rabbit reticulocytes. As noted earlier, these latter cells respond to low concentrations of dsRNA in a way that is seen in other animal cells only after interferon treatment.

2. Interferon-Induced Effects on RNA

Some of the cellular changes observed after interferon treatment appear to influence RNA stability. Lengyel's group (Sen *et al.*, 1976; Ratner *et al.*, 1977) reported an endonuclease activity in extracts of interferon-treated cells that was enhanced by dsRNA and ATP. Viral mRNA's, such as those of reovirus or vesicular stomatitis virus, were more rapidly degraded than ribosomal RNA and globin mRNA when added to extracts of interferon-treated Ehrlich ascites tumor cells, but reovirus dsRNA was not degraded. Protein synthesis inhibitors, such as sparsomycin or edeine, did not influence degradation of viral mRNA. The nuclease activity was found in the postribosomal supernatant and was enhanced by dsRNA and ATP.

Nuclease activity induced by interferon also may degrade tRNA, perhaps differentially with respect to different molecular species. Revel and co-workers (Zilberstein *et al.*, 1976a) found nuclease activity in extracts from interferon-treated L cells that was not dependent upon dsRNA or ATP. Inhibition of mRNA translation in extracts from these cells could be overcome by Leu-tRNA. One minor species of Leu-tRNA could restore translation when Mengo virus RNA was employed as mRNA; a different Leu-tRNA species was active in promoting translation of globin mRNA in the same system. A minor species of Lys-tRNA was reported to restore EMC RNA translation in extracts from interferon-treated Friend leukemia cells (Mayr *et al.*, 1977).

3. Protein Phosphorylation

The phosphorylation of the small subunit of eIF-2 and a protein with a molecular weight of 67,000 are two of the more striking differences between extracts from interferon-treated cells and extracts from untreated cells (Zilberstein *et al.*, 1976b; Lebleu *et al.*, 1976; Roberts *et al.*, 1976).

The situation appears to parallel that seen in extracts of reticulocytes treated with dsRNA, except that the response in the latter is not dependent upon pretreatment of the intact cells with interferon. The biochemical mechanism by which interferon sensitizes extracts from most treated cells to dsRNA or the reason why reticulocyte lysates respond to dsRNA without pretreatment with interferon is not known. Whether the synthesis of the oligonucleotide, the endonuclease activity, and the protein kinase activity are interrelated effects of dsRNA remains to be elucidated. Also, the relation of these biochemical phenomena to the interferon-induced "antiviral state" is not clear. Thus, a unifying hypothesis of interferon action is difficult to envision and probably will be very hard to prove.

B. Peptide Hormones and Cell Surface Factors

Early studies (summarized by Korner, 1970) on the effect of growth hormone on protein synthesis in the liver of hypophysectomized rats were interpreted to indicate that the hormone was active in regulating protein synthesis at the level of transcription and at a posttranscriptional site. The hormone caused a rapid, but relatively small, increase in the rate of protein synthesis, which was followed by a much larger effect that appeared to be associated with increased transcription. The latter, more sustained effect depended on stimulation of RNA synthesis, whereas the former did not. Widnell and Tata (1966) found that the two waves of response could be separated by administration of actinomycin D to block RNA synthesis.

The biochemical basis of these phenomena are not known. However, Feigelson and co-workers (Kurtz et al., 1978) have demonstrated what may be a similar effect on translation of a specific protein under more clearly defined conditions. They conclude that pituitary growth hormone is active at the translational level in the synthesis of α_{2u}-globulin in hepatic tissue of male rats. The protein is not formed in hypophysectomized animals, but is reinduced by the simultaneous administration of glucocorticoid, androgen, and thyroid hormones plus pituitary growth hormone. Administration of the hormone regimen lacking growth hormone induced the synthesis of α_{2u}-globulin mRNA to a normal level without detectable synthesis of the protein. Untranslated mRNA was associated with free polysomes and the nonribosomal fractions. Administration of growth hormone resulted in rapid synthesis of the protein and a shift of its mRNA into membrane-bound polysomes. A similar effect was seen if insulin was used in place of growth hormone. The authors suggest that hormone-induced attachment of polysomes to membranes might involve the amino-terminal portion of the nascent peptide, as suggested by Blobel and Dobberstein (1975) in what has come to be known as the signal hypothesis. This hypothesis holds that ribosomes are attached to membranes through a sequence of hydrophobic amino acids at the amino-terminal end of a nascent peptide destined to be completed in the endoplasmic reticulum. This hydrophobic tail is presumed to be bound to a hydrophobic region of the membrane early in the course of synthesis, thereby forming an attachment site for the ribosomal synthetic assembly. The nascent peptide itself is presumed to be channeled into the lumen of the endoplasmic reticulum as it is formed. Eventually the hydrophobic segment is lost during subsequent processing of the peptide.

Material on the outer surface of the cell membrane itself inhibits protein synthesis at the ribosomal level. Fisher and Koch (1977) found that material removed from the surface of HeLa cells by mild pronase treatment blocked translation in intact cells and cell-free systems. Inhibitory activity was associated with two fractions of heat-stable material, possibly glycoproteins, of 29,000 and 41,000 daltons. An intriguing, but entirely speculative, possibility is that material from the cell surface might be taken into the cell as part of the cap formed after the interaction of a hormone or protein factor with its cell surface receptor. This material or its degradation products might then affect translational control.

V. CONCLUSION

Translation of mRNA is sensitive to a wide range of conditions and specific or nonspecific factors that influence the metabolic economy of eukaryotic cells. Changes in the composition or concentration of intracellular salts, nucleoside triphosphates, or amino acids may have a dramatic effect on translation as well as the synthesis and degradation of RNA. Changes in the overall rate of protein synthesis and in the intracellular environment may lead to qualitative as well as quantitative changes in the proteins that are formed. Normally the α- and β-chains of hemoglobin are synthesized in equal amounts. However, excess β-chains are formed if the rate of synthesis is reduced so that peptide elongation rather than initiation becomes the rate-limiting step. Similarly, the relative proportion of virus and host cell proteins synthesized in infected cells or cell-free systems to which the respective mRNA's have been added may vary dramatically at different concentrations of KCl. These seemingly specific effects do not necessarily involve the action of a specific factor that will alter translation of a specific mRNA.

In principle, translational control might occur at any point in the reaction sequence by which proteins are synthesized. It is likely that different mechanisms will be found to operate in various systems and circumstances. For example, it appears that peptide elongation may be limited in some cases by deficiencies in minor species of tRNA's or as a result of the nuclease activity induced by interferon and dsRNA. Peptide elongation factor, EF-I, may be limiting in certain dormant forms such as brine shrimp cysts. However, at this time it seems that the most common points for translational regulation involve two steps of peptide initiation: (a) phosphorylation of the peptide initiation factor eIF-2, thus leading to a block in binding of Met-tRNA$_f$ to 40 S ribosomal subunits and (b) binding of mRNA to the 40 S ribosomal subunit · Met-tRNA$_f$ initiation complex. The former of these two mechanisms may be of special significance in that in some systems it involves multiple components that function in a cascade sequence of reactions that have the capacity to provide signal amplification. This sequence can be turned on in response to a variety of factors and stimuli. Of these, perhaps interferon is the most interesting at this time in that it may provide a model for the mechanism of action of certain peptide hormones and growth factors. These components share a common feature of being active in low concentrations by virtue of specific receptor sites on target cells. Will hormones and growth factors be found to control cell growth and differentiation by activation of cascade reaction sequences that will lead to translational restrictions? Will a direct, biochemical link be found by which these factors can control and regulate both transcription and translation in eukaryotic cells? These and related questions appear to provide the basis for extremely important and technically promising experimentation in the forthcoming decade.

REFERENCES

Adamson, S. D., Herbert, E., and Godchaux, W., III (1968). Factors affecting the rate of protein synthesis in lysate systems from reticulocytes. *Arch. Biochem. Biophys.* **125,** 671–683.
Adesnik, M., and Darnell, J. E. (1972). Biogenesis and characterization of histone messenger RNA in HeLa cells. *J. Mol. Biol.* **67,** 397–406.
Algranati, I. D., and Goldemberg, S. H. (1977). Translation of natural mRNA in cell-free systems from a polyamine-requiring mutant of E. coli. *Biochem. Biophys. Res. Commun.* **75,** 1045–1052.
Ash, J. F., Vogt, P. K., and Singer, S. J. (1976). Reversion from transformed to normal phenotype by inhibition of protein synthesis in rat kidney cells infected with a temperature-sensitive mutant of Rous sarcoma virus. *Proc. Natl. Acad. Sci. U.S.A.* **73,** 3603–3607.
Atkins, J. F., Lewis, J. B., Anderson, C. W., and Gesteland, R. F. (1975). Enhanced differential synthesis of protein in a mammalian cell-free system by addition of polyamines. *J. Biol. Chem.* **250,** 5688–5695.
Balkow, K., Mizuno, S., and Rabinovitz, M. (1973). Inhibition of an initiation codon func-

tion by hemin deficiency and the hemin-controlled translational repressor in the reticulocyte cell-free system. *Biochem. Biophys. Res. Commun.* **54**, 315–323.

Balkow, K., Hunt, T., and Jackson, R. J. (1975). Control of protein synthesis in reticulocyte lysates: The effect of nucleotide triphosphates on formation of the translational repressor. *Biochem. Biophys. Res. Commun.* **67**, 366–375.

Ball, L. A., and White, C. N. (1978). Oligonucleotide inhibitor of protein synthesis made in extracts of interferon-treated chick embryo cells. Comparison with mouse low molecular weight inhibitor. *Proc. Natl. Acad. Sci. U.S.A.* **75**, 1167–1171.

Baltimore, D. (1969). The replication of picornaviruses. *In* "The Biochemistry of Viruses" (H. Levy, ed.), pp. 101–176. Dekker, New York.

Baralle, F. E., and Brownlee, G. G. (1978). AUG is the only recognisable signal sequence in the 5′ non-coding regions of eukaryotic mRNA. *Nature (London)* **274**, 84–87.

Beavo, J. A., Bechtel, P. J., and Krebs, E. G. (1974). Activation of protein kinase by physiological concentrations of cyclic AMP. *Proc. Natl. Acad. Sci. U.S.A.* **71**, 3580–3583.

Ben-Hamida, F., and Beaud, G. (1978). *In vitro* inhibition of protein synthesis by purified cores from vaccinia virus. *Proc. Natl. Acad. Sci. U.S.A.* **75**, 175–179.

Bester, A., Kennedy, D., and Heywood, S. (1975). Two classes of translational control RNA: Their role in the regulation of protein synthesis. *Proc. Natl. Acad. Sci. U.S.A.* **72**, 1523–1527.

Beuzard, Y., Rodvien, R., and London, I. M. (1973). Effect of hemin on the synthesis of hemoglobin and other proteins in mammalian cells. *Proc. Natl. Acad. Sci. U.S.A.* **70**, 1022–1026.

Blair, G. E., Dahl, H., Truelsen, E., and Lelong, J. C. (1977). Functional identity of a mouse ascites and a rabbit reticulocyte initiation factor required for natural mRNA. *Nature (London)* **265**, 651–653.

Blobel, G. (1973). A protein of molecular weight 78,000 bound to the polyadenylate region of eukaryotic messenger RNAs. *Proc. Natl. Acad. Sci. U.S.A.* **70**, 924–928.

Blobel, G., and Dobberstein, B. (1975). Transfer of proteins across membranes. I. Presence of proteolytically processed and unprocessed nascent immunoglobulin light chains on membrane-bound ribosomes of murine myeloma. *J. Cell Biol.* **67**, 835–851.

Bollen, A., Heimark, R. L., Cozzone, A., Traut, R. R., Hershey, J. W. B., and Kahan, L. (1975). Cross-linking of initiation factor IF-2 to *Escherichia coli* ribosomal proteins with dimethyl suberimidate. *J. Biol. Chem.* **250**, 4310–4314.

Both, G. W., Furuichi, Y., Muthukrishan, S., and Shatkin, A. J. (1975a). Ribosome binding to reovirus mRNA in protein synthesis requires 5′ terminal 7-methylguanosine. *Cell* **6**, 185–195.

Both, G. W., Banerjee, A. K., and Shatkin, A. (1975b). Methylation-dependent translation of viral messenger RNAs *in vitro*. *Proc. Natl. Acad. Sci. U.S.A.* **72**, 1189–1193.

Brawerman, G. (1974). Eukaryotic messenger RNA. *Annu. Rev. Biochem.* **43**, 621–642.

Bruns, G. P., and London, I. M. (1965). The effect of hemin on the synthesis of globin. *Biochem. Biophys. Res. Commun.* **18**, 236–242.

Campbell, R. A., Morris, D., Bartos, D., Daves, G. D., Jr., and Bartos, F., eds. (1978). "Advances in Polyamine Research," Vol. 1. Raven Press, New York.

Carrasco, L., and Smith, A. E. (1976). Sodium ions and the shut-off of host cell protein synthesis by picornaviruses. *Nature (London)* **264**, 807–809.

Cimadevilla, J. M., and Hardesty, B. (1975). Evidence for a nonhemin regulated translational repressor in Friend leukemia virus transformed murine proerythroblasts. *Biochem. Biophys. Res. Commun.* **63**, 931–937.

Civelli, O., Vincent, A., Buri, J. F., and Scherrer, K. (1976). Evidence for a translational inhibitor linked to globin mRNA in untranslated free cytoplasmic mRNP complexes. *FEBS Lett.* **72**, 71–76.

Cohen, S. S. (1971). "Introduction to the Polyamines." Prentice-Hall, Englewood Cliffs, New Jersey.

Collett, M. S., and Erikson, R. L. (1978). Protein kinase activity associated with the avian sarcoma virus src gene product. *Proc. Natl. Acad. Sci. U.S.A.* **75**, 2021–2024.

Dasgupta, A., Roy, R., Palmieri, S., Das, A., Ralston, R., and Gupta, N. K. (1978). Protein synthesis in rabbit reticulocytes. XXII. A heat-stable dialysable factor (EIF-1*) modulates Met-tRNA$_f$ binding to EIF-1. *Biochem. Biophys. Res. Commun.* **82**, 1019–1027.

Datta, A., de Haro, C., Sierra, J. M., and Ochoa, S. (1977). Role of cAMP-dependent protein kinase in regulation of protein synthesis in reticulocyte lysates. *Proc. Natl. Acad. Sci. U.S.A.* **74**, 1463–1467.

Datta, A., de Haro, C., and Ochoa, S. (1978). Translational control by hemin is due to binding to cAMP-dependent protein kinase. *Proc. Natl. Acad. Sci. U.S.A.* **75**, 1148–1152.

Dworkin, M. B., Rudensey, L. M., and Infante, A. A. (1977). Cytoplasmic nonpolysomal RNP in sea urchin embryos and their relationship to protein synthesis. *Proc. Natl. Acad. Sci. U.S.A.* **74**, 2231–2235.

Ehrenfeld, E., and Hunt, T. (1971). Double-stranded poliovirus RNA inhibits initiation of protein synthesis by reticulocyte lysates. *Proc. Natl. Acad. Sci. U.S.A.* **68**, 1075–1078.

Ernst, V., Levin, D. H., and London, I. M. (1978a). Inhibition of protein synthesis initiation by oxidized glutathione: Activation of a protein kinase that phosphorylates the α subunit of eukaryotic initiation factor 2. *Proc. Natl. Acad. Sci. U.S.A.* **75**, 4110–4114.

Ernst, V., Levin, D. H., and London, I. M. (1978b). Evidence that glucose 6-phosphate regulates protein synthesis initiation in reticulocyte lysates. *J. Biol. Chem.* **253**, 7163–7172.

Farrell, P. J., Balkow, K., Hunt, T., Jackson, R., and Trachsel, H. (1977). Phosphorylation of initiation factor eIF-2 and the control of reticulocyte protein synthesis. *Cell* **11**, 187–200.

Farrell, P. J., Hunt, T., and Jackson, R. (1978). Analysis of phosphorylation of protein synthesis initiation factor eIF-2 by two-dimensional gel electrophoresis. *Eur. J. Biochem.* **89**, 517–521.

Filipowicz, W., Furuichi, Y., Sierra, J. M., Muthukrishnan, S., Shatkin, S., and Ochoa, S. (1976). A protein binding to methylated 5′ terminal sequence, m^{7}GpppN, of eukaryotic mRNA. *Proc. Natl. Acad. Sci. U.S.A.* **73**, 1559–1563.

Fisher, L. F., and Koch, G. (1977). Partial characterization and proposed mode of action of inhibitory HeLa surface polypeptides. *Biochim. Biophys. Acta* **470**, 113–120.

Giloh, H., and Mager, J. (1975). Inhibition of peptide chain initiation in lysates from ATP-depleted cells. I. Stages in the evolution of the lesion and its reversal by thiol compounds, cyclic AMP or purine derivatives and phosphorylated sugars. *Biochim. Biophys. Acta* **414**, 293–308.

Golini, F., Thach, S., Birge, C., Safer, B., Merrick, W., and Thach, R. E. (1976). Competition between cellular and viral mRNA's *in vitro* is regulated by a messenger discriminatory initiation factor. *Proc. Natl. Acad. Sci. U.S.A.* **73**, 3040–3044.

Grankowski, N., Kramer, G., and Hardesty, B. (1979). No effect of cAMP on protein synthesis in reticulocyte lysates. *J. Biol. Chem.* **254**, 3145–3147.

Grankowski, N., Lehmusvirta, D., Kramer, G., and Hardesty, B. (1980a). Partial purification and characterization of reticulocyte phosphatase with activity for phosphorylated peptide initiation factor 2. *J. Biol. Chem.* **255**, 310–317.

Grankowski, N., Lehmusvirta, D., Stearns, G. B., Kramer, G., and Hardesty, B. (1980b). The isolation and partial characterization of two substrate specific protein activators of the reticulocyte phosphoprotein phosphatase. *J. Biol. Chem.* (in press).

Grayzel, A. I., Hörchner, P., and London, I. M. (1966). The stimulation of globin synthesis by heme. *Proc. Natl. Acad. Sci. U.S.A.* **55,** 650–655.

Grollman, E. F., Lee, G., Ramos, S., Lazo, P. S., Kaback, H. R., Friedman, R. M., and Kohn, L. D. (1978). Relationships of the structure and function of the interferon receptor to hormone receptors and establishment of the antiviral state. *Cancer Res.* **38,** 4172–4185.

Groner, Y., Scheps, R., Kamen, R., Kolakofsky, D., and Revel, M. (1972a). Host subunit of Qβ replicase is translation control factor i. *Nature (London), New Biol.* **239,** 19–20.

Groner, Y., Pollack, Y., Berissi, H., and Revel, M. (1972b). Characterization of cistron specific factors for the initiation of messenger RNA translation in *E. coli. FEBS Lett.* **21,** 223–228.

Gross, M., and Mendelewski, J. (1978). Control of protein synthesis by hemin: An association between the formation of the hemin-controlled translational repressor and the phosphorylation of a 100,000 molecular weight protein. *Biochim. Biophys. Acta* **520,** 650–663.

Gross, M., and Rabinovitz, M. (1972a). Control of globin synthesis by hemin: Factors influencing formation of an inhibitor of globin chain initiation in reticulocyte lysates. *Biochim. Biophys. Acta* **287,** 340–352.

Gross, M., and Rabinovitz, M. (1972b). Control of globin synthesis in cell-free preparations of reticulocytes by formation of a translational repressor that is inactivated by hemin. *Proc. Natl. Acad. Sci. U.S.A.* **69,** 1565–1568.

Gross, M., and Rabinovitz, M. (1973). Partial purification of a translational repressor mediating hemin control of globin synthesis and implication of results on the site of inhibition. *Biochem. Biophys. Res. Commun.* **50,** 832–838.

Haigler, H., Ash, J. F., Singer, S. J., and Cohen, S. (1978). Visualization by fluorescence of the binding and internalization of epidermal growth factor in human carcinoma cells A-431. *Proc. Natl. Acad. Sci. U.S.A.* **75,** 3317–3321.

Held, W. A., Nomura, M., and Hershey, J. W. B. (1974). Ribosomal protein S21 is required for full activity in the initiation of protein synthesis. *Mol. Gen. Genet.* **128,** 11–18.

Henderson, A. B., and Hardesty, B. (1978). Evidence for an inhibitor of protein synthesis in rabbit reticulocytes activated by high pressure. *Biochem. Biophys. Res. Commun.* **83,** 715–723.

Henderson, A. B., Miller, A. H., and Hardesty, B. (1979). Multistep regulatory system for activation of a cyclic AMP-independent eukaryotic initiation factor 2 kinase. *Proc. Natl. Acad. Sci. U.S.A.* **76,** 2605–2609.

Hettinger, T., and Craig, L. (1970). Edeine. IV. Structures of the antibiotic peptides edeines A_1 and B_1. *Biochemistry* **9,** 1224–1232.

Hewlett, M. J., Rose, J. K., and Baltimore, D. (1976). 5'-terminal structure of poliovirus polyribosomal RNA is pUp. *Proc. Natl. Acad. Sci. U.S.A.* **73,** 327–330.

Hovanessian, A. G., and Kerr, I. M. (1978). Synthesis of an oligonucleotide inhibitor of protein synthesis in rabbit reticulocyte lysates analogous to that formed in extracts from interferon-treated cells. *Eur. J. Biochem.* **84,** 149–159.

Howard, G. A., Adamson, S. D., and Herbert, E. (1970). Studies on cessation of protein synthesis in a reticulocyte lysate cell-free system. *Biochim. Biophys. Acta* **213,** 237–240.

Huez, G., Marbaix, G., Burny, A., Hubert, E., Leclercq, M., Cleuter, Y., Chantrenne, H.,

Soreq, H., and Littauer, U. (1977). Degradation of deadenylated rabbit α-globin mRNA in *Xenopus* oocytes is associated with its translation. *Nature (London)* **266**, 473–474.

Hunt, T. (1976). Control of globin synthesis. *Br. Med. Bull.* **32**, 257–261.

Hunter, A., Jackson, R., and Hunt, T. (1977). The role of polyamines in cell-free protein synthesis in the wheat-germ system. *Eur. J. Biochem.* **75**, 149–159.

Hunter, T., Hunt, T., Jackson, R., and Robertson, H. D. (1975). The characteristics of inhibition of protein synthesis by double-stranded ribonucleic acid in reticulocyte lysates. *J. Biol. Chem.* **250**, 409–417.

Inouye, H., Pollack, Y., and Petre, J. (1974). Physical and functional homology between ribosomal protein S1 and interference factor i. *Eur. J. Biochem.* **45**, 109–117.

Insel, P. A. (1978). Membrane-active hormones; receptors and receptor regulation. *Int. Rev. Biochem.* **20**, 1–43.

Isono, S., and Isono, K. (1975). Role of ribosomal protein S1 in protein synthesis: Effects of its addition to *Bacillus stearothermophilus* cell-free system. *Eur. J. Biochem.* **56**, 15–22.

Jay, G., and Kaempfer, R. (1974). Host interference with viral gene expression: Mode of action of bacterial factor i. *J. Mol. Biol.* **82**, 193–212.

Jeffery, W. R. (1977). Characterization of polypeptides associated with messenger RNA and its polyadenylate segment in Ehrlich ascites messenger ribonucleoprotein. *J. Biol. Chem.* **252**, 3525–3532.

Kaempfer, R., Hollender, R., Abrams, W. R., and Israeli, R. (1978). Specific binding of mRNA and Met-tRNA$_f^{Met}$ by the same initiation factor for eukaryotic protein synthesis. *Proc. Natl. Acad. Sci. U.S.A.* **75**, 209–213.

Kaerlein, M., and Horak, I. (1978). Identification and characterization of ribosomal proteins phosphorylated in vaccinia-virus-infected HeLa cells. *Eur. J. Biochem.* **90**, 463–469.

Kamen, R., Kondon, M., Romer, W., and Weissmann, C. (1972). Reconstitution of Qβ replicase lacking subunit α with protein-synthesis-interference factor i. *Eur. J. Biochem.* **31**, 44–51.

Kenner, R. A. (1973). A protein–nucleic acid crosslink in the 30 S ribosomes. *Biochem. Biophys. Res. Commun.* **51**, 932–938.

Kerr, I. M., and Brown, R. E. (1978). pppA2′p5′A2′p5′A: An inhibitor of protein synthesis synthesized with an enzyme fraction from interferon-treated cells. *Proc. Natl. Acad. Sci. U.S.A.* **75**, 256–260.

Kerr, I. M., Brown, R. E., and Ball, L. A. (1974). Increased sensitivity of cell-free protein synthesis to double-stranded RNA after interferon treatment. *Nature (London)* **250**, 57–59.

Kerr, I. M., Brown, R. E., Clemens, M. J., and Gilbert, C. S. (1976). Interferon-mediated inhibitor of cell-free protein synthesis in response to dsRNA. *Eur. J. Biochem.* **69**, 551–561.

Kerr, I. M., Williams, B. R. G., Hovanessian, A. G., Brown, R. E., Martin, E. M., Gilbert, C. S., and Birdsall, N. J. M. (1979). Cell-free protein synthesis and interferon action: Protein kinase(s) and an oligonucleotide effector pppA2′p5′A2′p5′A. In "Modern Trends in Human Leukemia III" (R. Neth, R. Gallo, H.-P. Hofschneider, and K. Mannweiler, eds.), pp. 291–294. Springer-Verlag, Berlin and New York.

Kim, S. H., Quigley, G. J., Suddath, F. L., and Rich, A. (1971). High-resolution X-ray diffraction patterns of crystalline transfer RNA that show helical regions. *Proc. Natl. Acad. Sci. U.S.A.* **68**, 841–845.

Koch, G., Oppermann, H., Bilello, P., Koch, F., and Nuss, D. (1976). Control of peptide chain initiation in uninfected and virus infected cells by membrane mediated events. *In*

"Modern Trends in Human Leukemia II" (R. Neth, R. Gallo, K. Mannweiler, and W. C. Moloney, eds.), pp. 541–555. Lehmanns Verlag, Munchen.

Konecki, D., Kramer, G., Pinphanichakarn, P., and Hardesty, B. (1975). Polyamines are necessary for maximum in vitro synthesis of globin peptides and play a role in chain initiation. Arch. Biochem. Biophys. **169**, 192–198.

Korner, A. (1970). Insulin and growth hormone control of protein biosynthesis. In "Control Processes in Multicellular Organisms" (G. E. W. Wolstenholme and J. Knight, eds.), pp. 86–99. Churchill, London.

Kosower, N. S., Vanderhoff, G. A., Benerofe, B., Hunt, T., and Kosower, E. M. (1971). Inhibition of protein synthesis by glutathione disulfide in the presence of glutathione. Biochem. Biophys. Res. Commun. **45**, 816–821.

Kramer, G., Pinphanichakarn, P., Konecki, D., and Hardesty, B. (1975). Globin mRNA translation on Artemia salina ribosomes with components from Friend leukemia cells. Eur. J. Biochem. **53**, 471–480.

Kramer, G., Cimadevilla, J. M., and Hardesty, B. (1976). Specificity of the protein kinase activity associated with the hemin-controlled repressor of rabbit reticulocytes. Proc. Natl. Acad. Sci. U.S.A. **73**, 3078–3082.

Kramer, G., Henderson, A. B., Pinphanichakarn, P., Wallis, M. H., and Hardesty, B. (1977). Partial reaction of peptide initiation inhibited by phosphorylation of either initiation factor eIF-2 or 40 S ribosomal proteins. Proc. Natl. Acad. Sci. U.S.A. **74**, 1445–1449.

Kramer, G., Pinphanichakarn, P., and Hardesty, B. (1979a). Control of eukaryotic protein synthesis by phosphorylation. In "Modern Trends in Human Leukemia III" (R. Neth, R. Gallo, P.-H. Hofschneider, and K. Mannweiler, eds.), pp. 283–290. Springer-Verlag, Berlin and New York.

Kramer, G., Odom, O. W., and Hardesty, B. (1979b). Polyamines in eukaryotic peptide initiation. In "Methods in Enzymology" (K. Moldave and L. Grossman, eds.), Vol. 60, Part H, pp. 555–566. Academic Press, New York.

Krebs, E. G., and Beavo, J. A. (1979). Phosphorylation-dephosphorylation of enzymes. Annu. Rev. Biochem. **48**, 923–959.

Kruh, J., and Borsook, H. (1956). Hemoglobin synthesis in rabbit reticulocytes in vitro. J. Biol. Chem. **220**, 905–915.

Krystosek, A., Bitte, L. F., Cawthon, M. L., and Kabat, D. (1974). Phosphorylation of ribosomal proteins in eukaryotes. In "Ribosomes" (M. Nomura, A. Tissières, and P. Lengyel, eds.), pp. 855–870. Cold Spring Harbor Lab., Cold Spring Harbor, New York.

Kung, H.-F., Morrissey, J., Revel, M., Spears, C., and Weissbach, H. (1975). Studies on the lactose operon. The control of DNA-directed in vitro protein synthesis by interference factor 1-α. J. Biol. Chem. **250**, 8780–8784.

Kurtz, D. T., Chan, K.-M., and Feigelson, P. (1978). Translational control of hepatic α_{2u} globulin synthesis by growth hormone. Cell **15**, 743–750.

Lastick, S. M., Nielsen, P. J., and McConkey, E. H. (1977). Phosphorylation of ribosomal protein S6 in suspension cultured HeLa cells. Mol. Gen. Genet. **152**, 223–230.

Lawrence, C., and Thach, R. E. (1974). Encephalomyocarditis virus infection of mouse plasmacytoma cells. I. Inhibition of cellular protein synthesis. J. Virol. **14**, 598–610.

Leader, D. P., Rankine, A. D., and Coia, A. A. (1976). The phosphorylation of ribosomal protein S6 in baby hamster kidney fibroblasts. Biochem. Biophys. Res. Commun. **71**, 966–974.

Lebleu, B., Nudel, U., Falcoff, E., Prives, C., and Revel, M. (1972). A comparison of the translation of mengo virus RNA and globin mRNA in Krebs ascites cell-free extracts. *FEBS Lett.* **25**, 97–103.

Lebleu, B., Sen, G. C., Shaila, S., Cabrer, B., and Lengyel, P. (1976). Interferon, double-stranded RNA, and protein phosphorylation. *Proc. Natl. Acad. Sci. U.S.A.* **73**, 3107–3111.

Legon, S., Jackson, R., and Hunt, T. (1973). Control of protein synthesis in reticulocyte lysates by haemin. *Nature (London), New Biol.* **241**, 150–152.

Legon, S., Brayley, A., Hunt, T., and Jackson, R. (1974). The effect of cyclic AMP and related compounds on the control of protein synthesis in reticulocyte lysates. *Biochem. Biophys. Res. Commun.* **56**, 745–752.

Levin, D., Ernst, V., and London, I. M. (1979). Effects of the catalytic subunit of cAMP-dependent protein kinase (type II) from reticulocytes and bovine heart muscle on protein phosphorylation and protein synthesis in reticulocyte lysates. *J. Biol. Chem.* **254**, 7935–7941.

Levin, D., Ranu, R., Ernst, V., and London, I. M. (1976). Regulation of protein synthesis in reticulocyte lysates: Phosphorylation of methionyl-tRNA$_f$ binding factor by protein kinase activity of translational inhibitor isolated from heme-deficient lysates. *Proc. Natl. Acad. Sci. U.S.A.* **73**, 3112–3116.

Lin, A., and Wool, I. G. (1974). The molecular weights of rat liver ribosomal proteins determined by "three dimensional" polyacrylamide gel electrophoresis. *Mol. Gen. Genet.* **134**, 1–6.

Lodish, H. F. (1970a). Secondary structure of bacteriophage f2 ribonucleic acid and the initiation of *in vitro* protein biosynthesis. *J. Mol. Biol.* **50**, 689–702.

Lodish, H. F. (1970b). Specificity in bacterial protein synthesis: Role of initiation factors and ribosomal subunits. *Nature (London)* **226**, 705–707.

Lodish, H. F. (1971a). Thermal melting of bacteriophage f2 RNA and initiation of synthesis of the maturation protein. *J. Mol. Biol.* **56**, 627–632.

Lodish, H. F. (1971b). Alpha and beta globin messenger ribonucleic acid. Different amounts and rates of initiation of translation. *J. Biol. Chem.* **246**, 7131–7138.

Lodish, H. F. (1976). Translational control of protein synthesis. *Annu. Rev. Biochem.* **45**, 39–72.

Lodish, H. F., and Jacobsen, M. (1972). Regulation of hemoglobin synthesis. Equal rates of translation and termination of α- and β-globin chains. *J. Biol. Chem.* **247**, 3622–3629.

Majumdar, A., Roy, R., Das, A., Dasgupta, A., and Gupta, N. K. (1977). Protein synthesis in rabbit reticulocytes. XIX. EIF-2 promotes dissociation of Met-tRNA$_f$ · EIF-1 · GTP complex and Met-tRNA$_f$ binding to 40 S ribosomes. *Biochem. Biophys. Res. Commun.* **78**, 161–169.

Mayr, U., Bermayer, H.-P., Weidinger, G., Jungwirth, C., Gross, H., and Bodo, G. (1977). Release of interferon-induced translational inhibition by tRNA in cell-free extracts of mouse erythroleukemia cells. *Eur. J. Biochem.* **76**, 541–551.

Merrick, W. C. (1979). Evidence that a single GTP is used in the formation of 80 S initiation complexes. *J. Biol. Chem.* **254**, 3708–3711.

Morrison, M. R., Brinkley, S. A., Gorski, J., and Lingrel, J. B. (1974). The separation and identification of α- and β-globin messenger ribonucleic acid. *J. Biol. Chem.* **249**, 5290–5295.

Nomoto, A., Lee, Y., and Wimmer, E. (1976). The 5'-end of poliovirus mRNA is not capped with m⁷G(5')ppp(5')Np. *Proc. Natl. Acad. Sci. U.S.A.* **73**, 375–380.

Obrig, T., Irvin, J., Culp, W., and Hardesty, B. (1971). Inhibition of peptide initiation on reticulocyte ribosomes by edeine. *Eur. J. Biochem.* **21**, 31–41.

Odom, O. W., Kramer, G., Henderson, A. B., Pinphanichakarn, P., and Hardesty, B. (1978). GTP hydrolysis during methionyl-tRNA$_f$ binding to 40 S ribosomal subunits and the site of edeine inhibition. *J. Biol. Chem.* **253**, 1807–1816.

Orenstein, N. S., Dvorak, H. F., Blanchard, M. H., and Young, M. (1978). Nerve growth factor: A protease that can activate plasminogen. *Proc. Natl. Acad. Sci. U.S.A.* **75**, 5497–5500.

Petryshyn, R., Trachsel, H., and London, I. M. (1979). Regulation of protein synthesis in reticulocyte lysates: Immune serum inhibits heme-regulated protein kinase activity and differentiates heme-regulated protein kinase from double-stranded RNA-induced protein kinase. *Proc. Natl. Acad. Sci. U.S.A.* **76**, 1575–1579.

Pinphanichakarn, P., Kramer, G., and Hardesty, B. (1976). Partial reaction of peptide initiation inhibited by the reticulocyte hemin-controlled repressor. *Biochem. Biophys. Res. Commun.* **73**, 625–631.

Pinphanichakarn, P., Kramer, G., and Hardesty, B. (1977). Partial purification and characterization of a translational inhibitor from Friend leukemia cells. *J. Biol. Chem.* **252**, 2106–2112.

Quigley, G., Teeter, M., and Rich, A. (1978). Structural analysis of spermine and magnesium ion binding to yeast phenylalanine transfer RNA. *Proc. Natl. Acad. Sci. U.S.A.* **75**, 64–68.

Ranu, R. S., Levin, D. H., Delaunay, J., Ernst, V., and London, I. M. (1976). Regulation of protein synthesis in rabbit reticulocyte lysates: Characteristics of inhibition of protein synthesis by a translational inhibitor from heme-deficient lysates and its relationship to the initiation factor which binds Met-tRNA$_f$. *Proc. Natl. Acad. Sci. U.S.A.* **73**, 2720–2726.

Ranu, R. S., London, I. M., Das, A., Dasgupta, A., Majumdar, A., Ralston, R., Roy, R., and Gupta, N. K. (1978). Regulation of protein synthesis in rabbit reticulocyte lysates by the heme-regulated protein kinase: Inhibition of interaction of Met-tRNA$_f^{Met}$ binding factor with another initiation factor in formation of Met-tRNA$_f^{Met}$ · 40 S ribosomal subunit complexes. *Proc. Natl. Acad. Sci. U.S.A.* **75**, 745–749.

Ratner, L., Sen, G. C., Brown, G. E., Lebleu, B., Kawakita, M., Carber, B., Slattery, E., and Lengyel, P. (1977). Interferon, double-stranded RNA and RNA degradation. Characteristics of an endonuclease activity. *Eur. J. Biochem.* **79**, 565–577.

Revel, M., and Groner, Y., (1978). Post-transcriptional and translational controls of gene expression in eukaryotes. *Annu. Rev. Biochem.* **47**, 1079–1126.

Revel, M., Groner, Y., Pollack, Y., Cnaani, D., Zeller, H., and Nudel, U. (1973). Biochemical mechanism to control protein synthesis: mRNA specific initiation factors. *Acta Endocrinol. (Copenhagen), Suppl.* **180**, 54–74.

Roberts, W. K., Hovanessian, A., Brown, R. E., Clemens, M., and Kerr, I. M. (1976). Interferon-mediated protein kinase and low molecular weight inhibitor of protein synthesis. *Nature (London)* **264**, 477–480.

Rose, J., Trachsel, H., Leong, K., and Baltimore, D. (1978). Inhibition of translation by poliovirus. Inactivation of a specific initiation factor. *Proc. Natl. Acad. Sci. U.S.A.* **75**, 2732–2736.

Safer, B., and Jagus, R. (1979). Control of eIF-2 phosphatase in rabbit reticulocyte lysate. *Proc. Natl. Acad. Sci. U.S.A.* **76**, 1094–1098.

Salden, M., and Bloemendal, H. (1976). Polyamines can replace the dialyzable component from crude reticulocyte initiation factors. *Biochem. Biophys. Res. Commun.* **68**, 157–161.

Samuel, C. E. (1979). Mechanism of interferon action: Phosphorylation of protein synthesis initiation factor eIF-2 in interferon-treated human cells by a ribosome-associated kinase processing site specificity similar to hemin-regulated rabbit reticulocyte kinase. *Proc. Natl. Acad. Sci. U.S.A.* **76,** 600–604.

Schlessinger, J., Schechter, Y., Willingham, M. C., and Pastan, I. (1978). Direct visualization of binding, aggregation, and internalization of insulin and epidermal growth factor on living fibroblastic cells. *Proc. Natl. Acad. Sci. U.S.A.* **75,** 2659–2663.

Sehgal, P., Dobberstein, B., and Tamm, I. (1977). Interferon messenger RNA content of human fibroblasts during induction, shut off, and superinduction of interferon production. *Proc. Natl. Acad. Sci. U.S.A.* **74,** 3409–3413.

Sen, G. C., Lebleu, B., Brown, G. E., Kawakita, M., Slattery, E., and Lengyel, P. (1976). Interferon, double-stranded RNA, and mRNA degradation. *Nature (London)* **264,** 370–373.

Shafritz, D. A., Weinstein, J. A., Safer, B., Merrick, W. C., Weber, L. A., Hickey, E. D., and Baglioni, C. (1976). Evidence for role of $m^7G^{5'}$-phosphate group in recognition of eukaryotic mRNA by initiation factor IF-M_3. *Nature (London)* **261,** 291–294.

Shatkin, A. (1976). Capping of eukaryotic mRNA. *Cell* **9,** 645–653.

Shine, J., and Dalgarno, L. (1974). The 3'-terminal sequences of *Escherichia coli* 16 S ribosomal RNA: Complementarity to nonsense triplets and ribosome binding sites. *Proc. Natl. Acad. Sci. U.S.A.* **71,** 1341–1346.

Shine, J., and Dalgarno, L. (1975). Determinant of cistron specificity in bacterial ribosomes. *Nature (London)* **254,** 34–36.

Sonenberg, N., and Shatkin, A. (1977). Reovirus mRNA can be covalently crosslinked via 5' cap to proteins in initiation complexes. *Proc. Natl. Acad. Sci. U.S.A.* **74,** 4288–4292.

Sonenberg, N., and Shatkin, A. (1978). Non-specific effect of m^7GMP on protein–RNA interactions. *J. Biol. Chem.* **253,** 6630–6632.

Sonenberg, N., Morgan, M., Merrick, W., and Shatkin, A. (1978). A polypeptide in eukaryotic initiation factors that crosslinks specifically to the 5'-terminal cap in mRNA. *Proc. Natl. Acad. Sci. U.S.A.* **75,** 4843–4847.

Spirin, A. (1969). Informosomes. *Eur. J. Biochem.* **10,** 20–34.

Steitz, J. A. (1973). Discriminatory ribosome rebinding of isolated regions of protein synthesis initiation from the ribonucleic acid of bacteriophage R17. *Proc. Natl. Acad. Sci. U.S.A.* **70,** 2605–2609.

Steitz, J. A., and Jakes, K. (1975). How ribosomes select initiator regions in mRNA: Base pair formation between the 3' terminus of 16 S rRNA and the mRNA during initiation of protein synthesis in *Escherichia coli. Proc. Natl. Acad. Sci. U.S.A.* **72,** 4734–4738.

Stoltzfus, C. M., Shatkin, A. J., and Banerjee, A. B. (1973). Absence of polyadenylic acid from reovirus messenger ribonucleic acid. *J. Biol. Chem.* **248,** 7993–7998.

Szer, W., Hermoso, J. M., and Leffler, S. (1975). Ribosomal protein S1 and polypeptide chain initiation in bacteria. *Proc. Natl. Acad. Sci. U.S.A.* **72,** 2325–2329.

Trachsel, H., and Staehelin, T. (1978). Binding and release of eukaryotic initiation factor eIF-2 and GTP during protein synthesis initiation. *Proc. Natl. Acad. Sci. U.S.A.* **75,** 204–208.

Trachsel, H., Ranu, R. S., and London, I. M. (1978). Regulation of protein synthesis in rabbit reticulocyte lysates. Purification and characterization of heme-reversible translational inhibitor. *Proc. Natl. Acad. Sci. U.S.A.* **75,** 3654–3658.

Traut, R. R., Heimark, R. L., Sun, T. T., Hershey, J. W. B., and Bollen, A. (1974). Protein topography of ribosomal subunit from *Escherichia coli. In* "Ribosomes" (M. Nomura, A. Tissières, and P. Lengyel, eds.), pp. 271–308. Cold Spring Harbor Lab., Cold Spring Harbor, New York.

van Duin, J., and van Knippenberg, P. H. (1974). Functional heterogeneity of the 30 S ribosomal subunits of *Escherichia coli*. III. Requirement of protein S1 for translation. *J. Mol. Biol.* **84,** 185–195.

van Knippenberg, P. H., Hooykaas, P. J. J., and van Duin, J. (1974). The stoichiometry of *E. coli* 30 S ribosomal protein S1 on *in vivo* and *in vitro* polysomes. *FEBS Lett.* **41,** 323–326.

Vincent, A., Civelli, O., Buri, J. F., and Scherrer, K. (1977). Correlation of specific coding sequences with specific proteins associated in untranslated cytoplasmic mRNP complexes of duck erythroblasts. *FEBS Lett.* **77,** 281–286.

Wahba, A. J., Miller, M. J., Niveleau, A., Landers, T. A., Carmichael, G. G., Weber, K., Hawley, D. A., and Slobin, L. I. (1974). Subunit I of Qβ replicase and 30 S ribosomal protein S1 of *Escherichia coli*. Evidence for the identity of the two proteins. *J. Biol. Chem.* **249,** 3314–3316.

Wallis, M. H., Kramer, G., and Hardesty, B. (1980). Partial purification and characterization of a 90,000 dalton peptide involved in activation of the eIF-2α protein kinase of the hemin-controlled translational repressor. *Biochemistry* **19,** 798–804.

Walsh, D. A., Ashby, C. D., Gonzales, C., Calkins, D., Fischer, E. H., and Krebs, E. G. (1971). Purification and characterization of a protein inhibitor of adenosine 3′,5′-monophosphate-dependent protein kinases. *J. Biol. Chem.* **246,** 1977–1985.

Waxman, H. S., and Rabinovitz, M. (1966). Control of reticulocyte polyribosome content and hemoglobin synthesis by heme. *Biochim. Biophys. Acta* **129,** 369–379.

Widnell, C. C., and Tata, J. R. (1966). Additive effects of thyroid hormone, growth hormone, and testosterone on deoxyribonucleic acid-dependent ribonucleic acid polymerase in rat-liver nuclei. *Biochem. J.* **98,** 621–629.

Wittmann, H. G., Stöffler, G., Hindennach, I., Kurland, C. G., Randall-Hazelbauer, L., Birge, E. A., Nomura, M., Kaltschmidt, E., Mizushima, S., Traut, R. R., and Bickle, T. A. (1971). Correlation of 30 S ribosomal proteins of *Escherichia coli* isolated in different laboratories. *Mol. Gen. Genet.* **111,** 327–333.

Wool, J. G., and Stöffler, G. (1974). Structure and function of eukaryotic ribosomes. *In* "Ribosomes" (M. Nomura, A. Tissières, and P. Lengyel, eds.), pp. 417–460. Cold Spring Harbor Lab., Cold Spring Harbor, New York.

Wu, J. M., Cheung, C. P., and Suhadolnik, R. J. (1978). Stimulation and inhibition of the protein synthetic process by NAD$^+$ in lysed rabbit reticulocytes. *J. Biol. Chem.* **253,** 7295–7300.

Zilberstein, A., Dudock, B., Berissi, H., and Revel, M. (1976a). Control of mRNA translation by minor species of Leu-tRNA in extracts from interferon-treated L cells. *J. Mol. Biol.* **108,** 43–54.

Zilberstein, A., Federman, P., Shulman, L., and Revel, M. (1976b). Specific phosphorylation *in vitro* of a protein associated with ribosomes of interferon-treated mouse L cells. *FEBS Lett.* **68,** 119–124.

Zucker, W. V., and Schulman, H. M. (1968). Stimulation of globin-chain initiation by hemin in the reticulocyte cell-free system. *Proc. Natl. Acad. Sci. U.S.A.* **59,** 582–589.

3

Masked Messenger RNA and the Regulation of Protein Synthesis in Eggs and Embryos

Rudolf A. Raff

I. INTRODUCTION

The realization that informational macromolecules exist in the cytoplasm of eggs dates to observations made during the late nineteenth and early twentieth centuries on the development of certain marine embryos. These observations showed that morphogenetic information was present

107

in eggs and that various regions of the uncleaved egg were not equivalent with respect to their developmental potentials. This information was found to be cytoplasmic in localization, and it became clear that such information has a critical role in determination of nuclear fates. It was also realized that the early stages of development are generally independent of nuclear gene action until shortly before gastrulation. It has more recently been found that unfertilized eggs of sea urchins and other organisms possess a considerable store of messenger RNA in their cytoplasms. These mRNA species are generally present in the cytoplasm with little tendency to bind to ribosomes, which are also present, to initiate protein synthesis until the eggs have been fertilized. A variety of mechanisms have been proposed to explain these observations, with many investigators coming to favor the hypothesis of Spirin (1966) that mRNAs in eggs are "masked"; that is, unavailable for translation because of association with proteins to form stable messenger ribonucleoprotein particles (mRNPs).

This chapter will describe some of the phenomena in embryonic development in which masked forms of mRNA have been suggested to play a key role. The focus will be upon the function of masked mRNA and the direct demonstration of masking as a control mechanism for protein synthesis in eggs and embryos of the sea urchin and other organisms. Finally, space limitations have made it necessary to deal with a rather large body of material in a compact manner. While this has its advantages for the reader, it also means that much that is of interest is only briefly touched upon. I have fully documented key points and provided sufficient other references to allow the reader entry into the literature. However, this chapter is not an exhaustive review, and many interesting and pertinent papers are not cited.

II. MATERNAL EFFECTS AND THE EXISTENCE OF STORED mRNA IN EGGS

A variety of maternal effects has been noted in early embryos. These indicate that products of gene action during oogenesis are expressed in the embryo and control the events of early development and, in some cases, are responsible for differentiation of embryonic cells. Both quantitatively and qualitatively a large proportion of the oogenetic gene products are present as mRNAs.

A. Maternal Control of Early Development

Theodor Boveri (1902) established the need for a balanced set of chromosomes for normal development in his experiments on blastomeres from normal and dispermic sea urchin zygotes. He found that the dependence on a proper chromosome complement occurs only after the blastula stage is reached. Boveri's observations suggested that early development is dependent upon a "maternal" program in the egg and that control of development by the embryonic genome begins later.

The classic approach to the problem of time of onset of zygote gene action has been the use of cross-species hybrids. Many hybrids between sea urchin species eliminate all or part of the paternal chromosomes, but some hybrids retain both sets (Tennent, 1922). In such hybrids development strictly follows the maternal species pattern until the mesenchyme blastula stage, then paternal or hybrid characteristics may appear. For example, in crosses of *Cidaris* (♀) × *Lytechinus* (♂) studied by Tennent (1914), the maternal pattern is seen in timing of mesenchyme formation and invagination of the archenteron, but the paternal pattern is seen in the site of origin of primary mesenchyme cells. Similar results have been observed using biochemical rather than morphological criteria. Thus, while hybrids of *Dendraster* × *Strongylocentrotus* retain a hybrid genome and express both maternal and paternal histone H1, only the maternal forms of a variety of enzymes or protein antigens are expressed from cleavage through gastrula or pluteus (Whiteley and Whiteley, 1972; Ozaki, 1975; Badman and Brookbank, 1970; Easton *et al.*, 1974). Davidson (1976) proposed that these observations do not reflect a faulty paternal genome in these hybrids, but rather a mechanism of early development in which pregastrula events depend upon the mRNA stores of the egg, and not on new transcription.

B. Demonstration of Stored mRNAs in the Egg Cytoplasm

Two primary approaches have been used in the study of oogenetic mRNAs: first, the use of RNA synthesis inhibitors to prevent embryonic transcription and thus allow the observation of oogenetic mRNA function *in vivo* in embryos, and, second, direct characterization of mRNAs isolated from the egg cytoplasm.

1. Inhibitor Experiments

Brachet *et al.* (1963) and Denny and Tyler (1964) demonstrated protein synthesis in artificially activated sea urchin eggs which had been enucleated by centrifugation. This observation suggested the existence of mRNA in the unfertilized egg cytoplasm, but such activated, physically enucleated merogones have only a limited ability to cleave and exhibit little development. Most subsequent studies have instead used embryos in which transcription has been inhibited with actinomycin D following the initial observations of Gross and Cousineau (1963, 1964) that sea urchin embryos could be cultured in the presence of sufficient actinomycin D to block 95% of RNA synthesis during early development without preventing DNA replication and cleavage. Such embryos finally suffer developmental arrest at the hatched blastula stage, but prior to that continue to synthesize proteins at only slightly depressed rates (Gross and Cousineau, 1964; Greenhouse et al., 1971). Actinomycin D has no detectable direct effects on protein synthesis in sea urchin embryos. Sargent and Raff (1976) investigated the effects of actinomycin on protein synthesis in activated, enucleated sea urchin merogones and found no effect on level of protein synthesis over a 12-hr period or upon the spectrum of proteins made. Several specific proteins have been identified that are products of translation of oogenetic mRNAs: these include tubulins (Raff et al., 1971, 1972), hyaline membrane protein (Citkowitz, 1972), the major histones (Ruderman and Gross, 1974), and possibly the hatching enzyme (Barrett and Angelo, 1969).

Actinomycin experiments have also been carried out with other marine organisms, such as snails (Newrock and Raff, 1975) and tunicates (Whittaker, 1977), with results comparable to sea urchins. Similarly, mammalian embryos continue protein synthesis in the presence of doses of actinomycin D which greatly inhibit RNA synthesis (Tasca and Hillman, 1970; Manes, 1973), although it should be noted that actinomycin D is apparently toxic to mammalian embryos and causes rapid developmental arrest (Monesi et al., 1970; Golbus et al., 1973). However, α-amanitin, which also blocks mRNA synthesis in mammalian embryos, does not block cleavage nor inhibit protein synthesis (Golbus et al., 1973).

Actinomycin D has been injected into the large yolky eggs of insects and amphibians, but the results are difficult to interpret because of incomplete inhibition of RNA synthesis.

Stored mRNAs are not limited to animal systems. Actinomycin experiments with germinating plant seeds indicate that protein synthesis is directed by stored mRNA during the first hours of germination (see Dure, 1977). Similarly, protein synthesis during the early part of germination of the zoospores of the water mold, *Blastocladiella,* is unaffected by actinomycin D (Soll and Sonneborn, 1971; Leaver and Lovett, 1974; Silverman et al., 1974).

Taken together these observations provided strong evidence for the existence of stores of mRNA in eggs and similar systems, and for the crucial role these mRNAs play in providing the templates required for protein synthesis during early stages of development.

2. Direct Characterization

Slater and Spiegelman (1966) employed a cell-free system from *E. coli* to translate unfractionated RNA from unfertilized sea urchin eggs. By comparison with the amount of synthesis directed by viral RNA in their system, Slater and Spiegelman estimated that 4% of egg RNA represented mRNA. A comparable estimate of 3% of sea urchin egg RNA as mRNA was made by Jenkins et al. (1973) who used poly(A)$^+$ RNA to direct protein synthesis in a cell-free system from sarcoma-180 cells. No products were identified in either of the above studies. This was first accomplished by Gross et al. (1973) who showed that RNA extracted from 20–40 S particles from sea urchin eggs directed the synthesis of histones in vitro. Gabrielli and Baglioni (1977) obtained a similar result with RNA from 20–60 S particles extracted from 8-cell embryos of the clam, Spisula. Ruderman and Pardue (1977) analyzed the populations of mRNA present in sea urchin eggs and amphibian (Xenopus and Triturus) oocytes and ovaries by cell-free translation. Sea urchin eggs were found to contain poly(A)$^+$ mRNA, poly(A)$^-$ histone mRNA, and poly(A)$^-$ nonhistone mRNA. All were translatable, and there were some qualitative differences between products of poly(A)$^+$ and poly(A)$^-$ mRNAs. Amphibian oocytes lacked a significant amount of poly(A)$^-$ nonhistone mRNA, but possessed poly(A)$^+$ mRNA, poly(A)$^-$ histone mRNA, and poly(A)$^+$ histone mRNA.

Competition hybridization experiments performed by Farquhar and McCarthy (1973), Skoultchi and Gross (1973), and Lifton and Kedes (1976) demonstrated that a fraction of sea urchin egg RNA sequences competes with labeled histone mRNA for hybridization to histone DNA sequences. Histone mRNA sequences have also been detected in eggs by Woods and Fitschen (1978) who used a histone mRNA specific cDNA probe, and by Shepard (1977) who used a probe prepared from cloned sea urchin histone genes.

Lifton and Kedes (1976) isolated histone mRNAs from eggs and showed them to have electrophoretic mobilities identical to well-characterized embryonic histone mRNAs.

Sequence complexity analysis has been carried out in some detail for mature oocyte RNAs from sea urchins, Xenopus, and the echiuroid worm, Urechis. These studies have been reviewed by Davidson (1976). RNAs corresponding to the single copy portion of the genome are present in oocytes of these organisms in high complexity (30×10^6 to 40×10^6 nucleotides) equivalent to about 25,000 different mRNAs of 1500 nucleotides in length. These complex transcripts of unique DNA sequences make up a significant proportion of the oogenetic mRNA of eggs, but much oogenetic RNA consists of transcripts of repetitive sequences of the genome. Such transcripts were originally studied in amphibian oocytes (Crippa et al., 1967; Hough and Davidson, 1972; Hough et al., 1973), but are also present in a wide variety of other oocytes including sea urchins, mollusks, and tunicates (see Davidson, 1976). In sea urchin eggs, histone mRNA is a prominent member of this class of RNAs (Kedes and Birnsteil, 1971), but most repeat sequences are not structural genes, and their transcripts do not appear to be translated. The expression of such sequences in sea urchin egg RNAs has been investigated by Costantini et al. (1978), who have found that most of the repeat sequences of the genome are represented by transcripts in the egg. Some repeat families contribute up to 10^5 transcripts per egg. Surprisingly, both strands of the repeat families thus far examined are represented as transcripts. More recent data (W. Klein, personal communication) suggests that most oogenetic mRNAs are composed of single copy DNA transcripts linked to repeat sequence DNA transcripts. The function of these presumably nontranslated repeat sequences attached to the translated portions of the mRNAs of eggs is unknown.

C. Localization Phenomena

Localization of cytoplasmic information important in early determination events occurs in many animal phyla including ctenophores, nemerteans, annelids, mollusks, arthropods, echinoderms, and vertebrates. The most impressive examples of localization are provided by the highly mosaic development of many spiralian protostomes, such as annelids and snails. In some of these forms, such as the marine snail *Ilyanassa,* a spectacular cytoplasmic protuberance called the polar lobe forms shortly before the first mitotic division of the fertilized egg. This lobe is non-nucleated. Upon completion of cleavage the polar lobe is resorbed by one of the two daughter blastomeres. The polar lobe is easily removed, and embryos from which the lobe has been deleted continue to develop. However, although normal embryos differentiate into a complex larval form, the veliger, which possesses such complex structures as a foot, shell, ciliated velum, eye, statocyst, and internal organs, delobed embryos give rise only to a ciliated mass of cells. Newrock and Raff (1975) found that even at a stage prior to the beginning of morphogenesis, lobeless and normal embryos synthesize different proteins. These protein synthesis differences are also present between lobeless and normal embryos cultured continuously in the presence of a concentration of actinomycin D sufficient to abolish RNA synthesis. This result suggests that specific oogenetic mRNA sequences may be sequestered in the polar lobe, since actinomycin-treated embryos translate only the available stored mRNAs.

Similarly, Rodgers and Gross (1978) using a single-copy DNA probe found that in separated blastomeres of 16-cell stage sea urchin embryos mRNA sequences were not homogeneously distributed between cell types. The same result was obtained with embryos cultured in the presence of actinomycin D, indicating segregation of specific oogenetic mRNA sequences. This picture is complicated by the observation of Tufaro and Brandhorst (1979) that there are no differences between blastomeres in the synthetic patterns of the approximately 1000 proteins that can be resolved by two-dimensional gel electrophoresis. Rodgers and Gross (1978) may have detected segregation of mRNA sequences too rare to produce sufficient protein to be seen by Tufaro and Brandhorst (1979), or they may have detected sequences that do not function as mRNAs at the 16 cell stage.

In very different embryos, those of tunicates, localized stored mRNA is implicated in the development of alkaline phosphatase by endoderm cells, since histospecific appearance of this enzyme in the larva is not sensitive to actinomycin (Whittaker, 1977).

In the insect *Smittia*, Kalthoff and Sander (1968) found that ultraviolet irradiation of the animal pole of the egg produces an embryo in which head, thorax, and anterior abdominal segments are replaced by a mirror image duplication of the normal posterior end similar to the effect seen in the *bicaudal* mutant of *Drosophila*. Subsequent studies by Kalthoff (1971, 1973) and Kandler-Singer and Kalthoff (1976) have indicated that the anterior determining substance is RNA, possibly complexed with proteins.

Finally, polar granules localized at the vegetal pole of frog eggs and at the posterior pole of insect eggs have been shown to act as germ line determinants (reviewed by Davidson, 1976). Polar granules are sensitive to ultraviolet irradiation (Okada *et al.*, 1974; Warn, 1975) and appear from cytological staining to contain RNA (Mahowald, 1971), as well as (in *Drosophila*) at least one specific protein (Waring *et al.*, 1978).

D. Maternal Effect Genes

There are several maternal effect mutations, known from a variety of organisms, which specifically affect early developmental processes and provide further evidence for the storage of developmental information in the egg cytoplasm. In many cases, the molecular nature of the stored informational materials is not known, but in others proteins have been demonstrated, and in some RNA has been suggested. The major importance of these in the present context is to suggest that genetics may provide a tool for the study of stored mRNAs. The variety of maternal effect genes is indicated by the surprising range of early developmental processes affected.

One of the longest known cases is control of symmetry in the snail *Limnaea* (Morgan, 1927). The direction of spiral cleavage, and of the resulting symmetry of the adults, is apparently controlled by a pair of alleles of a single gene. Freeman (1977) has found that the effect of the allele for sinistral coiling can be reversed by injection of cytoplasm from dextral eggs into eggs from mothers homozygous for the sinistral allele. Other striking cases include the mutant *bicaudal* of *Drosophila* (Bull, 1966), in which embryos lack a head and thorax, but possess an abdomen at each end. The studies of Kalthoff on *Smittia* (cited in Section II,C) suggest that *bicaudal* may involve oogenetic mRNA. The *o* mutant of the axolotl produces embryos unable to gastrulate or develop further unless injected with a protein component from normal eggs (Briggs and Cassens, 1966; Briggs and Justus, 1968). Finally, the mutant *grandchildless* of *Drosophila*, in which homozygous flies are sterile, results from abnormal function of the polar plasm at the posterior pole of the egg, which is required for germ cell formation (Spurway, 1948; Fielding, 1967; Mahowald *et al.*, 1979).

III. PROTEIN SYNTHESIS FOLLOWING FERTILIZATION OF SEA URCHIN EGGS

Unfertilized sea urchin eggs synthesize a large spectrum of proteins (Brandhorst, 1976); however, this synthesis occurs at a comparatively low rate. Following fertilization the rate of protein synthesis begins to rise (Epel, 1967; Timourian and Watchmaker, 1970). By gastrulation the absolute rate is 100 times that of the unfertilized egg (Regier and Kafatos, 1977). As noted above (Section II,B,1), this rise is independent of mRNA synthesis by the zygote. The rise in synthesis is accompanied by an increase in the number of polysomes as oogenetic mRNA is recruited in a linear manner over a 2-hr period following fertilization (Rinaldi and Monroy, 1969; Humphreys, 1971; Dolecki *et al.*, 1977).

The presence of mRNAs and other components required for protein synthesis in sea urchin eggs taken together with the depressed stage of protein synthesis in eggs and the rapid rise following fertilization indicates that a significant translational control operates in this system. The proposed explanations for the dramatic postfertilization rise in protein synthesis in sea urchin eggs fall into three general classes: faulty or incomplete translation machinery, mRNA processing or modification, and mRNA unmasking. Unmasking appears, in fact, to be the primary mechanism.

The low rate of synthesis in eggs is not due to a lack of ribosomes, translation factors, or energy sources. Crude homogenates support *in vitro* translation of synthetic polyribonucleotides (Hultin, 1961; Stavy and Gross, 1967; Timourian, 1967), and concentrations of elongation factors, aminoacyl-tRNA synthetases, tRNAs, ATP, and GTP sufficient to support high levels of protein synthesis are present in eggs (Casteneda, 1969; Felicetti *et al.*, 1972; Ceccarini and Maggio, 1969; Molinaro and Farace, 1972; O'Melia and Villee, 1972; MacKintosh and Bell, 1969a; Zeikus *et al.*, 1969).

Metafora *et al.* (1971) and Gambino *et al.* (1973) demonstrated a translation inhibitor that could be removed from egg ribosomes with salt solutions. Addition of the inhibitor to an *in vitro* system containing embryo ribosomes suppressed the translation of poly(U) templates. Metafora *et al.* (1971) suggested that this inhibitor acts as a translation inhibitor in unfertilized eggs, but such a role was eliminated by the finding of Hille (1974) that the inhibitor could also be obtained from active embryo ribosomes. Moreover, egg ribosomes, which presumably possess the inhibitor, translate globin mRNA as well as embryo ribosomes do in a heterologous cell-free system (Clegg and Denny, 1974). A similar inhibitor has also been suggested to repress protein synthesis in brine shrimp cysts (Huang and Warner, 1974).

If binding of functional mRNAs with ribosomes were inhibited in the unfertilized egg, egg polysomes would be underloaded. Following fertilization, a significant shift in the polysome profile to a higher modal sedimentation value would be predicted (Lodish, 1971). No such shift has been observed (Humphreys, 1969; Brandis and Raff, 1978).

The second major hypothesis is that mRNA is modified or processed at fertilization (Greenberg, 1975; Perry, 1976). A common modification at the 5'-end of eukaryotic mRNAs is the inverted "cap" of 7-methylguanosine linked by a chain of three phosphates to the 5'-end of the next base ($m^7G^{5'}$ pppXp). The cap structure appears to be essential for translation (Both *et al.*, 1975; Muthukrishnan *et al.*, 1975a; Shafritz *et al.*, 1976). Hickey *et al.* (1976) tested for the presence of cap on extracted mRNAs from sea urchin eggs by translating this mRNA *in vitro* in the presence of S-adenosylhomocysteine to inhibit endogenous methylating enzymes of a wheat germ cell-free system. Since sea urchin egg mRNA stimulated protein synthesis in this system, which requires a cap structure on the mRNA, at least a significant portion of egg mRNAs already possess this structure.

It should be noted that incomplete capping may be responsible for translational repression in some systems. While the stored mRNAs of water mold zoospores and brine shrimp cysts, like sea urchin eggs, possess a methylated cap structure (Johnson *et al.*, 1977; Muthukrishnan *et al.*, 1975a), the stored mRNAs from the oocyte of the tobacco hornworm, *Manduca sexta,* have been reported to possess a nonmethylated cap with the structure $G^{5'}$pppXp (Kastern and Berry, 1976). This could provide a store of nontranslatable mRNAs, since mRNA bearing a 5'-terminal

$G^{5'}$pppXp structure are translationally inactive, but become translatable after methylation to m⁷GpppXp (Muthukrishnan *et al.*, 1975b).

A second modification of mRNAs are the poly(A) tracts found on the 3'-end of most mRNAs. The length of the poly(A) tracts of sea urchin oogenetic mRNAs increases twofold within about 2 hr after fertilization (Slater *et al.*, 1972; Wilt and Mazia, 1974). However, sufficient cordycepin to block 3'-polyadenylation of mRNA has no effect on the rise in protein synthesis (Mescher and Humphreys, 1974). Further, histone mRNA which lacks poly(A) tracts in sea urchins participates in the postfertilization rise in protein synthesis (Ruderman and Gross, 1974; Woods and Fitschen, 1978). As was noted above for cap structures, other dormant systems may differ from the sea urchin pattern with respect to the role of polyadenylation. Harris and Dure (1978) have noted that protein synthesis and germination of cotton seeds are unaffected by actinomycin D, but completely suppressed by cordycepin.

Other evidence suggests that the oogenetic mRNAs of sea urchin eggs are fully processed. Lifton and Kedes (1976) demonstrated that oogenetic histone mRNA is the same length to within 5 nucleotides, as embryonic histone mRNA.

The final hypothesis is that in eggs the protein synthetic machinery is competent and that mRNAs are potentially translatable, but that these mRNAs are sequestered or masked and thus unavailable for translation. An unmasking process is triggered by fertilization. Spirin (1966, 1969) presented detailed arguments for the role of masked forms of mRNA and predicted that these should exist as messenger ribonucleoprotein particles (mRNPs). Messenger RNPs and their role in masking are discussed in detail in Section IV.

IV. MASKING AND THE CONTROL OF PROTEIN SYNTHESIS IN EGGS AND EMBRYOS

A. Messenger RNP Structures in Unfertilized Eggs

Ribonucleoproteins may be defined as belonging to three broad classes on the basis of their intracellular origins: nuclear RNPs containing heterogeneous nuclear RNA (HnRNA) species, cytoplasmic unbound RNPs containing putative mRNAs, and functional mRNA-containing RNPs released from polysomes by EDTA or puromycin. Since it is clear that the bulk of oogenetic mRNAs of eggs is present in the cytoplasm, a search for masked mRNAs would logically concentrate upon free cytoplasmic RNPs. The study of cytoplasmic mRNPs in eggs has lagged behind the study of RNPs in nonrepressed cells, such as reticulocytes or cultured mammalian cells, because of the difficulty of labeling oogenetic mRNAs *in vivo*. Consequently, the detection of oogenetic mRNAs contained in presumptive RNP structures requires the use of either cell-free systems for translation of extracted mRNA or molecular hybridization to detect poly(A) tails or, if specific probes are available, the coding sequences of particular mRNA species.

As discussed in Section II,B,2, Gross *et al.* (1973) and Skoultchi and Gross (1973) using cell-free translation and competition hybridization, Woods and Fitschen (1978) using a histone-specific cDNA probe, and Shepard (1977) using a probe prepared from cloned histone genes have observed that histone mRNA sequences are present in 20–40 S structures in homogenates of sea urchin eggs. Poly(A)$^+$ mRNA has likewise been detected in cytoplasmic structures in sea urchin eggs and in insect and *Xenopus* oocytes by hybridization of [^{3}H]poly(U) to the poly(A) tails (Kaumeyer *et al.*, 1978; Lovett and Goldstein, 1977; Paglia *et al.*, 1976; Rosbash and Ford, 1974). These poly(A)$^+$ mRNA-containing structures range from 30 S to 70 S.

There are other developing systems in which repression of protein synthesis occurs and masked mRNA has been implicated in translational control. Encysted brine shrimp (*Artemia salina*) embryos possess mRNA in cytoplasmic structures, and poly(A)$^+$ mRNA-containing structures are present in the cytoplasm of wheat embryos, and in *Blastocladiella* zoospores. These systems will be considered in Section IV,B.

The mRNAs found in presumptive mRNPs in the cytoplasm exhibit S values higher than those of the same mRNAs after purification. Histone mRNPs typically have S values of 9–12 S, but sediment at 20–60 S in homogenates of eggs. Likewise, poly(A)$^+$ mRNAs, which have modal S values of 20–30 S when purified, are found to sediment at 30–70 S when detected in egg homogenates. These observations have been generally interpreted as indicating that the mRNAs in question are contained in mRNP structures in eggs.

Sedimentation data alone, however, are not sufficient to demonstrate the presence of mRNPs. This ambiguity arises from the ionic conditions under which mRNPs are studied. Messenger RNPs are generally prepared under ionic conditions which are adjusted to be similar to those used for isolation of intact polysomes in the system examined, and to prevent nonspecific complexing of protein with RNA. The critical concentration of monovalent cation to prevent formation of nonspecific complexes of RNA and protein is about 0.1 M (Baltimore and Huang, 1970). Echinoderm eggs have been reported to contain 0.2–0.4 M K$^+$ ion concentrations. Buffers with this range of K$^+$ ion concentrations are commonly used to yield intact

polysomes from sea urchin embryos, and similar buffers have been used
to prepare mRNPs from sea urchin eggs (Kaumeyer *et al.*, 1978; Jenkins *et
al.*, 1978; Ilan and Ilan, 1978; Shepard, 1977). Monovalent cation concentrations higher than 0.1 *M* were also used in the preparation of mRNPs
from insect oocytes, brine shrimp cysts, and wheat seeds. The difficulty in
interpretation of studies from which only sedimentation data are available
is that pure mRNAs aggregate under high salt conditions similar to those
used in sucrose gradient analysis of mRNPs (Nemer *et al.*, 1974; Haines *et
al.*, 1974; Bantle *et al.*, 1976; Kaumeyer *et al.*, 1978; Shepard (1977)). Thus
9–12 S histone mRNA sediments at 20 S in 0.35 *M* K^+, and 20 S poly(A)$^+$
mRNA at 30–70 S. This finding indicates that sedimentation analysis alone
is not sufficient to characterize mRNPs: buoyant density of fixed particles
on isopycnic gradients must also be measured to determine if protein is
indeed complexed with the mRNA. Density measurements have been
made on cytoplasmic mRNPs from some of the systems discussed above.
Paglia *et al.* (1976) found that the cytoplasmic poly(A)$^+$ mRNPs of
silkmoth oocytes had a buoyant density of 1.45 gm/cm^3 in CsCl (equivalent to approximately 80% protein).

The most extensive studies upon the buoyant density characteristics of
egg mRNPs have been carried out with sea urchin eggs. Shepard (1977)
isolated 20–30 S cytoplasmic structures containing the histone mRNA of
eggs: these banded at a density of 1.4–1.6 gm/cm^3 in Cs_2SO_4, indicating
that histone mRNAs in eggs are contained within mRNP particles composed of about 10–50% protein. Similarly, Kaumeyer *et al.* (1978) have
demonstrated that the poly(A)$^+$ mRNA of sea urchin eggs is contained
within mRNPs with a peak buoyant density in Cs_2SO_4 of 1.46 gm/cm^3
(approximately 45% protein). Kaumeyer *et al.* (1978) investigated the effects of monovalent cations and Mg^{2+} on the characteristics of poly(A)$^+$
mRNPs. The results of these studies are summarized in Table I. In the
presence of 0.35 *M* K^+, particles were stable to removal of Mg^{2+}. Particles
were also prepared in 0.35 *M* Na^+, since Na^+ has been used in some
studies of mRNPs, and because Na^+ was expected to be more effective
than K^+ in removing proteins from mRNP complexes. In the presence of
Na^+, deletion of Mg^{2+} resulted in loss of proteins. The presence of Mg^{2+}
has been observed in other studies to increase the resistance of *E. coli*
ribosomal and L cell and KB cell mRNP proteins to removal by high
monovalent cation concentrations (Spitnik-Elson and Atsmon, 1969; Perry
and Kelley, 1968; Kumar and Lindberg, 1972). Histone mRNPs are apparently unstable in Na^+ even in the presence of Mg^{2+} (Shepard, 1977).
Particles prepared in 0.05 *M* Na^+ (Table I) are very heterogeneous in
protein content and appear to have acquired proteins, confirming the need

TABLE I

Summary of Characteristics of mRNPs from Sea Urchin Eggs as a Function of
Isolation Conditions[a]

Class	Preparation buffer	Rate of sedimentation[b] (S)	Peak density in Cs_2SO_4 (gm/cm³)	Density range (gm/cm³)
Poly(A)⁺	0.35 M K⁺, 5 mM Mg²⁺	60–65	1.46	1.35–1.57
	0.35 M K⁺, 5 mM EDTA	60–65	1.46	1.35–1.57
	0.35 M Na⁺, 5 mM Mg²⁺	60–65	1.40	1.30–1.55
	0.35 M Na⁺, 5 mM EDTA	60–65	1.55	1.50–1.60
	0.05 M Na⁺, 5 mM EDTA	60–65	—	1.27–1.52
	Deproteinized poly(A)⁺mRNA	35–65 (20)	1.66	1.57–1.68
Histone	0.35 M K⁺, 5 mM Mg²⁺	19–30	—	1.41–1.60
[poly(A)⁻]	0.35 M Na⁺, 5 mM Mg²⁺	19–30	—	1.61–1.70
	Deproteinized histone mRNA	9–>30 (9–12)	1.65	1.57–1.70

[a] Data taken from Kaumeyer *et al.* (1978) and Shepard (1977).
[b] Sedimentation values in sucrose gradients containing 0.35 M K⁺ or Na⁺ except for
values in parentheses which were determined in sucrose gradients containing 70% for-
mamide to prevent aggregation of RNA.

for a monovalent cation concentration sufficient to avoid formation of
nonspecific RNP complexes during isolation.

The only data presently available on sea urchin egg mRNP proteins are
those of Peters and Jeffery (1978), who have reported that there are two
major high molecular weight proteins bound to the poly(A) tails of egg
mRNAs.

B. Messenger RNP Structures in Developing Embryos

There are two classes of mRNA present in developing embryos:
oogenetic mRNA stored in the egg and mRNA transcribed following fertilization. These mRNAs may be expected to be present in both free
cytoplasmic mRNPs and in polysomes. Because the focus of this chapter
is on masked forms of mRNA, I shall largely ignore the newly synthesized
mRNAs of actively developing embryos and concentrate upon oogenetic
mRNA which, as is shown in Section IV,C, exists in masked form in eggs.

Ribonucleoprotein particles containing oogenetic mRNAs have been
detected in the cytoplasm of developing embryos. For example, Gabrielli
and Baglioni (1977) demonstrated oogenetic histone mRNAs in 20–60 S

cytoplasmic particles from 8-cell embryos of the clam, *Spisula* (Section II,B,2). The mRNP identity of the particles was not established by buoyant density analysis, but the probability is good that histone mRNPs were present.

Young and Raff (1979) have investigated the characteristics of poly(A)⁺ mRNPs of oogenetic origin present in developing sea urchin embryos. Both free cytoplasmic mRNPs and mRNPs derived from polysomes were studied. While recruitment of mRNA from the pool of masked mRNA into functioning polysomes begins shortly following fertilization, a significant pool of oogenetic mRNA remains free in the cytoplasm (Hough-Evans *et al.*, 1977). Since recruitment is gradual (Dolecki *et al.*, 1977), it may be predicted that the free cytoplasmic mRNA remains in mRNPs similar or identical to those of the unfertilized egg. Embryos were cultured in the presence of actinomycin D so that only oogenetic mRNAs were present, and both free cytoplasmic mRNPs and polysomes were recovered. Cytoplasmic poly(A)⁺ mRNA was distributed in sucrose gradients over a range of 45 S to 80 S with a peak at 65–70 S, and exhibited a buoyant density of 1.45 gm/cm³ in Cs_2SO_4. These properties closely resemble those of the poly(A)⁺ mRNPs of the unfertilized sea urchin egg. Poly(A)⁺ mRNA released from polysomes was contained within mRNPs which sedimented at 55 S. The density of these particles was sensitive to the method of release; mRNPs released with puromycin had a density of 1.45 gm/cm³ in Cs_2SO_4, while release with EDTA caused a partial loss of protein. Cytoplasmic and polysomal mRNPs differed in ability to stimulate an *in vitro* protein synthesis system as discussed in Section IV,C.

Brine shrimp are unusual in that the cysts commonly called "eggs" are actually dessicated dormant embryos which are in a state of developmental arrest until rehydrated. This process, at least with respect to employment of masked forms of mRNA, bears a resemblance to the activation of protein synthesis at fertilization of eggs, although most of the masked mRNA of the cyst is probably of embryonic origin, since encysted embryos have largely reached the gastrula stage. Dormant brine shrimp embryos, like unfertilized eggs, contain a store of poly(A)⁺ mRNPs (Grosfeld and Littauer, 1975; Nilsson and Hultin, 1974, Slegers *et al.*, 1977; Felicetti *et al.*, 1975). These particles sediment from about 20 S to greater than 100 S and band at a buoyant density in sucrose at 1.27–1.28 gm/cm³ (approximately 80–90% protein) (Slegers and Kondo, 1977). The mRNAs in the cytoplasmic particles of dormant cysts shift to the polysomal fraction upon hydration of the cysts and resumption of development (Amaldi *et al.*, 1977; Sierra *et al.*, 1976).

Plant seeds resemble brine shrimp cysts in that embryonic development has begun prior to the onset of dormancy so that stores of mRNA maintained during the dormant state are apparently embryonic in origin (Dure, 1977). The mRNA of poly(A)$^+$ mRNPs can be labeled in wheat embryos, and shifts from the polysomes to free cytoplasmic mRNPs as the seed becomes dormant (Ajtkhozhin *et al.*, 1976). The buoyant density of the cytoplasmic mRNPs in CsCl is 1.45 gm/cm^3 (Ajtkhozhin *et al.*, 1973, 1976). Messenger RNA reappears in polysomes after germination (Ajtkhozhin and Akhanov, 1974; Hammett and Katterman, 1975). The stored poly(A)$^+$ mRNAs of the zoospore of the water mold, *Blasto-cladiella*, are present along with monosomes in a specialized organelle called the nuclear cap (Johnson *et al.*, 1977). The mRNAs sediment as 80 S particles and thus are either associated with monosomes or are present as 80 S mRNPs. Upon germination, stored mRNA enters polysomes and participates in the dramatic rise in protein synthesis that begins at about 20 min after the start of germination (Soll and Sonneborn, 1971; Leaver and Lovett, 1974; Silverman *et al.*, 1974).

C. Messenger RNPs as Masked mRNA

The existence of mRNPs in the cytoplasm of eggs and embryos, while consistent with the masked mRNA hypothesis, is by no means sufficient to substantiate the existence, and a developmental role of masked forms of mRNA. Several studies have shown that mRNPs isolated from polysomes or cytoplasm of various systems are as effective as the isolated mRNAs in stimulating heterologous *in vitro* protein synthesis systems (see Jenkins *et al.*, 1978, for references). Thus, demonstration of the validity of the masking hypothesis requires that egg mRNPs isolated in their native state be nontranslatable by an *in vitro* protein-synthesizing system until modified to allow the contained mRNA to be translated.

Direct tests of the translatability of presumed masked forms of mRNA have been made with mRNPs obtained from sea urchin eggs and embryos, brine shrimp cysts, and wheat seeds. The results of these studies are summarized in Table II.

Jenkins *et al.* (1978) and Young and Raff (1979) have examined the translatability of poly(A)$^+$ mRNPs isolated from the cytoplasm and released from polysomes. Poly (A)$^+$ mRNPs isolated from the cytoplasm of eggs in 0.35 M K$^+$ failed to stimulate translation in the wheat germ system (Fig. 1). However, mRNA extracted from these nontranslatable particles was as template active as mRNA extracted from whole eggs and directed the synthesis of the same spectrum of high molecular weight proteins. The lack of translation of the egg mRNPs was not due to the presence of an inhibitor, since addition of mRNPs to deproteinized mRNA had no effect on the efficiency of translation of the mRNA. These

TABLE II

Translatability *in Vitro* **of mRNPs Derived from Eggs and Embryos**[a]

Source	Cyto-plasmic mRNPs	Poly-somal mRNPs	*In vitro* system	References
Sea urchin egg	No	nd[b]	Wheat germ	Jenkins *et al.*, 1978
Sea urchin embryo	No	Yes	Wheat germ	Jenkins *et al.*, 1978; Young, 1978
Sea urchin egg	No	nd	Sea urchin egg	Ilan and Ilan, 1978
Brine shrimp cysts	No	nd	Wheat germ	Grosfeld and Littauer, 1975; Slegers *et al.*, 1977
Dry wheat embryos	Yes	nd	Wheat germ	Weeks and Marcus, 1971; Schultz *et al.*, 1972

[a] Under conditions optimal for translation of deproteinized mRNA, and using intact mRNPs. See text for description of translation experiments performed under alternate conditions or using modified mRNPs.
[b] nd, Not determined.

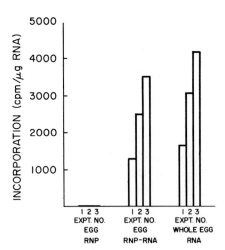

Fig. 1. Translation *in vitro* of egg mRNPs, mRNP-derived RNA, and total egg RNA from sea urchin eggs in the wheat germ cell-free system. Three separate experiments are plotted. Data are from Jenkins *et al.* (1978).

translation studies were carried out using 3.5 mM Mg^{2+} in the cell-free reactions. This is the optimal Mg^{2+} concentration for translation of mRNAs. Messenger RNPs were found to have an optimum of 3.0 mM Mg^{2+}, but even at this concentration efficiency of translation was very low. In contrast to the results obtained with cytoplasmic mRNPs, polysomal mRNPs isolated from embryos actively engaged in protein synthesis were translated as efficiently as deproteinized mRNA (Table III).

Masking appears to be labile, since egg cytoplasmic mRNPs prepared in the presence of 0.35 M Na$^+$ instead of K$^+$ were unstable and were readily translated *in vitro*.

Ilan and Ilan (1978) have confirmed the nontranslatability of egg cytoplasmic mRNPs using a cell-free system derived from sea urchin eggs, when ionic conditions optimal for mRNA translation were used. However, when the Mg^{2+} concentration in the cell-free system was raised to 12 mM, translation of a class of egg cytoplasmic mRNPs was observed. These particles differed from those studied by Jenkins *et al.* (1978) in that the particles of Ilan and Ilan sedimented at about 40 S and yielded mainly histones as products. These two sets of studies detected masked mRNA, and demonstrated that masking was sensitive to ionic conditions. Both poly(A)$^+$ and histone mRNAs occur in masked form in sea urchin eggs as mRNPs.

Evidence for the existence of masked mRNA in brine shrimp embryos comes from the observations of Grosfeld and Littauer (1975) and Slegers *et al.* (1977) that mRNA exists in a nontranslatable form in the cytoplasm of brine shrimp cysts. Grosfeld and Littauer (1975), Sierra *et al.* (1976),

TABLE III

Translation of Egg and Embryo mRNP and mRNA *in Vitro*[a]

Sample	Poly(A)$^+$ mRNA (ng/reaction mix)	Translational efficiency [cpm/ng of poly(A)$^+$ mRNA]
Egg RNA	471	66
Egg mRNP	444	15
Egg mRNP · RNA	162	157
Embryo cytoplasmic mRNP[b]	319	6
Polysomal mRNP (EDTA release)	147	111
Polysomal mRNP (puromycin release)	150	94

[a] Data of Young and Raff (from Young, 1978).
[b] Subpolysomal particles were prepared from embryos 10 min after fertilization. Polysomes were prepared from blastulae cultured with actinomycin D.

and Amaldi *et al.* (1977) found most of the poly(A)$^+$ RNA and mRNA
activity to be present in 40 S mRNPs. Grosfeld and Littauer (1975) demonstrated that the 40 S particles could not be translated in a wheat germ
cell-free system though deproteinized RNA isolated from the particles
was template active *in vitro*. Slegers and Kondo (1977) likewise observed
that the template-active poly(A)$^+$ mRNAs of the cyst were contained in
mRNPs with the somewhat different sedimentation value of 20–30 S.
These mRNAs were translatable *in vitro* in the wheat germ system. Slegers *et al.* (1977) present evidence that the actual translational inhibitor is
an RNA molecule which can be removed from the mRNA by poly(A)-
Sepharose chromatography in the presence of 10 mM EDTA. While there
is no reason to preclude a role for RNA as well as protein in masking, the
results of Slegers *et al.* (1977) are difficult to reconcile with those of other
workers who have reported brine shrimp mRNA to be translatable without such treatment (Grosfeld and Littauer, 1975; Felicetti *et al.*, 1975;
Sierra *et al.*, 1976; Amaldi *et al.*, 1977).

Unlike sea urchin eggs and brine shrimp cysts, the available evidence
suggests that the mRNPs of dormant seeds are translatable, and thus a
different translational control than masking may be present. Weeks and
Marcus (1971) reported that the mRNA of dry wheat embryos was present
in a membranous fraction which stimulated polysome formation and
amino acid incorporation in a wheat embryo cell-free system. Incorporation was abolished or reduced by treatments which destroyed or modified
either RNA or protein. In a similar study, Schultz *et al.* (1972) found that
template activity was present in 45–90 S particles. These were translatable
in a wheat embryo cell-free system without deproteinization of RNA.
Thus, either a translational repression system other than masking is present in seeds, or the wheat embryo cell-free system is capable of unmasking the mRNAs of the dry wheat embryo. This possibility could be tested
by use of a cell-free system of a different origin.

D. Unmasking and Quantitative Control of Translation

Humphreys (1969, 1971) pointed out that by determination of the ribosome transit times, polysome size, and change in overall rate of translation of oogenetic mRNA, a decision could be made between hypotheses in
which the postfertilization rise in protein synthesis in sea urchins results
from an increase in translational efficiency of ribosomes (more rounds of
translation per mRNA molecule) and those which predict a rise in synthesis because more translatable mRNA is made available. Ribosome transit
time (the time taken by a ribosome in traversing the length of an mRNA
molecule during translation) has been estimated for sea urchin eggs and

embryos. MacKintosh and Bell (1969b) calculated a transit time of about 6 min for both eggs and embryos, while Humphreys determined a transit time of approximately 1 min for both. These determinations differed greatly from one another and suffered from technical difficulties. Brandis and Raff (1978) using more reliable kinetic methods redetermined the transit times and observed that, in contrast to previous measurements, the average transit time for eggs at 16.5°C is 43 min. However, after fertilization the transit time falls to about 17 min. Hille and Albers (1979) have observed the same phenomenon at 12°C (72 min for eggs and 28 min for embryos).

The transit times reported by Brandis and Raff (1978) and Hille and Albers (1979) may appear at first to be unusually long when compared to the average transit times recorded for HeLa cells (1.8 min at 37°C) by Fan and Penman (1970) or to the rates of synthesis of specific proteins as globin (36 sec at 37°C) (Conconi et al., 1966) or ovalbumin (1.3 min at 41°C) (Palmiter, 1975). However, protein synthesis rates are strongly dependent upon temperature (Craig, 1975). A Q_{10} of 3.2 can be obtained from the data of Conconi et al. (1966), and preliminary measurements (J. W. Brandis and R. A. Raff, unpublished) suggest that a similar Q_{10} applies to protein synthesis in sea urchin eggs and embryos. Extrapolated elongation rates for globin and ovalbumin at 16.5°C, assuming a Q_{10} of 3.2 are 0.42 and 0.30 amino acids/sec. These elongation rates are similar to the average sea urchin elongation rates of 0.18 and 0.43 amino acids per second for eggs and embryos.

Thus, in contrast to the earlier reports, Brandis and Raff (1978) and Hille and Albers (1979) found that the transit time for eggs is long and that transit time decreases by about twofold following fertilization. Nevertheless, Humphreys' basic hypothesis (1969, 1971) that the rate of protein synthesis in eggs and zygotes is primarily controlled by availability of stored mRNA is correct. Most of the rise, as discussed below, is a consequence of unmasking of mRNPs present in the egg.

Sufficient data exist in the literature to allow the calculation of translational efficiencies (number of protein chains synthesized per active mRNA per minute) for eggs and embryos, and the actual number of mRNA molecules in active polysomes. Measurements of absolute rates of protein synthesis in sea urchin eggs and embryos indicate that embryos synthesize protein 30–100 times faster than do unfertilized eggs. Published rates of protein synthesis include cleavage and later stages and thus *may* overestimate the early rate of synthesis. These values are presented in Table IV along with the other parameters needed for the calculation of the values presented in Table V. The number of protein molecules made per minute per egg (S) is calculated from the mass of protein synthesized

TABLE IV

Parameters for Quantitation of Translation in Eggs and Embryos

Parameter	Value	Reference
Rate of protein synthesis, egg $(R_{\text{UF}})^a$	1.05×10^{-13} gm/egg/min	Regier and Kafatos, 1977
Rate of protein synthesis, embryo $(R_{\text{F}})^b$	2.86 to 11.7×10^{-12} gm/embryo/min	Regier and Kafatos, 1977; Fry and Gross, 1970; Seal and Aronson, 1973
Weight average molecular weight, protein chains, egg $(\overline{W}_{\text{UF}})$ and embryo $(\overline{W}_{\text{F}})^c$	50,000	Brandis and Raff, 1978
Polysome size, egg $(\overline{N}_{\text{UF}})$ and embryo $(\overline{N}_{\text{F}})^d$	10 ribosomes	Brandis and Raff, 1978
Transit time, egg $(\overline{t}_{\text{UF}})^e$	43 min	Brandis and Raff, 1978
Transit time, embryo $(\overline{t}_{\text{F}})^e$	17 min	Brandis and Raff, 1978

[a] Corrected to 16.5°C.

[b] Range of protein synthesis rates published for *Strongylocentrotus purpuratus* embryos corrected for temperature (16.5°C) and leucine content (7.5%).

[c] Determined by SDS-polyacrylamide gel electrophoresis of isotopically labeled nascent chains. It can be shown that the weight average molecular weight of completed chains (above) approximates 1.5 the weight average molecular weight of nascent chains.

[d] Weight average distribution determined from isokinetic gradients.

[e] Average of transit times determined by methods of Fan and Penman (1970) and Bremer and Yuan (1968).

TABLE V

Characteristics of Translation in Eggs and Embryos[a]

Parameter	Value
Rate of protein synthesis, egg (S_{UF})	1.3×10^6 molecules/egg/min
Rate of protein synthesis, embryo (S_F)	3.5×10^7 to 14×10^7 molecules/embryo/min
Translational efficiency, egg (T_{UF})	0.23 molecules/mRNA/min
Translational efficiency, embryo (T_F)	0.59 molecules/mRNA/min
Number of mRNAs on polysomes, egg (M_{UF})	5.7×10^6 molecules/egg
Number of mRNAs on polysomes, embryo (M_F)	5.9 to 24×10^7 molecules/embryo

[a] Calculated on basis of parameters set forth in Table IV.

per minute per egg (R) and the weight average molecular weight of the proteins ($\bar{W}$) as shown in Eq. (1).

$$S = (R/\bar{W})(6.023 \times 10^{23} \text{ molecules/mole}) \qquad (1)$$

Values of S for eggs and embryos are presented in Table IV. The ranges in the values reflect the wide range in published rates of protein synthesis in embryos.

Palmiter (1975) has defined translational efficiency, T, as a measure of how effectively an mRNA is being utilized. T can be calculated from Eq. (2).

$$T = \bar{N}/\bar{t} = I \qquad (2)$$

in which $\bar{N}$ is the average number of ribosomes per mRNA and $\bar{t}$ is the average transit time. Since we are dealing with the synthesis of a wide spectrum of proteins, $\bar{N}$ is a weight average for a heterogeneous population of polysomes. As discussed by Palmiter (1975), translational efficiency, T is equal to the rate of initiation, I. It can be seen that the translational efficiency increases after fertilization.

If translational efficiency, T, and rate of protein synthesis, S, are known, the number of mRNA molecules available for translation, M, can be calculated (Palmiter, 1975) from Eq. (3).

$$M = S/T \qquad (3)$$

Values of M for both eggs and early cleavage stage embryos are presented in Table V.

The increase in absolute rate of protein synthesis following fertilization is 30- to 100-fold, and there is an increase of from 10- to 40-fold in the number of mRNAs active in protein synthesis. At the stages used for these measurements, essentially all of the mRNA present in polysomes is oogenetic in origin and not newly transcribed. Messenger RNA availability accounts for most of the synthetic rate change. Thus, unmasking is responsible for major control over the rate of protein synthesis. Superimposed on this primary control is the change in transit time. Since the average size of polysomes, $\bar{N}$, in eggs and embryos is essentially the same, a shorter $\bar{t}$ requires that not only the rate of elongation, but also the rate of initiation [Eq. (2)], be faster in embryos than in eggs.

V. CONCLUDING REMARKS

It is clear that masked mRNAs, existing as mRNPs, provide the primary translational control in protein synthesis repression/activation in sea urchin eggs, and possibly in other embryonic systems such as brine shrimp cysts.

An overall scheme summarizing the events of mRNA recruitment in early stages of sea urchin development is presented in Fig. 2. A pool of masked mRNA is maintained in the egg. Upon fertilization, unmasking begins and translational efficiency rises. The result is the recruitment of mRNA into polysomes and the concomitant rise in protein synthesis. Recruitment is gradual, and the nonpolysomal pool of oogenetic mRNA persists for several hours. A few hours after fertilization significant synthesis of mRNA begins. The new mRNA is not masked, and rapidly enters polysomes. Polysomes translating oogenetic and new mRNAs are diagrammed separately in Fig. 2 to indicate the relative contributions of

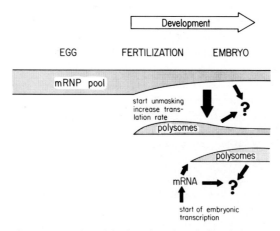

Fig. 2. Schematic representation of the function of masked and new mRNAs during early development of the sea urchin. The width of the stippled areas indicates the relative amounts of mRNA in the masked mRNA pool and in polysomes. See text for a discussion of the figure.

oogenetic and new mRNAs as development proceeds. The question marks (in Fig. 2) indicate destruction or inactivation of mRNA by as yet poorly defined processes.

The role of masking/unmasking is obviously to make available large amounts of mRNA for rapid development. However, the mechanism of unmasking remains obscure. In sea urchin zygotes this process is gradual and appears to involve modification or substitution of proteins in the mRNPs rather than simple removal. Masking is a coarse quantitative control rather than a qualitative one involving selection of specific mRNAs. Brandis and Raff (1979) have found that the trigger for unmasking can be separated from that for the change in transit time, since unmasking, but not the change in transit time can be stimulated by activation of eggs with NH_3. Preliminary experiments (J. W. Brandis and R. A. Raff, unpublished) with the Lilly calcium ionophore A23187 indicate that Ca^{2+} ions trigger both processes. The release of Ca^{2+} ions from intracellular stores that occurs shortly (<50 sec) following fertilization (Steinhardt *et al.*, 1977) may be the major initiator of metabolic events in fertilized eggs.

It is not yet clear how general masking is as a translation control mechanism. Masking appears to provide at least some eggs or dormant embryos with a means of wholesale repression and subsequent reactivation of protein synthesis. There are nonembryonic systems, such as cells in mitosis or meiosis, in which protein synthesis is inhibited. Masking provides a potential control mechanism in these systems. However, other mechanisms exist for inactivation of mRNAs or inhibition of translation so that masking should not be simply assumed in any uninvestigated system.

REFERENCES

Ajtkhozhin, M. A., and Akhanov, A. V. (1974). Release of mRNP- particles of the informo-some type from polyribosomes of higher plant embryos. *FEBS Lett.* **41**, 275–279.

Ajtkhozhin, M. A., Akhanov, A. V., and Doschanov, Kh. I. (1973). Informosomes of germinating wheat embryos. *FEBS Lett.* **31**, 104–106.

Ajtkhozhin, M. A., Doschanov, Kh. I., and Akhanov, A. V. (1976). Informosomes as a stored form of mRNA in wheat embryos. *FEBS Lett.* **66**, 124–126.

Amaldi, P. P., Felicetti, L., and Campioni, N. (1977). Flow of informational RNA from cytoplasmic poly(A)-containing particles to polyribosomes in *Artemia salina* cysts at early stages of development. *Dev. Biol.* **59**, 49–61.

Badman, W. S., and Brookbank, J. W. (1970). Serological studies of two hybrid sea urchins. *Dev. Biol.* **21**, 243–256.

Baltimore, D., and Huang, A. S. (1970). Interactions of HeLa cell proteins with RNA. *J. Mol. Biol.* **47**, 263–273.

Bantle, J. A., Maxwell, I. H., and Hahn, W. E. (1976). Specificity of oligo(dT)-cellulose chromatography in the isolation of polyadenylated RNA. *Anal. Biochem.* 72, 413–427.
Barrett, D., and Angelo, G. M. (1969). Maternal characteristics of hatching enzymes in hybrid sea urchin embryos. *Exp. Cell Res.* 57, 159–166.
Both, G. W., Banerjee, A. K., and Shatkin, A. J. (1975). Methylation dependent translation of viral messenger RNAs *in vitro. Proc. Natl. Acad. Sci. U.S.A.* 72, 1189–1193.
Boveri, T. (1902). On multipolar mitosis as a means of analysis of the cell nucleus. *In* "Foundations of Experimental Embryology" (B. H. Willier, and J. M. Oppenheimer, eds.), 1964, pp. 74–97. Prentice-Hall, Englewood Cliffs, New Jersey.
Brachet, J., Decroly, M., Ficq, A., and Quertier, J. (1963). Ribonucleic acid metabolism in unfertilized and fertilized sea-urchin eggs. *Biochim. Biophys. Acta* 72, 660–662.
Brandhorst, B. P. (1976). Two-dimensional gel patterns of protein synthesis before and after fertilization of sea urchin eggs. *Dev. Biol.* 52, 310–317.
Brandis, J. W., and Raff, R. A. (1978). Translation of oogenetic mRNA in sea urchin eggs and early embryos. Demonstration of a change in translational efficiency following fertilization. *Dev. Biol.* 67, 99–113.
Brandis, J. W., and Raff, R. A. (1979). Elevation of protein synthesis is a complex response to fertilisation. *Nature (London)* 278, 467–469.
Bremer, H., and Yuan, D. (1968). Chain growth rate of messenger RNA in *Escherichia coli* infected with bacteriophage T₄. *J. Mol. Biol.* 34, 527–540.
Briggs, R., and Cassens, G. (1966). Accumulation in the oocyte nucleus of a gene product essential for embryonic development beyond gastrulation. *Proc. Natl. Acad. Sci. U.S.A.* 55, 1103–1109.
Briggs, R., and Justus, J. T. (1968). Partial characterization of the component from normal eggs which corrects the maternal effect of gene *o* in the Mexican axolotl (*Ambystoma mexicanum*). *J. Exp. Zool.* 167, 105–116.
Bull, A. L. (1966). Bicaudal, a genetic factor which affects the polarity of the embryo in *Drosophila melanogaster. J. Exp. Zool.* 161, 221–242.
Castanada, M. (1969). The activity of ribosomes of sea urchin eggs in response to fertilization. *Biochim. Biophys. Acta* 179, 381–388.
Ceccarini, C., and Maggio, R. (1969). A study of aminoacyl transfer RNA synthetases by methylated albumin kieselguhr column chromatography in *Paracentrotus lividus. Biochim. Biophys. Acta* 190, 556–559.
Citkowitz, E. (1972). Analysis of the isolated hyaline layer of sea urchin embryos. *Dev. Biol.* 27, 494–503.
Clegg, K. B., and Denny, P. C. (1974). Synthesis of rabbit globin in a cell-free protein synthesis system utilizing sea urchin egg and zygote ribosomes. *Dev. Biol.* 37, 263–272.
Conconi, F. M., Bank, A., and Marks, P. A. (1966). Polyribosomes and control of protein synthesis: Effects of sodium fluoride and temperature in reticulocytes. *J. Mol. Biol.* 19, 525–540.
Costantini, F. D., Scheller, R. H., Britten, R. J., and Davidson, E. H. (1978). Repetitive sequence transcripts in the mature sea urchin oocyte. *Cell* 15, 173–187.
Craig, N. (1975). Effect of reduced temperatures on protein synthesis in mouse L cells. *Cell* 4, 329–335.
Crippa, M., Davidson, E. H., and Mirsky, A. E. (1967). Persistence in early amphibian embryos of informational RNA's from the lampbrush chromosome stage of oogenesis. *Proc. Natl. Acad. Sci. U.S.A.* 57, 885–892.
Davidson, E. H. (1976). "Gene Activity in Early Development," 2nd ed. Academic Press, New York.

Denny, P. C., and Tyler, A. (1964). Activation of protein biosynthesis in non-nucleate fragments of sea urchin eggs. *Biochem. Biophys. Res. Commun.* **14**, 245–249.

Dolecki, G. J., Duncan, R. F., and Humphreys, T. (1977). Complete turnover of poly(A) on maternal mRNA of sea urchin embryos. *Cell* **11**, 339–344.

Dure, L. S. (1977). Stored messenger ribonucleic acid and seed germination. *In* "The Physiology and Biochemistry of Seed Dormancy and Germination" (A. A. Khan, ed.), pp. 335–395. Elsevier, Amsterdam.

Easton, D. P., Chamberlain, J. P., Whiteley, A. H., and Whiteley, H. R. (1974). Histone gene expression in interspecies hybrid echinoid embryos. *Biochem. Biophys. Res. Commun.* **57**, 513–519.

Epel, D. (1967). Protein synthesis in sea urchin eggs: A "late" response to fertilization. *Proc. Natl. Acad. Sci. U.S.A.* **57**, 899–906.

Fan, H., and Penman, S. (1970). Regulation of protein synthesis in mammalian cells. II. Inhibition of protein synthesis at the level of initiation during mitosis. *J. Mol. Biol.* **50**, 655–670.

Farquhar, M., and McCarthy, B. J. (1973). Histone mRNA in egg and embryos of *Strongylocentrotus purpuratus*. *Biochem. Biophys. Res. Commun.* **53**, 515–522.

Felicetti, L., Metafora, S., Gambino, R., and DiMatteo, G. (1972). Characterization and activity of the elongation factors T1 and T2 in the unfertilized egg and in early development of sea urchins. *Cell Differ.* **1**, 265–277.

Felicetti, L., Amaldi, P. P., Moretti, S., Campioni, N., and Urbani, C. (1975). Intracellular distribution, sedimentation values and template activity of polyadenylic acid-containing RNA stored in *Artemia salina* cysts. *Cell Differ.* **4**, 339–354.

Fielding, C. J. (1967). Developmental genetics of the mutant *grandchildless* of *Drosophila subobscura*. *J. Embryol. Exp. Morphol.* **17**, 375–384.

Freeman, G. (1977). The transformation of the sinistral form of the snail *Lymnaea peregra* into its dextral form. *Am. Zool.* **17**, 946 (abstr.).

Fry, B., and Gross, P. R. (1970). Patterns and rates of protein synthesis. II. The calculation of absolute rates. *Dev. Biol.* **21**, 125–146.

Gabrielli, F., and Baglioni, C. (1977). Regulation of maternal mRNA translation in developing embryos of the surf clam *Spisula solidissima*. *Nature (London)* **269**, 529–531.

Gambino, R., Metafora, S., Felicetti, L., and Raisman, J. (1973). Properties of the ribosomal salt wash from unfertilized and fertilized sea urchin eggs and its effects on natural mRNA translation. *Biochim. Biophys. Acta* **312**, 377–391.

Golbus, M. S., Calarco, P. G., and Epstein, C. J. (1973). The effects of inhibitors of RNA synthesis (α-amanitin and actinomycin D) on preimplantation mouse embryogenesis. *J. Exp. Zool.* **186**, 207–216.

Greenberg, J. R. (1975). Messenger RNA metabolism of animal cells. Possible involvement of untranslated sequences and mRNA-associated proteins. *J. Cell Biol.* **64**, 269–288.

Greenhouse, G. A., Hynes, R. O., and Gross, P. R. (1971). Sea urchin embryos are permeable to actinomycin. *Science* **171**, 686–689.

Grosfeld, H., and Littauer, U. Z. (1975). Cryptic form of mRNA in dormant *Artemia salina* cysts. *Biochem. Biophys. Res. Commun.* **67**, 176–181.

Gross, K. W., Jacobs-Lorena, M., Baglioni, C., and Gross, P. R. (1973). Cell-free translation of maternal messenger RNA from sea urchin eggs. *Proc. Natl. Acad. Sci. U.S.A.* **70**, 2614–2618.

Gross, P. R., and Cousineau, G. H. (1963). Effects of actinomycin-D on macromolecular synthesis and early development of sea urchin eggs. *Biochem. Biophys. Res. Commun.* **10**, 321–326.

Gross, P. R., and Cousineau, G. H. (1964). Macromolecule synthesis and the influence of actinomycin on early development. *Exp. Cell Res.* **33**, 368–395.

Haines, M. E., Carey, N. H., and Palmiter, R. D. (1974). Purification and properties of ovalbumin mRNA. *Eur. J. Biochem.* **43**, 549–560.

Hammet, J. R., and Katterman, F. R. (1975). Storage and metabolism of poly(adenylic acid)-mRNA in germinating cotton seeds. *Biochemistry* **14**, 4375–4379.

Harris, B., and Dure, L. (1978). Developmental regulation in cotton seed germination: Polyadenylation of stored messenger RNA. *Biochemistry* **17**, 3250–3256.

Hickey, E. D., Weber, L. A., and Baglioni, C. (1976). Translation of RNA from unfertilized sea urchin eggs does not require methylation and is inhibited by 7-methylguanosine-5′monophosphate. *Nature (London)* **261**, 71–73.

Hille, M. B. (1974). Inhibitor of protein synthesis isolated from ribosomes of unfertilized eggs and embryos of sea urchins. *Nature (London)* **249**, 556–558.

Hille, M. B., and Albers, A. A. (1979). Efficiency of protein synthesis after fertilization of sea urchin eggs. *Nature (London)* **278**, 469–471.

Hough, B. R., and Davidson, E. H. (1972). Studies on the repetitive sequence transcripts of *Xenopus* oocytes. *J. Mol. Biol.* **70**, 491–509.

Hough, B. R., Yancy, P. H., and Davidson, E. H. (1973). Persistence of maternal RNA in *Engystomops* embryos. *J. Exp. Zool.* **185**, 357–368.

Hough-Evans, B. R., Wold, B. J., Ernst, S. G., Britten, R. J., and Davidson, E. H. (1977). Appearance and persistence of maternal RNA sequences in sea urchin development. *Dev. Biol.* **60**, 258–277.

Huang, F. L., and Warner, A. H. (1974). Control of protein synthesis in brine shrimp embryos by repression of ribosomal activity. *Arch. Biochem. Biophys.* **163**, 716–727.

Hultin, T. (1961). The effect of puromycin on protein metabolism and cell division in fertilized sea urchin eggs. *Experientia* **17**, 410–411.

Humphreys, T. (1969). Efficiency of translation of messenger RNA before and after fertilization of sea urchin eggs. *Dev. Biol.* **20**, 435–458.

Humphreys, T. (1971). Measurements of messenger RNA entering polysomes upon fertilization of sea urchin eggs. *Dev. Biol.* **26**, 201–208.

Ilan, J., and Ilan, J. (1978). Translation of maternal message ribonucleoprotein particles from sea urchin in a cell-free system from unfertilized eggs and product analysis. *Dev. Biol.* **66**, 375–385.

Jenkins, N. A., Taylor, M. W., and Raff, R. A. (1973). *In vitro* translation of oogenetic messenger RNA of sea urchins and picornavirus with a cell-free system from sarcoma-180. *Proc. Natl. Acad. Sci. U.S.A.* **70**, 3287–3291.

Jenkins, N. A., Kaumeyer, J. F., Young, E. M., and Raff, R. A. (1978). A test for masked message: The template activity of messenger ribonucleoprotein particles isolated from sea urchin eggs. *Dev. Biol.* **63**, 279–298.

Johnson, S. A., Lovett, J. S., and Wilt, F. H. (1977). The polyadenylated RNA of zoospores and growth phase cells of the aquatic fungus, *Blastocladiella. Dev. Biol.* **56**, 329–342.

Kalthoff, K. (1971). Photoreversion of UV induction of the malformaton "double abdomen" in the egg of *Smittia* spec. (Diptera, Chironomidae). *Dev. Biol.* **25**, 119–132.

Kalthoff, K. (1973). Action spectra for UV induction and photoreversal of a switch in the developmental program of the egg of an insect (*Smittia*). *Photochem. Photobiol.* **18**, 355–364.

Kalthoff, K., and Sander, K. (1968). Der Entwicklungsgang der Missbildung "Doppelabdomen" im partiell UV-bestrahlen Ei von *Smittia parthenogenica* (Dipt., Chironomidae). *Wilhelm Roux' Arch. Entwicklungsmech. Org.* **161**, 129–146.

Kandler-Singer, I., and Kalthoff, K. (1976). RNase sensitivity of an anterior morphogenetic determinant in an insect egg (*Smittia* spec., Chironomidae, Diptera). *Proc. Natl. Acad. Sci. U.S.A.* **73**, 3739–3743.

Kastern, W. H., and Berry, S. J. (1976). Non-methylated guanosine as the 5′-terminus of capped mRNA from insect oocytes. *Biochem. Biophys. Res. Commun.* **71**, 37–44.

Kaumeyer, J. F., Jenkins, N. A., and Raff, R. A. (1978). Messenger ribonucleoprotein particles in unfertilized sea urchin eggs. *Dev. Biol.* **63**, 266–278.

Kedes, L. H., and Birnstiel, M. L. (1971). Reiteration and clustering of DNA sequences complementary to histone messenger RNA. *Nature (London), New Biol.* **230**, 165–169.

Kumar, A., and Lindberg, U. (1972). Characterization of messenger ribonucleoprotein and messenger RNA from KB cells. *Proc. Natl. Acad. Sci. U.S.A.* **69**, 681–685.

Leaver, C. J., and Lovett, J. S. (1974). An analysis of protein and RNA synthesis during encystment and outgrowth (germination) of *Blastocladiella* zoospores *Cell Differ.* **3**, 165–192.

Lifton, R. P., and Kedes, L. H. (1976). Size and sequence homology of masked maternal and embryonic histone messenger RNAs. *Dev. Biol.* **48**, 47–55.

Lodish, H. F. (1971). Alpha and beta globin messenger ribonucleic acid. *J. Biol. Chem.* **246**, 7131–7138.

Lovett, J. A., and Goldstein, E. S. (1977). The cytoplasmic distribution and characterization of poly(A)+ RNA in oocytes and embryos of *Drosophila*. *Dev. Biol.* **61**, 70–78.

MacKintosh, F. R., and Bell, E. (1969a). Labeling of nucleotide pools in sea urchin eggs. *Exp. Cell Res.* **57**, 71–73.

MacKintosh, F. R., and Bell, E. (1969b). Regulation of protein synthesis in sea urchin eggs. *J. Mol. Biol.* **41**, 365–380.

Mahowald, A. P. (1971). Polar granules of *Drosophila*. IV. Loss of RNA from polar granules during early stages of embryogenesis. *J. Exp. Zool.* **176**, 345–352.

Mahowald, A. P., Caulton, J. H., and Gehring, W. J. (1979). Ultrastructural studies of oocytes and embryos derived from female flies carrying the *grandchildless* mutation in *Drosophila subobscura*. *Dev. Biol.* **69**, 118–132.

Manes, C. (1973). The participation of the embryonic genome during early cleavage in the rabbit. *Dev. Biol.* **32**, 453–459.

Mescher, A., and Humphreys, T. (1974). Activation of maternal mRNA in the absence of poly(A) formation in fertilized sea urchin eggs. *Nature (London)* **249**, 138–139.

Metafora, S., Felicetti, L., and Gambino, R. (1971). The mechanism of protein synthesis activation after fertilization of sea urchin eggs. *Proc. Natl. Acad. Sci. U.S.A.* **68**, 600–604.

Molinaro, M., and Farace, M. G. (1972). Changes in codon recognition and chromatographic behavior of tRNA species during embryonic development of the sea urchin *Paracentrotus lividus*. *J. Exp. Zool.* **181**, 223–232.

Monesi, V., Molinaro, M., Spalletta, E., and Davoli, C. (1970). Effect of metabolic inhibitors on macromolecular synthesis and early development in the mouse embryo. *Exp. Cell Res.* **59**, 197–206.

Morgan, T. H. (1927). "Experimental Embryology." Columbia Univ. Press, New York.

Muthukrishnan, S., Both, G. W., Furuichi, Y., and Shatkin, A. J. (1975a). 5′-Terminal 7′-methylguanosine in eukaryotic mRNA is required for translation. *Nature (London)* **255**, 33–37.

Muthukrishnan, S., Filipowicz, W., Sierra, J. M., Both, G. W., Shatkin, A. J., and Ochoa, S. (1975b). mRNA methylation and protein synthesis in extracts from embryos of brine shrimp, *Artemia salina*. *J. Biol. Chem.* **250**, 9336–9341.

Nemer, M., Graham, M., and Dubroff, L. M. (1974). Co-existence of non-histone messenger RNA species lacking and containing polyadenylic acid in sea urchin embryos. *J. Mol. Biol.* **89**, 435–454.

Newrock, K. M., and Raff, R. A. (1975). Polar lobe specific regulation of translation in embryos of *Ilyanassa obsoleta*. *Dev. Biol.* **42**, 242–261.

Nilsson, M. O. and Hultin, T. (1974). Characteristics and intracellular distribution of messenger-like RNA in encysted embryos of *Artemia salina*. *Dev. Biol.* **38**, 138–149.

Okada, M., Kleinman, I. A., and Schneiderman, H. A. (1974). Restoration of fertility in sterilized *Drosophila* eggs by transplantation of polar cytoplasm. *Dev. Biol.* **37**, 43–54.

O'Melia, A. F., and Villee, C. A. (1972). De novo synthesis of transfer and 5 S RNA[1f] in cleaving sea urchin embryos *Nature (London), New Biol.* **239**, 51–53.

Ozaki, H. (1975). Regulation of isozymes in interspecies sea urchin hybrid embryos. *In* "Isozymes" (C. L. Markert, ed.), Vol. 3, p. 543. Academic Press, New York.

Paglia, L. M., Kastern, W. H., and Berry, S. J. (1976). Messenger ribonucleoprotein particles in silkmoth oogenesis. *Dev. Biol.* **51**, 182–189.

Palmiter, R. D. (1975). Quantitation of parameters that determine the rate of ovalbumin synthesis. *Cell* **4**, 189–197.

Perry, R. P. (1976). Processing of RNA. *Annu. Rev. Biochem.* **45**, 605–629.

Perry, R. P., and Kelley, D. E. (1968). Messenger RNA–protein complexes and newly synthesized ribosomal subunits: Analysis of free particles and components of polyribosomes. *J. Mol. Biol.* **35**, 37–59.

Peters, C. and Jeffery, W. R. (1978). Postfertilization poly(A) · protein complex formation on sea urchin maternal messenger RNA. *Differentiation* **12**, 91–97.

Raff, R. A., Greenhouse, G., Gross, K. W., and Gross, P. R. (1971). Synthesis and storage of microtubule proteins by sea urchin embryos. *J. Cell Biol.* **50**, 516–527.

Raff, R. A., Colot, H. V., Selvig, S. E., and Gross, P. R. (1972). Oogenetic origin of messenger RNA for embryonic synthesis of microtubule proteins. *Nature (London)* **235**, 211–214.

Regier, J. C., and Kafatos, F. C. (1977). Absolute rate of protein synthesis in sea urchins with specific activity measurements of radioactive leucine and leucyl-tRNA. *Dev. Biol.* **57**, 270–283.

Rinaldi, A. M., and Monroy, A. (1969). Polyribosome formation and RNA synthesis in the early post-fertilization stages of the sea urchin egg. *Dev. Biol.* **19**, 73–86.

Rodgers, W. H., and Gross, P. R. (1978). Inhomogeneous distribution of egg RNA sequences in the early embryo. *Cell* **14**, 279–288.

Rosbash, M., and Ford, P. J. (1974). Polyadenylic acid-containing RNA in *Xenopus laevis* oocytes. *J. Mol. Biol.* **85**, 87–101.

Ruderman, J. V., and Gross, P. R. (1974). Histones and histone synthesis in sea urchin development. *Dev. Biol.* **36**, 286–298.

Ruderman, J. V., and Pardue, M. L. (1977). Cell-free translation analysis of messenger RNA in echinoderm and amphibian development. *Dev. Biol.* **60**, 48–68.

Sargent, T. D., and Raff, R. A. (1976). Protein synthesis and messenger RNA stability in activated, enucleate sea urchin eggs are not affected by actinomycin D. *Dev. Biol.* **48**, 327–335.

Schultz, G. A., Chen, D., and Katchalski, E. (1972). Localization of a messenger RNA in a ribosomal fraction from ungerminated wheat embryos. *J. Mol. Biol.* **66**, 379–390.

Seale, R. L., and Aronson, A. I. (1973). Chromatin-associated proteins of the developing sea urchin embryo I. Kinetics of synthesis and characterization of non-histone proteins. *J. Mol. Biol.* **75**, 633–645.

Shafritz, D. A., Weinstein, J. A., Safer, B., Merrick, W. C., Weber, L. E., Hickey, E. D., and Balgioni, C. (1976). Evidence for the role of $m^7G^{5'}$-phosphate group in recognition of eucaryotic mRNA by initiation factor IF-M_3. *Nature (London)* **261**, 291–294.

Shepard, H. M. (1977). Oogenetic histone messenger ribonucleic acid: Its form of storage in unfertilized eggs of the sea urchin, *Strongylocentrotus purpuratus*. Ph.D. Dissertation, Indiana University, Bloomington.

Sierra, J. M., Filipowicz, W., and Ochoa, S. (1976). Messenger RNA in undeveloped and developing *Artemia salina* embryos. *Biochem. Biophys. Res. Commun.* **69**, 181–189.

Silverman, P. M., Huh, M. M. O., and Sun, L. (1974). Protein synthesis during zoospore germination in the aquatic phycomycete *Blastocladiella emersonii*. *Dev. Biol.* **40**, 59–70.

Skoultchi, A., and Gross, P. R. (1973). Maternal histone messenger RNA: Detection by molecular hybridization. *Proc. Natl. Acad. Sci. U.S.A.* **70**, 2840–2844.

Slater, D. W., and Spiegelman, S. (1966). An estimation of the genetic messages in the unfertilized echinoid egg. *Proc. Natl. Acad. Sci. U.S.A.* **56**, 164–170.

Slater, D. W., Slater, I., and Gillespie, D. (1972). Post-fertilization synthesis of polyadenylic acid in sea urchin embryos. *Nature (London)* **240**, 333–337.

Slegers, H., and Kondo, M. (1977). Messenger ribonucleoprotein complexes of cryptobiotic embryos of *Artemia salina*. *Nucleic Acids Res.* **4**, 625–639.

Slegers, H., Mettrie, R., and Kondo, M. (1977). Evidence for a cytoplasmic translational inhibitor RNA in *Artemia salina* gastrula embryos. *FEBS Lett.* **80**, 390–394.

Soll, D. R., and Sonneborn, D. R. (1971). Zoospore germination in *Blastocladiella emersonii*: Structural changes in relation to protein and RNA synthesis. *J. Cell Sci.* **9**, 676–699.

Spirin, A. S. (1966). On "masked" forms of messenger RNA in early embryogenesis and in other differentiating systems. *Curr. Top. Dev. Biol.* **1**, 1–38.

Spirin, A. S. (1969). Informosomes. *Eur. J. Biochem.* **10**, 20–35.

Spitnik-Elson, P., and Atsmon, A. (1969). Detachment of ribosomal proteins by salt. *J. Mol. Biol.* **45**, 113–124.

Spurway, H. (1948). Genetics and cytology of *Drosophila subobscura*. IV. An extreme example of delay in gene action, causing sterility. *J. Genet.* **49**, 126–140.

Stavy, L., and Gross, P. R. (1967). The protein synthesis lesion in unfertilized eggs. *Proc. Natl. Acad. Sci. U.S.A.* **57**, 735–742.

Steinhardt, R., Zucker, R., and Schatten, G. (1977). Intracellular calcium at fertilization in the sea urchin egg. *Dev. Biol.* **58**, 185–196.

Tasca, R. J., and Hillman, N. (1970). Effects of actinomycin D and cycloheximide on RNA and protein synthesis in cleavage stage mouse embryos. *Nature (London)* **225**, 1022–1025.

Tennent, D. H. (1914). The early influence of the spermatozoan upon the characters of echinoid larva. *Carnegie Inst. Washington Publ.* **182**, 127–138.

Tennent, D. H. (1922). Studies on the hybridization of echinoids. *Carnegie Inst. Washington Publ.* **312**, 3–42.

Timourian, H. (1967). Protein synthesis in sea urchin eggs. I. Fertilization induced changes in subcellular fractions. *Dev. Biol.* **16**, 594–611.

Timourian, H., and Watchmaker, G. (1970). Protein synthesis in sea urchin eggs. II. Changes in amino acid uptake and incorporation at fertilization. *Dev. Biol.* **23**, 478–491.

Tufaro, F., and Brandhorst, B. P. (1979). Similarity of proteins synthesized by isolated blastomeres of early sea urchin embryos. *Dev. Biol.* **72**, 390–397.

Waring, G. L., Allis, C. D., and Mahowald, A. P. (1978). Isolation of polar granules and the identification of polar granule-specific protein. *Dev. Biol.* **66**, 197–206.

Warn, R. (1975). Restoration of the capacity to form pole cells in u.v.-irradiated *Drosophila* embryos. *J. Embryol. Exp. Morphol.* **33**, 1003–1011.

Weeks, D. P., and Marcus, A. (1971). Preformed messenger of quiescent wheat embryos. *Biochim. Biophys. Acta* **232**, 671–684.

Whiteley, A. H., and Whiteley, H. R. (1972). The replication and expression of maternal and paternal genomes in a blocked echinoid hybrid. *Dev. Biol.* **29**, 183–198.

Whittaker, J. R. (1977). Segregation during cleavage of a factor determining endodermal alkaline phosphatase development in ascidian embryos. *J. Exp. Zool.* **202**, 139–154.

Wilt, F. H., and Mazia, D. (1974). The stimulation of cytoplasmic polyadenylation in sea urchin eggs by ammonia. *Dev. Biol.* **37**, 422–424.

Woods, D. E., and Fitschen, W. (1978). The mobilization of maternal histone messenger RNA after fertilization of the sea urchin egg. *Cell Differ.* **7**, 103–114.

Young, E. M. (1978). Messenger ribonucleoprotein particle structures in developing sea urchin embryos. Ph.D. Dissertation, Indiana University, Bloomington.

Young, E. M., and Raff, R. A. (1979). Messenger ribonucleoprotein particles in developing sea urchin embryos. *Dev. Biol.* **72**, 24–40.

Zeikus, J. G., Taylor, M. W., and Buck, C. A. (1969). Transfer RNA changes associated with early development and differentiation of the sea urchin, *Strongylocentrotus purpuratus*. *Exp. Cell Res.* **57**, 74–78.

4

Structure and Function of Nuclear and Cytoplasmic Ribonucleoprotein Complexes

Terence E. Martin, James M. Pullman, and Michael D. McMullen

I. INTRODUCTION

Any attempt made at the present time to describe the role of ribonucleoprotein (RNP) complexes in the processing of nuclear RNA, the trans-

CELL BIOLOGY, VOL. 4

Copyright © 1980 by Academic Press, Inc.
All rights of reproduction in any form reserved.
ISBN 0-12-289504-5

port of mRNA to the cytoplasm, and the regulation of translation of that mRNA on polyribosomes must begin with a clear statement of our ignorance of the molecular details of these processes. The reader should be well aware that any reviewer will necessarily express both explicit and implicit prejudices when attempting to provide an overview of this relatively unmapped terrain. However, if we are to avoid simply providing a catalogue of published observations of nuclear and cytoplasmic RNP complexes, with all their bewildering complexity of descriptive data and relative lack of evidence of functional significance, then we must assume some point of view. The models illustrated in this chapter should therefore only be taken as ways of thinking about the structure and physiology of ribonucleoprotein complexes rather than as documented facts, or even generally recognized hypotheses.

While an obvious focus of interest in RNP complexes might center on the form in which mRNA is transported from the nucleus to the cytoplasm, the nature of that transport complex has been particularly elusive, and by far the majority of the studies we will describe have concerned themselves with the nature of either nuclear hnRNP or cytoplasmic mRNP complexes. The physical separation in eukaryotic cells of the events of RNA synthesis on cellular DNA and the final expression of genetic information by way of protein synthesis on cytoplasmic polyribosomes entails, in addition to the necessary transport processes, the possibilities of nuclear RNA processing and cytoplasmic mRNA storage or translational modulation.

The earliest observations of nonribosomal RNP complexes were made by electron microscope cytochemical techniques which determined that there were ribonuclease-sensitive granules on and near chromosomes in a number of different cell types (Gall, 1956; Swift, 1963), and that RNA was apparently transported from the nucleus to the cytoplasm through the nuclear pore in the form of RNP particles (Stevens and Swift, 1966). The present survey is primarily biochemical in its orientation but will draw some support from the morphological information obtained by the electron microscopy of spread chromatin and isolated RNP complexes. Current evidence suggests that newly synthesized or nascent RNA in its RNP form, as studied biochemically, is equivalent to the perichromatin fibrils observed cytochemically in the eukaryotic nucleus (Monneron and Bernhard, 1969), since it is these morphological entities which are most rapidly labeled in electron microscope radioautography studies (Fakan and Bernhard, 1971). Studies on puff regions of the giant polytene chromosomes of Diptera have provided striking cytochemical evidence for the association of newly made RNA with protein and for the passage of that RNA from nucleus to cytoplasm by way of the nuclear pore as RNP complexes (Stevens and Swift, 1966). Indeed the occurrence of giant RNA–protein granules in the Balbiani ring puffs of *Chironomus* has provided one of the most promising systems for the unification of the morphological and biochemical information on the accumulation and translocation of messenger-like RNA within cells (Daneholt *et al.*, 1978; Edström *et al.*, 1978).

The majority of researchers studying nuclear or cytoplasmic RNP complexes believe that the elucidation of the role of the proteins which bind RNA is essential for the understanding of the cellular processes involving these RNA molecules. It should be borne in mind that analysis of RNP has been crucially dependent on the evolution of our knowledge of the structure, synthesis, and turnover of eukaryotic RNA molecules as studied in their purified form. Thus, the still relatively primitive studies of nuclear RNP are best understood in the light of current knowledge of hnRNA and mRNA synthesis and processing (see relevant chapters in Volume 3 of this series; also Darnell, 1978 and Abelson, 1979).

In view of the relative infancy and the inherent difficulties of RNP studies, it should not surprise the reader to discover that the major source of controversy in the field concerns primarily the number and nature of the proteins involved in these structures. Conflicting evidence has accumulated for both nuclear and cytoplasmic complexes; thus, any comparison of the two is fraught with additional difficulty. The reason for the conflict is the lack of biologically relevant criteria for the purification of RNP complexes. Unfortunately there are as yet no self-evident means for the purification of RNP, and the situation is aggravated by the absence of functional tests for RNA binding proteins (that is, workers in this field do not have the catalytic or antigenic markers which have proven invaluable in the isolation and purification of other cellular components). This chapter will attempt to find some common ground between the various laboratories studying questions of RNP structure and function, will certainly not resolve all the conflicts inherent in the various viewpoints, and undoubtedly will imply an oversimplified view of the biological realities involved.

II. NUCLEAR RNP

A. Nuclear hnRNA Is Bound to Protein

While it should be pointed out that there is now considerable evidence that ribosomal RNA precursors in the nucleolus are bound to specific sets of proteins while undergoing maturation (Liau and Perry, 1969; Kumar and Warner, 1972; Matsumura *et al.*, 1974), most attention has been focused on the proteins associated with the rapidly synthesized and turning-over heterogeneous nuclear RNA (hnRNA) presumed to be the precursor of functional cytoplasmic mRNA. The first substantial biochemical evidence that hnRNA was associated with protein within the eukaryotic nucleus was provided by Georgiev and his co-workers (Samarina *et al.*, 1965), and indeed the model proposed by them for hnRNP structure in 1968 remains a valid basis for modification and extension (Samarina *et al.*, 1968). These experiments, which involved the leaching of pulse-labeled nuclear RNA from purified liver nuclei incubated in approximately isotonic salt solution at somewhat elevated pH (pH 8–9), suggested that newly synthesized RNA was not free within the nucleus, but rather associated with a protein complex consisting of multiple copies of a single polypeptide. Numerous subsequent experiments utilizing similar or different extraction procedures also led to the conclusion that newly synthesized nuclear RNA of the hnRNA class is indeed bound to proteins, although as we will see the nature of those proteins is debated. In many of the experiments, exogenous RNA or protein has been added in an attempt to determine whether the observed ribonucleoprotein complexes could be artifacts of the extraction conditions. This has been a serious consideration in view of the reports that certain cytoplasmic proteins could nonspecifically bind to naked RNA (Baltimore and Huang, 1970). Although these control experiments have generally indicated that such nonspecific binding does not occur, the possibility that the reorganization of the nuclear material during extraction leads to the unnatural association of certain nuclear proteins in close proximity to the RNA cannot be entirely eliminated. Although subject to similar reservations, the observation of protein bound to nascent nonribosomal RNA transcripts in chromatin spreads released from nuclei rapidly lysed on electron microscope grids lends graphic support to the notion that the majority of hnRNA is indeed in RNP complexes (Miller and Bakken, 1972; Franke *et al.*, 1976, 1978) (see also Fig. 1).

Although newly synthesized hnRNA is known to be of very high molecular weight, it is the common experience that the RNA in extracted RNP from nuclei of a great variety of cell types is fragmented to some degree. Georgiev and his colleagues (Samarina *et al.*, 1968) reported that the fragmentation of large complexes could be reduced by the addition of ribonuclease inhibitor to the extraction medium. However, it was the result of the action of either the endogenous nuclease activity or of added exogenous ribonuclease that initially led these authors to the conclusion that large hnRNP complexes consisted of long hnRNA molecules complexed to a series of globular protein complexes. Similar observations of the effect of endogenous or exogenous nucleases on chromatin made later

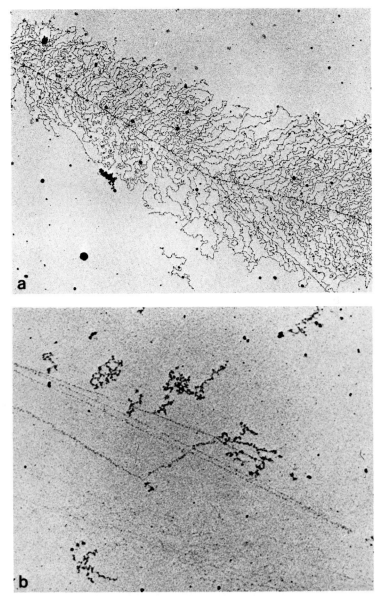

Fig. 1. Electron micrographs of transcription fibrils of nonribosomal genes in dispersed chromatin of (a) the lampbrush chromosomes of *Triturus viridescens* oocytes and (b) HeLa cells (Miller and Bakken, 1972). Nascent hnRNA chains extend from the central chromatin fiber, which in the case of the HeLa cell chromatin (b) can be seen to have the beaded "nucleosome" form. The attached RNA chains increase in length along the chromatin fiber; this polarity is most obvious for the densely packed amphibian oocyte transcription units (a). The nascent RNA is readily visualized only because of associated protein which imparts a particulate appearance to the fibril (b). (a) × 19,000. (b) × 28,000. (Micrographs courtesy of Drs. Aimee Bakken and Oscar Miller.) Excellent visualizations of the particulate nature of lampbrush transcription fibrils have also been presented by Angelier and Lacroix (1975).

by Hewish and Burgoyne (1973) have led to the nucleosome model for the substructure of the eukaryotic chromatin. The analogous "beads-on-a-string" model for hnRNP structure, in which the beads are sections of the giant hnRNA molecule associated with 30 S* globular protein complexes consisting of multiple copies of a single protein ("informatin"), has served as a reference or target for later experimentation.

The questions that have attracted most attention of workers studying RNP relate directly to the model: (a) Is there an inherent substructure to hnRNP and (b) is hnRNA associated with a simple and general, or a complex and specific set of polypeptides? In attempting to answer these questions the major concerns have been for the integrity and the purity of the complexes under study. Given the lack of direct functional tests the evidence accumulated has, for the most part, been circumstantial. Other recent reviews have also attempted to summarize and distill the increasing body of literature on the nature of hnRNP (Heinrich *et al.*, 1978; Van Venrooij and Janssen, 1978).

B. Methods of Isolation

The apparent structure and composition of hnRNP reported by various laboratories would seem to be a function of the procedure used to remove them from nuclei. The nuclear isolation procedure itself can be expected to contribute significantly to the yield and purity of the hnRNP complexes extracted. The stability and purity of isolated nuclei can vary greatly from one type of eukaryotic cell to another even if identical preparative procedures are employed. Most laboratories studying RNP complexes use aqueous sucrose–detergent methods for the preparation of nuclei from mammalian cells (Birnie, 1978), but the purity and integrity of the isolated nuclei are not always documented. The subsequent extraction of hnRNP from purified nuclei generally employs one of two general procedures, either isotonic extraction which maintains the nucleus intact, or nuclear disruption by sonication or high-pressure-induced cavitation which disperses the nuclear contents. Both methods require the release of hnRNP from any attachments that it may have to other nuclear components, such as RNA polymerase and the DNA template in the case of nascent RNA (Augenlicht and Lipkin, 1976), or from a possible association with nuclear skeletal elements for completed hnRNA molecules (Berezney and Coffey, 1974; Faiferman and Pogo, 1974; Herman *et al.*, 1978; Miller *et al.*, 1978).

---

* Sedimentation coefficients of 30–50 have been assigned to this apparent substructure of hnRNP. We have chosen to follow the original designation of Samarina and Georgiev in making general reference to this complex; the variation in its sedimentation behavior is discussed in the section on physical properties of hnRNP.

The isotonic extraction procedure introduced by Georgiev and co-workers is dependent on temperature, pH, divalent ion concentration, the number of extractions, and the duration of extraction (Samarina et al., 1968; Billings and Martin, 1978). Recovery of high molecular weight RNA by this procedure generally requires the presence of an RNase inhibitor (rat liver supernatant inhibitor is most often employed), but the effectiveness of the inhibitor may depend critically on the source of the nuclei. This procedure for the release of hnRNP from intact nuclei may, in fact, demand a limited activation of endogenous nucleases to detach the complex from internal nuclear structures. The extraction procedure can be very efficient, however, and it is possible to obtain greater than 50% of the total pulse-labeled RNA of a mammalian cell in the form of 30 S RNP sub-complexes by this technique (Martin et al., 1974).

While isotonic extraction procedures even at elevated pH require prolonged incubation times (particularly at low temperatures which favor the integrity of the RNA of the complex), nuclear disruption techniques allow the rapid release of RNP. Sonication has been the preferred method of this type, and cavitation procedures employing the French press or nitrogen bomb have been used by very few groups (for example, Faiferman et al., 1970). The use of ionic detergents as an aid in nuclear breakage once favored by some investigators is now generally avoided, since it may cause alterations in protein composition (Stevenin and Jacob, 1974). Likewise the use of DNase and high ionic strength to free hnRNP from contaminating chromatin has been criticized, since the procedure also releases histones and nonhistone chromosomal proteins thereby increasing the likelihood of artifactual associations (Georgiev and Samarina, 1971; Pederson, 1974).

Sonication and extraction each have relative advantages and disadvantages in the preparation of hnRNP, and some of these will be discussed in the context of composition, physical properties, and quaternary structure in the following sections. In general it would seem advisable in the isolation and purification of hnRNP to avoid extremes of ionic strength and the addition of protein or nucleic acid denaturants (see discussion by Billings and Martin, 1978).

C. Composition

Purified hnRNP are presumably composed entirely of RNA and protein; the various low levels of DNA reported for some preparations are most likely the result of chromatin fragmentation and release during extraction, although it cannot confidently be excluded that some DNA may arise from attachment sites for hnRNP in the nucleus. The buoyant densities of formaldehyde or glutaraldehyde fixed RNP particles centrifuged on isopycnic CsCl gradients are between 1.39 and 1.43 gm/cm^3 for all cell types investigated thus far, and appear to be independent of the isolation procedure employed (summarized in Heinrich et al., 1978). This density range corresponds to a composition of 75–80% protein and 20–25% RNA if the formula of Spirin (1969) is used, or slightly higher values for protein (85–90%) and correspondingly lower RNA content by the formula of Hamilton (1971). Direct measurements of RNA and protein content by the orcinol and Lowry methods have yielded significantly higher estimates of the protein–RNA ratio, viz., protein values of 90% or greater (Samarina et al., 1967a; Karn et al., 1977; Billings and Martin, 1978).

Similar density values are obtained upon isopycnic centrifugation in CsCl whether large hnRNP complexes or the 30 S substructures are used, indicating that there is no disproportionate loss of either RNA or protein during the cleavage of large hnRNP. This observation in itself, however, does not prove that large complexes are composed entirely of 30 S substructures.

Since isopycnic centrifugation in CsCl gradients requires the prior fixation of RNP complexes with aldehydes, it has not been possible to use this technique to purify native complexes for further studies. Attempts have been made to use gradients of the less ionically active compounds, such as metrizamide, for the analysis and purification of nuclear RNP complexes. Results obtained using metrizamide are conflicting, with some groups observing a buoyant density value comparable to that found on CsC₁ gradients (Karn et al., 1977) while others find particle dissociation (Gattoni et al., 1977). Although metrizamide may not disrupt RNP complexes by ionic means, the compound may interact with both RNA and protein in other ways which interfere with their association (Rickwood et al., 1974).

D. Physical Properties of hnRNP and Derived Substructures

As with most studies of nonribosomal RNP the question of biochemical purity of preparations is critical to attempts to determine size by analytical ultracentrifugation and structure by electron microscopy. While determination of relative sedimentation velocities by centrifugation of samples containing pulse-labeled RNA on density gradients circumvents most of the requirements for purity since the radioactivity provides the marker for the RNP, the method does not yield the precise information that can be obtained with purified preparations in the analytical ultracentrifuge. Unfortunately, there have been no precise sedimentation studies of large hnRNP and their relationship to substructures derived by nuclease action. The most simple view of this relationship has been provided by the sucrose gradient studies of Georgiev and his co-workers who reported that the large complexes were converted into 30 S substructures or "monomers" by very mild ribonuclease treatment (Samarina et al., 1968). In the results shown there appeared to have been a nearly quantitative conversion of the uv-absorbing material from the large heterogeneous region (up to 200 S or greater) of a gradient to the 30 S region. When large complexes were isolated from sucrose gradients and examined in the electron microscope, they appeared as chains of "monomers"; the number of particles in the chains was reported to be related to the position in the gradient (i.e., the sedimentation coefficient) and the length of RNA extractable from the fraction. The data presented were entirely consistent with a simple polymeric model for the large complexes in which hnRNA is bound to a series of identical protein particles ("informofers") forming a beads-on-a-string structure.

While various workers having very different views of hnRNP structure have subsequently reported the release of relatively homogeneous subcomplexes from large hnRNP by nuclease, the published data have not always been consistent with a quantitative conversion of heterogeneous large complexes into the smaller particles (Kinniburgh et al., 1976; Stevenin et al., 1977). Loss of material during this conversion raises the possibility of nonbeaded ribonucleoprotein regions in large hnRNP and lends some strength to the suggestion of Stevenin and Jacob (1979) that fragmentation of the RNA by nuclease may lead to protein rearrangements.

Whatever is the case, a major fraction of the nuclear hnRNA can be recovered in the form of derived substructures of hnRNP, and because of their relatively homogeneous sedimentation coefficients these can be easily purified for further study. As isolated, these complexes are thought to contain 700–1000 nucleotides of RNA and approximately 10^6 daltons of protein (Samarina et al., 1968; Billings and Martin, 1978), but these values are not as firmly established as one might wish. Indeed even the sedimentation coefficient of these subcomplexes has been variously reported as 30 (Samarina et al., 1968; Martin and McCarthy, 1972), 40 (Beyer et al., 1977), and 45 (Pederson, 1974). While there have been no detailed analytical studies published, the variations in the reported relative sedimentation values of the subcomplex (largely from sucrose gradient centrifugation experiments) are probably not merely the result of the imprecision of the method. The mouse tumor cell subcomplex isolated in our laboratory sediments distinctly more slowly than the 40 S ribosomal subunit (Martin and McCarthy, 1972). It has a modal value of 34 S determined in the analytical ultracentrifuge, but if the complex is pelleted and resuspended before analysis the value is increased to 40–45 S (Billings, 1979). Paradoxically, very mild ribonuclease digestion may also increase the sedimentation coefficient of the subcomplex, perhaps by removal of RNA tails which could cause frictional drag. The situation may be more complex in some tissues, since there is a report that rat brain large hnRNP yield two types of subcomplex upon fragmentation, and these may have different physical properties (Stevenin *et al.*, 1976, 1977).

There have been a limited number of electron microscope studies of shadowed or negatively stained hnRNP monomers. Early work reported particles roughly uniform in size, measuring about 20 × 20 nm in the plane of the specimen support and 10 nm in the third dimension (Samarina *et al.*, 1967a, 1968). The disk-like appearance may be an artifact of flattening of the particles on the supporting surface. More detailed analysis of micrographs reveals a variety of particle shapes, which cannot be explained as different two-dimensional projections of a single three-dimensional model (Martin *et al.*, 1978). The origin and significance of this heterogeneity has not been determined, but may be due to a microheterogeneity in protein or RNA composition (see below), or to the inherent dynamics of particle conformation in solution. The particles are at least in some sense labile, since they disintegrate unless fixed with formaldehyde or glutaraldehyde prior to their deposition on a supporting surface for microscopy (Georgiev and Samarina, 1971; Martin *et al.*, 1978).

Such flexibility in quaternary structure may explain the relatively broad optical density peaks on velocity gradient sedimentation, and the discrepancies among size measurements by different methods. Particles which sediment at 30 S have a computed molecular weight of approximately 1.0×10^6 as determined from the molecular weight of the RNA and the RNA : protein ratio (Samarina *et al.*, 1968). However, the particle dimensions as determined by electron microscopy show a particle comparable in size with the 60 S ribosomal subunit, which has a molecular weight of 2.5×10^6 (Van Holde and Hill, 1974).

The poly(A)-containing particle derived from hnRNP is a smaller complex. It sediments at 15 S, has a buoyant density slightly less than that of hnRNP complexes, and appears in electron micrographs as a uniform set of particles measuring 16.5 × 12.0 nm in projection (Martin *et al.*, 1978).

If, as labeling kinetics suggest, hnRNA as it is synthesized becomes almost immediately incorporated into the RNP form which yields the 30 S substructures, then we might expect that nascent RNA–protein fibrils observed by electron microscopy in spread chromatin would exhibit a particulate character somewhat similar to the beaded chromatin fibers. Because the single-stranded RNA transcript is more flexible than the double-stranded DNA of the chromatin template, it is not as easy to visualize precisely the linear path of the RNA in the transcription fibrils.

Many published electron micrographs show a particulate morphology of the presumed nascent hnRNP fibrils; suggestions of the particle substructure of hnRNP can be seen in the micrographs of the HeLa cell (Miller and Bakken, 1972) (Fig. 1b) and rat liver (Puvion-Dutilleul et al., 1977) chromatin.

Admitting that electron microscopy is as susceptible to artifacts as biochemical cell fractionation techniques, we nonetheless feel that there is growing evidence for a particulate substructure to both nascent and completed hnRNP. If that is so, we may reasonably ask what factors confer that substructure and how specific RNA sequences are organized in relationship to it. To approach these questions, the nature of both the RNA and protein components must first be examined in more detail.

E. Characteristics of Nuclear Ribonucleoprotein RNA

The RNA component of both large RNP and 30–40 S substructures has been identified as heterogeneous nuclear RNA by a number of criteria. It is similar to cellular DNA in base composition and hybridizes to homologous DNA in a manner similar to pulse-labeled hnRNA (Samarina et al., 1965; Parsons and McCarthy, 1968). The kinetics of synthesis and turnover of the pulse-labeled RNA of 30 S RNP substructures are consistent with the values obtained for hnRNA, and a majority of the nuclear pulse-labeled RNA can be obtained in the form of these 30 S complexes (Martin and McCarthy, 1972). Hybridization–competition experiments have indicated that the pulse-labeled RNA of 30 S subcomplexes contain nucleus-restricted sequences (Martin and McCarthy, 1972). More recent experiments in which 30 S RNP–RNA has been hybridized with complementary DNA (cDNA) transcribed from cytoplasmic poly(A)$^+$ mRNA suggest that 5–10% of hnRNP–RNA is homologous to cytoplasmic mRNA and that all mRNA sequences are represented in nuclear 30 S RNP structures (Kinniburgh and Martin, 1976a). These results are consistent with the notion that the majority of hnRNA is bound in these RNP complexes, but that only a fraction of this RNA is destined to reach the cytoplasm as mRNA. Labeling of nuclear ribonucleoprotein with RNA precursors is inhibited by compounds which interfere with hnRNA synthesis such as α-amanitin (Louis and Sekeris, 1976, Stunnenberg et al., 1978), but is not affected by low levels of actinomycin D which inhibit rRNA synthesis (Pederson, 1974).

While the size of the RNA contained in hnRNP has in some cases been shown to approach that of hnRNA extracted with phenol it is more usually found to be much smaller, particularly when attempts have been to purify the ribonucleoprotein complexes before RNA extraction (Samarina et al.,

1968; Pederson, 1974). Determinations of the composition of the 30 S hnRNP subcomplexes indicate that they should contain a total RNA length of 700–1000 nucleotides and indeed the largest molecules that can be obtained from purified 30 S RNP approach this size (Samarina *et al.*, 1968). More commonly, however, the RNA of highly purified 30 S RNP contains internal nicks, and the complexes can retain their structural integrity even when the majority of the RNA has been reduced to 50–100 nucleotide pieces (Kinniburgh and Martin, 1976a).

More recent studies have attempted to infer the location of specific RNA sequences within large hnRNP complexes. These studies which are in a very preliminary state will be discussed in Section II,I. For the moment it should be pointed out that the nuclear poly(A) of 200 nucleotides or greater in length, which is contained in large hnRNP, is in fact not isolated in the 30 S subcomplexes, but rather in a separate subcomplex with sedimentation coefficient of approximately 15 S (Quinlan *et al.*, 1974; Molnar and Samarina, 1975). The 15 S particle is, therefore, considered to be cleaved from large complexes by the same nuclease activity which yields the 30 S substructures. The poly(A) segment isolated in this distinctive RNP subcomplex remains intact during extraction and purification in the absence of ribonuclease inhibitor (Quinlan *et al.*, 1977), either as a result of more effective protection of the RNA by the proteins of the poly(A) nucleoprotein [poly(A)NP] complex, or because of the inherent ribonuclease resistance of poly(A). Buoyant density determinations made on CsCl gradients indicate that the poly(A)NP complex contains an even greater proportion of protein than does the 30 S hnRNP subcomplex (Quinlan *et al.*, 1977).

Some workers have presented data to suggest that hnRNP complexes contain, in addition to hnRNA, more stable species of nuclear RNA. For example, Sekeris and Niessing (1975) have argued that the 30 S RNP subcomplex contains in addition to rapidly turning-over RNA, stable species which may serve as a structural component of the particle. More recent experiments have been reported in which it appeared that small stable RNA's in the range of 4–7 S cosediment with hnRNP, particularly the large complexes, and that some of these species have been tentatively identified as certain of the small nuclear RNAs (snRNA) that have been known for many years but whose function is still obscure (Ro-Choi *et al.*, 1976; Deimel *et al.*, 1977; Northemann *et al.*, 1977; Howard, 1978). Although we feel that it is unlikely that these RNA species are required in the maintenance of RNP substructure (since purified 30 S subcomplexes do not contain them), it is certainly possible that interaction of these nuclear molecules with hnRNA bound in RNP complexes may have an important role in nuclear RNA processing. Low molecular weight RNAs have been found hydrogen-bonded to phenol-extracted hnRNA (Jelinek and Leinwand, 1978), and it has been suggested that snRNAs may be involved in specific splicing of nuclear pre-mRNA by way of base pairing with regions around the splice junctions (Murray and Holliday, 1979; Lerner *et al.*, 1979). As it stands at present however, the precise nature of the stable nuclear RNA species supposedly associated with hnRNP complexes remains to be determined (their identity with previously studied snRNA has not absolutely been proven), and in addition it must be conclusively demonstrated that these species are indeed physically bound to hnRNP complexes. The recent observations that human autoimmune sera include antibodies to 8–12 S RNP which contain snRNA (i.e., snRNP), provide the potential probes for further studies on the physiological roles of snRNA (Lerner and Steitz, 1979; and our unpublished results).

F. Characteristics of hnRNP Proteins

Currently, the most controversial area in the study of nuclear RNP complexes concerns the number and molecular weight distribution of their component proteins. The discrepancies which exist in the literature can partially be explained by the differences in the sources of RNP and in the analytical techniques employed. One generalization that can be made is that RNP prepared by the salt extraction technique generally have a much simpler protein composition on SDS-polyacrylamide gel electrophoresis than do RNP prepared by sonication of nuclei.

Employing salt extraction at elevated pH, Georgiev and his colleagues originally reported that both large hnRNP and the 30 S subcomplex contained only a single protein as determined by electrophoresis on urea gels in the presence of a reducing agent, although a more complex pattern was obtained if the reducing agent was omitted (Samarina *et al.*, 1968). They also stated but did not demonstrate that the protein migrated as a single component on SDS-acrylamide gels and had an apparent molecular weight of 40,000. Subsequent work in other laboratories lent support to the notion that the 30 S subcomplex possessed a relatively simple protein complement, although not as simple as had first been reported. Using basically similar techniques Martin *et al.* (1974), while demonstrating similar behavior of the 30 S RNP proteins on urea gels to those reported by Samarina *et al.*, found that SDS electrophoresis gave at least two distinct components in the molecular weight range of 35,000–40,000. The number of components detected in this molecular weight range has subsequently been shown to be critically dependent on the nature of the detergent used and the electrophoresis conditions (Billings and Martin, 1978). Thus, in later work from our own laboratory and from others, the reported number of polypeptides of this size class present in hnRNP subcomplexes ranges from 4 to 12 (Martin et al., 1978, 1979a; Beyer et al., 1977; Karn et al., 1977; Billings and Martin, 1978). As originally shown by Samarina et al. (1968), the majority of these proteins are mildly basic and have p*I*'s in the range of 8–9 (Beyer et al., 1977; Billings, 1979). The relative stoichiometry of these various polypeptides of 35,000–40,000 daltons varies from tissue to tissue and also apparently with the physiological state of the tissue (LeStourgeon et al., 1978; Billings, 1979). Thus, beginning with an apparently simple composition for the protein component of the 30 S hnRNP subcomplex, subsequent experimentation has revealed subtleties that require further discussion.

Comparative gel electrophoresis and amino acid analysis have demonstrated that the 30 S RNP proteins are a relatively conserved group among the vertebrates and possibly all higher eukaryotes (Martin et al., 1974;

TABLE I

Amino Acid Composition of Total 30 S RNP Proteins

Amino acid	Duck liver[a]	Mouse ascites[a]	HeLa cell[b]	Rat liver	
Lys	5.44	6.27	6.9	6.5[c]	7.71[d]
His	1.98	1.99	2.0	2.6	2.51
Arg	5.83	6.01	5.9	7.6	7.97
Asx	9.46	10.80	11.4	10.6	10.83
Thr	3.69	3.28	3.6	3.9	3.90
Ser	8.96	7.62	7.3	7.6	4.97
Glx	13.07	10.53	12.1	10.6	10.97
Pro	3.90	3.67	4.9	5.7	6.81
Gly	17.90	22.41	17.1	17.1	17.88
Ala	7.16	3.96	6.1	5.5	5.35
Cys	Trace	Trace	0.7	0.68	Trace
Val	4.35	4.75	4.1	4.1	5.68
Met	1.62	1.76	1.9	—	0.71
Ile	2.62	2.54	2.6	2.4	2.70
Leu	4.19	3.30	5.3	4.8	4.68
Tyr	3.73	5.64	3.7	4.5	3.33
Phe	4.09	5.42	3.9	4.2	3.26
X[e]	2.01	Trace	0.5	1.4	1.07

[a] Martin et al., 1974.
[b] Beyer et al., 1977.
[c] All data in this column from Krichevskaya and Georgiev, 1969.
[d] All data in this column from Karn et al., 1977.
[e] There are a number of modified amino acids, not all of which have been identified, but including N^G, N^G-dimethylarginine (Boffa et al., 1977; Beyer et al., 1977; Billings and Martin, 1978).

Billings and Martin, 1978); conservation of RNP protein structure is to be expected if they act as structural components in a way somewhat analogous to the role of histones in chromatin structure. Table I collates data from various laboratories comparing the amino acid composition of the major 30 S RNP subcomplex proteins from a variety of species, and in addition to the general similarities of composition, the data indicate the common finding of very high proportions of glycine, low cysteine content, and the presence of unusual amino acids, particularly N^G,N^G-dimethylarginine. Comparison of individual subspecies of the 30 S RNP protein is shown in Table II and demonstrates the general similarity of the individual polypeptides of this molecular weight class which have been analyzed thus far. Given the similarities in size, migration on urea gels, and amino acid composition it seems reasonable to suggest that these major RNP polypeptides may be closely related to one another, either as

TABLE II

Amino Acid Composition of Isolated 30 S RNP Proteins

Amino acid	Mouse ascites cell[a]		HeLa cell[b]	
	$M_r = 37,500$[c]	$M_r = 40,000$	$M_r = 34,000$	$M_r = 35,000$
Lys	4.95	5.11	7.7	6.0
His	2.07	2.24	1.5	1.6
Arg	4.86	4.28	3.3	5.3
Asx	9.83	9.21	7.6	10.3
Thr	3.27	3.17	3.5	2.7
Ser	7.04	8.20	13.0	10.4
Glx	11.76	12.97	12.6	10.7
Pro	3.59	3.52	3.5	4.0
Gly	23.27	25.56	23.2	24.9
Ala	6.56	5.67	8.1	5.8
Cys	Trace	Trace	Trace	Trace
Val	5.47	4.63	3.2	2.5
Met	1.09	1.04	0.7	1.5
Ile	2.73	2.45	3.3	2.7
Leu	4.41	4.32	4.8	4.3
Tyr	4.84	3.83	1.3	2.6
Phe	4.22	3.80	2.6	4.3
X	Trace	Trace	0.1	0.4

[a] Data from Billings and Martin, 1978.
[b] Data from Beyer et al., 1977.
[c] The apparent relative molecular weights are not strictly comparable between mouse and HeLa preparations because of the sensitivity of these polypeptides to the electrophoretic conditions (see Billings and Martin, 1978).

the products of a gene family or as modifications of the product of a single gene. There is no evidence that the smaller polypeptides are proteolytic products of the larger species. Unfortunately conventional sequencing techniques have proved unsuccessful, presumably because most of the species of the 30 S RNP complex contain blocked N-termini (Billings and Martin, 1978), and further discussion must await the analysis of peptide fragments from these proteins.

The size and composition of the 30 S RNP subcomplex suggests that each particle contains at least 800,000 daltons of protein, and that the polypeptide complexity and stoichiometry appear to be inconsistent with the notion that each RNP complex contains one copy of twenty or more polypeptides. One must, therefore, conclude that either all polypeptide species exist in varying proportions within a single 30 S subcomplex type or that cell nuclei contain a variety of different subcomplex classes each containing a single polypeptide type or a very limited subset of the total possible polypeptide species. Either possibility allows for changes in overall nuclear RNP composition as a function of physiological activity.

One immediate question that can be asked about this distinctive set of 34,000–40,000 dalton polypeptides of hnRNP, is whether the proteins turn-over at a rate comparable to the $t_{1/2}$ of 15–20 min for the RNA component. As expected, these proteins are not rapidly turned over, and in mouse tumor cells have half-lives approaching those of such metabolically stable polypeptides as histones and ribosomal proteins (Martin et al., 1979a). Thus, we may presume that these RNP polypeptides, at least, are reutilized many times in the nucleus. Since the proteins are insoluble in the test tube at physiological ionic strengths in the absence of RNA, questions then arise as to the nature and extent of the free protein pool, and the means by which the proteins reassociate with newly synthesized hnRNA. There is some preliminary evidence that the polypeptides may undergo posttranslational modifications other than the methylation of arginine mentioned above. The potential phosphorylation of RNP proteins (Martin et al., 1974; Blanchard et al., 1975; Gallinaro-Matringe et al., 1975; Karn et al., 1977) may be associated with recycling of the polypeptides, although roles in protein transport or in regulation of RNA processing remain equal possibilities at the present time.

While much of the existing information can be taken as support for a view of large hnRNP as a beads-on-a-string structure, in which the sub-units are composed of multiple copies of a limited set of conserved polypeptides, somewhat analogous to the nucleosome structure of nu-cleohistone in chromatin, not all researchers are willing to accept this idea in its most simple form. In fact, chromatin structure is obviously more complex than the nucleosome model suggests, and presumably includes a multitude of nonhistone proteins interacting with nucleosomes and the

DNA between them. Extending this analogy to RNP once more, one might anticipate that large hnRNP may contain polypeptides not characteristic of a repeating 30 S substructure. A more extreme view is that the sub-structures are themselves artifacts in which a limited subset of a complex protein complement of heterogeneous RNP complexes have aggregated specifically during the disruption of the large native structures (Stevenin and Jacob, 1979; Stevenin et al., 1979). We will return to the apparent subunit nature of hnRNP below, but some mention must be made here of the multiplicity of polypeptide components reported for hnRNP.

Since there is considerable agreement on the relatively simple composi-tion of the 30–40 S subcomplexes (whether they are considered artifacts or reality), discussion of other proteins is best considered in the context of large complexes. Unfortunately large hnRNP complexes are most readily obtained by means of sonication or cavitation, which disrupt a major part of the nuclear structure, and are thus more susceptible to contamination by fragments of chromatin, nuclear envelope (lamina) and intranuclear matrix. The protein compositions of the latter non-RNP nuclear compo-nents are only partly determined at the present time, so that the assign-ment of polypeptides observed on polyacrylamide gels to RNP or some other cosedimenting structure is often uncertain. Preparations of mamma-lian hnRNP obtained by nuclear disruption procedures have been re-ported to contain a very heterogeneous mixture of proteins, resolving into 50 or more bands of molecular weight 10,000–200,000 upon SDS gel elec-trophoresis (Pederson, 1974; Stevenin and Jacob, 1974). Tissue-specific differences have been observed in the complex polypeptide patterns of such preparations of hnRNP, and this implies that specific proteins may be associated with tissue-specific RNA sequences. Clearly, these results are in marked contrast to the report of Samarina and her colleagues (1968) that large complexes from rat liver nuclei essentially contain only the polypeptide(s) of the 30 S substructure. The question remains unresolved at the present time, and the truth is likely to lie between these extremes of complexity and simplicity.

It is certainly to be expected that proteins associated with the poly-(A)NP complex would be found in large hnRNP which contain poly(A). The 15 S poly(A)NP subcomplex isolated from mouse tumor cells appears to retain approximately five major associated polypeptides of apparent molecular weight 57,500, 63,000, 86,000, 120,000, and 140,000 (Quinlan et al., 1974; Billings and Martin, 1978); again there exist conflicting data, since Kish and Pederson (1977) report only 74,000 and 86,000 dalton polypeptides associated with nuclear poly(A) of the HeLa cells after pass-age over poly(U)-Sepharose.

Further, one would anticipate that enzymes of RNA processing may be found in hnRNP complexes, although one may not expect to detect by gel

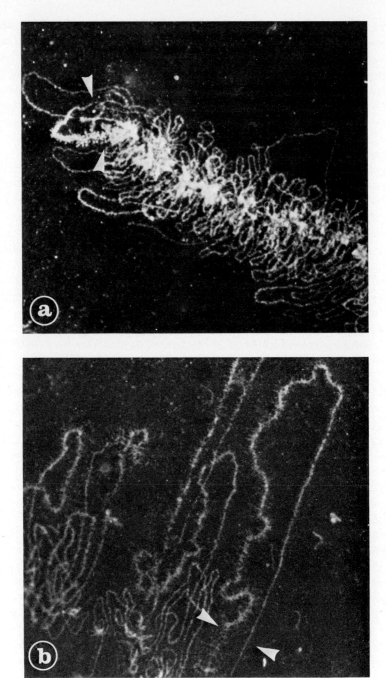

electrophoresis proteins such as methylases, nucleases, and splicing enzymes, which are required in catalytic amounts and need only be transiently associated with their RNA substrates. Such activities, as well as enzymes which may modify other RNP proteins, could be properties of the RNP "structural proteins" themselves. A variety of enzyme functions have been reported to be present in hnRNP preparations, including ribonuclease (Neissing and Sekeris, 1970), homopolymer synthetase (Neissing and Sekeris, 1973), and protein kinase (Martin *et al.*, 1974; Blanchard *et al.*, 1975; Gallinaro-Matringe *et al.*, 1975; Karn *et al.*, 1977). None of the reports confidently exclude the possibility of artifactual contamination of RNP preparations by these ubiquitous nuclear enzymes.

G. Intracellular Localization of hnRNP Proteins

As will be discussed below, there is considerable debate as to the nature of the proteins associated with mRNA in the cytoplasm of higher cells. While we may expect some nuclear polypeptides to accompany the mature mRNA during transport to the cytoplasm, such proteins have not yet been unequivocally identified with the possible exception of the poly(A) binding components. In the specific case of the major hnRNP polypeptides, i.e., those of 34,000–40,000 daltons, biochemical analysis has failed to detect their presence in significant quantities in cytoplasmic mRNP (see below).

Despite the conserved nature of the 34,000–40,000 dalton polypeptides, it has been possible to raise chicken antibodies against the mammalian proteins. Using such antisera Lukanidin *et al.* (1972) failed to detect these polypeptides on polyribosome mRNP.

More recently, we have used indirect immunofluorescence to examine the intracellular distribution of 30 S RNP antigens. These studies have indicated a primarily "nucleoplasmic" or "euchromatic" localization for these hnRNP proteins and have shown that they become distributed throughout the cell plasm during mitosis, reentering the nucleus as the nuclear envelope is reformed (Martin *et al.*, 1979a). Further experiments have provided perhaps the most direct evidence available that these proteins bind to nascent hnRNA. Figure 2 shows the reaction of chicken

Fig. 2. Immunofluorescent localization of hnRNP–protein antigenic determinants on the loops of *Triturus viridescens* lampbrush chromosomes. These loops, which contain very high densities of nascent RNP fibrils (see Fig. 1a), react strongly with antibodies raised against the purified 34,000–40,000 dalton polypeptide group of mouse 30 S hnRNP subcomplexes (a). Polarity of fluorescent loop width can be seen (between the arrowheads) and the fibrous character of the fluorescence extending from the loop core is sometimes apparent, particularly in well-spread giant loops (b). × 900. (Photographs courtesy of Carol Okamura.)

anti-mouse RNP γ-globulin with *Triturus* oocyte lampbrush chromosomes. Not only are the loops, which are known to contain high densities of nascent hnRNP fibrils (see Fig. 1a) highly fluorescent, but a polarity of width of the staining around the loop core is clearly visible in some cases (between arrows in Fig. 2a and b). Somewhat similar approaches in analyzing the distribution of proteins on lampbrush chromosomes have also been employed by Sommerville *et al.* (1978).

H. Protein–Protein Interactions: A Protein Core to hnRNP?

Models have been proposed in which the native hnRNP structure has been interpreted as either particulate or as a folded fibril. In the first model (Samarina *et al.*, 1968), nascent hnRNA is thought to complex with surface binding sites on a number of stable globular protein complexes ("informofers"). Cleavage between particles would yield individual informofers, each carrying a segment of RNA, and would thus explain the nature and origin of 30 S RNP. An alternative model implies that smaller protein complexes bind to the nascent RNA, forming an RNP fibril which is subsequently periodically folded into a more compact structure (Stevenin and Jacob, 1974) and possibly maintained in this conformation by protein–protein interactions of varying degrees of stability. These various alternatives for the organization of nascent RNP fibers have been discussed by Malcolm and Sommerville (1974) in the context of amphibian lampbrush loop fine structure.

In considering the interactions that maintain the integrity of the 30 S RNP subcomplex, several observations tend to diminish the importance of the RNA component. First, a large part of the RNA is relatively sensitive to RNase; this suggests, at least operationally, a surface location of the nucleic acid. At moderate salt concentrations, 30 S RNP proteins aggregate when the RNA is digested, implying a considerable tendency for protein–protein interactions. Second, the 30 S RNP contains approximately 1000 nucleotides of RNA (Samarina *et al.*, 1968; Martin *et al.*, 1974); however, whether this is relatively intact or internally cut into 50–100 nucleotide segments does not affect the integrity of 30 S RNP at physiological salt concentrations (Kinniburgh and Martin, 1976a). Thus, given the lack of a requirement for RNA continuity within the particle, our attention is focused on the protein component.

The concept of a reutilizable protein core complex [the informofer of Samarina *et al.* (1968)] is an attractive one and resembles in many ways the more recently evolved model for the nucleosome histone core in chromatin structure (reviewed by Kornberg, 1977). Major support for this model of the formation of hnRNP has come from the observation that chemically radioiodinated 30 S RNP protein cores, stripped of RNA and maintained in solution by high concentrations of salt, continue to sediment in approximately the same position in sucrose gradients as native 30 S RNP (Lukanidin et al., 1972b). The stability of these cores has suggested that the protein complexes could be reutilized intact in the nucleus. However, we have recently reexamined the nature of iodinated 30 S RNP and shown that the 34,000–40,000 dalton polypeptide components have been highly cross-linked by the iodination procedure (Martin et al., 1978). This cross-linking is responsible for the apparent salt stability of the "protein core," which, in fact; dissociates into polypeptide monomers or small oligomers in high salt if not iodinated. The iodination artifact, however, does give some impression of the intimate protein–protein contacts in the 30 S RNP subcomplex; the likely existence of free radical-mediated, zero-length cross-links between tyrosine or other amino acid residues implies that RNP structure is maintained by a net of polypeptide interactions.

While it would be easy in view of these findings to extend the analogy of nucleohistone structure to hnRNP and therefore to speak of "ribonucleosomes," some caution is necessary at the present time. First, we may recognize the possibility that the purified substructures studied so far may be the result of protein reorganization (Stevenin and Jacob, 1979). Second, hnRNP are much more dynamic structures than the bulk of the nucleosomes in a cell. There is insufficient evidence to decide whether the putative "protein cores" recycle as such or are dissociated and then reassembled on nascent hnRNA. The fact that not all of the "core" polypeptides have identical half-lives suggests that protein exchange may occur (Martin et al., 1979a). Further, we must expect a considerable degree of flexibility of structure to permit the complex events of RNA processing and maturation.

I. RNA–Protein Interactions: Preliminary Models for hnRNP Structure and Function

The question of the organization of particular RNA sequences of hnRNA in relationship to hnRNP proteins has thus far been approached in essentially three ways: (a) accessibility of the identifiable sequence to nuclease digestion, (b) association or not of the sequence with isolatable RNP substructures, and (c) cell-free binding of the sequence to purified RNP–proteins in "reconstitution experiments." Thus it can be shown that mRNA sequences are found in 30 S RNP substructures (Kinniburgh and Martin, 1976a) and also that purified mRNA has a high affinity for 30 S RNP–proteins (Martin et al., 1974). Double-stranded RNA (dsRNA) regions of hnRNA, first described by Ryskov et al. (1973), are not tightly associated with the 30 S RNP complexes (Kinniburgh *et al.*, 1976; Martin *et al.*, 1978), are relatively accessible to specific nuclease digestion (Calvet and Pederson, 1978), and have very low affinity for 30 S RNP–proteins *in vitro* (Martin *et al.*, 1978); thus, a particulate model for hnRNP structure would place such dsRNA sequences between particles or protruding from the surface of them (see Fig. 3). Similarly, the transcribed oligo(A) sequences of hnRNA are presumably only indirectly associated with the 30 S hnRNP subcomplex by way of covalent attachment to high affinity sequences (Kinniburgh and Martin, 1976b). Nuclear posttranscriptional poly(A), of course, is isolated in a distinctive 15 S RNP complex having none of the major 30 S RNP–proteins, and purified poly(A) has very low affinity for the 30 S RNP–proteins in cell-free binding assays (Martin *et al.*, 1974; Quinlan *et al.*, 1974). We may expect to find proteins other than those of the 30 S and 15 S complexes associated with certain specific sequences, such as oligo(A), 5'-capping groups, and some potential double-strand regions, as well as transiently associated processing and modification enzymes, perhaps accounting for some of the multiplicity of proteins reported for large hnRNP complexes.

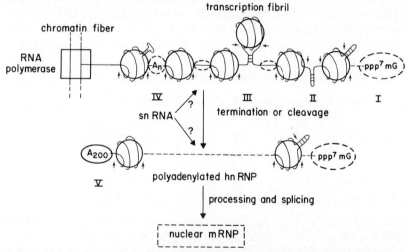

Fig. 3. A preliminary model for hnRNP structure illustrating the binding of nascent and newly terminated RNA molecules to the major hnRNP proteins which apparently fold the RNA chain to yield the 30 S RNP substructures. The binding leaves RNase-sensitive sites (small arrows) between and perhaps also within these substructures. The 5'-capping group (I) and double-strand (II and III) and oligo(A) (IV) regions are not tightly bound to the 30 S substructures, but may be associated with different proteins (broken lines). The 3'-terminal poly(A$_{200}$) is associated with a distinct class of polypeptides. The possible role of small nuclear RNA's (snRNA) in the processing of hnRNA in nuclear RNP is currently being studied.

Thus we can piece together a crude, circumstantial model for a major fraction of hnRNP structure (Fig. 3). The RNA contained in the structure proposed in this model would include nascent hnRNA, and recently completed molecules at early stages of processing. Therefore, while studies involving putative processing enzymes using naked RNA molecules are perfectly valid, as a first step, it is likely that the duplication of *in vivo* specificity will require the RNA to be in RNP form. This seems more probable when we consider that the splicing process apparently requires that sequences distant from each other along the hnRNA chain be brought into close proximity. The apparent ability to "reconstitute" RNP structures with exogenous RNA by alterations of ionic strength, etc., should allow the formation of suitable substrates for such experiments.

In the cell hnRNA sequence regions rapidly undergo maturation or degradation, yet many of the major hnRNP proteins are metabolically stable; we must therefore recognize nuclear hnRNP to be part of a highly dynamic system (Fig. 4). Based on our own studies, we believe that as the hnRNA strand is synthesized it becomes associated with a simple set of polypeptides, either in the form of an intact large protein complex or as small subunits, which fold the RNA strand into compact 30 S subcomplexes. The process of reutilization of the RNP–proteins upon turnover or transport of the RNA may involve such modifications of the

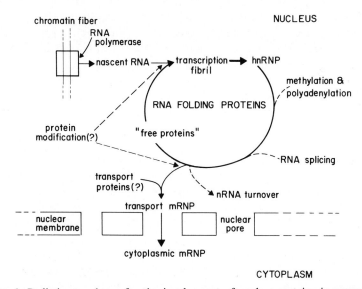

Fig. 4. Preliminary scheme for the involvement of nuclear proteins in processing of hnRNA to mRNA. The size of the "free protein" pool and the nature of the mRNA transport proteins are presently unknown.

polypeptides as phosphorylation, etc. Subtle alterations in the components of this hnRNP protein cycle may cause path switching between maturation and degradation of mRNA sequences, and, therefore, could act in a specific regulation mechanism. It is, however, idle to speculate further at the present time on the possible significance of such a control system given our continuing lack of (a) a precise structural model for hnRNP in which most components are chemically identified, (b) functional tests for RNP components, and (c) biochemical identification of the complex which transports mRNA from nucleus to cytoplasm.

One additional point which should be made is the need for a cell-free model for the *in vivo* transport of mRNA. There have been some reports of an ATP-dependent release of RNP from isolated nuclei (for example, Ishikawa *et al.*, 1969, 1970a,b); this has not been a universal finding, but the development of such systems deserves further consideration, so as to provide information linking nuclear and cytoplasmic RNP functions.

III. CYTOPLASMIC RNP CONTAINING mRNA

A. Free mRNP in the Cytoplasm

A most straightforward view of mRNA not bound to polyribosomes might be that it is newly made mRNA in transit from the nucleus, and therefore the study of the proteins associated with such "free" message would be expected to resolve some of the questions raised in Section II. Unfortunately it has become all too clear that the situation with regard to nonpolysomal mRNA in cells is not straightforward. Instead we are forced by currently available data to propose various functional classes of mRNP in the same cell, with different cell types perhaps having greatly differing levels of the individual classes.

The first suggestive evidence that nonpolysomal mRNA exists in the cytoplasm of eukaryotic cells and is complexed with protein was reported by Spirin and collaborators (reviewed by Spirin, 1969). In early stages of developing fish embryos, before ribosomal RNA synthesis is initiated, newly synthesized RNA could be found in structures that sedimented on sucrose gradients slower than polysomes or 80 S ribosomal monomers. The subribosomal material sedimented in structures of 20–75 S, and the authors termed these particles "informosomes." The RNP nature of this material was demonstrated by the procedure developed by Spirin and colleagues of buoyant density determination in CsCl equilibrium density gradients after fixation of these complexes with formaldehyde (Spirin *et al.*, 1965). They were able to show that the complexes had buoyant densities between those of free protein and free RNA, and distinct from ribosomes or ribosomal subunits. Following its introduction in these early experiments, the criterion of buoyant density of formaldehyde- or glutaraldehyde-fixed material in CsCl has frequently been used to define ribonucleoprotein complexes, and evidence has accumulated to support the idea that mRNA in the cytoplasm is always complexed with proteins.

The influential studies of Spirin and co-workers were of course carried out with cells from early embryos, and a variety of observations made over many years suggest that the embryo contains maternally inherited information stored, presumably in the cytoplasm, for programmed use after fertilization. This has led to the concept of "masked mRNA" and its possible identity with "informosomes" in developing systems. Since the question of the storage of mRNA in eggs and early embryos is discussed in detail in Chapter 5 of this volume, we will restrict our review to a brief outline of current knowledge of free mRNP in cells of adult tissues or late in the differentiation process.

Soon after the apparent discovery of free mRNP complexes in fish embryos, the presence of similar rapidly labeling, polydisperse RNA in subribosomal fractions was reported for cultured mouse L cells (Perry and Kelley, 1968), rat liver (Henshaw and Loebenstein, 1970), and HeLa cells (Spohr *et al.*, 1970). In all of these studies, the ribonucleoprotein nature of these complexes was demonstrated by CsCl density equilibrium centrifugation. Following the detection of poly(A) segments on most mRNA molecules, numerous reports of poly(A)$^+$ RNA in subribosomal fractions such as in mouse kidney (Ouellette *et al.*, 1976) and L cells (Perry and Kelley, 1976) have appeared. The fraction of the total cytoplasmic poly(A)$^+$ RNA of mouse liver found in free RNP is approximately 15%, while in tumor cells the level approaches 30% (Martin *et al.*, 1979b; McMullen *et al.*, 1979). Convincing evidence that the RNA in the nonpolysomal fraction is indeed mRNA has not always been provided. However, the case for a general occurrence of free cytoplasmic mRNP has been considerably strengthened by the detection of many specific mRNAs in subribosomal fractions: globin mRNA in rabbit (Lebleu *et al.*, 1971; Jacobs-Lorena and Baglioni, 1972; Olsen *et al.*, 1972) and duck (Spohr *et al.*, 1972), myosin mRNA (Buckingham *et al.*, 1976; Heywood *et al.*, 1975), actin mRNA (Bag and Sarkar, 1975), protamine mRNA (Gedamu *et al.*, 1977), and rat albumin mRNA (Yap *et al.*, 1978a,b).

As in the case of nuclear hnRNP, difficulties arise when one wishes to proceed beyond the simple detection of the complexes: the absence of proven purification methods and functional assays. The main hindrance to the characterization of free mRNP complexes is the difficulty in isolation of these complexes from other cytoplasmic proteins and from the ribosomal subunits. With sedimentation values of 20–75 S or greater, most free mRNP's cosediment with the ribosomal subunits. Free mRNPs can be separated from the ribosomal subunits on CsCl density gradients but only after fixation with formaldehyde, and this procedure renders the free RNP complexes relatively intractable for protein characterization, as the proteins have been cross-linked in the RNP by this procedure, and reversal of these cross-links is unreliable. A density separation method using Cs_2SO_4 has been proposed by Greenberg (1977), but the utility of the method for the isolation of native mRNP is as yet unproved.

Free mRNPs for four different specific mRNAs have been partially purified by sedimentation on sucrose gradients. Three of these free mRNPs are for duck globin mRNA (Gander et al., 1973), chicken actin mRNA (Bag and Sarker, 1975), and trout protamine mRNA (Gedamu et al., 1977); all of these mRNAs are relatively small, and their corresponding free mRNP particles sediment more slowly than the ribosomal small subunit. The specific free mRNP for chicken myosin heavy chain mRNA has been isolated by virtue of its large size (Bag and Sarkar, 1976). This free mRNP is found to sediment at about 120 S, in the range of small polysomes; therefore, fractions containing the complexes have been treated with EDTA to dissociate the small polysomes while leaving the myosin mRNP unaffected. The proteins of these partially purified free mRNP complexes have been characterized (Table III).

Other investigators have attempted to analyze the free mRNP proteins of the total postpolysomal mRNP in their cells. Liautard et al. (1976) have used the relatively high sensitivity to ribonuclease of free mRNP to release and characterize putative free mRNP proteins. They have reported a protein of MW 76,000, and four other proteins in the MW range of 16,000–40,000. Barrieux et al. (1975) used poly(U) immobilized on glass fiber filters to isolate free mRNP particles from Erlich ascites cells. Three

TABLE III

Published Protein Compositions for Specific Free mRNPs

mRNP	Polypeptide components (MW × 10^{-3})				
Protamine[a]		29,		73	
Globin[b]	15,15.5,17,19,21,22,24		51		
Actin[c]			44,49,	54,	75,80, 87,98
Myosin[d]					81,83,86,98

[a] Gedamu et al., 1977.
[b] Gander et al., 1973.
[c] Bag and Sarkar, 1975.
[d] Bag and Sarkar, 1976.

major polypeptides of MW 78,000, 52,000, and 34,000 were obtained along with four other relatively minor proteins. These latter studies, along with many others including an analysis of HeLa cell mRNA associated proteins (Kumar and Pederson, 1975), have taken advantage of the poly(A) tail of most mRNAs to employ poly(U) or oligo(dT) affinity methods to purify mRNP. Unfortunately the binding and release of the complexes has most often employed relatively harsh conditions and is open to criticism as a method for purifying RNP complexes (Billings and Martin, 1978). The more recent thermal elution procedure perhaps provides aid in circumventing some of the difficulties (Jain et al., 1979).

B. Polysomal mRNP

Henshaw (1968) and Perry and Kelley (1968) demonstrated that the rapidly labeling polysomal RNA (presumably mRNA) can be released from polysomes by EDTA treatment and that this mRNA is complexed with protein. The CsCl buoyant densities of these particles range from 1.35 to 1.55, and they are easily distinguished from ribosomal subunit densities. The values suggest a protein content of greater than 50%. The reports of protein composition of polysomal mRNP complexes are now too numerous to be fully covered in this chapter. A more detailed review has recently been published by Preobrazhensky and Spirin (1978). We will only summarize the general findings that have emerged from these studies.

The first reports of protein complements of polysomal mRNP were those for duck and rabbit globin mRNP. Morel et al. (1971) found two polypeptides of MW 73,000 and 49,000 associated with duck globin mRNA released from polysomes that had been dissociated with EDTA. Blobel (1972) reported two species with MW of 78,000 and 52,000 associated with globin mRNA which was released from polysomes of rabbit reticulocytes by the puromycin–high salt method. The globin mRNP from rabbit reticulocytes has become the most extensively characterized of polyribosome-associated mRNP. Bryan and Hayashi (1973), and Barrieux et al. (1975) have also reported polypeptides of molecular weight similar to those already mentioned, although in some cases additional minor proteins have been observed.

As was the case for free globin mRNP from reticulocytes, the 14 S polysomal mRNP has been isolated from ribosomal subunits by sucrose density gradient centrifugation. The method is obviously not satisfactory for larger mRNP in the range 30–80 S. Lindberg and Sundquist (1974) have developed the technique of oligo(dT)-cellulose affinity chromatography so that polysomal mRNPs could be isolated regardless of size.

They have reported that EDTA-released polysomal mRNP from KB cells isolated by this poly(A)-oligo(dT) affinity procedure contain four major polypeptides of MW 130,000, 78,000, 68,000, and 56,000. Polysomal mRNP from adenovirus infected KB cells have been reported to contain an additional protein of MW 110,000. Burns and Williamson (1975) used the oligo(dT)-cellulose method to isolate mouse globin polysomal mRNP and found two major proteins of MW 73,000 and 49,000, that is polypeptides with molecular weights close to that of the globin polysomal mRNP proteins found by Blobel and others. The oligo(dT)-cellulose method has been used to isolate polysomal mRNP from many cell types and lines. Examples include: HeLa (Kumar and Pederson, 1975), KB cells (Lindberg and Sundquist, 1974; Van der Marel *et al.*, 1975), Ehrlich ascites cells (Jeffery, 1977; Van Venrooij *et al.*, 1977), and mouse kidney (Irwin *et al.*, 1975). Although these investigators have reported different numbers of major proteins (3–6) and a variety of molecular weight distributions, a protein of MW 75,000–78,000 is a common finding. Whether the other differences reported represent real differences in protein composition or are isolation artifacts has yet to be determined. Philosophies of the experimentalists vary between the extremes of the idea of a limited general set of polypeptides bound to all mRNAs and the notion of mRNA-specific proteins.

C. Poly(A)–Ribonucleoprotein Subcomplex of mRNP

A majority of eukaryotic mRNAs contain a poly(A) region (reviewed by Greenberg, 1975). The resistance of poly(A) tracts to nucleases that solubilize the rest of the mRNA molecule allows a simple approach to the investigation of the proteins associated with this sequence. Kwan and Brawerman (1972) first suggested that the poly(A) region is associated with protein. The conclusion was based on the observation that poly(A) is released from polysomes by mild nuclease digestion in the form of a complex sedimenting at three to four times the rate of free poly(A), and that this poly(A) complex would bind to nitrocellulose filters at salt concentrations at which free poly(A) cannot bind but protein complexes can.

The first reported characterization of the protein associated with poly(A) was carried out by Blobel (1973), who reported a single polypeptide of MW 78,000. The poly(A) localization of proteins of similar size has been reported for other polysomal mRNPs (Barrieux *et al.*, 1975; Schwartz and Darnell, 1976; Jeffery, 1977). Some investigators find varying amounts of additional polypeptides associated with the poly(A) segment, and while there is general agreement that a 74,000–78,000 dalton polypeptide is most probably the poly(A) binding protein in mRNP, the case has yet to be proven and the function of the protein elucidated.

D. Possible Functions of Cytoplasmic mRNP

1. Transport of mRNA

The proteins of the free mRNP may function directly in the transport of mRNA from the nucleus to the cytoplasm, perhaps by presently un-defined interactions with the nuclear pore complex. Schwartz and Darnell (1976) have suggested that the protein of MW 74,000–78,000 found by themselves and many other laboratories to be a component of cytoplasmic mRNPs, may have a transport role. They find that the newest mRNA, that with the longest poly(A), has the largest amount of a 75,000 MW protein associated with its poly(A) region. They also report that blockage of new poly(A) synthesis leads to the virtual elimination of the 75,000 MW pro-tein as a poly(A)-particle protein. This is the strongest available evidence that a particular protein might be involved in the transport of mRNA, but more direct evidence is required to make the case convincing.

2. Storage and Protection of mRNA

Proteins associated with mRNA may simply serve to protect the mole-cule from nuclease action; however, a more specific storage function may be involved. As mentioned before, the storage of "masked" messenger RNA as found in embryonic systems is not covered by this chapter. The possibility exists, nonetheless, that some of the mRNA in any cell type may be in an homologous structure, specifically restricted from transla-tion by some mRNP component, possibly involving distinct proteins from those found in other free mRNP forms.

3. Translation Modulation

Not all of the free mRNP complexes in cells are in a sequestered or "masked" form unavailable for translation. Schochetman and Perry (1972) have demonstrated that when L cells are incubated at elevated temperatures, the initiation step of protein synthesis is inhibited and mRNA accumulates as free mRNP. Upon lowering the temperature to normal, the free mRNP quickly declines and polysomes accumulate utiliz-ing the original mRNA.

The distribution of the specific mRNA between free mRNP and poly-somes may be affected by the nutritional regime. Yap et al. (1978a,b) have shown that in starving rats up to 60% of the albumin mRNA in liver can be found in free mRNP, while in fed rats almost all of the liver albumin mRNA is found on membrane-bound polyribosomes. Refeeding amino acids to starved animals restores the mRNA to the polysomes. Also in rat liver, Zahringer *et al.* (1976) find that with a normal diet ferritin mRNA is about equally distributed between free mRNP and polysomes, but the addition of iron to the diet causes essentially all the ferritin mRNA to become associated with polysomes. In both of these situations it seems that the distribution of an mRNA species between free mRNP and polysomes can change very rapidly, and that all the mRNA for that species is available for translation.

The distribution of the generality of mRNAs in the cell between the free mRNP and polysome compartments has yet to receive detailed attention using sufficiently sensitive techniques. A nucleic acid hybridization and cell-free translation product analysis of the poly(A)$^+$mRNA of mouse ascites tumor cell-free mRNP and polysomes suggest a considerable degree of differential polarization of specific mRNAs between the two compartments; the mRNA of free mRNP appears to be a much simpler population of sequences than the functioning polysomal mRNA (McMullen *et al.*, 1979; Kinniburgh *et al.*, 1979).

A major question is whether any specific mRNA that is partly in a free mRNP form is readily exchangeable with the same mRNA species functioning on polysomes. If the mRNA molecules are exchanging between free mRNP and polysomes, then all of that mRNA species, whether in free

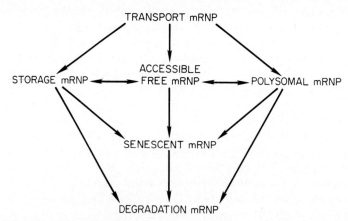

Fig. 5. Possible relationships between proposed cytoplasmic mRNP forms. The scheme is only intended to illustrate working concepts for free mRNP function, and it is not clear at the present time if all potential forms exist as structurally distinct complexes. A single cell type does not necessarily contain all possible mRNP forms. The model is shown only to make apparent ways of thinking about free cytoplasmic mRNP, rather than to represent any demonstrated physiological processes.

mRNP or polysomal mRNP, can be considered to be directly involved in protein synthesis.

Jacobs-Lorena and Baglioni (1972) have found that while about 20% of the total α-chain globin mRNA in rabbit reticulocytes is in free mRNP, essentially all of the β-globin mRNA is polysome bound. Lodish and Jacobsen (1972) have shown that β-chain mRNA is 40% more efficient in initiation of protein synthesis than α-chain mRNA. Instead of assuming that the 20% of the α-chain mRNA is actively "masked" by protein and unavailable for translation, we may consider the presence of the α-chain mRNA in free mRNP to result from a relatively lower initiation efficiency.

It is not at all clear what determines whether an mRNA is in the free mRNP or polysomes. It may result indirectly from the differential attachment of proteins to different mRNAs or directly from structural differences in the mRNAs themselves. Whatever the cause, the observation of differential polar distributions of mRNA species between free mRNP and polyribosomes raises the potential for translational regulation which may be mediated by modification of mRNA binding proteins.

4. Conclusion

Figure 5 illustrates the different possible forms of cytoplasmic mRNP and the possible routes that mRNA may follow in going from one form to another eventually leading to its degradation. We do not imply that all of these forms necessarily exist or that they exist in every cell type, but merely wish to formalize what appears to us to be the reasoning behind current experimentation in a field where, like that of nuclear RNP, the most interesting discoveries are yet to come. These will almost certainly involve the purification of mRNPs from different physiological states and the assay of functions *in vitro*. Preliminary observations indicate that polysomal mRNP are translatable in cell-free systems, while results differ using free mRNP (Olsen *et al.*, 1972; Sampson *et al.*, 1972; Nudel *et al.*, 1973; Civelli *et al.*, 1976). Improved purification and assay methods will certainly resolve some of the present confusion in this important area of research into biological control mechanisms.

REFERENCES

Abelson, J. (1979). RNA processing and the intervening sequence problem. *Annu. Rev. Biochem.* **48**, 1035–1069.

Angelier, N., and Lacroix, J. C. (1975). Complexes de transcription d'origines nucléolaire et chromosomique d'ovocytes de *Pleurodeles waltlii* et *P. poireti*. *Chromosoma* **51**, 323–335.

Augenlicht, L. H., and Lipkin, M. (1976). Appearance of rapidly labelled high molecular weight RNA in nuclear ribonuclear protein. Release from chromatin and association with protein. *J. Biol. Chem.* **251**, 2592–2599.

Bag, J., and Sarkar, S. (1975). Cytoplasmic nonpolysomal messenger ribonucleoprotein containing actin messenger RNA in chicken embryonic muscles. *Biochemistry* **14**, 3800–3807.

Bag, J., and Sarkar, S. (1976). Studies on a nonpolysomal ribonucleoprotein coding for myosin heavy chains from chick embryonic muscle. *J. Biol. Chem.* **251**, 7600–7609.

Baltimore, D., and Huang, A. S. (1970). Interaction of HeLa cell proteins with RNA. *J. Mol. Biol.* **47**, 263–273.

Barrieux, A., Ingraham, H. A., David, D. N., and Rosenfeld, M. G. (1975). Isolation of messenger-like ribonucleoproteins. *Biochemistry* **14**, 1815–1821.

Berezny, R., and Coffey, D. S. (1974). Identification of a nuclear protein matrix. *Biochem. Biophys. Res. Commun.* **60**, 1410–1417.

Beyer, A. L., Christensen, M. E., Walker, B. W., and LeStourgeon, W. M. (1977). Identification and characterization of the packaging proteins of core 40S hnRNP particles. *Cell* **11**, 127–138.

Billings, P. B. (1979). The proteins of ribonucleoprotein subcomplexes containing heterogeneous nuclear RNA. Ph.D. Thesis, University of Chicago, Chicago, Illinois.

Billings, P. B., and Martin, T. E. (1978). Proteins of nuclear ribonucleoprotein subcomplexes. *Methods Cell Biol.* **17**, 349–376.

Birnie, G. D. (1978). Isolation of nuclei from animal cells in culture. *Methods Cell Biol.* **17**, 13–26.

Blanchard, J. M., Ducamp, C., and Jeanteur, P. (1975). Endogenous protein kinase activity in nuclear RNP particles from HeLa cells. *Nature (London)* **253**, 467–468.

Blobel, G. (1972). Protein tightly bound to globin mRNA. *Biochem. Biophys. Res. Commun.* **47**, 88–95.

Blobel, G. (1973). A protein of molecular weight 78,000 bound to the polyadenylate region of eukaryotic messenger RNAs. *Proc. Natl. Acad. Sci. U.S.A.* **70**, 924–928.

Boffa, L. C., Karn, J., Vidali, G., and Allfrey, V. G. (1977). Distribution of N^G,N^G-dimethylarginine in nuclear protein fraction. *Biochem. Biophys. Res. Commun.* **74**, 969–976.

Bryan, R. N., and Hayashi, M. (1973). Two proteins are bound to most species of polysomal mRNA. *Nature (London), New Biol.* **244**, 271–274.

Buckingham, M. E., Cohen, A., and Gros, F. (1976). Cytoplasmic distribution of pulse-labelled poly(A)-containing RNA, particularly 26S RNA, during myoblast growth and differentiation. *J. Mol. Biol.* **103**, 611–626.

Burns, A. T. H., and Williamson, R. (1975). Isolation of mouse reticulocyte globin messenger ribonucleoprotein by affinity chromatography using oligo(dT)-cellulose. *Nucleic Acids Res.* **2**, 2251–2255.

Calvet, J. P., and Pederson, T. (1978). Nucleoprotein organization of inverted repeat DNA transcripts in heterogeneous nuclear RNA–ribonucleoprotein particles from HeLa cells. *J. Mol. Biol.* **122**, 361–378.

Civelli, O., Vincent, A., Buri, J.-F., and Scherrer, K. (1976). Evidence for a translational inhibitor linked to globin mRNA in untranslated free cytoplasmic messenger ribonucleoprotein complexes. *FEBS Lett.* **72**, 71–76.

Daneholt, B., Case, S. T., Derksen, J., Lamb, M. M., Nelson, L., and Weislander, L. (1978). The size and chromosomal location of the 75S RNA transcription unit in Balbiani ring 2. *Cold Spring Harbor Symp. Quant. Biol.* **42**, 867–876.

Darnell, J. E. (1978). Transcription units for mRNA production in eukaryotic cells and their DNA viruses. *Prog. Nucleic Acid Res. Mol. Biol.* **22**, 327–353.

Deimel, B., Louis, C., and Sekeris, C. (1977). The presence of small molecular weight RNAs in nuclear ribonucleoprotein particles carrying hnRNA. *FEBS Lett.* **73**, 80–84.

Edström, J.-E., Erickson, E., Lindgren, S., Lönn, U., and Rydlander, L. (1978). Fate of Balbiani-ring RNA *in vivo*. *Cold Spring Harbor Symp. Quant. Biol.* **42**, 877–884.

Faiferman, I., and Pogo, A. O. (1974). Isolation of a nuclear ribonucleoprotein network that contains heterogeneous RNA and is bound to the nuclear envelope. *Biochemistry* **14**, 3808–3816.

Faiferman, I., Hamilton, M. G., and Pogo, A. O. (1970). Nucleoplasmic ribonucleoprotein particles of rat liver, I. Selective degradation by nuclear nuclease. *Biochim. Biophys. Acta* **204**, 550–563.

Fakan, S., and Bernhard, W. (1971). Localization of rapidly and slowly labeled nuclear RNA as visualized by high resolution autoradiography. *Exp. Cell Res.* **67**, 129–141.

Franke, W. W., Scheer, U., Trendelenburg, M. F., Spring, H., and Zentgraf, H. (1976). Absence of nucleosomes in transcriptionally active chromatin. *Cytobiologie* **13**, 401–434.

Franke, W. W., Scheer, U., Trendelenburg, M., Zentgraf, H., and Spring, H. (1978). Morphology of transcriptionally active chromatin. *Cold Spring Harbor Symp. Quant. Biol.* **42**, 755–772.

Gall, J. G. (1956). Small granules in the amphibian oocyte nucleus and their relationship to RNA. *J. Biophys. Biochem. Cytol., Suppl.* **2**, 393–396.

Gallinaro-Matringe, H., Stevenin, J., and Jacob, M. (1975). Salt dissociation of nuclear particles containing DNA-like RNA. Distribution of phosphorylated and non-phosphorylated species. *Biochemistry* **14**, 2547–2554.

Gander, E. S., Stewart, A. G., Morel, C. M., and Scherrer, K. (1973). Isolation and characterization of ribosome-free cytoplasmic messenger ribonucleoprotein complexes from avian erythroblasts. *Eur. J. Biochem.* **38**, 443–452.

Gattoni, R., Stevenin, J., and Jacob, M. (1977). Metrizamide dissociates nuclear particles containing heterogeneous RNA. *Nucleic Acids Res.* **4**, 3931–3941.

Gedamu, L., Dixon, G. H., and Davies, P. L. (1977). Identification and isolation of protamine messenger ribonucleoprotein particles from rainbow trout testis. *Biochemistry* **16**, 1383–1391.

Georgiev, G. P., and Samarina, O. P. (1971). D-RNA containing ribonucleoprotein particles. *Adv. Cell Biol.* **2**, 47–110.

Greenberg, J. R. (1975). Messenger RNA metabolism of animal cells: The possible involvement of untranslated sequences and mRNA-associated proteins. *J. Cell Biol.* **64**, 269–288.

Greenberg, J. R. (1977). Isolation of messenger ribonucleoproteins in cesium sulfate density gradients: Evidence that polyadenylate and non-polyadenylate messenger RNAs are associated with protein. *J. Mol. Biol.* **108**, 403–416.

Hamilton, M. G. (1971). Isodensity equilibrium centrifugation of ribosomal particles: The calculation of the protein content of ribosomes and other ribonucleoproteins from buoyant density measurements. *In* "Methods in Enzymology" (K. Moldave and L. Grossman, eds.), Vol. 20, Part C, pp. 512–521. Academic Press, New York.

Heinrich, P. C., Gross, V., Northemann, W., and Scheurlen, M. (1978). Structure and function of ribonucleoprotein complexes. *Ergeb. Physiol., Biol. Chem. Exp. Pharmakol.* **81**, 101–134.

Henshaw, E. C. (1968). Messenger RNA in rat liver exists as ribonucleoprotein particles. *J. Mol. Biol.* **36**, 401–411.

Henshaw, E. C., and Loebenstein, J. (1970). Rapidly labeled, polydisperse RNA in rat liver cytoplasm: Evidence that it is contained in ribonucleoprotein particles of heterogeneous size. *Biochim. Biophys. Acta* **199**, 405–420.

Herman, R., Weymouth, L., and Penman, S. (1978). Heterogeneous nuclear RNA–protein fibers in chromatin depleted nuclei. *J. Cell Biol.* **78**, 663–674.

Hewish, D. R., and Burgoyne, L. A. (1973). Chromatin sub-structure. The digestion of chromatin DNA at regularly spaced sites by a nuclear deoxyribonuclease. *Biochem. Biophys. Res. Commun.* **52**, 504–510.

Heywood, S. M., Kennedy, D. S., and Bester, A. J. (1975). Stored myosin messenger in embryonic chick muscle. *FEBS Lett.* **53**, 69–72.

Howard, E. F. (1978). Small nuclear RNA molecules in nuclear ribonucleoprotein complexes from mouse erythroleukemia cells. *Biochemistry* **17**, 3228–3236.

Irwin, D., Kumar, A., and Malt, R. A. (1975). Messenger ribonucleoprotein complexes isolated with oligo(dT)-cellulose chromatography from kidney polysomes. *Cell* **4**, 157–165.

Ishikawa, K., Kuroda, C., and Ogata, K. (1969). Release of ribonucleoprotein particles containing rapidly labeled ribonucleic acid from rat liver nuclei. Effect of ATP and some properties of the particles. *Biochim. Biophys. Acta* **179**, 316–331.

Ishikawa, K., Kuroda, C., Ueki, M., and Ogata, K. (1970a). Messenger ribonucleoprotein complexes released from rat liver nuclei by ATP. I. Characterization of the RNA moiety of messenger ribonucleoprotein complexes. *Biochim. Biophys. Acta* **213**, 495–504.

Ishikawa, K., Kuroda, C., and Ogata, K. (1970b). Messenger ribonucleoprotein complexes released from rat liver nuclei by ATP. II. Characterization of the protein moiety. *Biochim. Biophys. Acta* **213**, 505–512.

Jacobs-Lorena, M., and Baglioni, C. (1972). Messenger RNA for globin in the post-ribosomal supernatant of rabbit reticulocytes. *Proc. Natl. Acad. Sci. U.S.A.* **69**, 1425–1428.

Jain, S. K., Pluskal, M. G., and Sarkar, S. (1979). Thermal chromatography of eukaryotic messenger ribonucleoprotein particles on oligo(dT)-cellulose. *FEBS Lett.* **97**, 84–90.

Jeffery, W. R. (1977). Characterization of polypeptides associated with messenger RNA and its polyadenylate segment in Ehrlich ascites messenger ribonucleoproteins. *J. Biol. Chem.* **252**, 3525–3532.

Jelinek, W., and Leinwand, L. (1978). Low molecular weight RNAs hydrogen-bonded to nuclear and cytoplasmic poly(A)-terminated RNA from cultured chinese hamster ovary cells. *Cell* **15**, 205–214.

Karn, J., Vidali, G., Boffa, L. C., and Allfrey, V. G. (1977). Characterization of the non-histone nuclear proteins associated with rapidly labelled heterogeneous nuclear RNA. *J. Biol. Chem.* **252**, 7307–7322.

Kinniburgh, A. J., and Martin, T. E. (1976a). Detection of mRNA sequences in nuclear 30S ribonucleoprotein subcomplexes. *Proc. Natl. Acad. Sci. U.S.A.* **73**, 2725–2729.

Kinniburgh, A. J., and Martin, T. E. (1976b). Oligo(A) and oligo(A)-adjacent sequences present in nuclear ribonucleoprotein complexes. *Biochem. Biophys. Res. Commun.* **73**, 718–726.

Kinniburgh, A. J., Billings, P. B., Quinlan, T. J., and Martin, T. E. (1976). Distribution of hnRNA and mRNA sequences in nuclear ribonucleoprotein complexes. *Prog. Nucleic Acid Res. Mol. Biol.* **19**, 335–351.

Kinniburgh, A. J., McMullen, M. D., and Martin, T. E. (1979). The distribution of cytoplasmic poly(A$^+$)RNA sequences in free mRNP and polysomes of mouse ascites cells. *J. Mol. Biol.* **132**, 695–708.

Kish, V. M., and Pederson, T. (1977). Heterogeneous nuclear RNA secondary structure: Oligo(U) sequences base-paired with poly(A) and their possible role as binding sites for heterogeneous nuclear RNA-specific proteins. *Proc. Natl. Acad. Sci. U.S.A.* **74**, 1426–1430.

Kornberg, R. D. (1977). Structure of chromatin. *Annu. Rev. Biochem.* **46**, 931–954.
Krichevskaya, A. A., and Georgiev, G. P. (1969). Further studies on the protein moiety in nuclear DNA-like RNA containing complexes. *Biochim. Biophys. Acta* **164**, 619–621.
Kumar, A., and Pederson, T. (1975). Comparison of proteins bound to heterogeneous nuclear RNA and messenger RNA in HeLa cells. *J. Mol. Biol.* **96**, 353–365.
Kumar, A., and Warner, J. R. (1972). Characterization of ribosomal precursor particles from HeLa cell nucleoli. *J. Mol. Biol.* **63**, 233–246.
Kwan, S.-W., and Brawerman, G. (1972). A particle associated with the polyadenylate segment in mammalian messenger RNA. *Proc. Natl. Acad. Sci. U.S.A.* **69**, 3247–3250.
Lebleu, B., Marbaix, G., Huez, G., Temmerman, J., Burney, A., and Chantrenne, H. (1971). Characterization of the messenger ribonucleoprotein released from reticulocyte polyribosomes by EDTA treatment. *Eur. J. Biochem.* **19**, 264–269.
Lerner, M. R., and Steitz, J. A. (1979). Antibodies to small nuclear RNAs complexed with proteins are produced by patients with systemic lupus erythematosus. *Proc. Natl. Acad. Sci. U.S.A.* **76**, 5495–5499.
Lerner, M. R., Boyle, J. A., Mount, S. M., Wolin, S. L., and Steitz, J. A. (1979). Are snRNPs involved in splicing? *Nature (London)* **283**, 220–224.
LeStourgeon, W. M., Beyer, A. L., Christensen, M. E., Walker, B. W., Poupore, S. M., and Daniels, L. P. (1978). The packaging proteins of core hnRNP particles and the maintenance of proliferative cell states. *Cold Spring Harbor Symp. Quant. Biol.* **42**, 885–898.
Liau, M. C., and Perry, R. P. (1969). Ribosomal precursor particles in nucleoli. *J. Cell Biol.* **42**, 272–283.
Liautard, J. P., Setyono, B., Spindler, E., and Köhler, K. (1976). Comparison of proteins bound to the different functional classes of messenger RNA. *Biochim. Biophys. Acta* **425**, 373–383.
Lindberg, U., and Sundquist, B. (1974). Isolation of messenger ribonucleoproteins from mammalian cells. *J. Mol. Biol.* **86**, 451–468.
Lodish, H. F., and Jacobsen, M. (1972). Regulation of hemoglobin synthesis. *J. Biol. Chem.* **247**, 3622–3629.
Louis, C., and Sekeris, C. E. (1976). Isolation of informofers from rat liver. Effects of α-amanatin and actinomycin D. *Exp. Cell Res.* **102**, 317–328.
Lukanidin, E. M., Olsnes, S., and Pihl, A. (1972a). Antigenic difference between informofers and protein bound to polyribosomal mRNA from rat liver. *Nature (London), New Biol.* **240**, 90–91.
Lukanidin, E. M., Zalmanzon, E. S., Komaromi, L., Samarina, O. P., and Georgiev, G. P. (1972b). Structure and function of Informofers. *Nature (London) New Biol.* **238**, 193–197.
McMullen, M. D., Shaw, P. H., and Martin, T. E. (1979). Characterization of poly(A+) RNA in free mRNP and polysomes of mouse Taper ascites cells. *J. Mol. Biol.* **132**, 679–694.
Malcolm, D. B., and Sommerville, J. (1974). The structure of chromosome-derived ribonucleoprotein in oocytes of *Triturus cristatus carniflex* (Laurenti). *Chromosoma* **48**, 137–158.
Martin, T. E., and McCarthy, B. J. (1972). Synthesis and turnover of RNA in the 30S nuclear ribonucleoprotein complexes of mouse ascites cells. *Biochim. Biophys. Acta* **277**, 354–367.
Martin, T. E., Billings, P. B., Levey, A., Ozarslan, S., Quinlan, T. J., Swift, H. H., and Urbas, L. (1974). Some properties of RNA:protein complexes from the nucleus of eukaryotic cells. *Cold Spring Harbor Symp. Quant. Biol.* **38**, 921–932.
Martin, T. E., Billings, P. B., Pullman, J. M., Stevens, B. J., and Kinniburgh, A. J. (1978). Substructure of nuclear ribonucleoprotein complexes. *Cold Spring Harbor Symp. Quant. Biol.* **42**, 899–909.

Martin, T., Jones, R., and Billings, P. (1979a). hnRNP core proteins: Synthesis turnover and intracellular distribution. *Mol. Biol. Rep.* **5,** 37–42.

Martin, T., McMullen, M., and Shaw, P. (1979b). Poly(A⁺) mRNA sequences in free mRNP and polysomes of mouse ascites cells. *Mol. Biol. Rep.* **5,** 87–90.

Matsumura, S., Morimoto, T., Tashiro, Y., Higashinakagawa, T., and Muramatsu, M. (1974). Ultrastructural and biochemical studies on the precursor ribosomal particles isolated from rat liver nucleoli. *J. Cell Biol.* **63,** 629–640.

Miller, O. L., Jr., and Bakken, A. H. (1972). Morphological studies of transcription. *Gene Transcription Reprod. Tissue, Trans. Karolinska Symp. Res. Methods Reprod. Endocrinol., 5th, 1972* pp. 155–167.

Miller, T. E., Huang, C., and Pogo, A. O. (1978). Rat liver nuclear skeleton and ribonucleoprotein complexes containing hnRNA. *J. Cell Biol.* **76,** 675–691.

Molnar, J. P., and Samarina, O. P. (1975). Purification of nuclear ribonucleoprotein complexes containing poly(adenylic) acid. *Acta Biochim. Biophys. Acad. Sci. Hung.* **10,** 263–266.

Monneron, A., and Bernhard, W. (1969). Fine structural organization of the interphase nucleus in some mammalian cells. *J. Ultrastruct. Res.* **27,** 266–288.

Morel, C., Kayibanda, B., and Scherrer, K. (1971). Proteins associated with globin messenger RNA in avian erythroblasts: Isolation and comparison with the proteins bound to nuclear messenger-like RNA. *FEBS Lett.* **18,** 84–88.

Murray, V., and Holliday, R. (1979). Mechanism for RNA splicing of gene transcripts. *FEBS Lett.* **106,** 5–7.

Neissing, J., and Sekeris, C. E. (1970). Cleavage of rapidly labeled DNA-like RNA by protein derived from nuclear particles. *Biochim. Biophys. Acta* **209,** 484–492.

Neissing, J., and Sekeris, C. E. (1973). Synthesis of polynucleotides in nuclear ribonucleoprotein particles containing heterogenous RNA. *Nature (London), New Biol.* **243,** 9–12.

Northemann, W., Scheurlen, M., Gross, V., and Heinrich, P. C. (1977). Circular dichroism of ribonucleoprotein complexes from rat liver nuclei. *Biochem. Biophys. Res. Commun.* **76,** 1130–1137.

Nudel, U., Lebleu, B., Zehavi-Willner, T., and Revel, M. (1973). Messenger ribonucleoprotein and initiation factors in rabbit-reticulocyte polyribosomes. *Eur. J. Biochem.* **33,** 314–322.

Olsen, S. D., Gaskill, P., and Kabat, D. (1972). Presence of hemoglobin messenger ribonucleoprotein in a reticulocyte supernatant fraction. *Biochim. Biophys. Acta* **272,** 297–304.

Ouellette, A. J., Kumar, A., and Malt, R. A. (1976). Physical aspects and cytoplasmic distribution of messenger RNA in mouse kidney. *Biochim. Biophys. Acta* **425,** 384–395.

Parsons, J. T., and McCarty, K. S. (1968). Rapidly labeled messenger ribonucleic acid protein complex of rat liver nuclei. *J. Biol. Chem.* **243,** 5377–5384.

Pederson, T. (1974). Proteins associated with heterogeneous nuclear RNA in eukaryote cells. *J. Mol. Biol.* **83,** 163–183.

Perry, R. P., and Kelley, D. E. (1968). Messenger RNA–protein complexes and newly synthesized ribosomal subunits: Analysis of free particles and components of polyribosomes. *J. Mol. Biol.* **35,** 37–59.

Perry, R. P., and Kelley, D. E. (1976). Kinetics of formation of 5′ terminal caps in mRNA. *Cell* **8,** 433–442.

Preobrazhensky, A. A., and Spirin, A. S. (1978). Informosomes and their protein components: The present state of knowledge. *Prog. Nucleic Acid Res. Mol. Biol.* **21,** 1–37.

Puvion-Dutilleul, F., Bernadac, A., Puvion, E., and Bernhard, W. (1977). Visualization of two different types of nuclear transcriptional complexes in rat liver cells. *J. Ultrastruct. Res.* **58,** 108–117.

Quinlan, T. J., Billings, P. B., and Martin, T. E. (1974). Nuclear ribonucleoprotein complexes containing polyadenylate from mouse ascites cells. *Proc. Natl. Acad. Sci. U.S.A.* **71**, 2632–2636.

Quinlan, T. J., Kinniburgh, A. J., and Martin, T. E. (1977). Properties of a nuclear polyadenylate–protein complex from mouse ascites cells. *J. Biol. Chem.* **252**, 1156–1161.

Rickwood, D., Hell, A., Birnie, G. D., and Gilhuus-Moe, C. (1974). Reversible interaction of metrizamide with protein. *Biochim. Biophys. Acta* **342**, 367–371.

Ro-Choi, T. S., Raj, N. B. K., Pike, L. M., and Busch, H. (1976). Effects of alpha-amanitin, cycloheximide, and thioacetamide on low molecular weight nuclear RNA. *Biochemistry* **15**, 3823–3828.

Ryskov, A. P., Saunders, G. F., Farashyn, V. R., and Georgiev, G. P. (1973). Double helical regions in nuclear precursor of mRNA. *Biochim. Biophys. Acta* **312**, 152–164.

Samarina, O. P., Asriyan, I. S., and Georgiev, G. P. (1965). Isolation of nuclear nucleoproteins containing messenger ribonucleic acid. *Dokl. Akad. Nauk SSSR* **163**, 1510–1513.

Samarina, O. P., Krichevskaya, A. A., Molnar, J., Bruskov, V. I., and Georgiev, G. P. (1967a). Nuclear ribonuclear proteins containing messenger RNA. Isolation and properties. *Mol. Biol. Moscow* **1**, 129–141.

Samarina, O. P., Molnar, J., Lukanidin, E. M., Bruskov, V. I., Krichevskaya, A. A., and Georgiev, G. P. (1967b). Reversible dissociation of nuclear ribonucleoprotein particles containing mRNA into RNA and protein. *J. Mol. Biol.* **27**, 187–191.

Samarina, O. P., Lukanidin, E. M., Molnar, J., and Georgiev, G. P. (1968). Structural organization of nuclear complexes containing DNA-like RNA. *J. Mol. Biol.* **33**, 251–263.

Sampson, J., Matthews, M. B., Osborn, M., and Borghetti, A. F. (1972). Hemoglobin messenger ribonucleic acid translation in cell-free systems from rat and mouse liver and Landschutz ascites cells. *Biochemistry* **11**, 3636–3640.

Schochetman, G., and Perry, R. P. (1972). Characterization of messenger RNA released from L cell polyribosomes as a result of temperature shock. *J. Mol. Biol.* **63**, 577–590.

Schwartz, H., and Darnell, J. E. (1976). The association of protein with polyadenylic acid of HeLa cell messenger RNA: Evidence for a "transport" role of a 75,000 molecular weight polypeptide. *J. Mol. Biol.* **104**, 833–851.

Sekeris, C. E., and Niessing, J. (1975). Evidence for the existence of a structural RNA component in the nuclear ribonucleoprotein particles containing heterogeneous RNA. *Biochem. Biophys. Res. Commun.* **62**, 642–650.

Sommerville, J., Crichton, C., and Malcolm, D. (1978). Immunofluorescent localization of transcriptional activity on lampbrush chromosomes. *Chromosoma* **66**, 99–114.

Spirin, A. S. (1969). Informosomes. *Eur. J. Biochem.* **10**, 20–35.

Spirin, A. S., Belitsina, N. V., and Lerman, M. I. (1965). Use of formaldehyde fixation for studies of ribonucleoprotein particles by caesium chloride density-gradient centrifugation. *J. Mol. Biol.* **14**, 611–615.

Spohr, G., Granboulan, N., Morel, C., and Scherrer, K. (1970). Messenger RNA in HeLa cells: An investigation of free and polyribosome-bound cytoplasmic messenger ribonucleoprotein particles by kinetic labeling and electron microscopy. *Eur. J. Biochem.* **17**, 296–318.

Spohr, G., Kayibanda, B., and Scherrer, K. (1972). Polyribosome-bound and free-cytoplasmic-hemoglobin-messenger RNA in differentiating avian erythroblasts. *Eur. J. Biochem.* **31**, 194–208.

Stevenin, J., and Jacob, M. (1974). Effects of sodium chloride and pancreatic ribonuclease on the rat brain nuclear particles: The fate of the protein moiety. *Eur. J. Biochem.* **47**, 129–137.

Stevenin, J., and Jacob, M. (1979). Structure of pre-mRNP: Models and pitfalls. *Mol. Biol. Rep.* 5, 29–35.

Stevenin, J., Divilliers, G., and Jacob, M. (1976). Size heterogeneity of the structural subunits of brain nuclear ribonuclear-protein particles. *Mol. Biol. Rep.* 2,˙385–391.

Stevenin, J., Gallinaro-Matringe, H., Gattoni, R., and Jacob, M. (1977). Complexity of the structure of particles containing heterogeneous nuclear RNA as demonstrated by ribonuclease treatment. *Eur. J. Biochem.* 74, 589–602.

Stevenin, J., Gattoni, R., Divilliers, G., and Jacob, M. (1979). Rearrangements in the course of ribonuclease hydrolysis of pre-messenger ribonuclear proteins. *Eur. J. Biochem.* 95, 593–606.

Stevens, B. J., and Swift, H. (1966). RNA transport from nucleus to cytoplasm in *Chironomus* salivary glands. *J. Cell Biol.* 31, 55–77.

Stunnenberg, H. G., Louis, C., and Sekeris, C. E. (1978). Depletion in nuclei of proteins associated with hnRNA, as a result of inhibition of RNA synthesis. *Exp. Cell Res.* 112, 335–344.

Swift, H. (1963). Cytochemical studies on nuclear fine structure. *Exp. Cell Res., Suppl.* 9, 54–67.

Van der Marel, P., Tasserson-deJong, J. G., and Bosch, L. (1975). The proteins associated with mRNA from uninfected and adenovirus type 5-infected KB cells. *FEBS Lett.* 51, 330–334.

Van Holde, K. E., and Hill, W. E. (1974). Physical properties of ribosomes. *In* "Ribosomes" (M. Nomura, A. Tissières, and P. Lengyel, eds.), pp. 53–92. Cold Spring Harbor Press, Cold Spring Harbor, New York.

Van Venrooij, W. J., and Janssen, D. B. (1978). hnRNP particles. *Mol. Biol. Rep.* 4, 3–8.

Van Venrooij, W. J., van Eekelen, C. A. G., Jansen, R. T. P., and Princen, J. M. G. (1977). Specific poly-A-binding protein of 76,000 molecular weight in polyribosomes is not present on poly A of free cytoplasmic mRNP. *Nature (London)* 270, 189–191.

Yap, S. H., Strair, R. K., and Shafritz, D. A. (1978a). Effect of a short term fast on the distribution of cytoplasmic albumin messenger ribonucleic acid in rat liver. *J. Biol. Chem.* 253, 4944–4950.

Yap, S. H., Strair, R. K., and Shafritz, D. A. (1978b). Identification of albumin in mRNPs in the cytosol of fasting rat liver and the influence of tryptophan or a mixture of amino acids. *Biochem. Biophys. Res. Commun.* 83, 427–433.

Zahringer, J., Baliga, B. S., and Munro, H. N. (1976). Novel mechanism for translational control in regulation of ferritin synthesis by iron. *Proc. Natl. Acad. Sci. U.S.A.* 73, 857–861.

5

Proteolytic Cleavage in the Posttranslational Processing of Proteins

D. F. Steiner, P. S. Quinn, C. Patzelt,
S. J. Chan, J. Marsh, and H. S. Tager

I. INTRODUCTION

The products arising from the linearly coded genetic units of eukaryotic organisms often undergo extensive editing, rearrangement, and condensation before appearing in a useful final form, for reasons that are not fully understood. Some of the many posttranscriptional and posttranslational modifications that have been discovered in the last decade appear to be dictated by the nature and properties of the physical pathway along which information is transferred from genes to specific cellular proteins, and which may impose many constraints on the structure, stability, and functionality of both messenger RNA molecules as well as the proteins derived from their translation. It indeed seems likely that secondary and tertiary

175

structure in both proteins *and mRNA* are critical determinants of their physical and functional properties, and therefore play a significant role in dictating their primary structures. Thus, the existence of any stable protein gene product presupposes the parallel evolution of a stable and compatible mRNA and gene structures.

The elaborate detail in the structural organization of most macromolecules, of course, reflects the cumulative refinements achieved through natural selection during evolution. Many stable and useful structures probably evolved very early and later were permuted in a variety of ways to expand their utility, e.g., by gene duplication (with or without fusion) followed by gradual divergence into families of related proteins, or by posttranslational modifications of many kinds. Neither of these transformations requires drastic revision of the original gene products; indeed, the perfection of particular sets of basic structures, or "folds," may have tended to reduce the probability that totally new, stable, and functional structures could be generated.

Some constraints upon the mutability of the pathway of gene expression might include the following: (a) introduction of new transcription initiation points would presumably require the formation or transfer of relatively large segments of specifically coded DNA, but might be accomplished *en bloc* if transposable elements indeed exist in higher eukaryotic organisms (Kleckner, 1977); (b) introduction of new initiation sites for protein synthesis would probably also require extensive reorganization of mRNA secondary structure in addition to the creation of appropriate sequences for ribosome and initiator tRNA binding (Kozak, 1978; Blumberg et al., 1979); (c) large deletions or insertions within protein domains would likely destabilize the final folded arrangement of the product, thus rendering it more susceptible to rapid turnover or degradation within the cell (Goldberg and Dice, 1974); and (d) size limitations in the translation products, i.e. the architecture of the ribosome–membrane junction may impose a lower limit on the size of secreted proteins, in the range of 70–80 residues (Patzelt et al., 1978a; Blobel and Sabatini, 1970). Because of these and other less obvious limitations, the development of new proteins and functions in the course of evolution has increasingly occurred through elaborations on preexisting structures and less often through the appearance of completely novel structures. It is thus inevitable that fragmentation (and other modifications) of gene products has become an important mechanism for the generation of additional bioactive configurations that cannot be produced more directly without drastic revisions in genetic "read out." An excellent example is the production of small peptide hormones by proteolytic cleavage of larger precursors which are largely nonfunctional.

This rather conservative evolutionary mechanism also can readily provide for the coordinated production of peptides with related functions, as, for example, the multihormone ACTH–endorphin precursor (Mains *et al.*, 1977; Roberts and Herbert, 1977), the more complex polyprotein precursors of virus capsules (Korant, 1975), or the precursor of some membrane-associated enzyme complexes (Poyton and McKemmie, 1979a). Some polyproteins, however, may have evolved by the serial fusion of several protein domains thereby achieving more efficient biosynthesis (Kirschner and Bisswanger, 1976) or regulation. Indeed, the ACTH–endorphin precursor contains homologous internal regions (the α-, β-, and γ-MSH regions) which presumably have arisen via gene duplication and fusion (Nakanishi *et al.*, 1979). This precursor thus may contain more than one structural domain, although it is clear that several distinct bioactive subspecies can reside within any one of its putative domains, e.g. the β lipotropin region (Hughes *et al.*, 1975). When viewed in this manner posttranslational processing is more readily comprehensible as an important mechanism of evolution, as well as a means for facilitating assembly, regulating the expression of activity, or any other more specialized functions of individual precursor forms, e.g., the facilitation of interchain disulfide bond formation by proinsulin (Steiner and Clark, 1968; Steiner, 1978). In view of the increasing appreciation of the importance of proteolysis in protein biosynthesis, this chapter will attempt to review our current knowledge of these mechanisms, and of the proteolytic enzymes involved and their subcellular localization(s).

II. EARLY CLEAVAGES—THE PRESECRETORY PROTEINS

Although it has been known for a long time that secretory proteins are synthesized in the rough endoplasmic reticulum (RER) by membrane-bound ribosomes (Palade, 1975), we have only recently begun to understand how the RER is formed and how mRNA for secretory proteins is made available exclusively to membrane-bound ribosomes. The development of cell-free systems that efficiently translate mRNA in the absence of an appreciable background of endogenous protein synthesis has been of great importance in resolving this problem (Roberts and Paterson, 1973; Pelham and Jackson, 1976). These systems not only faithfully translate mRNA molecules *in vitro* but are free of proteolytic enzymes that might break down or alter translation products as they are formed. When similar systems were used to study the cell-free translation of the mRNA for immunoglobin light chains extracted from mouse myelomas, it was found that an *in vitro* translation product was formed which was larger than the normal light chain (Milstein *et al.*, 1972). This observation was rapidly confirmed by other laboratories (Swan *et al.*, 1972; Mach *et al.*, 1973; Schechter, 1973). Additional amino acids were found exclusively at the

ProPTH (bovine)
Met-Met-Ser-Ala-Lys-Asp-Met-Val-Lys-Val-Met-Ile-Val-Met-Leu-Ala-Ile-Cys-Phe-Leu-Ala-Arg-Ser-Asp-Gly-Lys-

Proinsulin (rat)
I.   (Met)Ala-Leu-Trp-Met-Arg-Phe-Leu-Pro-Leu-Leu-Ala-Leu-Leu-Val-Leu-Trp-Glu-Pro-Lys-Pro-Ala-Gln-Ala-Phe-
II.   Ile   Ile
                                                              Phe Phe

Growth hormone (rat)
Met-Ala-Ala-Asp-Ser-Gln-Thr-Pro-Trp-Leu-Leu-Thr-Phe-Ser-Leu-Leu-Cys-Leu-Leu-Trp-Pro-Gln-Glu-Ala-Gly-Ala-Leu-

Prolactin (rat)
Met-Asn-Ser-Gln-Val-Ser-Ala-Arg-Lys-Ala-Gly-Thr-Leu-Leu-Leu-Leu-Met-Met-Ser-Asn-Leu-Leu-Phe-Cys-Gln-Asn-Val-Gln-Thr-Leu-

Lysozyme (chicken)
Met-Arg-Ser-Leu-Leu-Ile-Leu-Val-Leu-Cys-Phe-Leu-Pro-Leu-Ala-Ala-Leu-Gly-Lys-

Ovomucoid (chicken)
(Met)Ala-Met-Ala-Gly-Val-Phe-Val-Leu-Phe-Ser-Phe-Val-Leu-Cys-Gly-Phe-Leu-Pro-Asp-Ala-Ala-Phe-Gly-Ala-

Conalbumin (chicken)
Met-Lys-Leu-Ile-Leu-Cys-Thr-Val-Leu-Ser-Leu-Gly-Ile-Ala-Ala-Val-Cys-Phe-Ala-Ala-

Trypsinogen (dog)
I.   Ala-Lys-Leu-Phe-Leu-Phe-Leu-Ala-Leu-Leu-Ala-Leu-Tyr-Val-Ala-Phe-Val-
II.   Phe Pro

Myeloma L chain
(Mouse MOPC-41)
Met-Asp-Met-Arg-Ala-Pro-Ala-Gln-Ile-Fhe-Gly-Phe-Leu-Leu-Leu-Phe-Pro-Gly-Thr-Arg-Cys-Gln-
(Mouse MOPC-321)
Met-Glu-Thr-Asp-Thr-Leu-Leu-Leu-Trp-Val-Leu-Leu-Leu-Trp-Val-Pro-Gly-Ser-Thr-Gly-Gln-
(Mouse MOPC-104E)
Met-Ala-Trp-Ile-Ser-Leu-Ile-Leu-Ser- * -Leu-Leu-Ala-Leu-Ser-Ser-Gly-Ala-Ile-Ser-Gln-
(Mouse MOPC-315)
Met-Ala-Trp-Thr-Ser-Leu-Ile-Leu-Ser- * -Leu-Leu-Ala-Leu-Cys-Ser-Gly-Ala-Ser-Ser-Gln-

V_λ L chain gene (mouse)
Met-Ala-Trp-Thr-Ser-Leu-Ile-Leu-Ser-Leu-Leu-Ala-Leu-Cys-Ser-Gly-Ala-Ser-Ser-Gln-

Casein (ovine)
β
Met-Lys-Val-Leu-Ile-Leu-Ala- * -Leu-Val-Ala-Leu-Ala-Leu-Ala-Arg-
α_{s1}
Met-Lys-Leu-Leu-Ile-Leu-Thr- * -Leu-Val-Ala-Val-Ala-Leu-Ala-Arg-
α_{s2}
Met-Lys-Val-Leu-Met-Lys-Ala- * -Leu-Val-Ala-Val-Ala-Leu-Ala-Leu-Lys-
κ
Met- * -Lys- * -Ile-Leu-Leu-Val-Val- * -Ile-Leu-Ala-Leu- * -Leu-Pro- * -Leu-Ile- Ala-Lys-

Proalbumin (bovine)
Met-Lys-Trp-Val-Thr-Phe-Leu-Leu-Leu-Leu-Phe-Ile-Ser-Gly-Ser-Ala-Phe-Ser-Lys-

VSV glycoprotein
Met-Lys-Cys-Leu-Leu-Tyr-Leu-Ala-Phe-Leu-Phe-Ile-Gly-Val- * -Asn-Cys-Lys

Promellitin (honeybee)
Met-Lys-Phe-Leu-Val- * -Val-Ala-Leu-Val-Phe-Met-Val-Val-Tyr-Ile- * -Tyr-Ile-Tyr-Ala-Ala-

Lipoprotein (E. coli)
Met-Lys-Ala-Thr-Lys-Leu-Val-Leu-Gly-Ala-Val-Ile-Leu-Gly-Ser-Thr-Leu-Leu-Ala-Gly-Cys-

Penicillinase
(E. coli plasmid pBR322))
Met-Ser-Ile-Gln-His-His-Phe-Arg-Val-Ala-Leu-Ile-Pro-Phe-Phe-Ala-Ala-Phe-Cys-Leu-Pro-Val-Phe-Ala-His-

Ovalbumin (chicken)
Met-Gly-Ser-Ile-Gly-Ala-Ala-Ser-Met-Glu-Phe-Cys-Phe-Asp-Val-Phe-
-Lys-Glu-Leu-Lys-Val-His-His-Ala-Asn-Glu-Asn-Ile-Phe-Tyr-Cys-Pro-Ile-

amino-terminus in an extension that was estimated to be about 20 residues long. When rough microsomes from the same myelomas were incubated *in vitro* they synthesized only normal light chains, but when the microsomal membranes were lysed beforehand with detergents, the ribosomes liberated from them synthesized only the larger precursor chains. Milstein *et al.* (1972) suggested that the microsomes contained a protease(s) that normally rapidly processed the precursors to normal light chains and that the peptide extension might aid the light chains in gaining access to the microsomal cisternae (Milstein *et al.*, 1972).

Their results thus provided the first experimental support for the existence of a mechanism for segregation, first formulated in general terms by Blobel and Sabatini (1971) and later refined as the signal hypothesis by Blobel and Dobberstein (1975a). Blobel and Sabatini (1971) proposed that the amino-terminal regions of secretory proteins might contain sequences that enable them to interact with recognition sites on microsomal membranes, thus inducing the binding of polyribosomes to the membrane to give rise to the RER. They further suggested that the binding of large ribosomal subunits to the microsomal membrane might create pores or channels through which the nascent secretory products could be transferred into the cisternal spaces, thus leading to their rapid segregation. However, the Cambridge experiments (Milstein, *et al.*, 1972), as well as other studies on the cell-free synthesis of a variety of secreted proteins, rapidly led to the realization that the postulated amino-terminal recognition region is not present within the known sequence of most secretory proteins but is located in precursor sequences at their amino-termini. These presequences are cleaved off very rapidly after synthesis, and, under normal conditions, presecretory proteins are present only in very small amounts in intact cells synthesizing secretory proteins, or in rough microsomes extracted from them (Patzelt *et al.*, 1978a,b; Maurer and McKean, 1978; Habener and Potts, 1979). These precursors can be more readily identified, however, when mRNA molecules coding for secretory proteins are translated in nonprocessing cell-free systems. The N-terminal amino acid sequences of a number of known presecretory proteins are summarized in Fig. 1.

Fig. 1. Amino acid sequences and cleavage sites of various presecretory peptide extensions. Data sources are as follows: proPTH (Habener *et al.*, 1978); proinsulin (Chan *et al.*, 1976; Villa-Komaroff *et al.*, 1978; Chan *et al.*, 1979); growth hormone (Seeburg *et al.*, 1977); prolactin (McKean and Maurer, 1978); lysozyme, ovomucoid, and conalbumin (Thibodeau *et al.*, 1978); trypsinogen (Devillers-Thiery *et al.*, 1975); myeloma L chains (Burstein and Schechter, 1978); V_λ L chain gene (Tonegawa *et al.*, 1978); casein (Gaye and Gautron, 1977); proalbumin (Strauss *et al.*, 1978); VSV glycoprotein (Lingappa *et al.*, 1978a); promellitin (Suchanek *et al.*, 1978); lipoprotein (DiRienzo *et al.*, 1978); penicillinase (Sutcliffe, 1978); and ovalbumin (Palmiter *et al.*, 1978).

A further significant advance came from the work of Blobel and Dobberstein (1975b), who recombined a cell-free protein synthesizing system derived from wheat germ with dog pancreas microsomal vesicles stripped of ribosomes by treatment with EDTA and were able to demonstrate that presecretory peptides are both segregated and cleaved during synthesis by such systems. Moreover, these systems also will correctly cleave presecretory proteins from unrelated organs or species, but will not process or sequester the translation products of mRNA's for nonsecreted proteins such as globin (Blobel and Dobberstein, 1975b; Shields and Blobel, 1977; Lingappa et al., 1977). Boime and co-workers (1977) confirmed and extended these observations using a cell-free system derived from Krebs ascites cells to translate and process the mRNA for human placental lactogen. Other workers using similar systems have reported the correct processing of the precursors of proalbumin, proparathyroid hormone, and growth hormone (Strauss et al., 1978; Dorner and Kemper, 1978; Lingappa et al., 1978a), thus confirming the apparent universality of the recognition and cleavage mechanisms in the RER. The involvement of specific proteins in both of these functions is suggested by observations that smooth microsomes and other cellular membranes do not accept or cleave presecretory proteins (Blobel and Dobberstein, 1975b; Jackson and Blobel, 1977). Of interest in this connection are two unique RER proteins recently described by Sabatini and Kreibich (1976) and Kreibich et al. (1977) that are closely associated with attached ribosomes. Warren and Dobberstein (1978) have also recently shown that proteins removed from RER membranes by treatment with high salt concentrations are needed for the transfer of secretory proteins across the RER membrane. They have suggested that these proteins are probably bound to the cytoplasmic side of the membrane without spanning it but may be associated with proteins that do span the bilayer. Walter et al. (1979), on the other hand, have been able to remove translocation activity from microsomal membranes by tryptic digestion and have inferred that the trypsin-sensitive components are cytosolic portions of transmembrane proteins involved in ribosome or signal sequence recognition.

Cleavage of the prepeptide occurs very rapidly during or immediately after the nascent peptide chain enters the cisternae of the endoplasmic reticulum (Blobel and Dobberstein, 1975b; Boime et al., 1977; Strauss et al., 1978; Dorner and Kemper, 1978; Shields and Blobel, 1978; Albert and Permutt, 1979; Patzelt, et al., 1978b). Some preliminary efforts to characterize proteolytic enzymes associated with RER membranes in various tissues (Jackson and Blobel, 1977; Kaschnitz and Kriel, 1978) or in bacterial membranes (Chang et al., 1978) have been reported. The initial cleavage appears to be made by an endopeptidase (Jackson and Blobel, 1977), although there is little information currently available regarding its specificity, origin, or properties. Zimmerman *et al.* (1979) and Strauss *et al.* (1979) have suggested that the endopeptidase activity of microsomal preparations is due to a metalloprotease similar in properties to thermolysin. The latency of its activity has led to the assumption that the enzyme is located within the microsomes, possibly in association with the luminal surface of the limiting membrane (Jackson and Blobel, 1977).

Although the signal peptidase may act preferentially on amino acids having small neutral side chains, such as alanine, serine or cysteine (see Fig. 1), the presence of glycine at some of the cleavage sites indicates that other factors must also guide cleavage. Moreover, the ability of heterologous reconstituted systems consisting of wheat germ or reticulocyte ribosomes combined with dog pancreas or ascites cell microsomes to correctly cleave preproteins derived from totally unrelated organs or species militates against the possibility that a highly specific converting enzyme exists for each presecretory protein (Strauss *et al.,* 1978; Dorner and Kemper, 1978; Devillers-Thiery *et al.,* 1975; Shields and Blobel, 1977). In some instances it has been proposed that cleavage specificity may arise from conserved sequence elements among prepeptide substrates (Burstein and Schechter, 1978; Inouye and Halegoua, 1979). However, such similarities as have been observed tend to occur only within closely related subgroups, such as the immunoglobulin light chain precursors (Burstein and Schechter, 1978). Hence we must seek a proteolytic mechanism that always places the prepeptide in correct register with respect to a fixed protease (perhaps located on the inner surface of the microsomal membrane), which then accurately cleaves the peptide chain more or less irrespective of the nature of the amino acid side chains in the immediate vicinity of the cleavage site (*vide infra*).

Most of the prepeptide sequences contain a region rich in hydrophobic amino acids with large bulky side chains between positions -7 and -17 (Fig. 1). This region is followed by a region (between the cleavage site and position -7) of amino acids with small neutral side chains which would be more flexible and which, in many of the sequences, has a high β- turn-forming potential (Steiner *et al.,* 1979; Chan *et al.,* 1979). Observation of these structural features led us to postulate that the nascent prepeptide partitions directly into the bilayer rather than passing through a pore in the RER membrane into the cisternal space, and through interactions within, the nonpolar lipid phase assumes an orientation spanning the membrane and forming a looped structure at the site of entry of the nascent presecretory peptide chain into the luminal space, as outlined schematically in Fig. 2 (Steiner *et al.,* 1979). The hydrophobic central region of the prepeptide may form a β-pleated sheet structure with intrinsic membrane protein

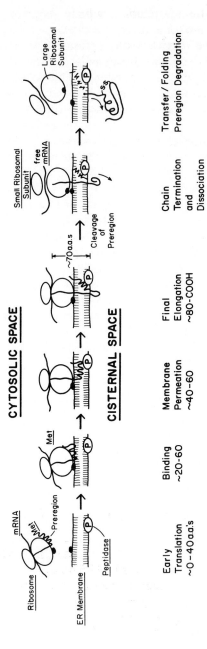

Fig. 2. Loop mechanism for the segregative transfer of presecretory proteins across the membrane of the rough endoplasmic reticulum (see text for details). ER, endoplasmic reticulum; a.a.'s, amino acids.

components, such as the ribophorins, which would further stabilize and orient the prepeptide with respect to a cleavage enzyme localized to the inner membrane. The formation of a transmembrane β-strand structure could also provide the necessary motive force for translocation of the nascent chain across the bilayer as shown schematically in Fig. 3. This model predicts that cleavages of presecretory proteins will be directed by their interactions with the hydrophobic core of the bilayer and protein components spanning it.

The ability of detergent-solubilized membrane preparations to correctly cleave some, but not all, presecretory proteins suggests that the signal peptidase is closely associated with those membrane components to which the presequence binds during segregative transfer, as indicated in Fig 3. A β-structure such as is postulated here can presumably form between any two (or three) peptide chains having the appropriate physical properties. In this case it would be defined by the dimensions of the nonpolar phase and oriented by the N-terminus and other charged residues on either side of the hydrophobic central region of the prepeptide. These polar regions might interact with the more polar membrane surface to add stability. This kind of receptor, involving mainly secondary structural interactions, could in theory accept and transfer a large variety of presecretory peptides provided that they all contained an appropriate hydrophobic segment.*

Evidence consistent with the loop models can be summarized as follows:

Lin *et al.* (1978) have shown that replacement of glycine with aspartic acid at a position 6 amino acids before the cleavage site of the *E. coli* lipoprotein (Inouye and Halegoua, 1979) results in inhibition of its cleavage but not of its transfer across the bacterial plasma membrane. Since the position of this mutation is just after the hydrophobic central region of the presequence, it would not be expected to interfere with the entry of the peptide into the lipid bilayer, but the additional charged group would probably alter the conformation of the putative peptide chain loop near the cleavage site which may be essential in orienting the peptide for cleav-

---

* The lipophilic central regions of most of the presequences have a high potential for forming either β-strands or α-helices, based on estimates using the rules of Chou and Fasman (1978). Although membrane-associated proteins have often been shown by physical studies to be rich in α-helices, the lengths of the lipophilic segments in many of the preproteins, if arranged in α-helices, would not be sufficient to span the bilayer without bringing hydrophilic residues from flanking regions into the apolar core of the membrane. The β-strand model has an additional advantage, beyond that of broad receptor specificity, of being more readily perturbed after completion of transfer so as to allow for rapid proteolysis of the residual prepeptide.

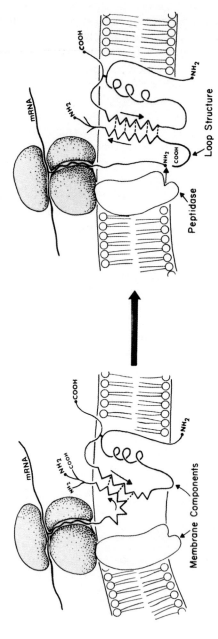

Fig. 3. More detailed scheme (β-strand model) proposed to account for transmembrane loop formation *via* progressive entry of hydrophobic central region of prepeptide into β-strand (or β-sheet) structure with one (or possibly two) extended peptide chains of preexisting transmembrane proteins. Such a receptor could accept almost any amino acid sequence having the requisite lipophilic and β-strand-forming properties.

age. Burstein and Schechter (1978) have found a nonproducing myeloma that synthesizes a smaller than normal light chain (C_K) which contains of the first 17 residues of the presequence attached to residue 109 of the constant portion of the chain. Despite the change in amino acid sequence at the cleavage site, the presecretory peptide is cleaved at position 20 in correct register with the first 17 residues. The three residues removed from the mature protein (Ala-Asp-Ala) differ from those normally present at these positions in the presequence (Ser-Thy-Gly). This finding is compatible either with the above model or with the proposal of Burstein and Schechter, 1978 that cleavage specificity for L chain precursors may require a glycine at position -4 and a pair of leucines at positions -8 and -9 in the presequence (see Fig 1). Palmiter et al. (1978) have observed that ovalbumin is secreted without a leader sequence. In this protein a hydrophobic segment is present near the amino-terminus, and the peptide is acetylated very rapidly during synthesis (Palmiter et al., 1978). It is thus possible that the blocked amino terminal region of ovalbumin contributes in some way to the transorption of the nascent peptide chain. Alternatively, formation of the postulated loop structure could conceivably occur at a more internal position in the nascent peptide chain, as suggested by recent results from Blobel's laboratory (Lingappa et al., 1979). Preliminary observations in our laboratory also provide further support for the loop models. We have detected fragments of the prepeptide containing the N-terminus in the external fluid medium surrounding dog pancreas microsomes during the segregation of secretory proteins in a cell-free system (Quinn, Ackerman and Steiner, unpublished results). These results indicate that a portion of the N-terminal region of the prepeptide remains outside the microsomal vesicles, and if blocked with N-formylmethionine, resists further proteolytic digestion.

In contrast with the proproteins (Section III,A) where peptide fragments are conserved and secreted, the prepeptide of preproinsulin is rapidly fragmented and degraded totally during short pulse-chase incubations of intact rat islets of Langerhans (Patzelt et al., 1978a,b). Presumably other membrane associated endo- and exopeptidases, must participate in this process, in addition to the postulated specific signal peptidase (Jackson and Blobel, 1977). Rapid degradation may serve to clear the membrane of the RER of residual dissolved peptide material which might disrupt or destabilize the bilayer if allowed to accumulate. Hence it is unlikely that prepeptides will be found in the secretions from cells or in the circulation.

It is interesting to note that N-terminal peptide extensions have been identified on the precursors of a number of exported prokaryotic proteins (Inouye and Halegoua, 1979; Inouye and Beckwith, 1977; Chang et al.,

1978; Mandel and Wickner, 1979; DiRienzo *et al.*, 1978). The N-terminal sequences of the precursors of penicillinase and lipoprotein are included in Fig. 1. These also appear to be composed of hydrophobic central sequences followed by a more flexible region with a high β-turn forming potential 3 to 6 residues from the cleavage site (Chan *et al.*, 1979). A signal peptidase activity has been detected in *E. coli* membranes (Mandel and Wickner, 1979; Chang *et al.*, 1978). Inouye and Halegoua (1979; DiRienzo *et al.*, 1978) recently independently proposed a model for the interaction of the prelipoprotein sequence with the bacterial cell inner membrane that is essentially similar to the loop model discussed above. Both models predict a transmembrane orientation of the prepeptide with the hydrophilic N-terminus remaining on the cytosolic side (Fig 2).

A difficult problem for any of the proposed models is the mechanism by which the nascent peptide chain continues to be driven across the membrane once the bilayer is spanned. In bacteria, the ribosomes synthesizing exported proteins appear not to be bound to the membranes by ionic interactions (Smith *et al.*, 1978), as is the case in eukaryotes (Sabatini and Kreibich, 1976), so that peptide chain elongation during protein synthesis cannot account for the continued passage of the nascent chain into the periplasmic space. However, we believe that translocation of the peptide chain could be completed by a combination of simple diffusion of the chain within the bilayer coupled with chain folding, causing the completed protein to accumulate on the luminal side of the R.E.R. (Chan *et al.*, 1979). Thus, the orderly folding of the peptide to a more stable conformation and perhaps related physicochemical processes, such as hydration and/or, in some cases, chemical modifications such as glycosylation (Lingappa *et al.*, 1978b; Leavitt *et al.*, 1977; Gibson *et al.*, 1979), may serve to make simple diffusion a unidirectional process. These attractive possibilities will require further investigation. In keeping with a diffusion mechanism for translocation, it is tempting to speculate also that regions of nascent peptides that are especially rich in hydrophobic residues may become fixed in the nonpolar phase of the membrane, thereby creating a permanent transmembrane insertion. The ''decision'' to stop transfer in such cases might also well depend on the structure and stability of the domain(s) already formed within the N-terminal segment of the protein on the luminal side of the membrane. If, on the other hand, the nascent peptide chain passes through a proteinaceous pore in the membrane as envisioned in the original signal hypothesis, then additional mechanisms that trigger the disassembly of the pore before completion of vectorial discharge must be invoked to account for insertion (Katz *et al.*, 1977).

III. POSTTRANSLATIONAL CLEAVAGES—INTRACELLULAR (THE PROPROTEINS)

It is now widely appreciated that a large number of mature proteins are produced by mechanisms involving intracellular proteolysis of larger precursor forms. Two main categories can be distinguished. The larger one, at present, includes many secreted proteins, among which are the small polypeptide hormones; the other group consists of a variety of cellular and organellar components, as well as many animal virus capsule proteins. Since the precursors of the first type all appear to be cleaved by a similar system of intracellular proteases, these will be examined in more detail in this chapter.

A. Precursors of Polypeptide Hormones and Other Secreted Proteins

Although not the best of terminology, the prefix "pro" has been generally used to designate many of the more slowly processed precursor forms that are cleaved intracellularly, while the prefix "pre" has been reserved for the presecretory proteins (Section II) that are rapidly processed by special microsomal proteases.

The discovery of proinsulin (Steiner and Oyer, 1967; Steiner et al., 1969) provided the first glimpse of the proteins comprising this group of precursors, the recognized numbers of which are rapidly enlarging. Proinsulin contains the A and B chains of insulin linked together in a single chain by a highly species variable segment containing at most 35 residues (Steiner et al., 1972, 1974, 1976). This connecting segment or C peptide, as it is usually designated, was originally thought to facilitate the formation of the correct disulfide linked three-dimensional structure of insulin. However, this role, while easily demonstrable (Steiner and Clark, 1968), cannot explain all of the structural features of the connecting segment, especially its great variability and its length, which exceeds by many times the 8–10 Å gap (Blundell et al., 1972) between the ends of the B and A chains in the final folded product—insulin. Moreover, the aforementioned function of the peptide of proinsulin can be reproduced by simple nonpeptide bifunctional cross-linking reagents (Brandenburg and Wollmer, 1973; Busse et al., 1974). For these and additional reasons summarized in greater detail elsewhere (Steiner, 1978), it now appears highly probable that the connecting segment also functions as a spacer that serves to enlarge the peptide chain to a length of 65–70 residues so that it can span the distance from the site of chain elongation between the large

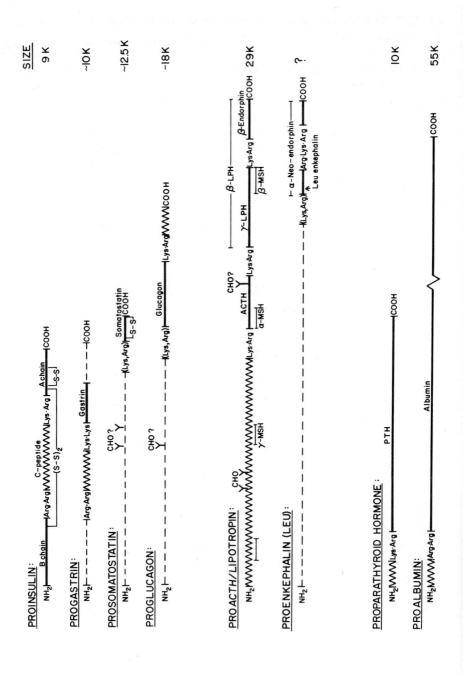

and small ribosomal subunits to the luminal side of the RER (Patzelt *et al.*, 1978a). According to this "minimum length hypothesis," many other small secreted peptides and hormones less than 50 residues in length would be expected to have similar nonfunctional and highly variable spacer regions included within larger precursor forms, and these proteins should be similar in overall size to proinsulin, or larger. Thus far, this simple idea seems to be supported by a constantly enlarging body of information on these precursor forms, as summarized in Fig. 4.

Since it is most useful for our purposes here to generalize, the various proproteins will not be considered in detail individually but instead will be considered from the standpoint of important structural similarities which point to the existence of basically similar processing systems for all of them. The data of Fig. 4 show that most proproteins studied thus far contain paired basic residues at the sites of cleavage and that arginine rather than lysine seems to be preferred on the carboxyl side of the pair. One exception to this is the Lys-Lys pair located just before the β-MSH sequence in the ACTH/endorphin precursor. However, much work now supports the view that this bond is cleaved only rarely during the maturation of the product (Chrétien *et al.*, 1977; Roberts *et al.*, 1978; Lowry *et al.*, 1977). The conversion of G34 gastrin to G17 gastrin also requires cleavage at a Lys-Lys pair, and seems to occur relatively slowly, allowing G34 to accumulate as an intermediate (Gregory and Tracy, 1975). Studies *in vitro* with proinsulin have shown that a mechanism involving both endo- and exopeptidases, e.g., a mixture of pancreatic cationic trypsin and carboxypeptidase B, readily reproduces the cleavage patterns seen *in vivo* and yields the correct products, i.e., insulin, C peptide, 3 Arg, and 1 Lys (Kemmler *et al.*, 1971). Similar considerations apply as well in proparathyroid hormone conversion (Habener *et al.*, 1977).

Conversion of proinsulin to insulin requires its prior transport to the Golgi area of the β cell (Steiner *et al.*, 1970, 1972), and similar requirements seem to exist for the processing of proparathyroid hormone (Habener *et al.*, 1977) and probably for proACTH/endorphin (Mains and Eipper, 1976; Eipper *et al.*, 1976) and proalbumin as well (Judah and Nicholls, 1971; Quinn and Judah, 1978). Thus, all of these precursors

Fig. 4. Structures of some of the known proproteins showing paired basic residues at cleavage sites. Heavy lines indicate regions which appear as biologically active or useful products. Unsequenced regions are indicated by the dashed lines. Data sources are as follows: proinsulin (Steiner *et al.*, 1972; Chance *et al.*, 1968); progastrin (Noyes *et al.*, 1979); prosomatostatin (Patzelt *et al.*, 1980); proglucagon (Tager and Steiner, 1973; Patzelt *et al.*, 1979); proACTH/lipotropin (Mains *et al.*, 1977; Nakanishi *et al.*, 1979; Hughes *et al.*, 1975); proenkephalin (Kanagawa and Matsuo, 1979); proparathyroid hormone (Hamilton *et al.*, 1974); and proalbumin (Russell and Geller, 1975).

exhibit a characteristically slow turnover with an initial delay of some 10–20 min postsynthesis, followed by the onset of cleavage with half-lives ranging from about 20 min to 1 hr or longer (Steiner *et al.*, 1967; Steiner, 1967). It is, therefore, likely that the cleavage enzymes are first brought into contact with the precursors as these pass into the Golgi apparatus and/or into newly formed (pro)secretory granules or condensing vacuoles (Steiner *et al.*, 1970; Howell, 1972; Jamieson and Palade, 1967a,b, 1968); Orei, *et al.*, 1971).

Smith and Farquhar (1966) first showed that lysosomal enzymes such as acid phosphatase might be cosegregated into newly forming secretion granules in the Golgi apparatus. Subsequent studies on β cells by Smith and VanFrank (1974) have provided additional direct evidence for the cosegregation into nascent β granules of a trypsin-like protease having some properties of cathepsin B. However, the studies of Quinn and Judah (1978; Judah and Quinn, 1978) on the conversion of proalbumin to albumin in isolated liver subcellular fractions suggest that the proteolytic enzymes acting on proalbumin must be added to preformed precursor-containing granules by a Ca^{2+}-dependent vesicle fusion process. These workers also have postulated a catheptic nature for these enzymes, presuming that they are related to cathepsin B. However, recent studies of MacGregor, *et al.*, (1979) indicate that cathepsin B does not have appropriate cleavage specificity to convert proparathyroid hormone. In view of the similarity of proparathyroid hormone to proalbumin it seems highly improbable that the latter would behave differently as a substrate for cathepsin B in liver.

Studies *in vitro* with isolated secretion granules containing prelabeled proinsulin have shown that normal processing activity is retained *in vitro* but only so long as the granules remain intact (Kemmler *et al.*, 1973; Sun *et al.*, 1973; Grant *et al.*, 1971; Sorenson *et al.*, 1972). Lysis of granules leads to a marked reduction in cleavage activity. Studies on such lysed granule preparations have failed to clearly demonstrate a trypsin-like activity, perhaps due to membrane association of the enzyme (Sun *et al.*, 1973), but have quite clearly disclosed the presence of a carboxypeptide B-like enzyme (Kemmler *et al.*, 1973, Zühlke *et al.*, 1976; Steiner *et al.*, 1975). Grant *et al.* (1971) detected membrane bound zymogen forms of both trypsin and chymotrypsin in codfish islet granule preparations. Other more recent studies on the nature of the proteases involved in proinsulin conversion have implicated catheptic proteases, rather than enzymes similar to the exocrine pancreatic proteases (Zühlke *et al.*, 1976; Zühlke and Steiner, 1977), but this problem is far from fully resolved. The identification of a protease within a tissue, or even within an appropriate tissue fraction, unfortunately, does not prove that the enzyme is involved in a particular processing event. Quite often in published studies cleavage products have been so poorly characterized that it is impossible to decide whether a described activity is even worthy of consideration as a possible participant in the *in vivo* cleavage mechanism (Sorenson *et al.*, 1972; Yip, 1971); Ole-Moi *et al.*, 1979). Moreover, it is necessary to show by appropriate fractionation studies with marker enzymes that the protease is localized within the cell at the site of cleavage.

Recently several families have been identified with high levels of circulating proinsulin (Gabbay *et al.*, 1976) or proalbumin (Brennan and Carrell, 1978). Evidently the affected individuals carry point mutations in sites of cleavage which result in the replacement of one of the basic residues by a neutral residue thereby lowering their susceptibility to cleavage (Brennan and Carrell, 1978). The proinsulin defect appears to involve only the cleavage site between the B chain and C-peptide (Gabbay *et al.*, 1979).

It is of interest that additional cleavages occasionally occur within precursors which clearly must be mediated by proteases other than those having trypsin-like or CPase B-like specificities. Thus in the rat, and several other species as well, about 20 percent of the proinsulin undergoes cleavage at a chymotrypsin-sensitive site located about two-thirds of the way along the C-peptide chain (Tager *et al.*, 1973), and the fragments thus generated are secreted along with the intact C peptide. In man these and other degradation products of the C peptide may complicate the interpretation of plasma C peptide levels dependent on the reactivity of the particular antiserum used in immunoassays (Kuzuya *et al.*, 1978). These results suggest that precursor cleavage patterns are determined both by the conformational availability and sensitivity of susceptible sites in the substrate proteins, as well as by the presence or absence of special proteases in the condensing vacuoles.

Additional processing of ACTH to α-MSH takes place in the pars intermedia of the pituitary in some species. To accomplish this conversion a sequence of 4 basic residues (Lys-Lys-Arg-Arg) is excised from positions 15 to 18 of ACTH to yield α-MSH-Gly (residues 1–14 of ACTH) which is then N-acetylated and amidated with loss of the glycine residue (Scott *et al.*, 1973; Chrétien *et al.*, 1977), while the C-terminal half of ACTH (residues 19–39) is left behind. The latter fragment has been designated CLIP (Scott *et al.*, 1973). In this case cleavage could be accomplished by the combined action of the same trypsin-like and carboxypeptidase B-like enzymes that evidently mediate the cleavage of the ACTH/endorphin precursor into ACTH, γ-lipotropin, and β-endorphin (see Fig. 3). What is unclear, however, is how this cleavage is prevented in the corticotrophic cells of the anterior pituitory, which do not process ACTH further. Little is known about the mechanism of N-acetylation of α-MSH. Presumably this is catalyzed by a specific transacetylase confined to the cells of the pars intermedia. The mechanism of the C-terminal amidation of α-MSH, as well as of mellitin, appears to require the presence of an additional C-terminal residue of glycine in the respective precursor forms (Scott *et al.*, 1973; Suchanek *et al.*, 1978) which, according to recent results of D. Smyth and co-workers (personal communication), is then decarboxylated and oxidized to the amide.

B. Other Posttranslationally Cleaved Precursor Forms

As shown by the compilation of Table I at least three additional broad categories of precursor forms in addition to those discussed above can be discerned. The viral precursors are generally polyproteins which are fragmented to produce both the structural (capsular) and some of the enzymic components required for virus production (Butterworth, 1976). In some instances, cellular proteases are required to carry out some maturation cleavages necessary for full viral infectivity (Scheid and Choppin, 1975), but evidence now indicates that some of the viruses also encode specialized proteases of their own (Korant *et al.*, 1979) or, as in the interesting case of Sindbis virus, the polyprotein precursor may cleave itself during biosynthesis (Schlesinger *et al.*, 1979).

TABLE I

A Classification of Precursor Proteins and Selected Examples

  I. Presecretory proteins—Early, rapid cleavage in RER or bacterial membrane
     Includes most secreted proteins, with rare exceptions (ovalbumin)
     Transmembrane proteins—VSV glycoprotein, probably many others
 II. Proproteins—Slow intracellular cleavage in Golgi/granules
     Prohormones—proinsulin, proPTH, progastrin, proglucagon, proACTH/endorphin
     Serum proteins—proalbumin
III. Virus precursors—Cleavage coordinated with assembly
     Capsule polyproteins—polio, EMC, etc. (may utilize virus coded proteases)
     Filamentous phage coat proteins—f1, M13
 IV. Membrane-associated proproteins—Posttranslational cleavage accompanying
     insertion/assembly
     Chloroplast—ribulosebiphosphate carboxylase subunit
     Mitochondria—subunits of ATPase and cytochrome oxidase
     Plasma membrane—none described
  V. Precursors cleaved during or after secretion
     Procollagen—cleavage facilitates assembly of quaternary structure
     Protoxins—diptheria, cholera, promelletin
     Vitellogenins—reptiles, insects
     Zymogens—blood clotting factors, fibrinolysins, digestive enzymes

---

Except for the membrane glycoproteins of some of the more complex animal viruses, which are inserted via signal sequences (Lingappa *et al.,* 1978b), the viral polyprotein precursors are cleaved mainly posttranslationally, but somewhat more rapidly than prosecretory proteins such as proinsulin or proACTH/endorphin. The role of the viral precursor forms can be seen to be one of coordination and regulation of assembly, and thus the precursor sequences may serve mainly to mask assembly sites until they are exposed at appropriate times during assembly of the virus capsular structures.

The precursors of class IV in Table I are concerned with the generation of membrane-associated protein assemblies, and this rapidly enlarging group includes both multisubunit polyproteins as well as precursors of individual subunits of membrane-localized enzymes. The first of this group of precursors to be identified was the nuclear-coded small subunit of the chloroplast enzyme ribulosebisphosphate carboxylase (Highfield and Ellis, 1978; Dobberstein *et al.,* 1977). The precursor is synthesized on free ribosomes in the cytoplasm and is then processed by a protease in the chloroplast envelope during its partition from the cytosol into the chloroplast stroma. Likewise, α, β, and γ subunit precursors of the yeast mitochondrial F1 ATPase have been identified by Schatz and his co-workers (Maccecchini *et al.,* 1979). These proteins are cleaved as they are taken up by mitochondria. Even more remarkable was the discovery of Poyton and McKemmie (1979a,b) that the four nuclear-coded subunits IV–VII of yeast cytochrome oxidase are synthesized initially in the cytoplasm as a 55,000-dalton polyprotein which is processed after its uptake into mitochondria into the four subunits and a small amount (8000 daltons) of residual peptide material. Uptake of the cytosolic polyprotein, moreover, appears to stimulate the intramitochondrial synthesis of subunits I–III of the cytochrome oxidase complex, thus providing an important clue to the mode of nuclear control of intrinsic mitochondrial protein synthesis (Poyton and Kavangh, 1976).

Little is known regarding the structures of the precursor regions of these proteins or of the properties and localization of their processing enzymes. Haas and Heinrich (1978) have recently reported the identification of a neutral protease in the inner membrane of rat liver mitochondria and have suggested that this enzyme may function in the conversion of enzyme precursors. Further information on the properties and specificity of this protease is eagerly awaited. Likewise, little information is available regarding the mechanisms for uptake of these precursors, or of other cytoplasmically synthesized organellar components that are not processed proteolytically during their transfer across membranes. It has been postulated that uptake may require the presence of specific recognition or carrier sites on organellar membranes (Highfield and Ellis, 1978; Maccecchini *et al.*, 1979). Proteolysis, where it occurs, may assist transfer via triggering a conformational change (Highfield and Ellis, 1978; Wickner, 1979) which exposes regions of the peptide which may interact with the membrane. Alternatively, observations that the intact cytochrome oxidase precursor protein enters the mitochondrial inner membrane prior to cleavage (Poyton and McKemmie, 1979b) suggests that proteolysis may be more relevent to assembly, in analogy to the role of the polyprotein viral capsule precursors discussed earlier. However, to the extent that the additional sequences of the precursors contribute to recognition or binding sites for the initiation of transfer, via membrane receptors, both mechanisms may actually be operative in guiding the products to their final destinations and coordinating the assembly of the active enzyme complex. Indeed as postulated earlier (Steiner *et al.*, 1972), it seems likely that similar mechanisms will prove to be involved in the formation of many other kinds of membrane-associated proteins including hormone receptor–effector complexes, transport systems and enzymes in both plasma membranes as well as in organelles.

It is beyond the scope of this chapter to discuss precursors and zymogen forms which are proteolytically processed after secretion from their cells of origin. Some of the proteins in this rather large and heterogeneous group are indicated in Table I, and extensive reviews on many of these are readily available elsewhere (c.f. Reich, *et al.* 1975; Pappenheimer, 1977; Fessler and Fessler, 1978).

ACKNOWLEDGMENTS

The authors are endebted to Ms. Sally Parks, Ms. Adelaide Jaffe, and Ms. Lise McKean for expert assistance in the preparation of this manuscript. Studies from this laboratory have been supported by grants from the United States Public Health Service (AM 13914, AM 18347, AM 20595) and in part by the Lolly Coustan Memorial Fund and the Kroc Foundation. H.S.T. is the recipient of a United States Public Health Service Research Career Development Award.

REFERENCES

Albert, S. G. and Permutt, A. M. (1979). Proinsulin precursors in catfish pancreatic islets. *J. Biol. Chem.* **254**, 3483–3492.
Blobel, G., and Dobberstein, B. (1975a). Transfer of proteins across membranes. I. Presence of proteolytically processed and unprocessed nascent immunoglobulin light chains on membrane bound ribosomes of murine myeloma. *J. Cell Biol.* **67**, 835–851.

Blobel, G., and Dobberstein, B. (1975b). Transfer of proteins across membranes. II. Reconstitution of functional rough microsomes from heterologous components. *J. Cell Biol.* **67**, 852–862.

Blobel, G., and Sabatini, D. D. (1970). Controlled proteolysis of nascent polypeptides in rat liver cell fractions. I. Location of the polypeptides within ribosomes. *J. Cell Biol.* **45**, 130.

Blobel, G., and Sabatini, D. D. (1971). Ribosome–membrane interactions in eukaryotic cells. *Biomembranes* **2**, 193–195.

Blumberg, B. M., Nakamoto, T., and Kézdy, F. J. (1979). Kinetics of initiation of bacterial protein synthesis. *Proc. Natl. Acad. Sci. U.S.A.* **76**, 251–255.

Blundell, T., Dodson, G., Hodgkin, D., and Mercola, D. (1972). Insulin: The structure in the crystals and its reflection in chemistry and biology. *Adv. Protein Chem.* **26**, 279–402.

Boime, I., Szczesna, E., and Smith, D. (1977). Membrane-dependent cleavage of the human placental lactogen precursor to its native form in ascites cell-free extracts. *Eur. J. Biochem.* **73**, 515–520.

Brandenburg, D., and Wollmer, A. (1973). The effect of a non-peptide interchain cross-link on the reoxidation of reduced insulin. *Hoppe-Seyler's Z. Physiol. Chem.* **354**, 613–627.

Brennan, S. O., and Carrell, R. W. (1978). A circulating variant of human proalbumin. *Nature (London)* **274**, 908–909.

Burstein, Y., and Schechter, I. (1978). Primary structures of N-terminal extra peptide segments linked to the variable and constant regions of immunoglobulin light chain precursors: Implications on the organizations and controlled expression of immunoglobulin genes. *Biochemistry* **17**, 2392–2400.

Busse, W. D., Hansen, S. R., and Carpenter, F. H. (1974). Carbonylbis (L-methionyl)insulin. A proinsulin analog which is convertible to insulin. *J. Am. Chem. Soc.* **96**, 5949–5950.

Butterworth, B. E. (1976). Proteolytic processing of animal virus proteins. *Curr. Top. Microbiol. Immunol.* Springer Verlag, New York.

Chan, S. J., Keim, P., and Steiner, D. F. (1976). Cell-free synthesis of rat preproinsulins: Characterization and partial amino acid sequence determination. *Proc. Natl. Acad. Sci. U.S.A.* **73**, 1964–1968.

Chan, S. J., Patzelt, C., Duguid, J., Quinn, P., Labrecque, A., Noyes, B., Keim, P., Heinrikson, R. L., and Steiner, D. F. (1979). Precursors in the biosynthesis of insulin and other peptide hormones. *Miami Winter Symp.* **16**, 361–378.

Chance, R. E., Ellis, R. M., and Bromer, W. W. (1968). Porcine proinsulin: Characterization and amino acid sequence. *Science* **161**, 165–167.

Chang, C. N., Blobel, G., and Model, P. (1978). Detection of prokaryotic signal peptidase in an *Escherichia coli* membrane fraction: Endoproteolytic cleavage of nascent f1 precoat protein. *Proc. Natl. Acad. Sci. U.S.A.* **75**, 361–365.

Chou, P. Y., and Fasman, G. D. (1978). Prediction of protein conformation. *Annu. Rev. Biochem.* **47**, 251.

Chrétien, M., Seidah, N. G., Benjannet, S., Dragon, N., Routhier, R., Motomatsu, T., Crine, P., and Lis, M. (1977). A LPH precursor model: Recent developments concerning morphine-like substances. *Ann. N.Y. Acad. Sci.* **297**, 84–105.

Devillers-Thiery, A., Kindt, T., Scheele, G., and Blobel, G. (1975). Homology in amino-terminal sequence of precursors to pancreatic secretory proteins. *Proc. Natl. Acad. Sci. U.S.A.* **72**, 5016–5020.

DiRienzo, J. M., Nakamura, K., and Inouye, M. (1978). The outer membrane proteins of gram-negative bacteria: Biosynthesis, assembly and functions. *Annu. Rev. Biochem.* **47**, 481–532.

Dobberstein, B., Blobel, G., and Chua, N.-H. (1977). *In vitro* synthesis and processing of a putative precursor for the small subunit of ribulose-1,5-bisphosphate carboxylase of *Chlamydomonas reinhardtii*. *Proc. Natl. Acad. Sci. U.S.A.* **74**, 1082–1085.

Dorner, A. J., and Kempner, B. (1978). Conversion of pre-proparathyroid hormone to proparathyroid hormone by dog pancreatic microsomes. *Biochemistry* **17**, 5550–5555.

Eipper, B. A., Mains, R. E., and Guenzi, D. (1976). High molecular weight forms of adrenocorticotropic hormone are glycoproteins. *J. Biol. Chem.* **251**, 1421–1426.

Fessler, J. H. and Fessler, L. I. (1978). Biosynthesis of Procollagen. *Annu. Rev. Biochem.* **47**, 129–62.

Gabbay, K. H., Deluca, K., Fisher, J. N., Mako, M. E., and Rubenstein, A. H. (1976). Familial hyperproinsulinemia: An autosomal dominant defect. *N. Engl. J. Med.* **294**, 911–915.

Gabbay, K. H., Bergenstal, R. M., Wolff, J., Mako, M. E., and Rubenstein, A. H. (1979). Familial hyperproinsulinemia: Partial characterization of circulating proinsulin-like material. *Proc. Natl. Acad. Sci.* **76**, 2881–2885.

Gaye, P., and Gautron, J. P. (1977). Amino terminal sequences of the precursors of ovine caseins. *Biochem. Biophys. Res. Commun.* **79**, 903–910.

Gibson, R., Schlesinger, S., and Kornfield, S. (1979). The nonglycosylated glycoprotein of vesicular stomatitis virus is temperature-sensitive and undergoes intracellular aggregation at elevated temperatures. *J. Biol. Chem.* **254**, 3600–3607.

Goldberg, A. L., and Dice, J. F. (1974). Intracellular protein degradation in mammalian and bacterial cells. *Annu. Rev. Biochem.* **43**, 835–869.

Grant, P. T., Coombs, T. L., Thomas, N. W., and Sargent, J. R. (1971). The conversion of [^{14}C]proinsulin to insulin in isolated subcellular fractions of fish islet preparations. *Mem. Soc. Endocrinol.* **19**, 481–495.

Gregory, R., and Tracy, H. (1975). *In* "Gastrointestinal Hormones" (J. Thompson, ed.), pp. 13–24. Univ. of Texas Press, Austin.

Haas, R., and Heinrich, P. C. (1978). The localization of an intracellular membrane-bound proteinase from rat liver. *Eur. J. Biochem.* **91**, 171–178.

Habener, J. F., and Potts, J. T., Jr. (1979). Subcellular distributions of parathyroid hormone, hormonal precursors, and parathyroid secretory protein. *Endocrinology* **104**, 265–275.

Habener, J. F., Chang, H. T., and Potts, J. T., Jr. (1977). Enzymic processing of proparathyroid hormone by cell-free extracts of parathyroid glands. *Biochemistry* **16**, 3910–3917.

Habener, J. F., Rosenblatt, M., Kemper, B., Kronenberg, H. M., Rich, A., and Potts, J. T., Jr. (1978). Pre-proparathyroid hormone: Amino acid sequence, chemical synthesis, and some biological studies of the precursor region. *Proc. Natl. Acad. Sci. U.S.A.* **75**, 2616–2620.

Hamilton, J. W., Niall, H. D., Jacobs, J. W., Keutmann, H. J., Potts, J. T., Jr., and Cohn, D. V. (1974). Amino terminal sequence of bovine proparathyroid hormone (calcemic fraction A). *Proc. Natl. Acad. Sci. U.S.A.* **71**, 653.

Highfield, P. E., and Ellis, R. J. (1978). Synthesis and transport of the small subunit of chloroplast ribulosebisphosphate carboxylase. *Nature (London)* **271**, 420–424.

Howell, S. L. (1972). Role of ATP in the intracellular translocation of proinsulin and insulin in the rat pancreatic B cell. *Nature (London), New Biol.* **235**, 85–86.

Hughes, J., Smith, T. W., Kosterlitz, H. W., Fothergill, L. A., Morgan, B. D., and Morris, H. R. (1975). Identification of two related pentapeptides from the brain with potent opiate agonist activity. *Nature (London)* **258**, 577–579.

Inouye, H., and Beckwith, J. (1977). Synthesis and processing of an *E. coli* alkaline phosphatase precursor *in vitro*. *Proc. Natl. Acad. Sci. U.S.A.* **74**, 1440–1444.

Jackson, R. C., and Blobel, G. (1977). Post-translational cleavage of presecretory proteins with an extract of rough microsomes from dog pancreas containing signal peptidase activity. *Proc. Natl. Acad. Sci. U.S.A.* **74,** 5598–5602.

Jamieson, J. D., and Palade, G. E. (1967a). Intracellular transport of secretory proteins in pancreatic exocrine cell. I. Role of peripheral elements of Golgi complex. *J. Cell Biol.* **34,** 577–596.

Jamieson, J. D., and Palade, G. E. (1967b). Intracellular transport of secretory proteins in pancreatic exocrine cell. II. Transport to condensing vacuoles and zymogen granules. *J. Cell Biol.* **34,** 597–615.

Jamieson, J. D., and Palade, G. E. (1968). Intracellular transport of secretory proteins in the pancreatic exocrine cell. IV. Metabolic requirements. *J. Cell Biol.* **39,** 589–603.

Judah, J. D., and Nicholls, M. R. (1971). Biosynthesis of rat serum albumin. *Biochem. J.* **123,** 649–655.

Judah, J. D., and Quinn, P. S. (1978). Calcium ion-dependent vesicle fusion in the conversion of proalbumin to albumin. *Nature (London)* **271,** 384–385.

Kangawa, K., and Matsuo, H. (1979). α-Neoendorphin: A "big" leuenkephalin with potent opiate activity from porcine hypothalami. *Biochem. Biophys. Res. Commun.* **86,** 153–160.

Kaschnitz, R., and Kreil, G. (1978). Processing of prepromelittin by subcellular fractions from rat liver. *Biochem. Biophys. Res. Commun.* **83,** 901–907.

Katz, F. N., Rothman, J. E., Lingappa, V. R., Blobel, G., and Lodish, H. F. (1977). Membrane assembly *in vitro:* Synthesis, glycosylation, and asymmetric insertion of a transmembrane protein. *Proc. Natl. Acad. Sci. U.S.A.* **74,** 3278–3282.

Kemmler, W., Peterson, J. D., and Steiner, D. F. (1971). Studies on the conversion of proinsulin to insulin. I. Conversion *in vitro* with trypsin and carboxypeptidase B. *J. Biol. Chem.* **246,** 6786–6791.

Kemmler, W., Steiner, D. F., and Borg, J. (1973). Studies on the conversion of proinsulin to insulin. III. Studies *in vitro* with a crude secretion granule fraction isolated from islets of Langerhans. *J. Biol. Chem.* **248,** 4544–4551.

Kirschner, K., and Bisswanger, H. (1976). Multifunctional proteins. *Annu. Rev. Biochem.* **45,** 143–166.

Kleckner, N. (1977). Translocatable elements in procaryotes (review). *Cell* **11,** 11–23.

Korant, B. D. (1975). Regulation of animal virus replication by protein cleavage. *In* "Proteases and Biological Control" (E. Reich, D. B. Ripkin, and E. Shaw, eds.), pp. 621–644. Cold Spring Harbor Lab., Cold Spring Harbor, New York.

Korant, B. D., Chou, N., Lively, M., and Powers, J. (1979). Virus-specified protease in poliovirus infected HeLa cells. *Proc. Natl. Acad. Sci. U.S.A.* **76,** 2992–2995.

Kozak, M. (1978). How do eucaryotic ribosomes select initiation regions in messenger RNA? *Cell* **15,** 1109–1123.

Kreibich, G., Grenbenau, R., Mok, W., Pereyra, B., Rodiriguez-Boulan, E., Ulrich, B., and Sabatini, D. D. (1977). Two membrane proteins of rat liver microsomes related to ribosome binding. *Fed. Proc., Fed. Am. Soc. Exp. Biol.* **36,** 656.

Kuzuya, H., Blix, P. M., Horwitz, D. L., Rubenstein, A. H., Steiner, D. F., Faber, O. K., and Binder, C. (1978). Heterogeneity of circulating human C-peptide. *Diabetes* **27,** Suppl. 1, 184–191.

Leavitt, R., Schlesinger, S., and Kornfeld, S. (1977). Impaired intracellular migration and altered solubility of nonglycosylated glycoproteins of vesicular stomatitis virus and Sindbis virus. *J. Biol. Chem.* **252,** 9018–9023.

Lin, J. J. C., Kanazawa, H., Ozols, J., and Wu, H. C. (1978). An *Escherichia coli* mutant with an amino acid alteration within the signal sequence of outer membrane prolipoprotein. *Proc. Natl. Acad. Sci. U.S.A.* **75,** 4891–4895.

Lingappa, V. R., Devillers-Thiery, A., and Blobel, G. (1977). Nascent prehormones are intermediates in the biosynthesis of authentic bovine pituitary growth hormone and prolactin. *Proc. Natl. Acad. Sci. U.S.A.* **74,** 2432–2436.

Lingappa, V. A., Katz, F. N., Lodish, H. F., and Blobel, G. (1978a). A signal sequence for the insertion of a transmembrane glycoprotein. *J. Biol. Chem.* **253,** 8667–8670.

Lingappa, V. R., Lingappa, J. R., Prasad, R., Ebner, K. E., and Blobel, G. (1978b). Coupled cell-free synthesis, segregation, and core glycosylation of a secretory protein. *Proc. Natl. Acad. Sci. U.S.A.* **75,** 2338–2342.

Lingappa, V. R., Lingappa, J. R., and Blobel, G. (1979). Chicken ovalbumin contains an internal signal sequence. *Nature (London)* **281,** 117–121.

Lowry, P. J., Silman, R. E., Hope, J., and Scott, A. P. (1977). Structure and biosynthesis of peptides related to corticotrophins and B-melanotropins. *Ann. N.Y. Acad. Sci.* **297,** 49–60.

Maccecchini, M.-L., Rudin, Y., Blobel, G., and Schatz, G. (1979). Import of proteins into mitochondria: Precursor forms of the extramitochondrially made F1-ATPase subunits in yeast. *Proc. Natl. Acad. Sci. U.S.A.* **76,** 343–347.

Macgregor, R. R., Hamilton, J. W., Kent, G. N., Shofstall, R. E., and Cohn, D. V. (1979). The degradation of proparathormone and parathormone by parathyroid and liver cathepsin B. *J. Biol. Chem.* **254,** 4428–4433.

Mach, B., Faust, C., and Vassalli, P. (1973). Purification of 14 S messenger RNA of immunoglobulin light chain that codes for a possible light-chain precursor. *Proc. Natl. Acad. Sci. U.S.A.* **70,** 451–455.

Mains, R. E., and Eipper, B. A. (1976). Biosynthesis of adrenocorticotropic hormone in mouse pituitary cells. *J. Biol. Chem.* **251,** 4115–4120.

Mains, R. E., Eipper, B. A., and Ling, N. (1977). Common precursors to corticotrophins and endorphins. *Proc. Natl. Acad. Sci. U.S.A.* **74,** 3014–3018.

Mandel, G., and Wickner, W. (1979). Translational and post-translational cleavage of M13 procoat protein: Extracts of both the cytoplasmic and outer membranes of *E. coli* contain leader peptidase activity. *Proc. Natl. Acad. Sci. U.S.A.* **76,** 236–240.

Maurer, R. A., and McKean, D. J. (1978). Synthesis of preprolactin and conversion to prolactin in intact cells and a cell-free system (communication). *J. Biol. Chem.* **253,** 6315–6318.

McKean, D. J., and Maurer, R. A. (1978). Complete amino acid sequence of the precursor region of rat prolactin. *Biochemistry* **17,** 5215–5219.

Milstein, C., Brownlee, G. G., Harrison, T. M., and Mathews, M. B. (1972). A possible precursor of immunoglobulin light chains. *Nature (London), New Biol.* **239,** 117–120.

Nakanishi, S., Inoue, A., Kita, T., Nakamura, M., Chang, A. C. Y., Cohen, S. N., and Numa, S. (1979). Nucleotide sequence of cloned cDNA for bovine corticotropinlipotropin precursor. *Nature (London)* **278,** 423–427.

Noyes, B. E., Mevarech, M., Stein, R., and Agarwal, K. L. (1979). Detection and partial sequence analysis of gastrin mRNA by using an oligodeoxynucleotide probe. *Proc. Natl. Acad. Sci. U.S.A.* **76,** 1770–1774.

Ole-Moi, Y. O., Pinkus, G. S., Spragg, J., and Austen, K. F. (1979). Identification of human glandular kallidrein in the beta cell of the pancreas. *New Engl. J. Med.* **300,** 1289–1294.

Orci, L., Lambert, A. E., Kanazawa, Y., Amherdt, M., Rouiller, C., and Renold, A. E. (1971). Morphological and biochemical studies of B cells of fetal rat endocrine pancreas in organ culture. Evidence for proinsulin biosynthesis. *J. Cell Biol.* **50,** 565–582.

Palade, G. (1975). Intracellular aspects of the process of protein synthesis. *Science* **189,** 347–358.

Palmiter, R. D., Gagnon, J., and Walsh, K. A. (1978). Ovalbubmin: A secreted protein without a transient hydrophobic leader sequence. *Proc. Natl. Acad. Sci. U.S.A.* **75,** 94–98.

Pappenheimer, A. M. (1977). Diphtheria toxin. *Annu. Rev. Biochem.* **46,** 69–94.

Patzelt, C., Chan, S. J., Duguid, J., Hortin, G., Keim, P., Heinrikson, R. L., and Steiner, D. F. (1978a). Biosynthesis of polypeptide hormones in intact and cell-free systems. *In* "Regulatory Proteolytic Enzymes and Their Inhibitors" (S. Magnusson, *et al.,* eds.), pp. 69–78. Pergamon, Oxford.

Patzelt, C., Labrecque, A. D., Duguid, J. R., Carroll, R. J., Keim, P., Heinrikson, R. L., and Steiner, D. F. (1978b). Detection and kinetic behavior of preproinsulin in pancreatic islets. *Proc. Natl. Acad. Sci. U.S.A.* **75,** 1260–1264.

Patzelt, C., Tager, H. S., Carroll, R. J., and Steiner, D. F. (1979). Identification and processing of proglucagon in pancreatic islets. *Nature* **282,** 260–266.

Patzelt, C., Tager, H. S., Carroll, R. J., and Steiner, D. F. (1979). Identification of prosomatostatin in pancreatic islets, *Proc. Nat'l. Acad. Sci. U.S.A.* (In press).

Pelham, H. R. B., and Jackson, R. J. (1976). An efficient mRNA-dependent translation system from reticulocyte lysates. *Eur. J. Biochem.* **67,** 247–256.

Poyton, R. O., and Kavangh, J. (1976). Regulation of mitochondrial protein synthesis by cytoplasmic proteins. *Proc. Natl. Acad. Sci. U.S.A.* **73,** 3947–3951.

Poyton, R. O., and McKemmie, E. (1979a). A polyprotein precursor to all four cytoplasmically-translated subunits of cytochrome *c* oxidase from *Saccharomyces cervisiae. J. Biol. Chem.* **254** (in press).

Poyton, R. O., and McKemmie, E. (1979b). Post-translational processing and transport of the polyprotein precursor to subunits IV–VII of yeast cytochrome *c* oxidase. *J. Biol. Chem.* **254** (in press).

Quinn, P. S., and Judah, J. D. (1978). Calcium-dependent Golgi-vesicle fusion and cathepsin B in the conversion of proalbumin into albumin in rat liver. *Biochem. J.* **172,** 301–409.

Quinn, P. S., Ackerman, E., and Steiner, D. F. (1979). Detection of an N-terminal fragment of pre-placental lactogen due to microsomal processing. *Proc. Conf. Precursor Process. Biosynth. Proteins, 1979* (in press).

Roberts, B. E., and Paterson, B. M. (1973). Efficient translation of tobacco mosiac virus RNA and rabbit globin 9 S RNA in a cell-free system from commercial wheat germ. *Proc. Natl. Acad. Sci. U.S.A.* **70,** 2330–2334.

Roberts, J. L., and Herbert, E. (1977). Characterization of a common precursor to corticotropin and β-lipotropin: Cell free synthesis of the precursor and identification of corticotropin peptides in the molecule. *Proc. Natl. Acad. Sci. U.S.A.* **74,** 4286–4830.

Roberts, J. L., Phillips, M., Rosa, P. A., and Herbert, E. (1978). Steps involved in the processing of common precursor forms of adrnocorticotropin and endorphin in cultures of mouse pituitary cells. *Biochemistry* **17,** 3609–3618.

Russell, J. H., and Geller, D. M. (1975). The structure of rat proalbumin. *J. Biol. Chem.* **250,** 3409–3413.

Sabatini, D. D. and Kreibich, G. (1976). *In* "Enzymes of Biological Membranes" (A. Martinosi, ed.), Vol. 2, pp. 531–579. Plenum, New York.

Schechter, I. (1973). Biologically and chemically pure mRNA coding for a mouse immunoglobulin L-chain prepared with the aid of antibodies and immobilized oligothymidine. *Proc. Natl. Acad. Sci. U.S.A.* **70,** 2256–2260.

Scheid, A., and Choppin, P. W. (1975). Activity of cell fusion and infectivity by proteolytic cleavage of sendai virus glycoprotein. *In* "Proteases and Biological Control" (E. Reich, D. B. Rifkin, and E. Shaw, eds.), Vol. 2, pp. 645–660. Cold Spring Harbor Lab., Cold Spring Harbor, New York.

Schlesinger, M. J., Schmidt, F. F. G., and Aliperti, G. (1979). Proteolytic cleavage and binding of fatty acids in the processing of Sindbis virus (SbB) and vesicular stomatitis virus (VSV) glycoproteins. *Proc. Conf. Precursor Process. Biosynth. Proteins, 1979* (in press).

Scott, A. P., Ratcliffe, J. G., Rees, L. H., Bennett, H. P. J., Lowry, P. J., and McMartin, C. (1973). Pituitary peptide. *Nature (London), New Biol.* **244,** 65–67.

Seeburg, P., Shine, J., Martial, J. A., Baxter, J. D., and Goodman, H. M. (1977). Nucleotide sequence and amplification in bacteria of the structural gene for rat growth hormone. *Nature (London)* **270,** 486–494.

Shields, D., and Blobel, G. (1977). Cell-free synthesis of fish preproinsulin and processing by heterologous mammalian microsomal membranes. *Proc. Natl. Acad. Sci. U.S.A.* **74,** 2059–2063.

Shields, D., and Blobel, G. (1978). Efficient cleavage and segregation of nascent presecretory proteins in a reticulocyte lysate supplemented with microsomal membranes. *J. Biol. Chem.* **253,** 3753–3756.

Smith, R. E., and Farquhar, M. G. (1966). Lysosome function in the regulation of the secretory process in cells of the anterior pituitary gland. *J. Cell Biol.* **31,** 319–347.

Smith, R. E., and VanFrank, R. M. (1974). Substructural localization of an enzyme in β-cells of rat pancreas with the ability to convert proinsulin to insulin. *Endocrinology* **94,** A190.

Smith, W. P., Tai, P.-C., and Davis, B. D. (1978). Nascent peptide as sole attatchment of polysomes to membranes in bacteria. *Proc. Natl. Acad. Sci. U.S.A.* **75,** 814–817.

Sorenson, R. L., Shank, R. D., and Lindall, A. W. (1972). Effect of pH on conversion of proinsulin to insulin by a subcellular fraction of rat islets. *Proc. Soc. Exp. Biol. Med.* **139,** 652–655.

Steiner, D. F. (1967). Evidence for a precursor in the biosynthesis of insulin. *Trans. N.Y. Acad. Sci.* [2] **30,** 60–68.

Steiner, D. F. (1978). On the role of the proinsulin C peptide. *Diabetes* **27,** Suppl. 1, 145–148.

Steiner, D. F., and Clark, J. L. (1968). The spontaneous reoxidation of reduced beef and rat proinsulins. *Proc. Natl. Acad. Sci. U.S.A.* **60,** 622–629.

Steiner, D. F., and Oyer, P. E. (1967). The biosynthesis of insulin and a probable precursor of insulin by a human islet cell adenoma. *Proc. Natl. Acad. Sci. U.S.A.* **57,** 473–480.

Steiner, D. F., Cunningham, D. D., Spigelman, L., and Aten, B. (1967). Insulin biosynthesis: Evidence for a precursor. *Science* **157,** 697–700.

Steiner, D. F., Clark, J. L., Nolan, C., Rubenstein, A. H., Margoliash, E., Aten, B., and Oyer, P. E. (1969). Proinsulin and the biosynthesis of insulin. *Recent Prog. Horm. Res.* **25,** 207–282.

Steiner, D. F., Clark, J. L., Nolan, C., Rubenstein, A. H., Margoliash, E., Melani, F., and Oyer, P. E. (1970). The biosynthesis of insulin and some speculations regarding the pathogenesis of human diabetes. *In* "The Pathogenesis of Diabetes Mellitus" (E. Cerasi and R. Luft, eds.), pp. 123–132. Almqvist & Wiksell, Stockholm.

Steiner, D. F., Kemmler, W., Clark, J. L., Oyer, P. E., and Rubenstein, A. H. (1972). The biosynthesis of insulin. *In* "Handbook of Physiology" (D. F. Steiner and N. Freinkel, eds.), Sect. 7, Vol. I, pp. 175–198. Williams & Wilkins, Baltimore, Maryland.

Steiner, D. F., Kemmler, W., Tager, H. S., and Peterson, J. D. (1974). Proteolytic processing in the biosynthesis of insulin and other proteins. *Fed. Proc. Fed. Am. Soc. Exp. Biol.* **33,** 2105–2115.

Steiner, D. F., Kemmler, W., Tager, H. S., Rubenstein, A. H., Lernmark, Å., and Zühlke, H. (1975). Proteolytic mechanism in the biosynthesis of polypeptide hormones. In "Proteases and Biological Control" (E. Reich, D. Rifkin, and E. Shaw, eds.), pp. 531–549. Cold Spring Harbor Lab., Cold Spring Harbor, New York.

Steiner, D. F., Terris, S., Chan, S. J., and Rubenstein, A. H. (1976). Chemical and biological aspects of insulin and proinsulin. In "Insulin" (R. Luft, ed.), pp. 53–107. A. Lindgren & Søner AB, Stockholm.

Steiner, D. F., Duguid, J. R., Patzelt, C., Chan, S. J., Quinn, P., Lernmark, Å., and Hastings, R. (1979). New aspects of insulin biosynthesis. Int. Symp. Proinsulin, Insulin, C-Peptide, 1978. Excerpta Medica Amsterdam/Oxford, pp. 9–19.

Strauss, A. W., Bennett, C. A., Donohue, A. M., Rodkey, J. A., Boime, I., and Alberts, A. W. (1978). Conversion of rat pre-proalbumin to proalbumin in vitro by ascites membranes. Demonstration by NH₂-terminal sequence analysis. J. Biol. Chem. 253, 6270–6274.

Strauss, A. W., Zimmerman, M., Alberts, A., and Mumford, R. (1979). Processing of pre-proalbumin and preplacental lactogen. Proc. Conf. Precursor Process. Biosynth. Proteins, 1979 (in press).

Suchanek, G., Kreil, G., and Hermodson, M. A. (1978). Amino acid sequence of honeybee prepromellitin synthesized in vitro. Proc. Natl. Acad. Sci. U.S.A. 75, 701–704.

Sun, A. M., Lin, B. J., and Haist, R. E. (1973). Studies on the conversion of proinsulin to insulin in the isolated islets of Langerhans in the rat. Can. J. Physiol. Pharmacol. 51, 175–182.

Sutcliffe, J. G. (1978). Nucleotide sequence of the ampicillin resistance gene of E. coli plasmid pBR322. Proc. Natl. Acad. Sci. U.S.A. 75, 3737–3741.

Swan, D., Aviv, H., and Leder, P. (1972). Purification and properties of biologically active messenger RNA for a myeloma light chain. Proc. Natl. Acad. Sci. U.S.A. 69, 1967–1971.

Tager, H. S., and Steiner, D. F. (1973). Isolation of a glucagon-containing peptide: Primary structure of a possible fragment of proglucagon. Proc. Natl. Acad. Sci. U.S.A. 70, 2321–2325.

Tager, H. S., Emdin, S. O., Clark, J. L., and Steiner, D. F. (1973). Studies on the conversion of proinsulin to insulin. II. Evidence for a chymotrypsin-like cleavage in the connecting peptide region of insulin precursors in the rat. J. Biol. Chem. 248, 3476–3482.

Thibodeau, S. N., Lee, D. C., and Palmiter, R. D. (1978). Identical precursors for serum transferrin and egg white conalbumin. J. Biol. Chem. 253, 3771–3774.

Tonegawa, J., Maxam, A. M., Tizard, R., Bernard, O., and Gilbert, W. (1978). Sequence of a mouse germ-line gene for a variable region of an immunoglobulin light chain. Proc. Natl. Acad. Sci. U.S.A. 75, 1485–1489.

Villa-Komaroff, L., Efstratidia, A., Broome, S., Lomedico, P., Tizard, R., Naber, S. P., Chick, W. L., and Gilbert, W. (1978). A bacterial clone synthesizing proinsulin. Proc. Natl. Acad. Sci. U.S.A. 75, 3727.

Walter, P., Jackson, R. C., Marcus, M. M., Lingappa, V. R., and Blobel, G. (1979). Tryptic dissection and reconstitution of translocation activity for nascent presecretory proteins across microsomal membranes. Proc. Natl. Acad. Sci. U.S.A. 76, 1795–1799.

Warren, G., and Dobberstein, B. (1978). Protein transfer across microsomal membranes reassembled from separated membrane components. Nature (London) 273, 569–571.

Wickner, W. (1979). The assembly of proteins into biological membranes: The membrane trigger hypothesis. Annu. Rev. Biochem. 48, 23–45.

Yip, C. C. (1971). A bovine pancreatic enzyme catalyzing the conversion of proinsulin to insulin. *Proc. Natl. Acad. Sci. U.S.A.* **68**, 1312–1315.

Zimmerman, M., Ashe, B. M., Alberts, A. W., Pierzchala, P. A., and Mumford, R. A. (1979). Proteases present in membranes which process human preplacental lactogen. *Proc. Conf. Precursor Process. Biosynth. Proc. Proteins, 1979* (in press).

Zühlke, H., and Steiner, D. F. (1977). Metabolism of proinsulin and insulin in islets of Langerhans. *Ergeb. Exp. Med.* **28**, 53–61.

Zühlke, H., Steiner, D. F., Lernmark, Å., and Lipsey, C. (1976). Carboxypeptidase B-like and trypsin-like activities in isolated rat pancreatic islets. *Ciba Found. Symp.* **41** (new ser.), 183–195.

6

Principles of the Regulation of Enzyme Activity

Peter J. Roach

I. INTRODUCTION

Analysis of the functional significance of biological structures has long been an important aspect of biology. The last half-century, in fact, has provided so many advances in our knowledge of the structure of organisms, their cells, and cellular components that we must now assess the physical and chemical properties of individual molecular species in relation to physiological function. This chapter, then, will deal with the interpretation of the chemical properties of enzymes within the context of

203

the functioning of the cell or organism, and particularly, with the role of enzymes in the regulation of cellular processes. The discussion is founded on the assertion that evolution has fashioned the properties of individual proteins to be advantageous for the efficient functioning, and thus survival, of the cell or organism, and in this strict biological context, we will talk of the function or purpose of given physical and chemical properties of cellular components (Krebs, 1954; Pittendrigh, 1958; Mayr, 1961; Atkinson, 1970, 1977). Though our knowledge is, in many respects, still very immature, enzyme regulation is a flourishing theme in biochemistry so that limitations of space would not permit a complete, detailed survey of the subject. This chapter will attempt to respect its title, then, and concentrate on principles rather than an exhaustive coverage of this rapidly expanding field.

Implicit in the concept of enzyme regulation is the contention that enzymic properties, other than basic catalytic function, have been selected during evolution. Let us examine this statement. The cell contains hundreds of linked* chemical reactions that underlie its maintenance, growth, and function. It is by no means self-evident, from our knowledge of chemistry, that the order and coherence that characterize the life of the cell could emerge, even given appropriate and highly specific catalysts, without mechanisms for the coordination of the various reaction chains. Acceptance of such an intuitive, and perhaps *a posteriori,* argument for the existence of evolved molecular control mechanisms is possibly aided by our contemporary familiarity with numerous complex regulated devices. It should be added, though, that the concept of regulation is much older than this century, and that homeostatic mechanisms in physiology have been recognized for many years as biological control systems (Cannon, 1926, 1929).

Besides suggestive arguments, however, recent biochemical studies have provided many examples of the existence of highly specialized properties that are almost impossible to rationalize without invoking the evolution of regulatory function. Among the most definitive examples are the identification of allosteric ligand-binding sites, the discovery of converting enzymes that covalently modify other enzymes, and the elucidation of the complex mechanisms of the induction and repression of enzyme synthesis. These phenomena, which will be treated again later, involve a high degree of evolutionary specialization and, especially in the last two cases, genes other than those coding exclusively for catalytic function.

---

* Linked because nearly all metabolites participate in at least two reactions (formation and degradation), many in several reactions, and a smaller number in many reactions (see Section IV,A,1).

II. SOME BASIC CONCEPTS OF BIOCHEMICAL REGULATION

A. Metabolic Regimes and Metabolic Control

No cell has a completely fixed metabolic regime; the fluxes of reactants through the various reaction chains, the activities of enzymes, and even the set of enzymes present all change, to a greater or lesser extent, with time. The metabolic activity of the cell may be considered as a continuously self-adjusting response to the environment to permit the fulfillment of its evolved function. The mechanistic strategy of the response may differ from organism to organism, and this is, of course, the outcome of evolution at the level of the species. Some aspects of metabolism and its control are remarkably constant throughout nature, and others are extremely variable.

One generalization that will recur is that the efficient coupling of the catabolic sequences that furnish metabolic energy and intermediates with the biosynthetic and other energy-requiring processes must have been an important evolutionary pressure. *How* efficient a coupling is difficult to say. Systems will often be encountered that are very costly in terms of energy and whose value to the cell can best be rationalized in terms of improved regulatory characteristics. It is evident, though, that too ineffective a coupling between catabolism and anabolism is unlikely to have conferred selective advantages to an organism. This basic viewpoint tempers much current thinking about metabolic regulation.

Avoidance of apparently "wasteful" energy consumption is especially relevant to metabolic sequences that must be traversed in both directions depending upon the momentary metabolic needs. Often this involves simply the reversal of a thermodynamically reversible reaction; in some important instances, however, different reactions are invoked, and this can be understood as the need to ensure that a metabolic pathway is exergonic. For example, the glycolytic pathway involves two particularly exergonic, ATP-consuming reactions catalyzed by hexokinase (or glucokinase) and phosphofructokinase. The reverse pathway of gluconeogenesis involves hydrolysis of the sugar phosphates at these points by fructose-1,6-diphosphatase and glucose-6-phosphatase. On paper, then, metabolism abounds with enzyme couples (such as glucokinase and glucose-6-phosphatase) that could cycle between substrates and products with the consumption of ATP. One basic pressure in evolution must have been to suppress such cycles *unless* some metabolic advantage was derived from their operation. This point will emerge several times later.

The response of unicellular organisms to their surroundings is normally to maximize their rate of growth and cell division. In the absence of essential nutrients, either minimal maintenance is effected (for example, bacteria in the stationary phase) or else specialized survival programs are adopted as in the case of spore formation by some bacteria (Hanson *et al.*, 1970) or slug formation by slime molds (Newell, 1971). The cells of multicellular organisms, on the other hand, contribute in a much more complex and indirect way to the survival of the organism. Differentiation leads to cells competent in their basic maintenance and duplication, but also equipped for specific individual roles within the organism. The environment of the individual cell is no longer, in most cases, that of the whole organism; it represents not only a local nutritional status but contains also many intercellular control signals, all of which reflect indirectly the response of the whole organism to *its* environment.

How, then, do cells respond metabolically to their immediate surroundings? First, the environment may dictate which enzymes, and hence which metabolic pathways, are present. Bacteria, for example, are often very versatile in their ability to tailor both biosynthetic and catabolic patterns to the availability of nutrients, thus eliminating the synthesis of some unnecessary enzymes (Magasanik, 1976). Similar phenomena occur in higher organisms. Liver cells, for example, are able to induce the production of the appropriate metabolizing or detoxifying enzymes in response to the presence of various substances (Schimke, 1973; Gillette *et al.*, 1972). Hormones, also, may lead to radical alterations in the pattern of gene expression.

An enzyme, though, is characterized not only by its effective absence or presence but also by its concentration. Enzyme concentration depends on the balance between the rates of both synthesis and degradation, and so may involve controls at the level of transcription, translation, or proteolytic degradation. The molecular mechanisms regulating enzyme concentration are not well understood in all cases, although the observation of changes in various cell types in response to changing environmental conditions is well established. It is clear, then, that the enzyme level can be an important element in determining the characteristics of metabolism, in particular, fixing the maximum catalytic capacity for a given interconversion.

Finally, though, the built-in autoregulatory properties of individual enzymes must ensure the detailed balance and control of metabolism, whatever the prevailing extracellular environment and the actual enzyme complement. Such control is rapid and on the time scale of molecular diffusion within the cell. In this way, available nutrients are channeled through existing pathways to supply the energy and structural components necessary for cell function. Sometimes, and especially for higher organisms, the cell's environment may contain specific signals, for example, hormones, capable of altering the intrinsic properties of some enzymes, often through covalent modification. In this way, the regulatory requisites of the whole organism are imposed upon and integrated with the basic autonomous control mechanisms of the individual cell. This brief introductory survey should emphasize the importance of enzymes as units of regulation for the function of the cell.

B. Classifications of Enzymic Reactions

1. Regulatory and Nonregulatory Enzymes

It has become customary to distinguish regulatory from nonregulatory enzymes, a distinction that, at least at the present level of knowledge, has an experimental basis. It is reasonable, for example, to recognize the differences in sophistication between, say, a simple nonregulatory enzyme in a specialized biosynthetic pathway and regulatory enzymes such as phosphofructokinase (Bloxham and Lardy, 1973; Mansour, 1972) or bacterial glutamine synthetase (Stadtman and Ginsburg, 1974), which have particularly refined kinetic properties. Moreover, such "regulatory" enzymes are often located at seemingly strategic points in metabolism as, for instance, at the first committed steps of reaction sequences. However, this distinction, though often convenient, must be appreciated as quantitative rather than qualitative. So long as we accept that the results of natural selection are manifested at all hierarchical levels of organisms down to individual proteins, it follows that probably every enzyme will have precisely evolved specifications and will contribute to the efficiency of metabolic regulation. What may be true is that some enzymes have been subject to more severe selective pressures than others. Ultimately, though, efficient metabolism as a whole is what is important and an enzyme with complex control features in one cell type may have a less sophisticated counterpart in another. For example, the major point of control of bacterial glycogen synthesis is the formation of ADPglucose (Preiss, 1973); in higher organisms and even yeast, the regulation appears concentrated at the level of glycogen formation from, in this case, UDPglucose (Larner and Villar-Palasi, 1971).

2. Equilibrium and Nonequilibrium Reactions

A second type of classification of enzymic reactions, relevant to regulatory function, is of equilibrium and nonequilibrium reactions. When the potential "activity" (see below) of an enzyme is high relative to the other enzymes of a metabolic pathway, the reaction catalyzed will be close to apparent equilibrium; in other words, the ratios of concentrations between substrates and products are close to those predicted from equilibrium thermodynamics for a closed system. The *absolute* concentrations, though, are determined by the kinetic properties of the reaction chain. In the opposite situation of relatively low enzymic "activity," substrates and products will be displaced from equilibrium ratios. Since the effective catalytic activity of an enzyme in the cell depends on the combination of its intrinsic kinetic properties, the levels of all possible modifiers of activity and its concentration, whether or not the reaction catalyzed by an enzyme is close to equilibrium is not just a property of that enzyme but of the whole of metabolism. Furthermore, the effective activity of an enzyme may vary with time, since neither enzyme nor modifier concentrations need remain constant and covalent modification can alter the intrinsic kinetic properties of an enzyme.

So-called equilibrium enzymes, then, have usually been regarded as unlikely sites for regulation per se, since massive changes in kinetic properties would be required to influence the flux through the reaction chain. A role for such enzymes in regulation has, however, been suggested by Krebs (1969) in that they determine the concentration ratios between certain metabolites which in their turn may act as important control signals. Nonequilibrium enzymes, on the other hand, contribute directly to determining metabolic fluxes and are potential sites of direct regulation. This leads to a very important concept in discussions of the metabolic process, that of rate-determining reactions or, by association, rate-determining enzymes. Unfortunately, some treatments of this topic are misleading, as explained below.

3. Rate-Limiting Steps

In studies of chemical reaction mechanisms, it is often useful to recognize the rate-determining step of a reaction chain (see, for example, Jencks, 1969). Consider the following sequence of two first-order reactions.

$$A \xrightarrow{k_1} B \xrightarrow{k_2} C$$

The appearance of C is given by the expression (see Laidler, 1965):

$$c = \frac{a_0}{(k_2 - k_1)} [k_2(1 - e^{-k_1 t}) - k_1(1 - e^{-k_2 t})] \tag{1}$$

where a_0 is the initial concentration of A, and c the concentration of C. If one reaction is inherently much slower than the other ($k_1 \ll k_2$, or $k_2 \ll k_1$), that reaction may be unequivocally identified as rate limiting. The rate of formation (v) of the product, C, then becomes independent of the inherent kinetics of the non-rate-determining step ($\partial v/\partial k \to 0$). This may easily be checked by letting either k_1 or k_2 tend to large values in Eq. (1). If, however, the rate constants of the two reactions do not differ greatly, there is no unique rate-determining step (see Jencks, 1969). Algebraically, this means that in the expression for the rate of appearance of C both kinetic constants are significant. Hypothetical changes in either k_1 or k_2 would alter the reaction rate (neither $\partial v/\partial k_1$ nor $\partial v/\partial k_2$ equals zero). It is true that the rate might be more sensitive to variation in the kinetics of one stage than the other; this could be evaluated by the derivative $\partial v/\partial k$. The same is true if a steady state is assumed for B. For this discussion, though, the critical point is to appreciate that the flux through a reaction pathway may depend on the kinetics of *more than one* individual step.

In passing to the complex reaction chains of metabolism, little of the above argument is modified in principle for enzymes operating below saturation by substrates. Statements quite commonly found to the effect that "it is of course a truism to say that every metabolic pathway can and must have only one rate-limiting step" are simply unfounded unless the system involves a saturated enzyme (see below). What *is* true generally is that certain reactions, and hence enzymes, will exert quantitatively more profound influences on metabolic fluxes than others, and these will be the nonequilibrium enzymes of the discussion above.

For example, in the glycolytic pathway, measurement of metabolite levels in several tissues indicates that only the reactions catalyzed by phosphoglucomutase, phosphoglyceroisomerase, and enolase are clearly close to equilibrium (Newsholme and Start, 1973). Therefore, several of the enzymes are "nonequilibrium" as defined above, and thus might be expected to influence the rate of glycolysis. It often happens that a metabolic pathway is dominated by the properties of a single enzyme; this, however, is not the consequence of general *a priori* principles but of evolutionary design.

The only situation in which a single reaction of a simple linear pathway is rate-limiting *by definition* is when the corresponding enzyme is operating at its maximal velocity. Thus, the smallest V_{max} of a sequence does determine the maximum possible flux, just as the slowest ship of the convoy limits the speed. The extent to which metabolism is controlled by enzymes working at saturation is not known, but this is probably not a general occurrence. Certainly much of the sophisticated design of regulatory enzymes seems directed to operation below saturation. Returning to glycolysis, for example, the glycolytic rates in normal and even some abnormal conditions are less than the lowest V_{max}'s of the individual enzymes of the pathway (hexokinase, phosphofructokinase, or aldolase) by severalfold in most cases (Newsholme and Start, 1973; Larner, 1971).

Some confusion in this area may have arisen from too imprecise a use of the term "activity." Rate-limiting metabolic steps are often stated to be associated with low enzyme "activity" which is usually based on the V_{max}. In fact, unless we are treating a system limited by a saturated enzyme, the appropriate activity measure to judge rate-determining steps is some analogue of the kinetic constants of the chemical scheme above. For a Michaelis–Menten enzyme, one approximation is V_{max}/K_m. In general, V_{max} is only *one* parameter contributing to the kinetic criteria relevant to evaluating the degree of rate limitation exerted by a given enzyme on a metabolic pathway.

A further point concerning the rate-determining metabolic steps is that the factors determining effective enzyme properties in the cell are subject to considerable variation. Therefore, the extent to which a given enzyme contributes to determining metabolic fluxes may likewise vary with time. In other words, the set of enzymes that most influences the regulation of metabolism may change with cellular conditions.

4. Reversible and Irreversible Reactions

The final division of enzymic reactions that will be briefly discussed is between reactions that are exergonic or endergonic in the normal metabolic direction. From a general metabolic point of view, the principal thermodynamic constraint is that the overall conversion of cellular nutrients to products be exergonic; life does not contravene thermodynamics. Through metabolism, though, this globally exergonic process is broken down into a large number of smaller steps, few of which, if any, have standard free energies outside the range of, perhaps, -10 to $+10$ kcal/mole. The evolution of various coupling factors, such as the ATP/ADP/AMP system, has contributed to keeping standard free energy changes within these limits.

A priori, the standard free energy change associated with a biochemical reaction does not dictate the direction of the metabolic flux within the cell, since evolved kinetic factors may lead to metabolite concentrations that favor the reaction in the opposite sense of its standard ΔG^0. Often quoted examples are malate dehydrogenase and aldolase. Nonetheless, most biochemical reactions do proceed in the direction of standard free energy decrease, as must be the case if metabolism is as a whole exergonic. The thermodynamics of a reaction can of course determine completely the concentration ratios of metabolites linked by an equilibrium enzyme as defined above. In such cases, extremely exergonic or endergonic reactions could lead to correspondingly large or small metabolite concentrations ($\Delta G^0 = -10$ kcal/mole corresponds to $K_{eq} = 1.2 \times 10^8$; for a reaction A $\rightleftharpoons$ B, this means a difference in equilibrium concentrations of

1.2×10^8 between A and B), and this will have had significant consequences during evolution (see Atkinson, 1969).

Quite frequently, regulatory enzymes are associated with irreversible reactions, but this does not seem a strong generalization. Some irreversible reactions are catalyzed by enzymes that do not appear to have especially refined regulatory roles, as for example the reduction of pyruvate to lactate by lactate dehydrogenase ($\Delta G^0 = -6$ kcal/mole). On the other hand, several thermodynamically reversible reactions are associated with regulatory enzymes and an effective unidirectional catalysis. A notable example is glycogen phosphorylase.

C. Signals, Sensitivity, and Time Scales

1. Signals

Any self-regulating device is dependent on some form of signal, carrying information about the state of the system, that can be translated into appropriate regulatory action. As an example, the "cruise-control" mechanism fitted on some automobiles will be taken: the maintenance of speed by such a device relies on the appropriate response of the fuel supplying systems to a signal bearing information about the road speed (Fig. 1). There is a specially designed interaction between the active element, accelerator pedal–carburetor, and the signal, which in this case is electrical. Note that the signal has a functional significance beyond its own physical nature that can only be expressed in the context of this control system: electrical signals in other devices will usually have nothing to do with road velocities. The type of regulatory communication and control action described above is, in principle, of the same nature as must occur in the cell, except that the latter is of an exceptional level of complexity by comparison. There is one basic difference between such a system and the biochemical analogue, however. The cruise-control setting is determined by the driver, or perhaps indirectly by the federal government; identification of the analogous constraints in a cellular system is more difficult. Various environmental conditions, such as nutrient levels and intercellular signals, will influence the metabolic regime, but at least, as yet, the interpretation of what constitutes the "setting" of a metabolic regulatory system is only possible in a very qualitative sense.

The basis of nearly all biochemical processes is the specific interaction of two or more molecules. Enzyme catalysis involves the binding of substrate molecules to the active site of the enzyme; hormone action requires the recognition by the hormone of specific receptor sites and so on. The interactions between proteins and small molecules appear to be especially

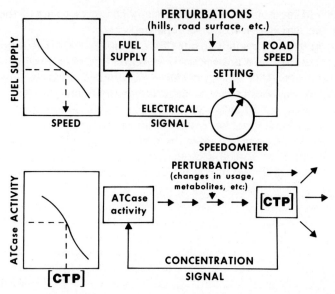

Fig. 1. Examples of control circuits. The upper portion shows a schematized circuit for an automobile "cruise-control" system whereby the supply of fuel is subject to negative feed-back by the road speed. Various external factors (such as the incline of the road) will modify the rate of fuel supply needed to maintain a fixed speed. The lower portion is a very schematic model of CTP synthesis incorporating the biochemical negative feedback loop exerted by CTP concentration on an enzyme regulating its own rate of synthesis, aspartate transcarbamylase (ATCase). For both examples, the panels on the left represent the de-signed ("cruise-control": Detroit; ATCase response: evolution) response between the con-trol element and the signal.

important and are certainly the best studied. Interactions between other classes of cellular components are undoubtedly significant in many in-stances, as for example, between proteins and other proteins, nucleic acids, or lipids. For the analysis of the control patterns of cellular chemis-try, these interactions are the basis of regulatory communication analo-gous to that between the fuel supply system and the signal indicating road speed. In many cases, proteins may be recognized as the active, func-tional elements, and those compounds that interact with proteins, for this discussion often metabolites, as the signals. The protein reacts to a pre-vailing set of signal levels with a particular, evolved response.

Let us take the example of a culture of *Escherichia coli* growing on lactose; the disaccharide may obviously be considered as a nutrient (rather than a metabolite in this case), a source of carbon and energy. In the analysis of the complex regulatory system by which lactose governs the synthesis of those enzymes necessary for its utilization, lactose, or in fact a derived metabolite, may additionally be seen as part of the signal chain for the expression of the *lac* operon.

Indeed, in some cases, substances have been discovered that appear to have no other role than as mediators of regulatory phenomena.* A good example is cyclic AMP. Metabolically, the synthesis and degradation of cyclic AMP would seem a rather elaborate and energetically wasteful way of hydrolyzing ATP to AMP plus PP_i; the cyclic AMP level, though, carries much information, largely of hormonal origin in higher organisms, that is expressed through its interaction with cyclic AMP-dependent protein kinase (Section III,A,3). The hormones themselves are another class of compounds whose existence appears exclusively related to bearing information. Biochemical signals, then, are often associated with the concentrations of small molecules, but this is by no means obligatory. Strictly, some peptide hormones would not usually be classed as small molecules. The control of gene expression by steroid hormones is thought to involve the interaction of the hormone–receptor complex with the genome. For the process of transcription, the receptor, a protein, is clearly acting as a signal. In fact, identifying only certain elements of the cell as being associated with regulatory signals is probably wrong and is just a convenient classification that is limited by our knowledge of, and ability to understand, the cell. The concentrations or states of many cellular components probably serve in some way as signals contributing to the regulation of metabolic and other processes.

2. Response of Regulated Element

To be effective, a signal must not only carry information but must also modify some property of the element with which it interacts. In an enzymic example, the reaction velocity may be a function of the concentrations of substrates and other modifiers of activity (effectors); the extent of binding of these compounds is related, in a more or less complex way mechanistically, to the functional activity. More generally, the degree of biological activity of cellular components is often determined through the formation of binary or higher-order complexes whose concentrations are in turn dependent on the concentrations of signal metabolites. The mechanisms of such control of enzyme activity are discussed later (Section III). Empirically, a curve of enzyme activity versus effector concentration may be viewed, for the purpose of analyzing regulatory function, as the response curve of a controlling element to a signal, irrespective of the under-

---

* Tomkins (1975) made a special classification of such compounds which he termed "symbols" associated with a "metabolic code." I would emphasize the possible "symbolic" significance of many cellular components for regulation, although some compounds such as cyclic AMP or ppGpp can be viewed as a definable class (see Section IV,A,1).

lying mechanism (Atkinson and Walton, 1965; Stadtman, 1970). For such curves, it is useful to consider the sensitivity of the response to the signal (see also Fig. 7 and Section III), and this is of course related to the gradient of the response curve. If an enzyme is sensitive to a certain metabolite concentration, small changes in the level of the latter will cause large changes in reaction rate. In general, one might anticipate a finer regulation to be correlated with greater sensitivity to signals; if the response is unchanged by a signal, no regulation is possible.

The analogy with the sensitivity of nonbiological control systems is qualitatively clear but is difficult to quantitate. If v and x represent rate (response) and effector (signal) concentration, respectively, sensitivity as defined here is dv/dx, that is, absolute rate change for absolute change in effector concentration. In some cases, though, it might seem more reasonable to use fractional changes in defining sensitivity, such as $dv/d \ln (x)$. The question remains, what is more indicative or relevant for the cell, fractional or absolute changes in rates and concentrations? Occasionally, I will distinguish absolute sensitivity (dv/dx) from what will be termed relative sensitivity $dv/d \ln (x)$, but I do not wish to be polemic about their relative importance (for further discussion of definitions of sensitivity and amplification, see Stadtman and Chock, 1978; Savageau, 1976; Newsholme and Crabtree, 1978).

3. Specificity

It is worth emphasizing again how the specificity of interactions is central to regulatory communication in biochemical systems. Nonspecific interactions would constitute a form of background noise. Chemically, specificity will correlate with a high affinity between target and signal compounds relative to all other potential binding species. This is illustrated, for an equilibrium system, in Fig. 2. What level of preferential binding is necessary to yield a useful signal–target couple is hard to judge; apparent K_m's of enzymes for substrates are mostly in the range micromolar to millimolar, which may be a crude indication of the range of dissociation constants that have been evolved between proteins and small ligands. Some systems of much higher affinity have also been described: hormone–receptor interactions are good examples here. Examples of weak but meaningful interactions by their nature probably elude present day experimentation. Of course, the extent of interaction (that is, the concentration of the signal–target complex) depends on concentration as well as affinity. Simple binding by a ligand to a protein may be described by

$$PX/P_T = X/(X + K) \doteq 1/(1 + K/X) \tag{2}$$

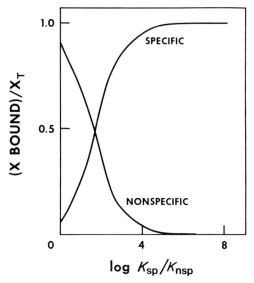

Fig. 2. Specificity of binding as a function of affinity. The model system comprises 20 protein species and a ligand, X. The concentration (bound plus free) of each species, as well as the ligand, is 1 arbitrary unit. One protein is identified as a potential target for ligand binding and has an affinity K_{sp} for X. The other 19 proteins, which will represent nonspecific binding in the model, have each an affinity $K_{nsp} = 1$ for X. The curves represent, then, the proportion of X bound specifically to the target protein or else nonspecifically to the interfering proteins as a function of the ratio of affinities [precisely $\log(K_{sp}/K_{nsp})$]. Note that K_{sp} and K_{nsp} are *association* constants here. After Roach (1976).

where PX, P_T, and X are the concentrations of bound ligand, total protein, and total ligand, respectively, and K is the dissociation constant. The ligand is assumed to be present in large excess over the protein. The degree of saturation of the protein by the ligand X is clearly determined only by the ratio of the ligand to the dissociation constant. In other words, many aspects of a system with high affinity and low ligand concentration would be comparable to a lower affinity system with a correspondingly higher ligand level. Evolution will have operated on both affinities and concentrations, either simultaneously or successively.

4. Time Scales

A controlled device requires also that the time scales of regulatory processes be geared to the needs of the system. For example, the ability of *E. coli* to sense the presence of lactose in the gut and effect the expression of the lac operon must match the frequency of appearance of the disaccharide; otherwise this control mechanism could not have conferred survival advantage on the organism. Obviously, only biological control sys-

TABLE I

Time Scales of Some Biologically Relevant Events

Process or parameter	Value	Reference[a]
Maximum first-order rate constant ($\Delta G\ddagger = 0$)	10^{13} sec^{-1}	1
Diffusion-controlled second-order[b] rate constant for typical protein and small ligand association	$\sim 10^{8}$ M^{-1} sec^{-1}	1
Observed second-order rate constant for substrate and enzyme association	10^{7}–10^{8} M^{-1} sec^{-1}	1, 2
Observed second-order rate constant for protein–protein associations	$\sim 5 \times 10^{5}$ M^{-1} sec^{-1}	1
Time constants for protein[c] conformational changes	10^{-4}–10^{-1} sec	2
Diffusion times[d]		
$\quad 10^{-6}$ m (bacterial cell)	5–50 msec	
$\quad 20 \times 10^{-6}$ m (higher organism cell)	2–20 sec	
Enzyme turnover numbers	1–500 sec^{-1}	2
Turnover of adenine in bacterial adenylate pool (growing bacteria)	~ 40 sec	3
Turnover of ATP in bacterial adenylate pool	0.1–1 sec	3
Turnover of rat liver proteins	10 min–20 days (average 2–3 days)	4
Rate of synthesis of an average protein molecule		
$\quad$ Vertebrate	~ 5 min	5
$\quad$ Bacterial	~ 30 sec	
Time scales of hormone action		
$\quad$ Catecholamines, glucagon	seconds–minutes	6
$\quad$ Insulin, ACTH, cortisone	minutes–hours	
$\quad$ Testosterone, estrogens, GH, TSH	hours–days	
Propagation of action potential in nerves	0.5–50 m/sec	7
"Cycle time" for human blood[e]	0.2–1 min	
Cell doubling times		
$\quad$ Bacteria	20 min to hour	8
$\quad$ Amoeba	30–60 hr	
$\quad$ Sea urchin eggs (2-cell stage)	~ 1 hr	
$\quad$ Sea urchin eggs (200-cell stage)	2.5 hr	
$\quad$ Cultured animal cells	12–26 hr	
$\quad$ Mouse intestinal epithelium	13–25 hr	
$\quad$ Mature neurons	∞	

[a] Key to references: (1) Gutfreund (1972); (2) Hammes and Schimmel (1970); (3) Chapman and Atkinson (1977); (4) Schimke (1973); (5) Mahler and Cordes (1966); (6) Larner (1971); (7) Davson (1951); (8) Novikoff and Holtzman (1970).

[b] This depends on the diffusion coefficient assumed for the species.

[c] This does not intend small fluctuations in protein structure.

[d] t, taking the diffusion coefficient $D = 10^{-6}$ to 10^{-7} cm^2 sec^{-1} and using the expression $\overline{x^2} = 2Dt$.

[e] Based on a 5.6 liters blood volume and heart output of 5 to 25 liters/min.

tems that have met such criteria have survived. It is interesting, however, that biochemical regulatory mechanisms span an enormously wide range of time scales, from seconds to months, even years (Table I). While no very severe restriction on the time scale of regulatory events is thus evident, different mechanisms appear to have evolved to match different temporal requirements. The regulation of enzyme activity through ligand binding, for example, is relatively rapid, the rates of formation of protein–ligand complexes being of the order of 10^{-4} to 10^{-5} sec. Control of enzyme activity through changes in enzyme concentration occurs generally on the scale of minutes or longer. A number of hormonal effects are measured in days or even months. Other long-term controls, such as gene expression in development, must pose different types of design necessities. Such phenomena are not well understood at the molecular level, but it is tempting to imagine that very tight interactions are involved that will be impervious to the changes in the metabolic state occurring on a short time scale.

At least in a general way, there must be a trade-off in design characteristics between specificity and a rapid time scale of regulation. Greater affinity will correlate generally with an improved specificity, but will normally imply a slower reversal time, unless other more specialized mechanisms come into play. This is because, for a simple equilibrium,

$$A + B \underset{k_2}{\overset{k_1}{\rightleftharpoons}} AB$$

increases in the affinity imply an increased ratio k_1/k_2. Mechanistically, k_1 can be increased or k_2 decreased. There are ultimate physical limits on the maximum value of k_1, but in any case, high-affinity binding will tend to correlate with lower values for k_2. Therefore, the evolution of specificity and rapid reversibility for such simple binary systems must represent a compromise. The apparent conflict, however, is not insuperable if more complex mechanisms are involved, such as interaction of a third component whose binding could alter the value of k_2. The role of elongation factors EF-Tu and EF-Ts in prokaryotic protein synthesis may provide a related example (Fig. 3) (Haselkorn and Rothman-Denes, 1973; Lucas-Lenard and Beres, 1974). The complex EF-Tu–GDP has a dissociation constant of $3 \times 10^{-9} M$ (Miller and Weissbach, 1970), and the release of GDP from this complex could pose a kinetic barrier to the elongation cycle. However, exchange between GDP and GTP is accelerated by the presence of another elongation factor, EF-Ts, so that the formation of the ternary complex EF-Tu–EF-Ts–GDP may provide a mechanism to overcome a potential limitation on the system. The role of this type of negative cooperative effect (see Section III,A) in a ternary complex as a mecha-

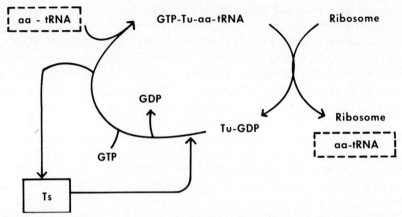

Fig. 3. Scheme of the elongation phase of prokaryotic protein synthesis. The diagram illustrates the cycle whereby an aminoacyl-tRNA (aa-tRNA) is transferred to the ribosome with the hydrolysis of a molecule of GTP. For the present discussion, the cyclic role of EF-Ts is the point of interest; binding of this component to EF-Tu–GDP destabilizes the binding of the nucleotide thus facilitating its release from EF-Tu. Otherwise, the very high stability of the EF-Tu–GDP complex could place a kinetic constraint on the process because of the very low k_{off} for GDP. As drawn, EF-Ts is released before interaction with the aminoacyl-tRNA; in fact the order of these events is not clearly established. See Haselkorn and Rothman-Denes (1973) and Lucas-Lenard and Beres (1974).

nism of rate facilitation has been stressed by G. Romero, V. Chau, and R. L. Biltonen (unpublished).

III. MECHANISMS OF CONTROLLING ENZYME ACTIVITY

A. Noncovalent Mechanisms

1. Direct Active Site Interactions and Cooperative Substrate Binding

a. Substrates. One of the simplest ways in which the modification of enzyme activity might be envisaged is through direct interactions at the active site of the enzyme. The number of natural substances capable *in vivo* of interacting specifically with the active site may be quite limited. Such compounds would, obviously, include substrates, products, possibly some compounds related chemically to these, and, in certain cases, some simple species such as metal ions.

Complete descriptions of enzyme kinetics are outside the scope of this chapter, and the reader is referred to many treatments of this topic (for example, see Gutfreund 1972, 1975; Cleland, 1970). Some fundamental points, however, are pertinent to discussions of regulation. First, two basic classes of enzyme kinetic behavior are usefully distinguished: what might be termed simple, hyperbolic kinetics and complex, cooperative kinetics (see later). The simple type, often called Michaelis–Menten kinetics, describes velocity–substrate concentration curves that are hyperbolic (see Fig. 7) and follow the rate law

$$v = e_T k_{cat} s/(K_m + s) \tag{3}$$

where the product of e_T, the total enzyme concentration, and k_{cat}, the intrinsic kinetic coefficient associated with the catalytic step, is commonly designated V_{max}. The substrate concentration, s, is assumed to be much greater than e_T.

The importance of protein–ligand interactions has been repeatedly stressed, and the chemistry of substrate binding is intimately involved in enzyme kinetic behavior. The simple ligand binding curve [Eq. (2)] is algebraically identical to a Michaelis–Menten rate equation, for example. Nonetheless, the catalytic act encompasses more than ligand binding, and so the description of the kinetic phenomenon may be correspondingly more complex. In particular, while the thermodynamics of substrate binding contributes to K_m, the K_m is not necessarily the measure of an equilibrium constant (see Dalziel, 1962; however, see also Atkinson, 1977), being a function of the reaction mechanism and the relative values of the rate constants for various individual steps. Thus, indiscriminate identification of K_m with K_{diss} is not valid *a priori* and reflects an expectation that the reaction of a simple enzyme–substrate complex to form products is unequivocally rate-limiting.

Having said this, however, the lack of chemical rigor in taking K_m measurements as a rough index of substrate affinity probably does not generate grave misconceptions in most discussions of metabolism, especially given the many imprecisions that must be tolerated anyway in this area. Generally, where it is preferable to avoid mechanistic connotations, one may adopt operational parameters, such as $S_{0.5}$ or $M_{0.5}$, the substrate or modifier concentration corresponding to half-maximal rate or effect (Koshland, 1970).

Two extremes of the simple kinetic behavior will be noted. First, at very low substrate concentration ($s \ll K_m$), the rate law tends to the simple first-order expression

$$v = e_T (k_{cat}/K_m) s \tag{4}$$

The rate is thus susceptible to hypothetical variation through changes in either of the intrinsic properties (K_m and k_{cat}) or through altered enzyme or substrate levels. The sensitivity of the reaction to substrate concentration is related to the quantity $e_T(k_{cat}/K_m)$.

The second extreme is of saturation ($s \gg K_m$) in which situation neither K_m nor the substrate concentration can readily influence the reaction rate

$$v = e_T k_{cat} \tag{5}$$

Control would have to be exerted through changes in the intrinsic catalytic capacity or in the enzyme concentration. There is, of course, a graded transition from the first to the second extreme with increasing substrate concentration, reduced sensitivity to K_m and substrate level following increasing saturation of the enzyme.

The kinetic properties of even "humble" Michaelis–Menten enzymes have most probably been subject to evolutionary design. It seems a fairly good generalization that evolution has led to substrate levels around or less than their apparent K_m's for the enzymes with which they interact (Atkinson *et al.*, 1975; Trevelyan, 1958). That is, most enzymes operate far from saturation [where Eq. (4) is not too incorrect for Michaelis–Menten enzymes]. This permits changes in the flux through a reaction by variations in substrate levels or substrate affinities, a situation not possible for an enzyme working at saturation. This is a very basic example of an expenditure of cellular resources in order to ensure metabolic and regulatory flexibility, in this case by the energetic outlay of using enzymes at only some fraction of their potential maximum catalytic activity. This may, in a certain sense, seem strange in view of the high degree of catalytic proficiency displayed generally by enzymes; based on very thorough kinetic studies of triosephosphate isomerase, Albery and Knowles (1976) have emphasized how close to theoretical perfection this enzyme, at least, has evolved as a catalyst. Why such refined properties if only a fraction of the catalytic capacity is to be utilized? An evolving cell has two principal means of increasing catalytic capacity: the evolution of a more efficient catalyst or increased enzyme concentration. The latter is an expenditure of energy for protein synthesis, and the former will be limited at some stage by the chemistry of the situation. However, accepting the metabolic benefits of running enzymes below maximum capacity, the evolution of enzymes of as high catalytic potential as possible is not at all contradictory since this permits savings in the number of enzyme molecules necessary to sustain a given reaction rate.

Another general evolutionary pressure on the design of enzyme kinetic properties is related directly to the limited solvent capacity of the cell. Atkinson (1969) has pointed out that the available cell water can maintain only a certain amount of material in solution and argued that the restriction of metabolite concentrations will have been a critical factor in evolution. The same pressures will have tended to increase the affinities of metabolites for enzymes. To allow a given reaction rate at a given substrate level, a high K_m would have to be offset either by increased enzyme concentration or increased k_{cat} [see Eq. (3)]. The latter, as noted above, has chemically determined limits, while the former itself leads to increased solute concentration. This is a general consideration, and just as some metabolites are present at relatively high levels, so some enzymes have high K_m's. An example is glucokinase, which has a K_m for glucose at about 10 mM, tuned as it were to the similarly high levels of glucose encountered by liver cells (Weinhouse, 1976; Walker, 1966).

   b. **Products.** A second class of simple active site effects is mediated by product binding. It should be added immediately that, from a purely chemical point of view, the enzyme sees no difference between substrates and products. The terms "substrate" and "product," with their implied directionality, are supplied by us, either because of the presumed physiological direction of the reaction or because of our design of kinetic assays.

   Product binding in the vicinity of the active site will almost always cause inhibition of the net forward reaction. The flux through any reaction is the difference between the forward and the back reaction rates; increased product levels, then, reduce the reaction flux simply by the mass action effect of promoting the backward reaction rate. In a simple closed system, a reaction will continue until the forward and backward rates are equal, that is, equilibrium is attained. Thermodynamics requires that the rate is zero at equilibrium and that the reaction always tends toward equilibrium. Note that the temporal nature of the approach to equilibrium is, of course, determined by kinetic factors and, in the case of enzyme-catalyzed reactions, there are enough kinetic variables to permit considerable flexibility in the relation between reaction rate and extent of reaction. Examples are known where the kinetic properties of the enzyme can render a thermodynamically reversible reaction effectively unidirectional (Sly and Stadtman, 1963; Ramos et al., 1967; LeJohn, 1968; Barnes et al., 1972).

   In the cell, the situation is further complicated because the reactants linked by a given enzyme will in general participate in other interconversions. Unlike what happens in a closed system in vitro, the enzyme will not usually determine completely the relative concentrations of products and substrates present. For multiproduct enzymes, then, it may happen that one product concentration is very low and cannot contribute to a substantial rate of the back reaction, but that another product is maintained at a sufficiently high level to bind to the enzyme and inhibit the forward rate. Experimentally, this is the situation studied when initial rates are measured in the presence of one (out of two or more) products. There may be a simple competition of product for substrate binding (no change in V_{max}) but the kinetic mechanism may lead to changes of both K_m and/or V_{max} (Cleland, 1970). Presumably in most simple cases evolutionary constraints will have operated against excessive inhibition and, in fact, where strong product inhibition is encountered, one may wonder whether any functional advantage derives from such properties. Product inhibition may sometimes be rationalized as a short feedback loop, but to what extent such inhibition is a selected response and not a concomitant of the evolution of catalytic activity is not generally clear.

In this context, it is interesting to compare the kinetic properties of hexokinase and glucokinase, both of which catalyze the phosphorylation of glucose to glucose 6-phosphate (glucose-6-P). Two striking differences in kinetic properties exist; glucokinase has a high apparent K_m for glucose [~10 mM (Weinhouse, 1976)] compared with hexokinase [~0.03 mM (Walker, 1966; Colowick, 1973)], and only hexokinase is subject to significant product inhibition by glucose-6-P. Thus, the design of two glucose-phosphorylating enzymes occurred with and without strong product inhibition. This leads to the suspicion that the glucose-6-P inhibition of hexokinase might have some functional significance. The two enzymes do seem to have somewhat different roles. Hexokinase is present in almost all cell types, whereas glucokinase is limited to the liver of some species, in which its high K_m value for glucose seems fitted to the purpose of responding to changes in the high levels of glucose that prevail in the portal vein. Glucokinase appears to function in the mass storage of blood glucose by the liver. Hexokinase, the sole glucose-phosphorylating agent in other tissues, is associated with the regulated utilization of blood glucose by peripheral cells, a process under a variety of hormonal and local metabolic controls.

Product inhibition may be a specially evolved feature in some other situations. A notable exception to the occurrence of substrates at levels around K_m values is provided by metabolites that act as "coupling agents," that is, metabolites that exist as two or more interconvertible forms in a relatively stable metabolic pool. Examples are the NAD$^+$/ NADH system and the ATP/ADP/AMP system (see Section IV,A,3). Such metabolites or cofactors enter into many reactions and are functionally quite distinct from simple metabolites that are involved in two or a few reactions. This distinction has been extensively discussed by Atkinson (1977). It also turns out to be a generalization that the apparent affinities of enzymes for such cofactors are usually high in comparison with their concentrations. Thus, enzymic sites for these cofactors will normally be saturated. For example, the nucleotide-binding site of a dehydrogenase will tend to be occupied by either NAD$^+$ or NADH, or of a kinase by ATP, ADP, or AMP. Frequently, the affinity for the "product" cofactor is greater than that for the "substrate" cofactor, another incidence of an apparent sacrifice of catalytic activity. Analysis of such systems, though, shows that strong product inhibition can lead to a useful response of enzyme activity to the composition of the cofactor pool (Fig. 4), namely, a sensitive variation of reaction rate with changes in that composition. Thus, the simple evolutionary manipulation of the relative affinities for the two cofactor forms can generate sensitive response curves. It should be noted that the above system provides an example where the more relevant cellular signal is probably the ratio of metabolite concentrations and not their absolute levels; the proportion of the saturated binding site bearing the substrate cofactor is a function of the mole fraction of that form of the cofactor.

The discussion in this section was initiated for enzymes with simple Michaelis–Menten kinetic behavior, but the general principles are equally relevant to enzymes with more complex properties. This latter class of enzymes will now be considered from the point of view of the dependence of reaction rate on substrate level; allosteric effects are treated later.

c. **Cooperativity.** The discovery of enzymes with more complex kinetic behavior and the correlation of such properties with enzyme structure have had far-reaching influence in biochemistry. An important historical example is aspartate transcarbamylase (Fig. 5) (Gerhart, 1970) whose substrate saturation curve is sigmoid instead of hyperbolic. Many en-

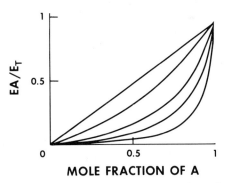

MOLE FRACTION OF A

Fig. 4. Response of enzymic rate to the composition of a cofactor pool. A simple Michaelis–Menten enzyme interacts with a cofactor pool (substrate, A, and product, B) and the proportion of EA complex, and hence v/V_{max}, is shown as a function of the mole fraction of A $\{[A]/([A] + [B])\}$. The pool of A + B saturates the enzyme, with the apparent dissociation constant for A fixed at 0.05 times the pool concentration. The different curves represent the effect of changing the apparent affinity for B (that is increasing the relative affinity for product compared with substrate) from 0.05, uppermost curve, to 0.025, 0.01, 0.005, and 0.0025 times the pool size. From Atkinson (1977).

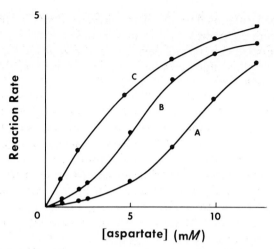

Fig. 5. Kinetic behavior of aspartate transcarbamylase. Curve B shows the sigmoid shape of the plot of enzymic activity against the concentration of the substrate, aspartate. The presence of CTP, 20 mM, causes inhibition and the response is shifted to that represented by curve A. Curve C shows the acquisition of simple Michaelis–Menten kinetics if the enzyme is treated with a mercurial. See Gerhart (1970).

zymes have since been found to have similar kinetics (see Table II) and most also have a multisubunit structure, so that each enzyme molecule contains more than one active site. The most frequently observed stoichiometries of subunits in enzymes are 2, 4, and 8 subunits per molecule (Klotz *et al.*, 1975).

A detailed presentation of the models proposed to account for cooperativity will not be made here, but some basic principles essential for the present discussion are noted. The most important rationalization of nonhyperbolic kinetic behavior is based on the concept of cooperativity, which may be defined operationally as the binding of ligand influencing the subsequent binding of more ligand. In the case of enzymes the multisubunit structure usually found for enzymes displaying "cooperative" kinetics is correlated with the binding of substrate at one site effectively* altering the affinity of other sites for the substrate. Most experimental data concern kinetic measurements and the extrapolation to ligand binding assumes that the reaction rate is simply proportional to the concentration of substrate bound to the enzyme. In a much smaller number of examples, however, binding itself has been measured.

Interpretations of the phenomenon are founded largely on the existence

---

* Note the distinction between the intrinsic affinity of a site for ligand and the effective, apparent affinity of a population of sites for ligand.

of multiple conformational states of the protein (Fig. 6). The first model proposed to treat cooperative substrate kinetics (Monod et al., 1965) postulated a preexisting equilibrium between two enzyme conformations, one of which (T_n) displays a greater affinity for the ligand than the other (R_n).

$$R_n \rightleftharpoons T_n \rightleftharpoons T_nS \rightleftharpoons T_nS_2 \quad \text{etc.}$$

The two species are assumed to be symmetrical, which implies that all n binding sites are of equal affinity in a given conformational state. In other words, there is an "all-or-none" transition between enzyme with low-affinity sites (R_n) and enzyme with high-affinity sites (T_n). Further, the equilibrium favors the low-affinity species in the absence of ligand. Binding of the ligand preferentially to the high affinity (T_n) form will thus increase the number of subunits with high affinity for ligand. Such a system would generate sigmoid substrate-binding curves similar in form to the kinetic curves in Fig. 5, since the greater the substrate concentration, the greater the concentration of high-affinity sites.

The model proposed by Monod, Wyman, and Changeux is notable for its essential simplicity. The later formulation of Koshland et al. (1966) in a sense exchanges simplicity for generality, basically by allowing for a greater number of distinguishable enzyme states and relinquishing the dictate of symmetry. In addition, there is no requirement for a preexisting equilibrium between enzyme forms. The enzyme can exist in states containing a mixture of different subunit forms (see Fig. 6). The distribution among the different enzymic species as a function of ligand concentration is described thermodynamically in terms of an intrinsic affinity of a given subunit form for the ligand, the equilibrium constant for the conversion of one subunit form to the other, and a term accounting for the variation in subunit interactions. In the simple example of Fig. 6, three such interactions (circle–circle, circle–square, and square–square) are possible. With more than two subunits, the system is more complex and will depend also on the geometric arrangement of the subunits. Analysis of such systems (see Koshland, 1970) can generate equations that predict cooperative binding.

Much effort has been expended in trying to establish which model is more accurate. Whether either need by universally true or untrue is of doubtful importance, certainly not for this discussion. The essential points here are (1) the considerable implications of cooperativity in metabolic control and (2) the concomitant alterations in the form of velocity–substrate curves.

Both of the formalisms above can account for sigmoid ligand-binding curves which correspond to "positive cooperativity," that is, effective affinity increasing with ligand binding. One point of difference between the

TABLE II

Some Enzymes with Cooperative Kinetics or Allosteric Effectors[a]

Enzyme	Substrate cooperativity	Effectors	References[b]
Phosphofructokinase	ATP [0] F6P [+]	High ATP (I) Citrate (I) AMP, ADP, FDP, P_i (A)	1
Pyruvate kinase (L type)	PEP [+] ADP [0]	FDP (A) ATP, Ala (I)	2
Glyceraldehyde-3-P dehydrogenase (muscle)	NAD$^+$ [−]		3
Pyruvate carboxylase			4
Mammalian	ATP, pyruvate, HCO$_3^-$ [0]	Acetyl-CoA (A) [+]	
Prokaryote	ATP, pyruvate, HCO$_3^-$ [0]	Acetyl-CoA (A) Asp (I) [+] or [−]	
PEP carboxylase			5
E. coli	PEP [0]	Acetyl-CoA (A) Asp (I)	
S. typhimurium	PEP [+]	Acetyl-CoA (A) FDP (I)	
Glycogen phosphorylase	P_i [0] Glycogen [0]	AMP (A) [+] ATP, G6P, UDPG, Glucose (I)	6
Glycogen synthase	UDPG [−] Glycogen [−]	G6P (A) [+] ATP, ADP, AMP, UDP (I)	7
Isocitrate dehydrogenase (NAD$^+$; yeast)	Isocitrate [+] NAD$^+$ [+]	AMP (A) [+]	8
ADPG pyrophos- phorylase	GIP [−] ATP [−]	FDP, NADPH, PLP (A) AMP (I)	9
Aspartokinase I	Asp [+]	Threonine (I) [+]	10
CTP synthetase	ATP [+] UTP [+]	GTP (A) [+]	11
Aspartate trans- carbamylase	Asp [+] Carbamyl P [+]	CTP (I) [−] ATP, dATP (A)	12
Glutamine PRPP amidotransferase	PRPP [+]	AMP, ADP, GMP, usually GDP, IMP (A) [+]	13
Deoxycytidylate deaminase	dCMP [+]	dTTP (I) dCTP (A)	14

[a] The table lists (by no means comprehensively) several enzymes that either have allosteric effectors or display cooperative kinetic behavior. Only in some cases has the cooperativity been verified by direct ligand binding studies. The inherent complexity of the enzymes makes the succinct compilation of such a table difficult since usually substrate and effector parameters are highly interdependent. Thus, only some general

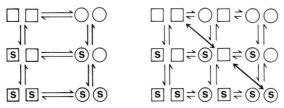

Fig. 6. Formulations of cooperative enzyme kinetic behavior. A dimeric enzyme is considered, whose subunits can exist in two states, designated by the squares or circles, with different affinity for a ligand. The proposal of Monod *et al.* (1965), on the left, requires that all (two in this case) subunits must exist in the same form in the enzyme molecule. The treatment of Koshland *et al.* (1966) allows the subunits individually to change from one form to another, thus introducing *a priori* a larger number of potential enzyme species. In the simplest case, progressive ligand binding would follow the diagonal.

two models, however, is that the opposite situation, decreased affinity with ligand binding, cannot be explained by the Monod–Wyman–Changeux model. This is called "negative cooperativity" (Levitski and Koshland, 1976) and has been documented in several instances, the best-studied probably being the binding of NAD^+ to muscle glyceraldehyde-3-P dehydrogenase (Levitski and Koshland, 1976; Harris and Waters, 1976). A phenomenon that may be related to negative cooperativity is "half-of-the-sites" activity (Lazdunski, 1972; Mathews and Bernhard, 1973; Levitski and Koshland, 1976), where only one-half of the total number of sites of an enzyme molecule are able to bind ligand. This could be rationalized as an extreme form of negative cooperativity (Levitski and Koshland, 1976), mediated through changes in subunit interactions. Alternatively, as proposed by Bernhard and his colleagues (MacQuarrie and Bernhard, 1971; Mathews and Bernhard, 1973), this could arise from preexisting asymmetry in the enzyme molecule. In this case, two types of site with intrinsically different properties would exist, and the phenome-

properties have been noted. Square brackets [ ] refer to cooperativity, of binding or kinetic behavior: + or −, formal positive or negative cooperativity; 0, hyperbolic. For effectors, A and I indicate activator and inhibitor respectively; however, realization of the effect may depend on the presence of other effectors. Most effectors operate on the $S_{0.5}$ for substrate or $M_{0.5}$ for other effectors.

[b] Key to references: (1) Mansour (1972), Bloxham and Lardy (1973), Ramaiah (1974); (2) Kayne (1973);   (3) Levitski and Koshland (1976), Harris and Waters (1976); (4) Scrutton and Young (1972), Utter *et al.* (1975);   (5) Utter and Kolenbrander (1972); (6) Fischer *et al.* (1970), Fischer *et al.* (1971);   (7) Larner and Villar-Palasi (1971), Roach and Larner (1977);   (8) Barnes *et al.* (1971), Kuehn *et al.* (1971);   (9) Preiss (1973);   (10) Truffa-Bachi (1973);   (11) Koshland and Levitski (1974);   (12) Gerhart (1970), Jacobson and Stark (1973);   (13) Wyngaarden (1972);   (14) Maley and Maley (1972).

non would not depend on changes in subunit interactions. For completeness, it is noted that some other mechanisms have been proposed to account for sigmoid velocity–substrate curves (Weber and Anderson, 1965; Rabin, 1967; Sweeney and Fisher, 1968; Griffin and Brand, 1968; see also Section III,A,3). Again, no model for cooperativity need be universal, since selection will have operated on function rather than the mechanism. Thus, we might anticipate any chemically feasible mechanism, providing it is associated with physiologically appropriate performance.

A parameter often found in discussions of cooperative kinetics is the Hill coefficient, m, which is the slope of the plot, for kinetic data, of log $[v/(V_{max} - v)]$ against log s. This derives from the equation

$$v/V_{max} = s^m/(K + s^m) \tag{6}$$

where K is the dissociation constant for $E + mS \rightleftharpoons ES_m$. Equation (6) would describe an infinitely cooperative system with m interacting sites, assuming that enzyme is at much lower concentration than substrate and that reaction rate is simply proportional to the concentration of bound substrate. On either the Monod or the Koshland model, this would imply that the only species existing in practice are free enzyme and completely saturated enzyme (m substrate molecules bound). Then, the Hill coefficient is equal to the number of binding sites. In practice, this is not normally true; the Hill plot need not be linear, and the Hill coefficient underestimates the real number of binding sites. Nonetheless, a Hill coefficient greater than unity *operationally* defines a positively cooperative situation, while a value less than 1 corresponds to formal negative cooperativity. For Michaelis–Menten kinetics, the Hill coefficient is 1.

d. **Functional Significance of Cooperativity.** From the viewpoint of regulation and enzyme design, what advantages derive from cooperative substrate kinetics? The most common answer centers on the increased sensitivity of the enzymic rate to substrate concentration. This may be expressed as the ratio of substrate concentrations leading to, say, 10 and 90% saturation, respectively; for simple, hyperbolic kinetics this ratio is 81, but can be much less for a positively cooperative enzyme (see Koshland, 1970). The point may also be seen from Fig. 7, where Michaelis–Menten and cooperative kinetics are compared. In fact, two sigmoid curves are shown. One (A) corresponds to an enzyme with the same inherent affinity for substrate, before substrate binding, as the simple enzyme. The other (B) shares with the Michaelis–Menten enzyme the same $S_{0.5}$ value, and thus initially has binding sites of lower affinity. From the point of view of evolutionary protein design, it is not obvious which is the more appropriate comparison. In any case, however, around the $S_{0.5}$ value for substrate, sigmoid curves are always more sensitive to variations in substrate concentration.

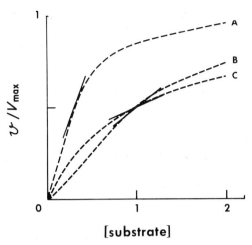

[substrate]

Fig. 7. Comparison of positive cooperative with Michaelis–Menten enzyme kinetics. The kinetic behavior of a simple Michaelis–Menten enzyme with $S_{0.5} = 1$ is shown for reference in curve C. The other curves correspond to an empirical positive cooperative behavior with a dimeric enzyme in which binding of substrate causes a tenfold increase in the affinity of the other site for substrate. In curve A, the initial affinity for substrate equals that of the simple Michaelis–Menten enzyme. In curve B, the initial affinity (0.316) is chosen so that the cooperative and simple enzymes share the same $S_{0.5}$. The tangents drawn at the $S_{0.5}$ values for the three hypothetical enzymes are to give an indication of the sensitivity of the reaction rate to changes in substrate concentration.

Although negative substrate cooperativity probably requires no less sophisticated protein design than positive cooperativity, the functional advantages of the former are less immediately obvious. Some possibilities, however, have been discussed (Levitski and Koshland, 1976). A negatively cooperative system has a reduced sensitivity to changes in substrate concentration compared with hyperbolic kinetics but a correspondingly wider range of response. The increased sensitivity of positive cooperativity is, incidentally, at the expense of having a narrower range of dependence on ligand concentration. The low sensitivity associated with negative cooperativity, then, might be viewed as a damping mechanism, a means to attenuate the effects of large fluctuations of ligand concentration on enzyme activity (Levitski and Koshland, 1976). In other words, enzyme sensitivities might be fitted to the magnitude of signal variations, large fluctuations being linked to negative cooperativity and small fluctuations with positive cooperativity. Conceivably circumstances exist where reduced sensitivity is beneficial although, if we jump ahead to think of allosteric effectors (Section III,A,2) as opposed to substrates, the most obvious value of negative cooperativity is in an allosteric system.

   e. Cyclical Logic. The rates of a vast number of metabolic reactions are sensitive, to greater or lesser extents, to variations in substrate and product concentrations, fundamentally through interactions at active sites. Conversely, the concentrations of most metabolites are sensitive to variations in enzymic activities. The necessarily cyclic nature of cause and effect arguments in discussing such a complex phenomenon as metabolism is, as Atkinson (1977) has stressed, an essential feature of the beast [and is related to the impossibility of deciding what sets metabolic control settings (Section II,C)]. It will often not be clear whether substrate levels control reaction rates or vice versa. Nonetheless, many regulatory connections can be identified and the analysis of the kinetic performance of individual enzymes gives us insight into the nature of metabolic control. The type of conceptual difficulty noted above may be responsible for slightly insipid concepts, such as the control of enzyme activity by substrate availability, sometimes invoked as a category of control mechanism. In a reasonable sense, the fluxes through all enzyme-catalyzed reactions not running at saturation and not catalyzed by so-called equilibrium enzymes are limited in rate by substrate availability, and this must include a very large number of enzymes.

2. Allosteric Interactions

a. **Basic Phenomenon.** The discovery of allosteric interactions has had a profound effect on our conception of metabolic regulation. The cooperative binding or kinetic properties described above derived from a dependence of effective ligand binding affinity on the degree of saturation of the protein by the same ligand. In an allosteric system, the extent of ligand binding may influence not only the affinity for the ligand itself (homotropic effect) but also the properties of binding by other ligands (heterotropic effect). In the most common situation, one of the sites involved is the active or substrate-binding site, and the effect of ligand (effector or modifier) bound at an allosteric site is to alter enzymic activity (Fig 5); either activation or inhibition may be observed (Table II). A large number of allosteric enzymes have now been recognized, and Table II lists several examples.

Both heterotropic and homotropic effects involve a dependence of binding affinities on the degree of ligand saturation, and this effectively means a communication of the fact of ligand binding to physically distinct sites. Allosteric effects are readily incorporated into the two formalisms for dealing with cooperativity noted above. On the Monod model, an effector would bind preferentially to one or the other form of the enzyme, thus promoting the presence of high (activitor) or low (inhibitor) affinity sites of the enzyme. On the Koshland scheme, inhibitor binding would stabilize enzyme species with lower effective affinity for the substrate, and activator binding, the opposite. That conformational changes accompanying ligand binding are actually transmitted through space to alter the properties of other binding sites directly is not required on either model, although this is not inconceivable. However, for the present discussion, it is not inexact to speak functionally of the binding of one ligand as modifying the effective affinity for another. Both positive and negative cooperativity are known for allosteric effectors (Table II), and some enzymes interestingly display both types of behavior. CTP synthetase (Koshland and Levitski, 1974) shows positive cooperativity toward the substrates ATP and UTP, but negative cooperativity for the binding of GTP, an activator. Aspartate transcarbamylase similarly exhibits negative cooperativity with CTP and positive cooperativity with aspartate (Jacobson and Stark, 1973; Gerhart, 1970).

b. **Nonidentical Subunits.** As for homotropic interactions, a multisubunit structure is usually associated with heterotropic interactions. In this latter case, however, an additional feature is sometimes observed; some enzymes are composed of more than one species of subunit, one of which bears the active site while allosteric sites are located on a different type of subunit [see also cyclic AMP-dependent protein kinase (Section III,A,3)]. Aspartate transcarbamylase again serves as an example; the *E. coli* enzyme (Jacobson and Stark, 1973; Gerhart, 1970) contains two types of subunit, R (regulatory) and C (catalytic), and exists in the stoichiometry R_6C_6. The C subunits have the binding sites for the substrates aspartate and carbamyl phosphate and the R subunits possess sites specific for the allosteric inhibitor CTP. In fact, abolition of kinetic cooperativity by chemical modification of the enzyme (desensitization) with the retention of catalytic activity (see Fig. 5) was important in formulating the concept of distinct inhibitor and active sites.

c. **Functional Significance of Allosteric Sites.** At nonsaturating substrate levels, curves of enzyme activity versus the concentration of an allosteric effector will usually be nonhyperbolic, as a result of homotropic and/or heterotropic effects. Thus, the same advantages in the design of enzyme–response curves are possible as in the case of cooperative substrate kinetics discussed earlier, namely, a greater flexibility in the relative sensitivity of activity to effector levels. In addition, the evolution of allosteric systems probably heralded a significant advance in the design of regulatory mechanisms, since it permitted the interaction of chemically unrelated substances to determine a reaction rate. In terms of regulation, the notion of "communication" between binding sites mentioned above is certainly much more than the gratuitous borrowing from the vocabulary of another discipline. It is to emphasize this basic mechanism of metabolic communication that heterotropic phenomena have been placed in a separate section from cooperative substrate kinetics. This distinction, which from a chemical point of view is not perhaps the most logical, has definite grounds for the discussion of metabolism: the design of response curves versus the integration of metabolic signals.

Allosteric enzymes may possess several ligand binding sites, as is evident from Table II; glutamine synthetase, an enzyme well-known in this respect, has more than ten effectors of likely physiological significance (see Table IV and Section III,B,2). Besides the type of direct effector–substrate interactions, such as between CTP and substrates in aspartate transcarbamylase, allosteric enzymes permit more subtle interrelations among different effectors. An example is provided by phosphofructokinase (Mansour, 1972; Bloxham and Lardy, 1973; Ramaiah, 1974). ATP, a substrate, inhibits at high concentrations, but its effect may be annulled by AMP. Glycogen synthase, in its phosphorylated form, is inhibited by several metabolites, including P_i, adenine nucleotides, and UDP, but their effects may be counteracted by the activator glucose-6-P (Larner and Villar-Palasi, 1971). Thus, some enzymes are highly sophisticated regulatory units that are capable of integrating numerous metabolic signals to determine a reaction rate. The rate will respect, in this way, the constraints imposed by the rest of metabolism, in this case, through metabolite levels and ligand-binding interactions.

3. Protein–Protein Associations and Other Interactions

a. **Changes in Subunit Aggregation.** Of course, the allosteric and cooperative types of behavior just discussed derive from interactions of polypeptide chains; this section will begin with some examples of protein–protein interactions that are usually placed outside the formalisms of allosteric systems even though there is an underlying chemical similarity.

Frieden (1971) has emphasized how effectors can modify the aggregation state of an enzyme and has suggested the possible physiological significance of this phenomenon. For example, GTP and GDP decrease the extent of aggregation of bovine liver glutamate dehydrogenase, while ADP increases it. In the presence of NADH, the binding of GTP follows a sigmoid curve whose exact shape depends on enzyme concentration. In general, if different multimers have different kinetic properties, then any compound whose binding differentially stabilizes a particular aggregation state will be an effector of the enzyme (compare with the Monod–Wyman–Changeux model above). Moreover, such systems may have non-hyperbolic kinetic behavior, that is, empirically cooperative kinetics. Whether or not such changes in the aggregation state of identical subunits are of physiological significance is not known with certainty.

However, current feeling would almost certainly agree that some types of modification of subunit composition are biologically relevant. Protein kinase from muscle probably exists as a tetramer of two regulatory (R) and two catalytic (C) subunits (Walsh and Krebs, 1973; Rubin and Rosen, 1975):

$$R_2C_2 + 2 \text{ cAMP} \rightleftharpoons R_2cAMP_2 + 2 \text{ C}$$

Upon binding of cyclic AMP, the complex dissociates and liberates free C. Since only the free catalytic subunit is enzymatically active, cyclic AMP activates this enzyme. The binding of cyclic AMP to one site ultimately influences the activity of the catalytic site and so the system is quite analogous to the allosteric enzymes discussed above. Note that cyclic AMP exerts a negative heterotropic effect upon the binding of R to C; cyclic AMP binding effectively destabilizes the association of R and C.

A somewhat similar system is found for RNA polymerase (Chamberlin, 1974a,b). The bacterial enzyme exists in two catalytically active forms, characterized by subunit structures $\sigma\beta\beta'\alpha_2$ (holoenzyme) and $\beta\beta'\alpha_2$ (core). Only the holoenzyme, however, has a high specificity in starting RNA synthesis at the correct promoter sites on the DNA molecule. The σ-subunit regulates, in conjunction with other protein factors, the initiation of RNA synthesis, and once begun, the σ factor is released.

Another nonenzymic regulatory protein is associated with mammalian lactose synthetase. Lactose synthetase A protein (Ebner, 1973; Hill and Brew, 1975) exists in the liver and mammary gland during pregnancy and is involved in glycoprotein biosynthesis, catalyzing the formation of N-acetyllactosamine from UDPgalactose and N-acetylglucosamine. However, after parturition, the mammary gland synthesizes a second protein, α-lactalbumin (B protein) which, upon association with the A protein, leads to an enzyme of modified catalytic properties. The lactose synthetase now catalyzes the synthesis of lactose from UDPgalactose and glucose.

An especially interesting example of a nonenzymic regulatory protein is provided by the Ca^{2+}-dependent modulator protein (Wang, 1977; Kretsinger, 1976), a small polypeptide of about 17,000 daltons. This species is found in many mammalian cell types and can interact, in the presence of Ca^{2+}, with several enzyme systems including adenylyl cyclase, phosphodiesterase, and myosin light chain kinase. The modulator protein binds Ca^{2+} and confers a Ca^{2+} activation to the enzyme systems with which it interacts. A positive heterotropic effect is formally exerted by Ca^{2+}, since its presence favors modulator protein binding. Recently, this polypeptide has been identified in muscle phosphorylase kinase, another important Ca^{2+}-requiring enzyme (Cohen et al., 1978).

Several other examples of this kind are known [see also the P_{II} protein

of the glutamine synthetase system (Section III,B,2)]. Two general points
are worth making. First, only relatively recently are specific functions
being found for a growing number of noncatalytic proteins—enzyme activity has long been the traditional marker for protein purification. Second,
for polypeptides such as the protein kinase regulatory subunit, σ factor,
the lactose synthetase B protein, and Ca^{2+}-dependent modulator protein,
the line between enzyme modifier and subunit becomes blurred, any distinction reflecting primarily the relative stability of subunit complexes
through purification procedures.

   b. **Multienzyme Complexes.** In several cases, complexes of proteins
have been isolated that contain several functionally related enzyme activities. What might be termed classic examples are the pyruvate dehydrogenase complex (Reed and Cox, 1970) and the "fatty acid synthetase"
complex (Volpe and Vagelos, 1973, 1976; Bloch and Vance, 1977). The
pyruvate dehydrogenase complex from E. coli, with a molecular weight
around 4×10^6, comprises three enzymatic activities: pyruvate dehydrogenase, dihydrolipoyl transacetylase, and dihydrolipoyl dehydrogenase,
which are responsible for the conversion of pyruvate plus CoA to acetyl-
CoA. A similar large complex has been isolated from mammalian tissues
(Reed, 1974), but in this case includes also protein kinase and phosphatase
activities (see Table IV). Likewise, stable enzyme complexes have been
obtained from yeast and from avian and mammalian tissues that catalyze
the synthesis of fatty acids from malonyl-CoA and acetyl-CoA. The complexes apparently contain the necessary enzymic activities, as well as a
carrier protein which was postulated to bear the growing fatty acid residue
as a thioester. While the original conception was of separate active sites
associated with distinct components, the complex is now believed to be
composed of multifunctional elements in which more than one active site
is associated with a single polypeptide chain. The current idea is that the
enzyme activities of eukaryotic fatty acid synthetase (type I) are distributed between only two distinct, multifunctional polypeptides (see Bloch
and Vance, 1977). Both of these types of complexes may be obtained as
fairly well-defined, although certainly very complicated, associations of
polypeptides with relatively fixed stoichiometries. Such assemblies seem
to represent units of structural organization beyond the individual enzyme. Functionally, some advantages of this metabolic arrangement may
be postulated (Reed and Cox, 1970; Srere and Mosbach, 1974; Kempner,
1975; Gaertner, 1978). For example, intermediates of no metabolic consequence other than in the pathway in question might be generated in close
proximity to the next enzyme in the sequence so that a low cellular concentration of the intermediate could be associated with an effectively high
local concentration. Further, protein–protein interactions throughout the

complex present, at least in theory, the possibility of an amazing number of subtle controls of the pathway.

To be known, a multienzyme complex must survive purification, to some degree. Indeed, reports range from the two examples cited to less defined and even to quite speculative complexes. Fischer and his colleagues (Meyer et al., 1970) have isolated from muscle a complex which contains all of the enzymes of glycogen metabolism associated with glycogen. In fact, several other enzymes are present in lesser amounts, and the complex can, in the presence of appropriate cofactors, catalyze the production of lactate from glycogen. The evidence strongly favors the existence of this entity, but, as isolated, it does not appear to have a well defined stoichiometry. A particularly interesting but still speculative complex that might be associated with the glycolytic pathway has been obtained from E. coli (Mowbray and Moses, 1976). Similarly, there is evidence for an association between enzymes of the pathway of "de novo" purine biosynthesis in pigeon liver (Rowe et al., 1978). How many enzyme–enzyme interactions exist within the cell is a fascinating question which cannot be answered at present. It should be appreciated, though, that the term "complex" in this context is related largely to stability during and after purification. Not all interactions that are strong intracellularly need be so in vitro, and weaker associations may themselves have functional roles in the cell.

c. **Multifunctional Enzymes.** Fatty acid synthetase exemplifies what could be considered an extreme as regards the stability of multienzyme complexes, namely, multiple enzyme activities residing in the same polypeptide chains (Table III) (Kirschner and Bisswanger, 1976; Stark, 1977). A few of these enzymes have been known for some time (glycogen debranching enzyme, tryptophan synthase, aspartokinase I–homoserine dehydrogenase I), but the number of examples is steadily growing. A major problem in such studies is the need to avoid even limited proteolysis, which can artificially generate species that are no longer multifunctional. With awareness of the existence of these systems, one might certainly anticipate that, through appropriate experimentation, new examples of multifunctional enzymes will continue to be detected.

Structurally, one current hypothesis is that different contiguous stretches of the polypeptide chain would fold to form somewhat autonomous, globular units associated with a given function. In other words, the continuity of the peptide chain would, from a simplistic point of view, connect two functional units much as noncovalent forces do in multisubunit enzymes. Of course, the relation between domains and functional sites may be less direct, and ligand binding sites could be located between domains.

TABLE III

Multifunctional Proteins[a]

Enzyme system	Number of catalytic activities	Metabolic relationship
Anthranilate synthase (M)	2	A → (B) → C
P-ribosylanthranilate isomerase: indoleglycerol-P synthase (M)	2	A → B → C
Tryptophan synthase (M)	3	A ────→ C ⟍ ↗ (B)
Chorismate mutase: prephenate dehydrogenase (M)	2	A → (B) → C
Chorismate mutase: prephenate dehydratase (M)	2	A → B → C
Aspartokinase I: homoserine dehydrogenase I (M)	2	A → B ; C → D
Histidinol dehydrogenase: imidazole acetol-P-aminotransferase (M)	2	A → B ; C → D
Acetyl-CoA carboxylase (A)	2	A → (B) → C
Fatty acid synthetase (A, Y)	8	Complete pathway
Glycogen debranching (A) enzyme	2	A → B → C
Carbamyl-P synthetase (A): aspartate transcarbamylase: dihydro-orotase[b]	3	A → B → C → D

[a] The table is largely taken from Kirschner and Bisswanger (1976). (M) microorganisms; (A) animals; (Y) yeast. The column "metabolic relationship" identifies which reactions of a metabolic sequence are catalyzed by the multifunctional enzyme, and parentheses indicate where there is evidence for a covalently attached or channeled intermediate. Note that the eight activities associated with fatty acid synthesis are distributed among two multifunctional polypeptides.

[b] See Stark (1977).

As to physiological relevance, what was said of multienzyme complexes is applicable to multifunctional enzymes. Curiously, at least of the known examples, the enzyme activities housed by the polypeptide are not always strictly sequential in terms of a metabolic pathway (Table III). This is less easy to rationalize than those cases where a multifunctional enzyme does contain the activities necessary for the operation of a metabolic sequence. The latter would indeed appear as a "higher-order" functional unit compared with simpler monofunctional enzymes.

   d. Interaction of Proteins with Other Classes of Cellular Constituents. Many proteins are associated in some way or another with the lipids of membranes; the mode in which lipid–protein interactions (Gennis, 1977) modify or contribute to the functional activities of proteins is not well understood. With the extensive studies of membrane receptors, ion pumps, transport mechanisms, etc., in progress, though, advances may be expected in this area.

Adenylyl cyclase is taken as a prototype of a membrane-associated enzyme system that introduces some new concepts of enzymic control mechanisms. This enzyme catalyzes the formation of cyclic AMP from ATP, and of special interest is its activation by several hormones (Section IV,B,2). The ensuing increase in cellular cyclic AMP concentration is thought to mediate, at least in part, the action of glucagon, postaglandins, β-adrenergic agonists* (e.g., isoproterenol) and other hormones (Section IV,B,2). The mechanism of the coupling of hormone interactions outside the cell with the activation of the center catalyzing cyclic AMP formation has commanded a large amount of study but, as yet, has not reached the same level of biochemical understanding as some of the enzymes cited earlier. In large part this is due to the difficulties in working with a complex, membrane-associated system present at low concentration in cells. Nonetheless, significant advances have been made. We will discuss mainly β-adrenergic receptor-stimulated adenylyl cyclase (Maguire et al., 1977; Lefkovitz et al., 1976; Haber and Wrenn, 1976), but this probably serves as a good model for cyclase stimulated by other hormones (Cuatrecasas, 1974; Braun and Birnbaumer, 1975; Helmreich et al., 1976).

Membrane fragments can be purified containing adenylyl cyclase that retains the property of being stimulated by isoproterenol or other β-adrenergic agonists. The additional presence of a nucleotide such as GTP or ITP is necessary for this activation and indeed may quantitatively modulate the relationship between hormone binding and enzyme activation. Insight into the composition of the system has come largely from two experimental approaches: first, conventional separation techniques applied to detergent extracts of cyclase-containing membranes, and, second, studies of cultured mouse and rat cell lines carrying genetic defects in the cyclase system. In the latter approach, the ability to reconstitute a hormone-sensitive cyclase *in vitro* from separated components has allowed a form of complementation analysis of the genetic variants (see Gilman et al., 1979).

Current evidence argues for a minimum of three definable components necessary for the integrity of the hormone-sensitive enzyme (Fig. 8). The

---

* Physiological effects of catecholamines are conveniently categorized, on the basis of pharmacological properties, as mediated by α-adrenergic or β-adrenergic receptors (see Innes and Nickerson, 1975). Epinephrine can act through both receptor types, the relative importance of each depending on the tissue. Isoproterenol is a drug acting almost completely through β-adrenergic receptors.

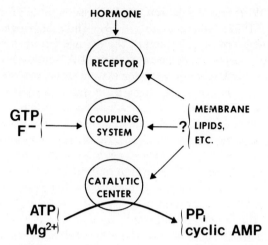

Fig. 8. Hormone-sensitive adenylyl cyclase. Current knowledge of the system is summarized in the model which comprises three distinguishable components: a hormone receptor, a coupling or regulatory component which is the site of binding of guanine nucleotides and F⁻, and a catalytic center. In fact, the enzyme may have an even more complex structure. Although, of course, the idea is that hormone binding at the exterior face of the cell membrane is transmitted to an activation of the catalytic center inside the cell, the components are not shown in any particular relative configuration since the physical nature of this relay and the exact arrangement of the components are unknown. Likewise, the interrogative expresses uncertainty as to the precise role of lipids or other membrane components in the function of this enzyme.

β-adrenergic receptor has been resolved from the catalytically active moiety both physically (Limbird and Lefkovitz, 1977; Haga *et al.*, 1977) and genetically (Gilman *et al.*, 1979). Furthermore, catalytic activity itself appears to require two distinguishable components, one bearing the active site of cyclic AMP production and the other, a regulatory or coupling factor that is the likely site of interaction of GTP and F⁻, activators (Gilman *et al.*, 1979). A specific GTP-binding protein that may be analogous to the factor (G/F) defined by Gilman and his associates has in fact been isolated from avian erythrocyte adenylyl cyclase (Pfeuffer, 1977). In addition, there is evidence for a GTPase activity in avian erythrocytes (Cassel and Selinger, 1976, 1977) that is stimulated by β-adrenergic agonists and whose operation has been postulated as an integral part of cyclase function. Cholera toxin has been known for some time to cause an irreversible activation of cyclase (see Gill, 1977) associated probably with a modification of the coupling between the receptor and the cyclase activity. Recent evidence suggests that the toxin provokes the ADP-ribosylation of the same GTP-binding component discussed above (Moss and Vaughan, 1977; Gill and Meren, 1978; Cassel and Pfeuffer, 1978; G. L. Johnson et al., 1978; Gilman et al., 1979). Thus, although the coupling mechanism is still not clear in detail, the linkage between the binding of hormone to the receptor and the regulation of the catalytic center probably involves a separate component (or conceivably, more than one) (see Gilman et al., 1978). Figure 8, then, provides a minimal model of the β-adrenergic receptor-linked adenylyl cyclase.

For our purposes, the system provides a somewhat novel example of the transmission of binding at one site to modify the activity of a catalytic center. In purely mechanistic terms, analogy with allosteric binding to enzymes is evident. The cyclase system may be more complex, however, and future information on the physical interactions involved in linking hormone binding to the cyclase activity will be of great interest. One wonders how the lipids of the membrane interact with this system? Stoichiometrically, how many receptors feed onto each catalytic subunit, and are different receptor types coupled to the same or different catalytic components? In Section IV,B,2, we trace some of the intracellular effects of increased cyclic AMP concentration.

Another major category of macromolecular interactions is between protein and nucleic acid. A large proportion of cellular nucleic acids is associated with proteins (Jovin, 1976; von Hippel and McGree, 1972) either in a relatively stable manner, such as in ribosomes, or in a possibly more intermittent fashion, such as between polymerases and nucleic acids during nucleic acid synthesis. The expression of the *lac* operon in *E. coli,* for example, involves very specific interactions between RNA polymerase core enzyme, σ factor, CAP protein, and maybe other factors with particular stretches of the DNA molecule. These proteins clearly have evolved the ability to interact not only with proteins and, where relevant, small molecules, but also with specific nucleic acid sequences.

B. Covalent Modification of Proteins

1. General Comments

A number of enzymes are known to be subject to covalent modification catalyzed by other enzymes (converting enzymes), and those for which sufficient information is available appear to possess especially complex and interesting regulatory properties. The list of such enzymes or proteins is growing steadily, and the phenomenon may be more widespread than had been supposed (Tables IV and V).

One of the most frequent forms of modification is the phosphorylation of

TABLE IV

Some Interconvertible Enzymes[a]

Enzyme	Activity change after modification	Possible initiators	Modifying enzyme	Reference[b]
Phosphorylation				
Phosphorylase *b*	Activated	Ca²⁺ Cyclic AMP	Phosphorylase Kinase	1
Phosphorylase kinase	Activated	Cyclic AMP	Cyclic AMP PK	2
Glycogen synthase	Inactivated	Cyclic AMP ? Ca²⁺	Cyclic AMP PK IND PK's Phosphorylase kinase	3
Pyruvate dehydrogenase	Inactivated	Metabolites	IND PK	4
Phosphofructokinase	Activated	?	IND PK	5
Fructose-1,6-diPase	Activation	Cyclic AMP	Cyclic AMP PK	6
Pyruvate kinase (L type)	Inactivated	Cyclic AMP	Cyclic AMP PK	7
Hormone-sensitive lipase	Activated	Cyclic AMP	Cyclic AMP PK	8
Fatty acid synthetase	Inactivated	?	?	9
Acetyl-CoA carboxylase	Inactivated	?	IND PK ?Cyclic AMP PK	10
Tyrosine aminotransferase	?	?	?	11
Phenylalanine hydroxylase	Activated	Cyclic AMP	Cyclic AMP PK	12
RNA polymerase (yeast)	?	?	?	13
Myosin light chain kinase	Inactivated	Cyclic AMP	Cyclic AMP PK	14
Hydroxymethyl-glutaryl CoA reductase	Inactivated	?	Reductase kinase (IND PK)	15
Reductase kinase	Activated	?	IND PK	15
Nucleotidylylation				
Glutamine synthetase (adenylylation)	Inactivation	Metabolites	ATase + P_II	16, 17
P_II regulator protein (uridylylation)	Specifies different reaction	Metabolites	UR and UT activities	16, 17
ADP-ribosylation				
Elongation factor 2	Inactivation		Diphtheria toxin A fragment	18
Adenylyl cyclase component	"Activation" (see text)		Cholera toxin fragment A₁	19
RNA polymerase	Specificity Altered	T4 infection		20

TABLE V

Some Specific Proteolytic Modifications of Proteins

Precursor	Product	Agent
Trypsinogen	Typsin	Trypsin or enterokinase
Chymotrypsinogen	Chymotrypsin	Trypsin
Procarboxypeptidase	Carboxypeptidase	Trypsin
Pepsinogen	Pepsin	Pepsin, gastric proteases
Blood coagulation factors	Usually protease	See text
Complement factors	Usually protease	See text
Angiotensinogen	Angiotensin II	Renin, converting enzyme
Proinsulin	Insulin	Trypsin-like activity
Diphtheria toxin	A fragment	Protease
Kininogen	Bradykinin	
	Kallidin	Kallikrein
Plasminogen	Plasmin (fibrinolysin)	Proteases ? other factors

a serine (or occasionally a threonine) residue in a protein; the phosphate derives from ATP. Adenylylation and uridylylation of tyrosine residues, consuming ATP or UTP, although as yet firmly established in fewer instances, form another important mechanism of modification. A third mechanism involves the attachment of an ADP-ribosyl group, deriving from NAD^+, to an arginine or other residue of a protein. Specific proteolytic cleavage of polypeptide chains is a further general class of modification of proteins (Table V). Methylation or acylation of proteins, though less well understood so far, may also be of importance in enzyme regulation.

[a] The table lists a number of enzymes subject to covalent phosphorylation. The column "modifying enzyme" for the phosphorylated enzymes identifies the protein kinase(s) involved; the reverse reactions are catalyzed by protein phosphatase(s). Cyclic AMP PK represents cyclic AMP-dependent protein kinase, and IND PK signifies cyclic AMP-independent protein kinase but does not imply that only one such kinase exists. The "possible initiators" refer to small molecules, where known, that may control the activity of the modifying enzymes.

[b] Key to references: (1) Fischer et al. (1970, 1971), Nimmo and Cohen (1977); (2) E. G. Krebs et al. (1964);   (3) Friedman and Larner (1965), Nimmo and Cohen (1977); Roach et al. (1978);   (4) Reed (1974);   (5) Brand and Söling (1975), Brand et al. (1976), Hofer and Fürst (1976);   (6) Riou et al. (1977);   (7) Engström (1978); (8) Steinberg et al. (1975);   (9) Qureshi et al. (1975);   (10) Carlson and Kim (1973), Seubert and Hamm (1975), Hardie and Cohen (1978);   (11) Lee and Nickol (1976); (12) Milstein et al. (1976);   (13) Bell et al. (1976);   (14) Adelstein et al. (1978); (15) Ingebritsen and Gibson (1980);   (16) Stadtman and Ginsburg (1974);   (17) Adler et al. (1975);   (18) Pappenheimer (1977);   (19) Moss and Vaughan (1977), Gill and Meren (1978), Cassel and Pfeuffer (1978), G. L. Johnson et al. (1978);   (20) Goff (1974), Zillig et al. (1977).

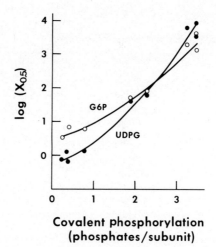

Covalent phosphorylation
(phosphates/subunit)

Fig. 9. Dependence of enzyme kinetic properties on covalent phosphorylation state. The example is taken of muscle glycogen synthase and the figure shows the variation of two kinetic parameters, the $S_{0.5}$ of the substrate UDP glucose and the $M_{0.5}$ for the activator glucose-6-P, as a function of the average number of covalent phosphate residues per enzyme subunit. The $S_{0.5}$ is calculated in millimolar and the $M_{0.5}$ in micromolar. Note that the ordinate has a logarithmic scale. Plotted from data in Roach *et al.* (1976).

These systems have certain features in common: all involve the action of an enzyme on a protein, usually another enzyme, and at least where the physiological role is understood, the modification causes significant changes in the protein, usually of enzyme kinetic properties (Fig. 9). On the other hand, it is useful to distinguish irreversible (such as associated with specific proteolysis) (Table V) from reversible systems (Table IV). The physiological events associated with proteolytic control can, of course, be reversed; blood clots eventually dissolve, for example. The next round of activation of the coagulation system, however, will involve now proenzyme molecules. This contrasts with reversible modifications where the covalently attached group (such as phosphate) will turn over many times during the lifetime of the enzyme molecule.

2. Covalent Modification Systems: Enzymic Cascades

a. Phosphorylation. As is evident from Table IV, phosphorylation accounts for the majority of the reported covalent modifications but this might just reflect experimental approaches rather than a true generalization. Besides those listed, the phosphorylation of many unidentified cellular components has been noted. For instance, the incubation of isolated cells with ^{32}P leads to many electrophoretically distinct, radioactively labeled proteins (for example, Avruch *et al.,* 1976; Garrison, 1978). This obviously gives a minimal estimate of the number of cellular phosphoproteins.

Each site of phosphorylation is associated with a protein kinase, normally utilizing Mg-ATP as cosubstrate, and a protein phosphatase responsible for the hydrolysis of the phosphoprotein. On the whole, there seem to be fewer kinases and phosphatases than phosphorylation sites, suggesting multiple specificity of these converting enzymes. In general, the relation of specificity between sites and converting enzymes poses interesting questions both for the chemistry of action and for regulatory function. This point will emerge again later.

Enzyme regulation by covalent phosphorylation will be largely exemplified by a discussion of muscle glycogen metabolism. The individual components will be described, gradually building up a picture of the complicated regulatory connections between these. It may be useful to refer to Fig. 10, which gives an overall summary of the interconversions involved. Glycogen metabolism is one of the best documented and most complicated systems regulated through phosphorylation but its choice as an example should not obscure the importance of other systems, further details of which are to be found in the references to Table IV.

Glycogen phosphorylase (Fischer *et al.,* 1970, 1971) catalyzes the phosphorolysis of α-1,4-linkages in glycogen to yield glucose-1-P. Degradation of the polysaccharide beyond the α-1,6-linkages associated with its

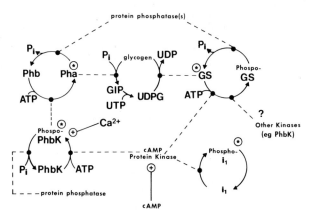

Fig. 10. Enzyme interconversions in glycogen metabolism. The major interconversions, which are all phosphorylation–dephosphorylation cycles, of the glycogen metabolism cascade are illustrated. More active (see text) enzyme species are marked by an asterisk; in the case of the phosphatase inhibitor i_1, this means greater effectiveness as an inhibitor. Enzymes are linked to the reactions they catalyze by the filled circle. The noncovalent inputs of cyclic AMP and Ca^{2+} are indicated by plus sign in circle. GS, glycogen synthase; Ph, phosphorylase; PhbK, phosphorylase kinase.

branched structure requires the participation of a debranching enzyme; control, however, seems to be most directed at phosphorylase. Glycogen phosphorylase exists in two forms, phosphorylase b which is not phosphorylated and phosphorylase a which is phosphorylated at one serine (Ser-14) per subunit. The two forms differ in a number of molecular properties and, of particular interest, as to kinetic properties. The b form is essentially inactive in the absence of AMP and is generally more susceptible to inhibition by metabolites such as glucose-6-P, ATP, and UDPglucose. The combination of these allosteric modifiers would, on the basis of probable levels in the cell, render phosphorylase b inactive but leave phosphorylase a active *in vivo* (Morgan and Parmeggiani, 1964). The phosphorylase interconversion *in vivo* should thus correlate with a change in effective activity.

Recently, considerable advances in resolving the crystal structure of phosphorylase have been made (see Fletterick and Madsen, 1977; L. N. Johnson *et al.*, 1978), and a precise, physical basis has been given to many of the known kinetic properties through the identification of several effector binding sites. Interestingly, the active centers of dimers are located at the subunit interfaces and are composed of residues from both. This observation explains the earlier finding that the isolated subunits were inactive (Feldman *et al.*, 1972). The high resolution structure of the b form (only the a form is yet known at 3 Å resolution) will be very important to our understanding of the chemical mechanisms by which covalent modification influences enzyme activity. The comparison of phosphorylase a and b at low resolution has already indicated a large shift in the position of the amino-terminal portion (of 20 amino acids) of the molecule (Fletterick *et al.*, 1976).

The phosphorylation of phosphorylase b is catalyzed by phosphorylase kinase which is itself a complex enzyme, with subunit structure $(ABC)_4$, that undergoes phosphorylation (see Nimmo and Cohen, 1977). All forms of this kinase require Ca^{2+} ions for activity, half-maximal activation occurring in the range 10^{-7} to 10^{-6} M (Ozawa *et al.*, 1967; Brostrom *et al.*, 1971). The native, nonphosphorylated enzyme displays lower activity at pH 6.8, around the physiological value, than at pH 8.2. Phosphorylation of the enzyme by cyclic AMP-dependent protein kinase results in an activation correlating with a large differential increase in the activity at pH 6.8 (increasing the pH 6.8 : pH 8.2 activity ratio) (E. G. Krebs *et al.*, 1964). The activation follows from an increase in the apparent affinity for phosphorylase. Cyclic AMP-dependent protein kinase phosphorylates one site on each of the A and B subunits, the B subunit reacting some four to five times faster than the A subunit (Hayakawa *et al.*, 1973; Cohen, 1973). Cohen and Antoniw (1973) suggested that modification of the B subunit is related to enzyme activity changes, while phosphorylation of the A subunit predisposes the B subunit to a more rapid dephosphorylation. There is evidence for the phosphorylation of these cyclic AMP-dependent kinase-specific sites *in vivo* following the administration of epinephrine to rabbits (Yeaman and Cohen, 1975). As noted earlier, the latest evidence (Cohen *et al.*, 1978) suggests that this kinase may contain a fourth subunit, namely, the Ca^{2+}-dependent modulator protein. The hypothesis is that binding of Ca^{2+} to this component would mediate the sensitivity of the enzyme to Ca^{2+}.

Glycogen synthase (Larner and Villar-Palasi, 1971) catalyzes the addition of a glucose moiety from UDPglucose, by an α-1,4-linkage, to a growing glycogen molecule. Branches, determined by α-1,6-linkages, are introduced by the branching enzyme. Synthase too is subject to interconversion; the nonphosphorylated (I form) is generally more active, while the phosphorylated enzyme (D form) has low activity unless the effector glucose-6-P is present. The phosphorylated enzyme is also more susceptible to inhibition by metabolite effectors such as ATP, ADP, AMP, and P_i (Larner and Villar-Palasi, 1971; Roach and Larner, 1976), and this form of the enzyme is unlikely to be active *in vivo* (Piras *et al.*, 1968). Again the covalent modification is associated with an alteration of the enzymic activity expressed under given levels of effectors, but especially for glycogen synthase, the correlation between phosphorylation and activity is complex (Fig. 10). The enzyme appears to exist as multimers of a single subunit which can each be phosphorylated at several sites; there has been some controversy over the number of phosphorylation sites, but the value is probably between 4 and 6 per subunit (see Roach and Larner, 1977). There is some evidence also that the enzyme *in vivo* may contain from 1 to 3 phosphates per subunit (Roach *et al.*, 1977; Nimmo and Cohen, 1977).

Unlike phosphorylase, muscle glycogen synthase can be phosphorylated by at least three different protein kinases: cyclic AMP-dependent protein kinase (Friedman and Larner, 1965), phosphorylase kinase (DePaoli-Roach *et al.*, 1979), and what has been termed "cyclic AMP-independent synthase kinase" (see Nimmo and Cohen, 1977). The last-named "kinase" probably comprises two or more distinguishable kinases, the enzymology of which remains to be worked out in detail. The cyclic AMP-dependent protein kinase and phosphorylase kinase preferentially phosphorylate different sites or sets of sites on the synthase molecule (Roach *et al.*, 1978), and both phosphorylations result in an inactivation of the enzyme. Currently, there is good evidence for a physiological role of cyclic AMP-dependent protein kinase to phosphorylate synthase. The importance of the other phosphorylations remains to be established, although the involvement of phosphorylase kinase, with a potential control through Ca^{2+}, is a provocative hypothesis.

The phosphatases (see Curnow and Larner, 1979) involved in glycogen metabolism have proved more difficult to study. Often, multiple peaks or fractions of phosphoprotein phosphatase are encountered during purification, and the assignment of such activities to specific dephosphorylation reactions has not been simple, due partly perhaps to a relative nonspecificity of action. There is fairly good evidence that the same enzyme is responsible for the dephosphorylation of both glycogen synthase and phosphorylase (see Curnow and Larner, 1979). Cohen and co-workers (Antoniw and Cohen, 1975) also suggest that this same enzyme dephosphorylates the phosphorylase kinase B subunit, while a separate phosphatase acts on the A subunit. Other workers have reported that phosphorylase phosphatase does not act on phosphorylase kinase (Gratecos *et al.*, 1974).

Phosphorylase phosphatase activity from liver (Brandt *et al.*, 1974) or muscle (Huang and Glinsmann, 1975, 1976) may be associated with inhibitory protein factors or subunits. In the case of muscle, the inhibition by one factor (i_1) is greater if the protein factor has itself been phosphorylated by cyclic AMP-dependent protein kinase (Huang and Glinsmann, 1975). Interestingly, this phosphorylation is one of the few examples of attachment to a threonine residue (Cohen *et al.*, 1977). Phosphatase activity toward glycogen synthase or histone is unaltered by this phosphorylation. A second protein inhibitor, i_2, that inhibits the dephosphorylation of phosphorylase, synthase, and phosphorylase kinase and that is not subject to covalent modification (Huang and Glinsmann, 1976) has also been described.

The numerous interconversions and regulatory interactions of muscle glycogen metabolism are displayed in Fig. 10. Although the scheme is very complex, our knowledge is clearly far from complete, and the diagram is merely a "status report." Nonetheless, this system provides an almost unique opportunity to relate molecular events with physiological function. Inspection of Fig. 10 will demonstrate how an alteration in cyclic AMP concentration will exert, through cascaded enzyme interconversions, a coordinated influence on both glycogen synthase and phosphorylase activities. Increased cyclic AMP would correlate with increased glycogen degradation and decreased glycogen synthesis, and this provides a mechanistic description for the hormonal control of glycogen metabolism by, for example, glucagon or epinephrine. A more detailed discussion of the integrated control of this system is given in Section IV,B,2.

Phosphorylation of a single residue per subunit is clearly capable, chemically, of exerting a profound effect upon enzymatic properties, and phosphorylase may be taken as a good example of this. As more systems are studied, it is becoming clear that, at least *in vitro*, multiple phosphorylation of enzymes, i.e., phosphorylation of two or more distinguishable sites per molecule, can occur. We have already noted two examples, glycogen synthase and phosphorylase kinase. Another example is pyruvate dehydrogenase, which can be phosphorylated at three separate sites in the α-subunit (Reed *et al.*, 1976), although only two have so far been implicated in altering enzyme activity. For muscle glycogen synthase, however, evidence is good for an influence of more than one phosphorylation site on activity *in vitro*, as just discussed. If this is true also *in vivo*, it would represent a very interesting, and unique, situation whereby different controls, acting to phosphorylate different sites through different kinases, could be integrated physically at the level of the glycogen synthase molecule (see Roach and Larner, 1977).

    b. Adenylylation and Uridylylation. Glutamine synthetase catalyzes the formation of glutamine from glutamate and ammonia with the concomitant hydrolysis of ATP to ADP plus P_i. The enzyme is thus critically associated with nitrogen metabolism in many cells. In fact, the *E. coli* enzyme is subject to a complex set of feedback inhibitions by several metabolites that are indirectly end products of glutamine synthesis; these include ATP, CTP, glucosamine-6-P, histidine, carbamyl-P, glycine, tryptophan, and alanine (Stadtman and Ginsburg, 1974; Holzer and Duntze, 1971). The inhibition has been termed "cumulative" (see Stadtman and Ginsburg, 1974) because each metabolite alone exerts maximally only partial inhibition; acting together, the effects are additive* and, in the presence of all of the effectors, strong inhibition of synthetase activity can be achieved. Subsequently, some interactions between effectors were observed, but the behavior is probably close to cumulative at physiological metabolite levels.

    Besides this rather complex inhibition pattern, however, it was discovered that glutamine synthetase could be adenylylated at one tyrosine residue in each of its 12 subunits (Kingdon *et al.*, 1967; Shapiro and Stadtman, 1968) (Fig. 11). The nonadenylylated form of the enzyme is more sensitive to metabolite inhibition and is activated by Mn^{+2} ions; the adenylylated enzyme is less sensitive to inhibition and is activated by Mg^{2+}. The covalent modification is catalyzed, in *both* directions, by adenylylate transferase (ATase). However, the specificity of ATase for adenylylation or deadenylylation is modulated both by metabolite effectors, including α-ketoglutarate, ATP, glutamine, and P_i, and a regulatory protein called P_{II}. This protein is likewise covalently modified, in this instance by uridylylation of a tyrosine residue. The nonmodified form of P_{II} (P_{IIA}) specifies adenylylation by ATase; the uridylylated form, P_{IID},

---

  * In the sense that, if two effectors reduce activity to $P\%$ and $q\%$ of the uninhibited rate, respectively, then their cumulative action would correspond to $(p \times q)/100\%$ of the uninhibited activity.

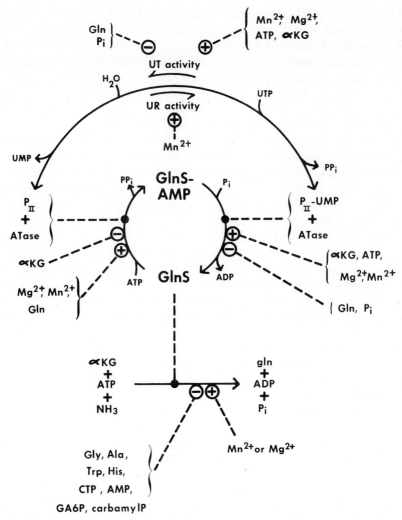

Fig. 11. Enzyme interconversions in the control of *E. coli* glutamine synthetase. The basic interconversion of glutamine synthetase (GlnS) by adenylylation and deadenylylation is shown in the center. The filled circle links enzymes to the reactions they catalyze; minus sign in circle and plus sign in circle denote noncovalent inhibition or activation, respectively. For further explanations, see text. Taken from the work of the laboratories of Stadtman and of Holzer.

stimulates the adenylyl-removing activity of ATase. More recent evidence (Adler *et al.*, 1975) further suggests that the converting activities acting on P_{II}, the uridylyl-removing activity (UTase) and the deuridylylating activity (UR-enzyme) are associated with the same enzyme or enzyme complex.

The specification of UTase or UR activity arises from the concentrations of metabolites, including P_i, glutamine, ATP, α-ketoglutarate, and UTP, as well as metal ions.

The glutamine synthetase system is, thus, exceedingly complicated and embodies several important regulatory features other than covalent modification. Particularly interesting are the two "bidirectional" converting enzyme activities whose directionality is subject to control. Taking the system as a whole, other points of interest are the recurrence of some metabolites or ion interactions at several levels of the cascade and the complementary actions of some effectors on converting enzyme activity. For example, glutamine influences both uridylylation of P_{II} and adenylylation of synthetase; in addition, glutamine inhibits the adenylyl-removing activity of ATase but activates the adenylylation reaction. By way of allosteric or small molecule interactions at various points of the glutamine synthetase system, then, in addition to the numerous direct effectors of the target enzyme, a quite prodigious number of indirect effectors is possible (Table VI). A bewildering complexity of control is thus channeled into determining the rate of glutamine formation.

c. Proteolysis. Proteolysis is a common form of covalent modification (Table V) that may have a variety of functional implications. Protein turn-

TABLE VI

Effectors of Glutamine Synthetase Activity[a]

Direct		
Alanine	CTP	
Glycine	AMP	
Serine	GMP	
Histidine	IMP	
Tryptophan	Glutamate	
Carbamyl-P	Mg^{2+}	
Glucosamine-6-P	Mn^{2+}	
ATP	NH_4^+	
Indirect		
Glutamine	3-Phosphoglycerate	Malate
α-Ketoglutarate	Fructose-6-P	Glycerol
Phosphate	Fructose-1,6-diP	PP_i
Glutamate	Phosphoenolpyruvate	Mg^{2+}
Tryptophan	CoA	Mn^{2+}
Methionine	Oxaloacetate	CMP
ATP	Citrate	UMP
UTP	Pyruvate	GMP
NH_4^+	Succinate	IMP
	Fumarate	

[a] From Stadtman and Chock (1978).

over, of course, involves proteolytic degradation; in addition, the cleavage of peptide bonds may be associated with production of active protein species from inactive precursors (see Steiner, Chapter 5). This latter may be in response to precise signals or as "routine" posttranslational processing. Relatively specific proteolytic action is usually involved in such processes. In addition, the literature abounds with examples where limited proteolytic activity *in vitro* may cause significant changes in enzymic properties without destroying basic catalytic capacity.

A number of enzymes (and some other proteins) are first synthesized as inactive zymogens or proenzymes (Table V). This phenomenon has been noted particularly for proteolytic enzymes such as trypsin, chymotrypsin, and carboxypeptidase. Storage of such enzymes as zymogens may be a form of "safety" mechanism, since the uncontrolled presence of nonspecific proteolytic activity could evidently be disruptive. On the other hand, not all degradative hydrolytic activities are thus protected, not is specific cleavage restricted to proteases. Proinsulin, from which active insulin is produced by the excision of an internal stretch of polypeptide chain (the connecting polypeptide), is really a biosynthetic intermediate of insulin synthesis (in fact, the processing is even more complex: Steiner, Chapter 5).

Blood coagulation is a cascade system in which proteolytic modification plays a central role (Fig. 12) (Davie and Kirby, 1973; Davie and Fujikawa, 1975). The so-called intrinsic system involves, after initial activation of factor XII by contact with collagen, glass or similar surface, a cascade of some eight or nine levels culminating in the activation of thrombin and a cross-linking transglutaminase activity, factor XIII, thus giving rise to an insoluble, cross-linked fibrin clot. At least eight proteolytic activities are involved in the scheme and specific proteolysis, and sometimes the combination with other components, always correlates with the appearance of an enzymic activity, which is itself usually a proteolytic activity. The extrinsic system, initiated by the combined action of tissue factor (a lipoprotein) and factor VII, feeds in at the level of the modification of factor X, thus bypassing the earlier stages of the intrinsic system. However, both the extrinsic and intrinsic systems appear physiologically important and may operate simultaneously *in vivo*. Besides proteolytic modification, several regulatory interactions between blood clotting factors have been suggested, and Ca^{2+} is necessary at four or so of the stages. The complexity of the system is evident from Fig. 12; for further details of the properties of the individual components, recent reviews may be consulted (Davie and Kirby, 1973; Davie and Fujikawa, 1975). Within the connections of the coagulation scheme it is interesting to note several regulatory configurations.

INTRINSIC

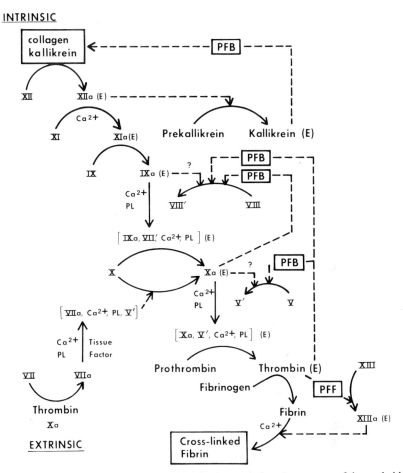

Fig. 12. The enzymic cascade involved in blood coagulation. A summary of the probable mechanism for blood clotting by either the intrinsic or extrinsic systems is shown. The Roman numerals refer to blood clotting factors; (E) indicates an enzymatically active form (usually a protease); curved arrows denote an enzymatic conversion; vertical arrows denote combination; dashed arrows visualize the multiple interactions of some components of the system; PL is phospholipid; PFB and PFF identify some potential positive feedback and positive feedforward loops, respectively, in the system. Taken largely from Davie and Fujikawa (1975).

Kallikrein stimulates the production of factor XIIa but activated factor XII (i.e., XIIa) itself generates kallikrein through the cleavage of the proenzyme prekallikrein. This constitutes a positive feedback mechanism, since factor XIIa action accelerates its own formation. Similarly, the action of thrombin to activate transglutaminase may be considered a positive feedforward. The possible effect of thrombin to feedback and activate factors V and VIII, on the other hand, would represent positive feedbacks on thrombin generation. Interwoven within the cascade, then, there may be other classes of regulatory connection, all of which seem geared to an explosive autocatalytic, all-or-none activation of blood clot formation.

Serum also contains the inactive precursors of another cascaded defense system, the complement (C), which is responsible for the lysis of microorganisms and erythrocytes (Müller-Eberhard, 1975; Fothergill and Anderson, 1978). Two pathways exist for the production of an active cytolytic agent (Fig. 13). The classic route, involving some nine complement glycoproteins, is activated by immune complexes containing certain immunoglobulins. The alternative pathway, whose initial phases are less well understood, may be triggered by immunoglobulins, as well as by some lipopolysaccharides and polysaccharides associated with microbial invasion. The two modes converge at the level of component C3. Again, specific proteolysis is central to the activation of the complement components. For the present discussion, the overall cascaded pattern and regulatory configuration is emphasized (more details concerning this complex process may be found in the citations above).

As with the coagulation scheme, this system has some interesting internal structure. At two points, an enzymically active species combines with the product of its reaction to constitute the active components responsible for the next stage of the cascade (C3 convertase and C3b; C423b and C5b). The precise functional meaning of this unusual arrangement is hard to judge at present but it would tend to suppress spurious triggering at intermediate levels of the cascade, accentuating only that deriving from earlier stages.

Another feature of the alternative pathway is an autocatalytic loop at the level of the formation of the "C3 convertase" activity. The complex C3b–Bb itself catalyzes the formation of C3b from C3, so that the reaction product forms part of the producing enzyme (See Fig. 13). Again we have a very complex, cascade system that appears designed for "failsafe" all-or-none operation such as for the destruction of unwanted cells invading the body.

d. Miscellaneous. The ADP-ribosylation of a number of proteins has now been reported (Hayaishi and Ueda, 1977). One of the most interesting of these is elongation factor 2 (EF-2) necessary for the transfer of the growing polypeptide from the acceptor to the donor site of the ribosome during eukaryotic protein synthesis. ADP-ribosylation of this protein leads to inactivation, although the identification of the site of attachment has not been possible so far. Interestingly, the stimulation of this modifica-

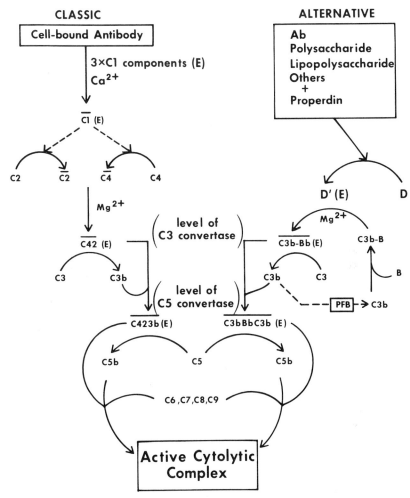

Fig. 13. The complement system. The possible mechanism for the cascade-linking interaction with immunoglobulin and the formation of active cytolytic complexes is shown for either the "classic" or "alternative" pathway. Symbols as for Fig. 12. Composed from Müller-Eberhard (1975), and Fothergill and Anderson (1978).

tion appears to be the molecular basis of the action of diphtheria toxin (see Pappenheimer, 1977). Cholera toxin may similarly provoke an ADP-ribosylation reaction of a component of adenylyl cyclase (Section III,A,3). T4 infection of *E. coli* leads to ADP-ribosylation of the α-subunit of the *E. coli* RNA polymerase (Goff, 1974; Zillig *et al.,* 1977), and some changes in transcription specificity may result from this modification. ADP-ribosylation is also reported for a mitochondrial protein, and poly(ADP)-ribosylation of nuclear proteins, probably histones, has been observed. Histones, in fact, are subject to a variety of chemical modifications besides this, including phosphorylation (Rubin and Rosen, 1975; Langan, 1973), acetylation (Allfrey *et al.*, 1964; Gallwitz, 1970) and methylation (Cantoni, 1975; Paik and Kim, 1975). In many of these cases, however, the biological implications of covalent modification are as yet unclear.

3. Functional Significance of the Covalent Modification of Proteins: Enzymic Cascades

a. General Comments. Enough information is available, at least for a few well-studied examples, to ask seriously what is the physiological significance of covalent modification (Fischer *et al.*, 1971; Davie and Kirby, 1973; Roach, 1977; Stadtman and Chock, 1978). Of course, perhaps not all new examples of covalently modified enzymes will be confirmed, and some well-established modifications have yet to be implicated in metabolic control. Having made this necessary preface, however, I suspect that the number of enzymes known to be subject to covalent control will increase. Furthermore, there is no particular reason to believe that all classes of modification have yet been identified in chemical terms.

Covalent modification systems involving enzymes as the target proteins are often called "enzymic cascades." Each interaction of this type is a "stage" of the cascade; the blood clotting sequence comprises eight or nine such stages. Cascaded activation of a target enzyme seems often associated with a complete switching on (or off) of activity; this correlates most simply with the complete inactivity of converting enzymes prior to stimulation. Recently (Stadtman and Chock, 1978), more emphasis has been placed on the possibly dynamic nature of the interconversion; in other words, the simultaneous activity of both enzymes of a converting enzyme couple (such as a protein kinase and phosphatase). This is of course only relevant to those modifications defined earlier as reversible. A dynamic modification would, thus, involve turnover of modified protein with the consumption of metabolic energy, usually as ATP. Such a phenomenon is an example of a "futile cycle" (See Section IV,B), although, at least in this case, it may not be futile (see below).

A number of model configurations of cascade systems have been analyzed on the basis of a steady state* turnover of the target enzyme (Stadtman and Chock, 1978) and the expected performance of such sys-

---

* Not to be confused with the steady state assumption in enzyme kinetics. In this case, it is the concentration of covalently modified enzyme, not an enzyme–substrate complex, that would be in a steady state.

tems discussed. Many of these predictions will be mentioned below. Experiments with purified interconvertible enzyme systems have illustrated that such steady states may be set up *in vitro*. Figure 14 shows such an experiment with glutamine synthetase where the same steady state level of adenylylation results whether the enzyme has a low or high initial modification state. Similar steady states have been established with pyruvate dehydrogenase (Pettit *et al.*, 1975; Reed *et al.*, 1973). The situation remains suggestive rather than proven, but it is important to appreciate the possibility of dynamic interconversions *in vivo,* even if it does not occur in all cases and even if perfect steady states are not attained.

b. **Sensitivity and Amplification.** Previously, cooperative enzyme kinetics were discussed as a means of varying the relative sensitivity of an enzymic rate to an effector or substrate. For a positively cooperative system, there are presumably limits to this increased sensitivity imposed by the chemical properties of single proteins. Cascaded enzyme systems

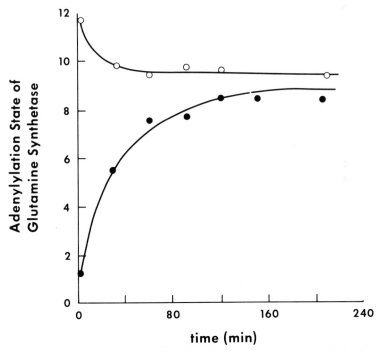

Fig. 14. Steady state covalent modification of an enzyme. The example is glutamine synthetase incubated with ATase, P_{II}- protein, and appropriate effectors. The glutamine synthetase tends to the same steady state level of adenylylation (expressed as AMP per molecule in the ordinate) whether it starts with a low (closed circles) or a high (open circles) level of covalent modification. After Stadtman and Chock (1978).

can have sensitivities much greater than are possible for single enzymes. This, as well as some other points made in the present section, are often quite intuitive; the approach of Stadtman and Chock (1978) to analyze cascade systems with the steady state assumption, however, gives an objective base upon which to conceptualize about performance characteristics.

From the analysis of a number of configurations of cascade systems, where effector molecules interacted with various combinations of converting enzymes, it is a good generalization that the compound system can correlate with vast reductions in the $M_{0.5}$ of an effector for the response of the target enzyme compared with its $M_{0.5}$ for a converting enzyme. This means an increase in absolute sensitivity. Interestingly, in the simpler systems analyzed, where an effector enters only at one level of the cascade, the *shape* of the curves of target enzyme activity against effector concentration is not subject to great change and, in particular, no increases in sigmoidicity are noted. The possibility of a more sigmoid response curve, that is, increased relative sensitivity, is not excluded by this analysis, and relatively simple modifications could accommodate this. An effector that entered at more than one level or stage of a cascade, for instance, could lead to an increased apparent kinetic order (increased sigmoidicity) of the curve of target enzyme activity against effector concentration. The same outcome would result from nonlinear relationships between target enzyme activity and covalent modification, as might be true for glycogen synthase (Fig. 9) (Roach et al., 1976). Thus, the configuration of cascade systems can allow for a greatly reduced $M_{0.5}$ for effectors, as well as manipulation of the shape of the response curve between effector and target enzyme activity.

As the response of the target enzyme becomes more and more sensitive to an effector, one can arrive at the situation where a physiologically feasible change in effector concentration can provoke the complete "all-or-none" activation or inactivation of the target enzyme. For example, synthesis–degradation couples, such as the glycogen synthase–phosphorylase couple, are unlikely to be useful if they catalyze constant cycling between, in this case, glycogen and sugar phosphate at the expense of ATP. Thus, one might anticipate a coordination of the activities of the two enzymes so that only one of the enzymes is significantly active at a given time. This would require an "all-or-none" inactivation of the appropriate enzyme when either synthesis or degradation is signaled.

Related with the idea of increased sensitivity is the fact that cascaded systems can enable one molecule of effector to activate more than one molecule of target enzyme. This is a true "molecular amplification" and is built into the system because enzymes act on other enzymes. Thus, through a cascade, cyclic AMP, present at less than 1 μM, can govern the activity of muscle phosphorylase, present at 0.1 mM. The allosteric effector of phosphorylase, AMP, which contributes to controlling the same enzyme, is present at around 0.1 mM. However, cascade systems are unlikely to have evolved exclusively to cater for low effector concentrations. Generally, effector and binding-site concentrations will have evolved alongside one another, and to a certain extent, increased affinity can compensate lower effector levels (see Section II,C). Where converting enzymes appear to have evolved to link incompatible concentrations of effector and target enzyme, one may suspect that other pressures in some way constrained this disparity of concentrations. Cyclic AMP-mediated controls, for example, might have been a later overlay on existing cellular metabolism (Tomkins, 1975; Roach, 1977) so that the inclusion of a new "messenger" metabolite at high levels would have been impossible in an already saturated cell (Atkinson, 1969). In fact, converting enzymes are not always present at concentrations greatly inferior to those of the target enzyme (Burchell *et al.*, 1976), suggesting that "molecular amplification" of the type described above is not an indispensable aspect of cascade systems. Nonetheless, it is easy to show that great increases in sensitivity to effectors are equally possible for cascades with relatively high levels of converting enzymes (for example, Roach 1977).

c. Integration and Coordination. An important aspect of the function of cascade systems is undoubtedly their capacity to integrate a large number of regulatory signals (Stadtman and Chock, 1978; Roach, 1977). Some individual enzymes are capable of sensing a large number of metabolite concentrations (see, for example, Section III,A); nonetheless, one may generally anticipate that the number of potential small molecule interactions, and hence binding sites, increases with the number of proteins involved in the system. However, not only would the multienzyme cascade thus be able to respond to more small molecule concentrations than a single enzyme, but would permit new and subtle ways in which these very signals might interact among one another to give an integrated output in the form of the activity of the target enzyme.

It should be evident from the examples quoted that exceedingly complex interactions do exist in cascade systems, not all of which we are able to rationalize in terms of a specific regulatory feature. One aspect of the integrative control that we can appreciate is the possibility of coordinating different enzyme activities. In noncovalent systems, as will be discussed in Section IV, the interaction of a common effector with several enzymes can contribute to this type of coordination, which is also true for the cascade systems. However, the latter present a new feature, since the converting enzymes can guide the coordination. For example, the multiple substrate specificity of cyclic AMP-dependent protein kinase will ensure the coordinated modification of a number of enzymes or proteins following hormonal stimulation, which in the already mentioned case of glycogen synthesis will lead to reciprocal effects on synthase and phosphorylase activities. In addition, for hormonally stimulated changes in covalent modification state, the resulting alterations in the intrinsic kinetic properties of the affected enzymes will permit not only the integration of different hormonal controls, but also their integration with localized intracellular regulation.

When discussing the extent to which enzymic cascades lead to the coordination of vast amounts of metabolic information, no example is better than glutamine synthetase (Table V and Fig. 11). Evidently, the concentrations of many metabolites are sensed and processed into the expression of the rate of glutamine formation.

C. Regulation of Enzyme Concentration

Variation in enzyme activity through changes in enzyme concentration is of considerable importance for cellular regulation and has been observed in response to numerous stimuli, including nutritional, developmental, circadian, and hormonal factors. The concentration of any protein will depend on the balance between synthesis and degradation. Both are complex processes and so a number of controls may determine the concentration of a given enzyme, as is discussed in detail in other chapters of this series (see Vol. III; see other chapters of this volume; see also Pine, 1972; Schimke, 1973, 1975; Goldberg and Dice, 1974; Katanuma, 1975, 1977; Holzer et al., 1975; Goldberg and St. John, 1976; Lodish, 1976; Segal, 1976). Changes in enzyme concentration in vivo have not usually been postulated to alter the intrinsic properties of an enzyme, which have been emphasized in this chapter. The regulation of enzyme concentration is not covered in any detail in this presentation, and the reader is referred to the citations above.

We should note, however, that variation in protein concentration could modify any complex formation involving proteins, and particularly any process in which protein–protein interactions occur (see Section III,A), so that some kinetic properties might be a function of enzyme concentration. The extent to which such changes in protein association may be operative in vivo, however, has not been extensively documented, but, for example, the levels of some of the noncatalytic proteins discussed in Section III,A,3 could clearly be of regulatory significance.

Brief mention will be made of the induction and repression of the synthesis of bacterial enzymes, since these phenomena enter into the discussion of patterns of metabolic control in Section IV (see Chapter 4 by Zubay in Vol. III of this series; also, Epstein and Beckwith, 1968; Magasanik, 1976; Goldberger et al., 1976). Mention has been made of the control of expression of the lac operon, an example of induction of the enzymes necessary for lactose utilization by the organism in response to the presence of lactose in the medium. Transcription is normally blocked by the binding of a specific "repressor" protein (Müller-Hill, 1975; Bourgeois and Pfahl, 1976) to a region of DNA designated the "operator"; in the presence of lactose, a small quantity of the disaccharide enters the cell and its metabolite, allolactose, can bind the repressor molecule and weaken its binding to the DNA. The operon is now free for transcription by RNA polymerase, providing that positive signals in the form of the CAP protein plus cyclic AMP are present. These compounds relay signals about the presence or absence of other carbon sources, such as glucose, that could be "preferred" over lactose (Magasanik, 1962, 1976; Carpenter and Sells, 1975; Goldberger et al., 1976). This latter phenomenon, termed catabolite repression, is mediated by the CAP protein, which binds cyclic AMP whose concentration is lowered during growth on glucose, thus suppressing the transcription of the lac operon, and other operons, when glucose is available.

Slightly different is the induction of the arabinose operon (Englesberg and Wilcox, 1974) since the "repressor" here, once freed from binding to the operator by interaction with arabinose, then functions in a positive sense. The "repressor," along with RNA polymerase, the CAP protein, and cyclic AMP, is needed for the initiation of transcription. A number of other inducible systems have been studied in some detail (see review articles cited above).

A related phenomenon is termed repression in which the presence of a compound acts as a negative signal to switch off the production of the enzymes necessary for its own biosynthesis. Thus, for example, the appearance of a necessary amino acid such as arginine in the medium will suppress the expression of the relevant biosynthetic pathway. The mechanism is not understood in all cases, although the classic hypothesis is for the binding of the metabolite with an "aporepressor," which is then able to interact specifically with the genome to block RNA synthesis, much as in the case of the lac operon. In some cases, different mechanisms may be involved (Calhoun and Hatfield, 1975; Brenchley and Williams, 1975). Independent of the mechanism, however, the phenomenon of "feedback repression" fits elegantly into our concepts of metabolic control (see Section IV).

IV. INTEGRATION OF METABOLIC CONTROLS

A. Classification of Metabolites

1. Functional Distinctions between Metabolites

The idea that molecules can act as signals has been emphasized here. Thus, in metabolism there exists a regulatory connectedness between metabolites and enzymes, different from, and probably much more complicated than, the network of chemical conversions. Much of what we think we know about metabolic regulation depends upon the identification of metabolites, or other compounds, as carrying particular regulatory signals. For this type of analysis, it is therefore useful, and perhaps in one or two cases critical, to recognize different broad classes of cellular intermediates. One such categorization is made in Table VII, where an attempt is made to distinguish different functional roles for some metabolites. I has-

TABLE VII

Classification of Some Metabolites and Other Compounds on the Basis of Function[a]

Class	Examples	Typical concentration	Number of interactions with enzymes
Simple	Intermediate in biosynthesis	Very low	Small number typically 2
Storage	Glycogen, triglycerides, starch, polyphosphate, poly-β-hydroxybutyrate	Variable, can be high	Small number
Regulatory			
"General"	Glucose-6-P, citrate, α-keto-glutarate, fructose-1,6-diP	Moderate	Multiple
"Specific"	End products, e.g., amino acid feedback inhibitors	Moderate	Small number
Pure regulatory	Cyclic AMP ? cyclic GMP, hormones ? ppGpp, ? pppGpp	Usually very low can be high	Limited number
Coupling agents	NTP/NDP/NMP NDP-sugars, folate, S-adenosyl-methionine, NAD$^+$/NADH, NADP$^+$/NADPH, acetyl-CoA/CoA	Often high	From several to hundreds

[a] The table attempts to make some distinctions between metabolites based on metabolic function. In fact, the categories are probably not all mutually exclusive. NTP, NDP, etc., refers to unspecified nucleoside triphosphate, diphosphate, etc.

ten to add that few classifications, and certainly not this one, are absolute. Included in Table VII are, for the most part, what we may consider metabolizable species, those most clearly implicated as "signals" rather than the "control elements" that we would identify frequently as proteins.

Insofar as the classification is meaningful, the entry under "number of interactions" is one of the most important. It is quite striking to compare "coupling agents" such as adenine nucleotides, which interact with hundreds of enzymes, with a metabolite in a biosynthetic sequence interacting only with the enzyme of its synthesis and subsequent transformation (not that biosynthetic intermediates cannot be of great regulatory importance: some are involved in sophisticated feedback controls). On the whole, coupling agents may be expected to have a specially evolved regulatory function simply on the basis of the large number of reactions in which they enter. Changes in the concentrations of such compounds would clearly impinge upon all manner of metabolic sequences and, at least intuitively, one would expect carefully evolved responses of the appropriate enzymes to shield against chaotic metabolic consequences. The number of enzymes known with potential control by, for example, adenine nucleotides, in part bears out this logic. PP_i is included as a coupling agent because there is growing evidence that this compound, long regarded simply as a product of nucleoside triphosphate hydrolysis to nucleoside monophosphate, may sometimes substitute ATP as a mediator of metabolic energy in certain microorganisms (Wood, 1977).

A category in Table VII lists "regulatory metabolites," and this entry is somewhat fancifully divided into "general" and "specific." By "general" are intended a number of metabolites, from the "central" areas of metabolism, which appear to interact with a considerable number of enzymes in an apparently regulatory fashion. Compounds such as citrate, α-ketoglutarate, fructose-1,6-diP, and so on come readily to mind in this context. Presumably, the concentrations of such species reflect information on the progress and proportioning of intermediary metabolism that is transmitted to a number of relevant enzymes. The class of "specific" regulatory metabolites is only marginally distinct, the intent being to distinguish those which feedback on a relatively small number of enzymes directly related to their biosynthesis.

Storage metabolites do represent a fairly definable class and in some cases, such as starch, fat, or glycogen, can be correlated well with the gross physiological life of the cell. Depletion of such stores will normally follow starvation or other energetic demands upon the cell. The poly-β-hydroxybutyrate and polyphosphate deposits, as well as some other types of granular deposits [e.g., sulfur and proteinaceous granules (Dawes and Senior, 1973)] of microorganisms, probably represent the same strategy of laying down reserves in times of nutritional plenty for periods of deprivation.

One of the most interesting classes of metabolites is what I have designated "pure regulatory," compounds whose sole function, as known so far, is related to relaying information. The classic example of cyclic AMP has already been noted. Cyclic GMP (Goldberg and Haddox, 1977) is included although the status of this compound as an unequivocal mediator of regulatory events is not as well established as cyclic AMP. Hormones, though not classic intracellular metabolites, are included on the basis that they too function exclusively as information carriers. There is some evidence also to place the guanosine polyphosphates ppGpp and pppGpp in this category, in that their accumulation in some bacteria in response to various nutritional limitations may coordinate the inhibition of several areas of biosynthesis. Neurotransmitters also are concerned with information transfer and might be considered for this class of compounds. Some, such as acetylcholine, probably have no other role than as transmitters. Others, however, such as glycine or adenosine, do have other roles in other contexts. The important point, though, is simply that a number of known compounds have functions, at least in specified locales, related entirely to the transfer of information.

While Table VII may not be definitive, it does note the type of distinctions that can be made. It is very probable that in 50 years time, an updated version of the table will include, as self-evident, not only newly discovered compounds but also newly recognized categories.

2. Intracellular Diffusion and Compartmentation: Implications for Metabolic Regulation

If interactions between various cellular components are the basis for communication within the cell, to what extent will their freedom to interact influence this role? Here one might distinguish interactions that are essentially fixed, such as subunit interactions, from the binding of a metabolite or regulatory protein to an enzyme. For a metabolite to carry a signal about the cellular state, some degree of diffusion within the cell is necessary. On the other hand, the environment within the cell is clearly not homogeneous and, without doubt, rather different from the dilute aqueous solutions in which most experiments are conducted *in vitro* (see Sols and Marco, 1970; Srere and Mossbach, 1974; Masters, 1977).

Many cells, especially of higher organisms, contain organelles that may provide physical compartmentation of both enzymes and substrates. Many metabolites do not diffuse freely across the boundary membranes of such organelles, and the control of their passage may be an important feature of metabolism. For example, a variety of mechanisms may be operative for the transport of metabolites into and out of the mitochondrion (see Stadtman, 1970). Furthermore, the metabolite concentrations within such a compartment need not necessarily reflect those of the cytoplasm. Experimentally, this is a difficult problem to approach, and until recently most estimates of intramitochondrial concentrations, for example, have been indirect, based on the assumption of equilibrium for specific enzymic reactions (see Gumaa et al., 1971). For more detailed discussion of physical compartmentation, the reader is referred to Gumaa et al. (1971), Srere and Mossbach (1974), and Ottoway and Mowbray (1977).

Spatial heterogeneity of concentrations within a given compartment does not require the type of physical barriers noted above. Many enzymes probably are not floating freely in solution but are bound to or associated with cellular structures or other proteins. The multienzyme complexes described earlier could be taken as examples of this. This view, then, would replace the cell as a sack of freely diffusable enzymes and metabolites with a cell containing a degree of structure, with more or less stable organization of enzymes and other macromolecular components. Rigorous experimental verification of this contention is not really available, but the idea is quite feasible and in some ways attractive. Note that a possible, though not obligatory, consequence of spatially organized metabolic pathways is that some metabolite concentrations might not be homogeneous within the cell. In other words, diffusion might limit the mixing of some metabolic intermediates. There is reasonable evidence, mostly from radioactive tracer experiments, for distinguishable pools of a number of metabolites (see Srere and Mossbach, 1974), although this could result from any form of compartmentation within the cell.

In this context, there has been a considerable discussion as to whether diffusion of metabolic intermediates would be sufficiently rapid for an unstructured, homogeneous cytoplasm to support observed metabolic rates (see Srere and Mossbach, 1974; Ottoway and Mowbray, 1977). Making simple calculations for diffusion times and the average separation of enzymes in a homogeneous cell, many results suggest that diffusion would not be rate-limiting, except possibly in large cells. However, the difficulty of such calculations makes it hard to have a definitive answer. The outcome rests heavily on the models analyzed, and particularly on the value taken for the diffusion coefficient, whose value in vivo is simply not known.

Of course, whether or not diffusion limitation dictates such a need does not preclude the existence of a structured cell in any case, and indeed some potential advantages of organized multienzyme complexes have been noted (Section III,A,3). If we accept some degree of internal cellular organization, what implications are there for metabolic regulation? First, many more macromolecular interactions might be present, with possible control features, than can be judged by studies of isolated components. Second, in enzymic chains, the local concentration of an intermediate might be higher than the cellular average, leading to the more efficient channelling of the intermediate through a pathway. An extreme, in fact, has been discussed where the intermediates of fatty acid biosynthesis are covalently linked to the ACP protein (see also Section III,A,3).

What are the implications of a heterogeneous, ordered cell for the evaluation of cellular control? A major point would be an added strain on the already difficult extrapolation from *in vitro* to *in vivo*. Obviously, if metabolite concentrations are spatially heterogeneous, even the best measure of such levels in cell populations will give only an average cellular concentration (without accounting for any cell-to-cell variation), with no indication of intracellular distribution. Perhaps this is a problem that can be resolved by improved technology. The methods of "in vivo NMR," for example, may in the future be able to provide very precise information on the state of a given metabolite *in vivo*. A second question raised is that another current approach, the study of enzymic properties *in vitro,* usually respects neither the physiological concentration of enzymes nor all the potential protein–protein interactions of an ordered cell. Clearly one must not lose sight of these problems.

A third prediction for a highly structured, diffusion-limited cell might be that a metabolite would be unable to carry a signal, at least in the sense stressed so far in this chapter. The logic here is that for any metabolite concentration to carry information reflecting the operation of metabolism as a whole, it must have mixed, to some extent, in a pool of the same species. This point could be very important, and, in fact, such a general diffusion limitation would alter many of our concepts of regulation. I do not think, though, that this can be generally true. Certainly, there is no evidence for such a strict curtailment of the diffusion of metabolites in general. More importantly, however, the documented existence of specific allosteric sites on proteins and other very specific "noncatalytic" interactions must be viewed as strong evidence for a degree of free diffusion of some compounds within the cell, or at least, a relevant region of the cell. Were this not so, one would have to find a new rationalization for such noncatalytic properties which, it can hardly be denied, must reflect a high degree of evolutionary specialization.

In summary, structure within the cell may provide for a number of metabolic advantages, such as an increased number of regulatory interactions and the possibility of "channeling" certain metabolites through metabolic pathways (perhaps mostly the type of compound classed as simple in Table VII). On the other hand, it is improbable that all metabolites, or even proteins, be subject to such strict limitation of diffusion, and the concept of a compound bearing a signal through its concentration must remain, at least for now, central to our thinking of cellular regulation.

3. Metabolic Coupling Agents

Coupling agents (see Table VII) were designated as certain metabolites, or really interconvertible pools of metabolites, that enter into large numbers of reactions and hence serve as an evolved "common currency" for the stoichiometric coupling of diverse areas of metabolism. The adenylate system, ATP/ADP/AMP, is one of the most important here and essentially links the production of useful cellular energy with its utilization for various cellular requirements. Other important systems are $NAD^+/NADH$ and $NADP^+/NADPH$. These are also related with energy metabolism and act as intermediaries for oxidation–reduction reactions. Such metabolites clearly have a role distinguishable from the simple metabolites involved in linear reaction sequences.

The central importance of ATP, ADP, and AMP in metabolism and its regulation is beyond doubt, and the inputs of these compounds to inhibit or activate enzymes have been recognized as a possible means of controlling the metabolic energy status of the cell. One formulation of interest is that of Atkinson (1968, 1977), who defines a special parameter to characterize the ATP/ADP/AMP system. This proposal has, in fact, provoked considerable controversy. However, I consider it instructive to discuss some of the principles leading to the definition by Atkinson of what he terms the "energy charge" of the adenine nucleotide pool, namely, $([ATP] + \frac{1}{2}[ADP])/([ATP] + [ADP] + [AMP])$.

A fundamental point to note is that a coupling agent is constantly being interconverted between different forms which, however, constitute a relatively stable pool. In other words, taking the ATP/ADP/AMP system as an example, one expects that the rates of interconversion of ATP, ADP, and AMP will be more rapid than the flux of the adenine nucleus into and out of the pool. For example, in growing bacteria, the turnover time of the adenine nucleus is of the order of 40 sec, whereas the turnover of ATP occurs generally within 1 sec (see Chapman and Atkinson, 1977). As a consequence, the distribution of adenylates among the different forms, ATP, ADP, and AMP, is likely to be a more pertinent index of the instantaneous balance between ATP consumption and utilization than the individual absolute concentrations. Phrased differently, on the short term, changes in ATP concentration must be accompanied by variation of the ADP and/or AMP concentrations. The total pool concentration ([ATP] + [ADP] + [AMP]), however, will be relatively constant on the same time scale. This argument, then, would lead one to seek a measure of the relative concentrations of adenine nucleotides as a measure of the cellular energy status.

Turning now to studies *in vitro* of purified enzymes, we note that the activities of many enzymes are modified by adenine nucleotides (see Table II), either as reactants, simple inhibitors, or allosteric effectors, and many enzymes interact with more than one adenine nucleotide. Therefore, since adenine nucleotide binding sites will tend to be saturated *in vivo* (see Section III,A,1), the ratios of concentrations of adenine nucleotides will be important in describing the kinetics of such enzymes. The "energy charge," then, is proposed as a parameter to describe the kinetic properties of certain enzymes as a function of the metabolic energy status. Enzymes in biosynthetic sequences consuming metabolic energy should be activated at high energy-charge; enzymes contributing to the supply of metabolic energy should be stimulated by low energy-charge.

The arguments above determine the general form of the "energy charge," which is a modified mole fraction. For a two-component system, matters are simpler: the absolute concentrations of two cofactor forms, A and B, may be transformed to the total pool concentration ([A] + [B]) and a mole fraction {[A]/([A] + [B])} to describe completely the system. The mole fraction indexes relative concentrations. Description of a three-component system, with a single variable reflecting relative concentrations, is less simple. Atkinson, therefore, assumes the reaction catalyzed by adenylate kinase (2 ADP $\rightleftharpoons$ AMP + ATP) to be at equilibrium, thus reducing to two the independent variables needed to define the adenylate system. These two variables are chosen to be the total pool concentration ([ATP] + [ADP] + [AMP]) and the energy charge as defined above. This latter is proposed as the parameter most relevant to correlating the evolved kinetic properties of enzymes with the energetic state of the cell. The coefficient of $\frac{1}{2}$ for ADP simply reflects the approximate difference, in terms of metabolic energy, between ATP and ADP, and also normalizes so that an energy charge of 1 would correspond to pure ATP.

For a number of enzymes studied *in vitro* whose metabolic sense (i.e., in energy-consuming or energy-producing pathways) can be fairly unequivocally rationalized, the predicted response to energy charge was indeed observed (see Atkinson, 1977). Of course, many regulatory inputs besides the energy charge will contribute to the control of a given enzyme. Another feature of the hypothesis is that the value of the energy charge *in vivo* would be stabilized by the evolved responses of the enzymes, and hence pathways, influenced by the energy charge. A lowering of the value would stimulate ATP production; an increase would provoke greater ATP consumption. This is similar to, but more complex than, the simple feedback loops in Fig. 1. Measurements in a number of cell types do suggest that the numerical value of the energy charge is relatively stable (see Atkinson, 1977).

This interpretation of the energy charge is mine, and those interested in a more complete exposition should consult the work of Atkinson and his colleagues (see Atkinson, 1977).

Several potential problems with the hypothesis should be noted. For the cells of higher organisms the compartmentation of adenine nucleotides between the cytoplasm and the mitochondrion may be important, and their concentrations need not be equal in these different locales. In other words, no unique cellular energy charge may exist. Another question is whether the total adenylate pool is unimportant in describing the kinetics of the relevant enzymes, and the answer rests essentially on the degree to which binding sites would be truly saturated by adenine nucleotides. Several other criticisms have been voiced, at least some deriving from a reaction to what is viewed as the "unscientific nature" of the energy charge. Whatever one believes, the energy charge must not be interpreted as a mystical, supernatural index of metabolism; it is, quite the contrary, a highly pragmatic, working parameter whose exact definition is sanctified only by the generalizations and premises upon which it is based; consideration of the evolved biological role of adenine nucleotides and the observed kinetic properties of enzymes. It serves to link these two factors. Of course, in focusing upon a single measure of a three-component system, instead of three separate variables, a unification and verbal simplification is introduced. Whether or not the energy charge hypothesis is "true" or "untrue," however, is not, in my view, essential to its significance. As a hypothesis, it addresses certain basic aspects of the control of metabolism in an important way, focusing upon the evolved functionality of both regulated enzymes and such fundamental metabolites as the adenine nucleotides.

4. Metal Ions of Modifiers of Enzyme Activity

The activities of a large number of enzymes depend on or are modified in some way by metal ions (Mildvan, 1970). In some instances, the metal is an intrinsic structural component of the protein; in others, the metal may enter critically into the formation of the enzyme substrate complex at the catalytic center of the enzyme. Given the importance of metals and metal ions in determining enzymic activity, it has been natural to ask whether metal ions contribute to metabolic regulation. In other words, are metal ion involvements in enzymic catalysis merely the exploitation of available chemical tools for catalysis or has their role been further refined to include control? No general answer is possible at present, but two particular ions will be taken as examples, Mg^{2+} and Ca^{2+}.

Magnesium ions influence the activities of a large number of enzymes, probably the largest class of which is the kinases (Morrison and Heyde, 1972) that utilize ATP-Mg as the normal substrate (Mildvan, 1970). This ion, however, is a modifier of many other enzymes besides kinases. Indeed, changes in intracellular magnesium would exert a profound effect on metabolism through effects on a number of susceptible enzymes. Actual intracellular free Mg^{2+} levels and physiologically significant changes in these levels, however, are not well documented. One problem is the fact that the Mg^{2+} is almost certainly partitioned among its various binding sites; indeed, it has been suggested that the large number of Mg^{2+}-binding sites would tend to buffer the free Mg^{2+} level against changes (Veloso *et al.*, 1973). Thus, while Mg^{2+} is of undoubted significance for the metabolic process, it is probably fair to say that no example of a specific, precise regulatory role for this cation has yet been demonstrated. This does not mean that no such role exists, however.

The second cation to be mentioned is Ca^{2+}. Though associated probably with fewer enzymic reactions than Mg^{2+}, there is growing evidence that Ca^{2+} is critically involved in several physiological control processes, including muscle contraction, various secretory phenomena, blood coagulation, ion transport, cell division, and so on (Berridge, 1975; Larner, 1977). In fact, current thought would tend to class Ca^{2+} in many such cases as a second messenger, somewhat analogous to cyclic AMP; indeed, the two appear often to interact in a complex way, interlocking to determine cellular events in response to external stimuli. Much current research is directed at the physiological role of Ca^{2+}, and a recent review (Kretsinger, 1976) lists some 70 known Ca^{2+}-binding proteins. How many of these represent specific control mechanisms awaits further study.

A few general comments concerning metal ion modification of enzyme activity are worthwhile. First, from a regulatory point of view, the simple requirement of a metal ion for biological activity should be distinguished from those processes in which variations of metal ion level could exert a reversible influence on the activity of a target component. Second, metal ions are not metabolized in the normal sense of the word. Indirectly, of course, changes in the concentrations of metabolites to which the metal ions bind could lead to changes in free ion concentration. This would be relevant to Mg^{2+} and ATP, say, although in this case cellular ATP levels do not appear to vary greatly. Otherwise, for a metabolically inert compound to carry a signal, concentration changes must be associated with variations in the spatial distribution of that compound. Accordingly, almost all models of possible Ca^{2+}-mediated controls are based either on exchange between intracellular and extracellular Ca^{2+}, or else partitioning of the ion between the cytoplasm and an intracellular reserve (Berridge, 1975).

B. Patterns of Metabolic Regulation

1. Feedback Inhibition and Other Metabolic Strategies

a. **Pyrimidine and Arginine Biosynthesis.** Aspartate transcarbamylase was described earlier in relation to its allosteric inhibition by CTP. The *E. coli* enzyme is placed in its metabolic context in Fig. 15. Aspartate transcarbamylase catalyzes the reaction that commits carbamyl phosphate toward the biosynthesis of pyrimidines, and, by virtue of the allosteric properties of the enzyme, the rate of this reaction will depend not only on the concentrations of substrates but also on the concentration of an end product of the metabolic pathway, CTP. The CTP concentration, then, serves as a negative signal for its own synthesis and this regulatory configuration is called negative feedback (see also Fig. 1). CTP itself is substrate for various reactions, such as RNA synthesis and CDPsugar formation, and hence is subject to variations in its rate of utilization. Thus, increased drain on CTP for synthetic purposes, through an initial depletion of the CTP concentration, would signal an increase in its production rate by alleviating the feedback inhibition. This simple type of logic is the prototype for much of our reasoning about the control of metabolism.

In fact, the biosynthesis of pyrimidines is probably under more intricate feedback control and is further coordinated (a) with purine synthesis and (b) with the production of arginine, for which carbamyl phosphate is also a

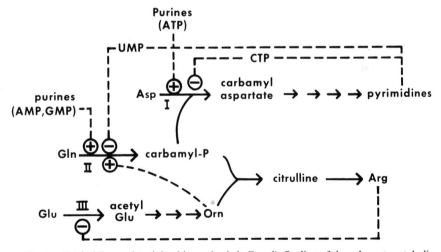

Fig. 15. Pyrimidine and arginine biosynthesis in *E. coli*. Outline of the relevant metabolic pathways with major noncovalent controls indicated: minus sign in circle, inhibition; plus sign in circle, activation; I, aspartate transcarbamylase; II, carbamyl phosphate synthetase; III, acetylglutamate synthetase. Modified from Pierard (1966).

precursor. Carbamyl-phosphate synthetase is inhibited by UMP, and to a lesser extent by UDP and UTP (Pierard, 1966). Thus, pyrimidines in general exert a negative feedback on two of the first enzymes of their biosynthetic sequence. Carbamyl-phosphate synthetase is activated by AMP and GMP, aspartate transcarbamylase by ATP, and CTP synthetase by GTP, so that purines have an overall stimulatory influence on pyrimidine biosynthesis. Balanced by the negative feedback inputs of the pyrimidines themselves, it has been suggested that this might provide a means to coordinate the relative rates of purine versus pyrimidine biosynthesis. Remember that for growing bacteria, precise relative amounts of the bases are utilized in nucleic acid synthesis.

Arginine is another end product of carbamyl phosphate metabolism. If there is a sufficiency of carbamyl phosphate, arginine itself is able to exert an autoregulatory feedback on its production by inhibiting acetylglutamate synthetase (Vyas and Maas, 1963) and reducing the production of its other precursor, ornithine. Arginine does not control the supply of carbamyl phosphate directly, however. Instead, ornithine is able to override the inhibitory effects of pyrimidine nucleotides on carbamyl-phosphate synthetase (Pierard, 1966). Thus, in situations where the arginine level is reduced and hence is signaling its own production, any build-up of ornithine would ensure production of the other requisite for arginine synthesis, carbamyl phosphate. The scheme provides an elegant model for the controlled partitioning of carbamyl phosphate between two pathways in which it enters. High arginine and pyrimidine would close off both pathways; high arginine and low pyrimidine levels would channel carbamyl phosphate into pyrimidines; low arginine and high pyrimidine would funnel the intermediate into arginine synthesis.

b. Purine Interconversion. The synthesis of purines (Magasanik, 1962) also provides a system in which feedback controls may establish a balance between two products deriving from a common precursor (Fig. 16). In this instance, *de novo* purine biosynthesis yields IMP from which divergent

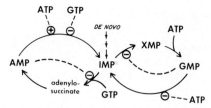

Fig. 16. Purine nucleotide interconversion. The pathways allowing for the balance of the availability of adenine and guanine nucleotides are illustrated together with some potential regulatory connections. Symbols as for Fig. 15. After Magasanik (1962) and Stadtman (1970).

pathways lead to either AMP (and subsequently ADP and ATP) or GMP (and GDP and GTP). In either case, metabolic energy is consumed for the conversion of IMP to GMP (ATP consumed) or AMP (GTP consumed). Furthermore, there exist return reactions from GMP or AMP to IMP (see Fig. 16). Two points will be made from this example: (a) the nature of the controls suggests a regulation of the relative production of AMP and GMP and (b) appropriate controls must exist to prevent energy-wasting metabolic cycles (e.g., IMP → XMP → GMP → IMP). With regard to the fate of newly synthesized IMP, both AMP and GMP exert feedback inhibitions on their respective synthetic pathways. In addition, there is a "crossed" requirement of GTP for AMP synthesis and ATP for GMP synthesis. Both of these features will proportion synthesis to the purine present in metabolically lower concentration. Besides *de novo* synthesis through IMP, however, the system also allows redistribution of existing purine nucleus between adenine and guanine nucleotides. Thus, following through the scheme in Fig. 16, it can be seen that high GTP concentration will push toward AMP production, and high ATP toward GMP. We can see, furthermore, that the nucleoside triphosphate involvements tend to oppose the wasteful cycling mentioned earlier. For example, as a substrate, ATP favors the sequence IMP → XMP → GMP but hinders the completion of the cycle by inhibiting the conversion of GMP to IMP. The question of these potentially "futile cycles" will be raised in a little more detail later.

c. **Biosynthesis of Lysine, Methionine, Threonine, and Isoleucine.** The synthesis of lysine, methionine, threonine, and isoleucine (Fig. 17) (Truffa-Bachi, 1973; Truffa-Bachi and Cohen, 1973; Stadtman, 1970) introduces yet another regulatory feature, the existence of isoenzymes with different regulatory properties, in particular different sensitivities to feedback inhibitors. In addition, the system involves multifunctional enzymes of which mention was made in Section III,A,3. Here, the "traffic control" problem is clearly more complex than the examples noted above since four different amino acids, needed for protein synthesis, derive from a common precursor, aspartate, and share different common segments of the biosynthetic chain.

The first committed step for the biosynthesis from aspartate of all four amino acids is common, but is catalyzed by three distinct aspartokinases with different regulatory characteristics. Aspartokinase III is inhibited by lysine; aspartokinase II is repressed by methionine; aspartokinase I is inhibited by threonine and homoserine, and also repressed in the presence of threonine and isoleucine. In this way the total flow of carbon through aspartyl-P is geared to the requirements for all four amino acids; the separate controls ensure that feedback by one amino acid does not have detrimental repercussions for the production of the others. However, any

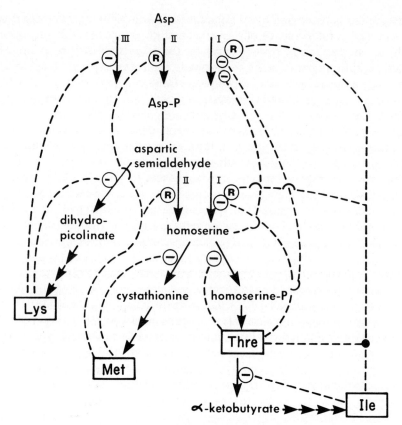

Fig. 17. Biosynthesis of lysine, methionine, threonine, and isoleucine in *E. coli*. The complex regulation of the biosynthesis of these amino acids is shown using the symbolism of Fig. 16. In addition, Roman numerals identify the three isoenzymes of aspartokinase and the two isoenzymes of homoserine dehydrogenase; R in circle denotes feedback repression; the solid circle joining feedbacks from threonine and isoleucine indicates the simultaneous requirement of these two amino acids to effect the repression of the indicated enzymes. Based on the model of Stadtman (1970).

negative effect on the total common flux through aspartyl-P provoked by a sufficiency of a given amino acid must still divert flow away from its specific synthetic path. Inspection of Fig. 17 will show that invariably this is achieved. For example, excess lysine will inhibit aspartokinase III, thus reducing aspartyl-P formation, but will additionally inhibit the first step committed specifically to its own biosynthesis. Aspartic semialdehyde is also a common intermediate for methionine, threonine, and isoleucine synthesis. Again, multiple enzymes exist to handle the common transformation of aspartic semialdehyde to homoserine. One homoserine dehydrogenase is repressed by methionine; one is inhibited by threonine and repressed in the presence of threonine and isoleucine. Partitioning occurs, then, on the same principle as at the aspartokinase reaction. The homoserine branch point is further controlled by feedbacks from methionine and threonine. Finally, the last linear portion from threonine to isoleucine is regulated through a feedback inhibition on α-ketobutyrate formation. In fact, careful study of Fig. 17 will probably reveal the elegance of the regulation better than verbal description. The evolution of separate enzymes to respond to separate controls might be viewed as an alternative to evolving new allosteric sites on a single enzyme. A final feature of interest in the above system is that the active sites corresponding to aspartokinase I and homoserine dehydrogenase I, as well as aspartokinase II and homoserine dehydrogenase II, are carried in single polypeptide chains. These are examples of multifunctional enzymes. Precise rationalization of the advantages of this arrangement is not really possible, although some of the ideas expressed in Section III,A,3 may be valid. A clear consequence, though, is that the two enzyme activities will always be produced at a fixed ratio.

  d. Pyruvate Branch Point. As a final example of the sort of logic used in trying to understand metabolic regulation, a somewhat larger and more central area of metabolism, the Krebs cycle, will be viewed (Fig. 18) (see Atkinson, 1977). The Krebs cycle furnishes both metabolic energy and

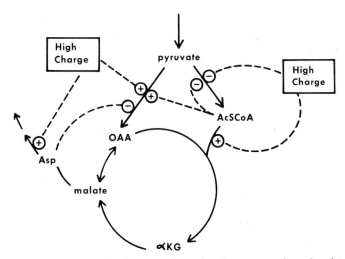

Fig. 18. Pyruvate branch point in yeast. Control at the pyruvate branch point in yeast grown on glucose is shown using the symbols used in Fig. 15. High charge refers to the adenylate energy charge (see text). OAA, oxalacetic acid, AcSCoA, acetyl-SCoA. Modified from Miller and Atkinson (1972).

intermediates for biosynthesis. The example in question, of yeast growing on glucose, expresses this divergence at the level of pyruvate. Entry into the cycle via acetyl-CoA will contribute both to energy production and the supply of Krebs cycle intermediates to be drained as biosynthetic precursors. On the other hand, pyruvate carboxylase action produces oxalacetic acid, which may be considered as replenishing carbon taken out of the cycle for biosynthesis. For such a basic branch point as this, it is reasonable that the cellular energy status, as indexed for example by the adenylate energy charge, is an important control. Adequate metabolic energy will tend to favor oxalacetic acid synthesis from pyruvate over acetyl-CoA production, commensurate with a greater commitment to biosynthesis. Also, the feedback activation of oxalacetic acid formation by acetyl-CoA will provide a further control on the partition of pyruvate to ensure that the cycle is not limited by the supply of oxalacetic acid. The negative feedback by aspartate may be an indicator of biosynthetic adequacy signaling a reduction in the need for anabolic precursors.

e. **Futile or Substrate Cycles.** A coherent metabolism was stressed on the outset as being an intuitive requisite for the cell. By this is meant that an organism that was extremely inefficient in coupling the metabolic energy produced to the optimal regime of energy utilization (biosynthesis, motion, transport, etc.) is unlikely to have competed successfully in the course of evolution. Inherent in this idea is that opposed reaction sequences that, on paper, could cycle with the consumption of ATP or another mediator of metabolic energy would in normal circumstances be controlled to obviate what have been termed "futile cycles." Examples have just been noted in relation to purine interconversions. This point has been discussed at some length [Stadtman (1970) lists a number of examples], and in general such cycling is probably limited; otherwise it is difficult to envisage any effective utilization of metabolic energy. For example, the reactions catalyzed by phosphofructokinase and fructose-1,6-diphosphatase constitute such a couple and would formally permit a cycling between fructose-6-P and fructose-1,6-diP, with the hydrolysis of ATP to ADP (Fig. 19). The necessity for both enzymes is rationalized in the need, at different times, for either glycolysis or gluconeogenesis. Thus, one envisages that cellular controls, for example, via metabolite concentrations, would have evolved to allow kinetically only one direction of the metabolic flux at a given time. In the example, the effects of ATP, AMP, and fructose-1,6-diP would indeed appear to suppress futile cycling.

Having said this, however, though the argument above must be to some extent valid, it is difficult to be quantitative. In other words, how inefficient must the coupling between energy production and utilization be before metabolism becomes chaotic? We have noted several situations

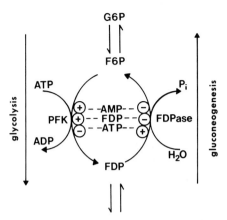

Fig. 19. A potential futile cycle. The figure shows the opposing reactions catalyzed by phosphofructokinase (PFK) and fructose-1,6-diphosphatase (FDPase). This illustrates, in the first place, a potential "futile cycle" that unchecked would cycle between fructose-6-P (F6P) and fructose-1,6-diphosphate (FDP) with the hydrolysis of ATP to ADP and P_i. In addition are shown noncovalent interactions by AMP, fructose-1,6-diP and ATP with these enzymes which had previously been felt to suppress the operation of the system as a cycle (see text).

where control has apparently been achieved through a considerable expenditure of metabolic energy. Thus, regulatory features have been evolved despite a high metabolic cost. Recently, this same type of reasoning has led to an interest in the regulatory potential of so-called "futile cycles" (Katz and Rognstad, 1976; Newsholme and Crabtree, 1976, 1978; Newsholme and Start, 1973). It can be shown simply that an effector that interacts with two oppositely directed enzymes of a cycle can exert a more profound influence on the flux through the chain than through interaction with a single enzyme, but only if the cycling rate is high. The most appealing example of this phenomenon is the phosphofructokinase/fructose-1,6-diphosphatase system in insect flight muscle. Lardy and his colleagues presented elegant evidence (Clark *et al.*, 1973) to correlate the rate of cycling at the fructose-6-P/fructose-1,6-diP level with the need to generate heat in order to maintain a temperature compatible with flight. Thus, the ATP hydrolysis associated with the operation of this cycle appears to have a defined functional role, namely, the production of heat. Another example of cycling has been discussed in relation to covalently modified enzymes (see Section III,B). Here, the simultaneous action of the opposite converting enzymes (e.g., a protein phosphatase and kinase) would form such a cycle and a number of potential advantages, in terms of the response properties of the system, can be postulated (Stadtman and Chock, 1978). Note, however, that here the substrate involved in the cycle is an enzyme and hence generally at a relatively low concentration. Thus, even rapid cycling would consume much less ATP than a system involving a metabolite present at, say, millimolar concentrations. Substrate cycling as a more general regulatory phenomenon has its protagonists, and time will be needed to evaluate more fully this idea. For the present, we may propose such systems as potential examples of an improved regulatory response achieved at a certain metabolic price; at the same time, the general theme that, for most of metabolism, the uncoordinated and futile consumption of ATP is avoided, was probably one of the major evolutionary pressures.

2. Control of Metabolic Pathways by Extracellular Signals

It is artificial to separate metabolic regulation into intracellular and extracellular components since the metabolism of virtually any cell is in some way geared to its environment. Nonetheless, many of the controls of the preceding section seem more fashioned to an intracellular partitioning of the available resources, that is, the exercise of a more localized, autonomic control of cellular metabolism. In addition, and this becomes especially complex for higher organisms, we can usefully identify mechanisms whereby extracellular signals are translated into specific, intracellular regulatory action. Essentially, the several means of regulation of enzyme activity discussed earlier can enter into this type of control. Indeed, it was noted (Section III,C) how enzyme concentrations are often a function of the presence of nutrients or hormones or other agents in both higher and lower organisms. Discussion in this section, however, will focus on the involvement of other means of modulating enzyme activity, ligand binding, and covalent modification in response to external stimuli. Again examples will be selective and not comprehensive.

a. Control by ppGpp and pppGpp in Bacteria. Stringent strains of *E. coli* and some other bacteria accumulate guanosine polyphosphates (ppGpp and pppGpp) in response to various types of nutritional deprivation (Cashel, 1975), such as starvation for a required amino acid. These nucleotides, which can attain concentrations of several millimolar, appear to be synthesized on the ribosome, and a stimulus for their formation is the presence of uncharged tRNA. Thus, a diminished supply of an amino acid, reflected in an increase in the concentration of uncharged tRNA, could provoke the production of ppGpp and pppGpp. Associated with amino acid starvation is the decreased synthesis of RNA, carbohydrates, nucleotides, and lipids. An attractive hypothesis, then, is that these guanine nucleotides function as a coordinating signal for this general metabolic suppression. It has been shown that ppGpp inhibits acetyl-CoA carboxylase (Polakis *et al.,* 1973) and glycerol-3-P acyltransferase (Lueking and Goldfine, 1975) from *E. coli,* potentially controlling fatty acid and phospholipid synthesis (see also Bloch and Vance, 1977). The inhibition of the latter enzyme occurs only when palmitoyl-CoA and not palmitoyl-ACP is the substrate. Thus, ppGpp control of phospholipid biosynthesis from *de novo* fatty acid production (supplied as the ACP derivative) might be directed at acetyl-CoA carboxylase, but when the fatty acids derive from phospholipid turnover (supplied as acyl-CoA), the control could be effected through the acyltransferase. There are also reports of an inhibition by ppGpp of the membrane-associated phosphoribosyltransferase (Hochstadt-Ozer and Cashel, 1972) as well as adenylosuccinate synthetase and IMP dehydrogenase (Gallant *et al.*, 1971). The former system is reponsible for the uptake of purine bases and the latter for the synthesis of AMP and GMP from IMP.

Although matters may be more complicated than the simple description above, there are some grounds to view the production of these guanine nucleotides as a specialized response of the cell to altered nutritional regimes. The ppGpp and pppGpp might, thus, provide a coordinating mechanism for inhibiting the several biosynthetic processes that are shut down as a result of nutritional restriction. This would predict a number of specific interactions of ppGpp and pppGpp with enzymes involved in these various areas of metabolism, such as noted above. Taking this kind of view, ppGpp and pppGpp could be tentatively classified as "pure regulatory" within the framework of Table VII (see also Tomkins, 1975). These nucleotides, however, can be present at concentrations several orders of magnitude greater than, for example, cyclic AMP or hormones in higher organisms. In addition, their synthesis is at the expense of ATP and, moreover, the half-life of these compounds *in vivo* is 30 sec or less (Cashel, 1969; Lund and Kjeldgaard, 1972). Thus, the maintenance of ppGpp and pppGpp at relatively high levels represents a not inconsiderable expenditure of energy. One might, therefore, wonder, as did Cashel (1975), whether these compounds have metabolic roles besides the purely regulatory function implied above, a question that must await further study.

b. Hormones. As a rather naive simplification one might view many hormonal signals as providing a coordinated operation of diverse cell types in the interests of the organism's survival. In some cases, we can trace this type of rationalization to the molecular properties of the enzymes themselves. Most hormones circulate at exceedingly low concentrations (10^{-9}–10^{-11} M; Tepperman, 1973), and yet a brief review of some of their effects on cells demonstrates the amazing potency of these compounds to modify cellular metabolism (and of course to cause numerous other dramatic changes). Most current thinking about hormonal control seeks some very specific locus of interaction of the hormone with the cell, the result of which can provoke, with amplification where needed, the biochemical effects of the hormone. The "locus of interaction" is the receptor, which in the best-characterized cases seems likely to be protein or protein-containing in nature. The cell specificity of hormone action can thus be correlated with the presence of the appropriate receptor types in the target cell.

Mechanistically, the action of several hormones, notably the steroid and thyroid hormones, involves interaction with a receptor and the subsequent migration of the hormone–receptor complex to the nucleus where controls are exerted on the expression of the genetic material. In other cases, including catecholamines and probably some actions of peptide hormones, the hormone exerts its action without entry into the cell. Here, hormone binding is able to trigger internal biochemical events that lead to the appropriate metabolic adjustments. Expression of the hormone binding requires an intracellular signal; the best understood of these is cyclic AMP, discovered by Rall and Sutherland (1958), which is a so-called "second messenger" for a number of hormones. Growing evidence points also to Ca^{2+} in this role. Table VIII lists some hormones, their metabolic

TABLE VIII

Some Metabolic Effects of Insulin, Glucagon, and Catecholamines

Hormone	Mediators	Increased	Decreased
Insulin	?	Glucose, nucleotide and amino acid uptake, glycogen synthesis, protein synthesis, nucleic acid synthesis, fatty acid synthesis, pentose pathway, glycolysis, cell growth	Glycogenolysis, protein degradation, lipolysis
Glucagon	Cyclic AMP	Glycogenolysis, gluconeogenesis, protein degradation, lipolysis, ketogenesis, fatty acid release	Glycogen synthesis, fatty acid synthesis
Catecholamine (β)	Cyclic AMP	Glycogenolysis, gluconeogenesis, lipolysis, ketogenesis, fatty acid release	Glycogen synthesis, fatty acid synthesis, protein synthesis
Catecholamine (α)	? Ca^{2+}	Glycogenolysis, gluconeogenesis	Glycogen synthesis

effects, and some of the intracellular signals potentially involved in their action.

c. Control of Glycogen Metabolism. To exemplify this type of hormonal control, we will return to glycogen metabolism and its hormonal control, especially by epinephrine, glucagon, and insulin. Even though this is probably one of the best understood and important examples of hormonally controlled systems, knowledge is far from complete.

Some of the noncovalent interactions involved in the control of glycogen metabolism are illustrated in Fig. 20. As noted earlier, several of the enzymes have sophisticated regulatory properties as regards ligand binding. Inspection of Fig. 20, in the vein of our discussion of intracellular metabolic controls in Section IV,B,1, will show that most of the effector interactions display a logical strategy. The differential influence of P_i and glucose-6-P on the degradative and synthetic arms of glycogen metabolism would tend to suppress potential futile cycling between sugar phosphate and glycogen. Glucose-6-P activation of glycogen synthase is an example of a positive feed-forward control; if glucose-6-P concentration indexes a plentiful supply of glucose to the cell, then this link would tend to stimulate a reaction "downstream" leading to the storage of glucose in the form of glycogen. Imposed on all this control are the hormonal inputs (see Larner, 1977), which in this case find expression mainly in changes in the covalent phosphorylation states of several key enzymes. In terms of the metabolic flow, the activities of glycogen synthase and phosphorylase are the targets of hormone action.

The enzymology of the various components was discussed in Section III,B,2; these elements may now be placed in their metabolic context in relation to hormonal control (Fig. 21). Epinephrine and glucagon provoke

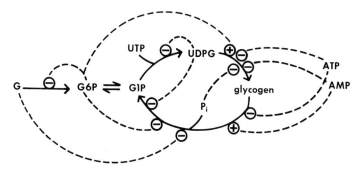

Fig. 20. Noncovalent controls of glycogen metabolism. Using the symbols as described in Fig. 16, a number of possible noncovalent interactions relevant to the regulation of glycogen metabolism are illustrated. The feedback of glucose-6-P (G6P) on its formation would relate to hexokinase and not glucokinase action.

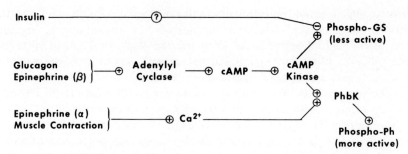

Fig. 21. Covalent controls of glycogen metabolism. Extracellular inputs, on the left, are connected, with intervening mechanistic steps where known, with the activities of glycogen synthase and phosphorylase. The minus or plus here denote either changes in concentration or enzyme activity. Epinephrine enters twice: its "traditional" action is thought to be mediated by β-adrenergic receptors, but in some tissues (e.g., rat liver) its primary control of glycogen metabolism seems through α-adrenergic receptors and possibly Ca^{2+}. Insulin action remains one of the most important open questions in the field and only its very well-established action to inactivate glycogen synthase is noted here. Phospho-GS, phosphorylated glycogen synthase; PhbK, phosphorylase kinase; Phospho-Ph, phosphorylated phosphorylase.

the mobilization of cellular glycogen reserves. Liver glycogen contributes to the maintenance of blood glucose levels; muscle glycogen usage is related primarily to supplying energy for muscular activity, such as is often correlated with the increased release of epinephrine following, for example, fright. Insulin, as regards glycogen metabolism, opposes both of the above hormones, signalling the deposition of glycogen when blood sugar levels are elevated. Not all of the details of the mechanisms of action of the three hormones are known yet, but our knowledge is sufficient to postulate a rather elegant link between the gross, physiological properties of the hormones and the metabolic consequences of their action.

Epinephrine (as a β-adrenergic agonist) and glucagon interact with specific cell-surface receptors that are able to translate this binding into an activation of adenylyl cyclase. The result is an increased intracellular cyclic AMP concentration that causes an activation of the cyclic AMP-dependent protein kinase. The cyclic AMP signal, then, is translated into the phosphorylation of the various substrates for cyclic AMP-dependent protein kinase. For our present discussion, the principal substrates are glycogen synthase, phosphorylase kinase, and possibly the phosphatase inhibitor, i_1. Action on the first leads to the inactivation of glycogen synthesis. Phosphorylase kinase, on the other hand, is activated and this leads in turn to the phosphorylation and activation of phosphorylase.

Generation of active phosphatase inhibitor would tend to reinforce the stimulation of kinases. Glycogen synthesis is suppressed and its degradation is favored. The result is an increased supply of sugar or sugar-phosphate to serve as a metabolic fuel, commensurate with the believed physiological roles of glucagon and epinephrine.

Another important input into this system is mediated by Ca^{2+}. The best studied tissue is muscle, in which contraction is associated with increased cytoplasmic Ca^{2+} concentration that stimulates phosphorylase kinase, in turn provoking the activation of phosphorylase and hence glycogen degradation (Ozawa et al., 1967; Brostrom et al., 1971; Ebashi, 1974). This may be rationalized with the increased requirement for metabolic energy in the face of muscular work. Recently, Ca^{2+} has also been invoked as a potential messenger for epinephrine action on rat liver cells, where the catecholamine appears to function through α-adrenergic, and not β-adrenergic, receptors to control glycogen metabolism (Hutson et al., 1976; Cherrington et al., 1976).

The detailed mechanism of action of insulin (Larner and Haynes, 1975; Czech, 1977; Larner et al., 1978) is not known with the same certainty, although its action to dephosphorylate and inactivate glycogen synthase is well established (Villar-Palasi and Larner, 1960; Roach et al., 1977). Most conceivable sites of action for insulin have been proposed at one time or another; it is now fairly well accepted that direct effects on the cyclic AMP concentration are not primary to the action of the hormone. In terms of glycogen synthase, insulin definitely opposes the action of epinephrine and glucagon, and in some instances, also their effects on phosphorylase. Of the endocrine signals, then, insulin appears alone in balancing the glycogenolytic stimuli of several hormones, including epinephrine and glucagon.

In addition to this poised endocrine control, note that hormonal controls will themselves be integrated with the internal cellular regulation. The "activation" and "inactivation" of glycogen synthase and phosphorylase discussed above are in fact contingent on the cellular metabolite concentrations. The hormones modify, in this case, the phosphorylation states of enzymes and lead to changes in the intrinsic properties of the regulatory enzymes. Under permissive metabolic conditions, the import of the hormonal signals will be realized. This may be viewed as an integration of local, metabolic controls with more widespread hormonal regulation than is expressed in the covalent phosphorylation of the enzymes (for example, Roach and Larner, 1976). Generally, one might envision the needs of the cell as subservient to the demands of the whole organism, for which reason hormonal metabolic controls might tend to overrride the local cellular ones. Clearly, however, the individual cells must be autonomous to some degree in conducting their metabolism and further may be capable of resisting the hormonal imperatives when necessary.

The overall picture of the control of glycogen metabolism, even allowing for our incomplete knowledge of some aspects, is one in which several types of regulatory mechanism are exploited in a complex control system that integrates different hormonal stimuli as well as local metabolic conditions to determine the rate of glycogen synthesis or degradation.

   d. **General Metabolic Interrelations of Insulin, Glucagon, and Epinephrine.** Although the control of glycogen metabolism has been isolated for more detailed discussion, it is important to note that hormonal effects on metabolism are much more widespread. Often, the mechanism is known less completely, but this section will conclude with a short discussion of some of the general metabolic results of the action of insulin, catecholamines, and glucagon. It will be seen that here too the hormonal inputs can be viewed as counterbalancing stimuli whose resultant is an important determinant of the metabolic regime (see Table VIII).

   Insulin action may be characterized to a large extent as anabolic, stimulating both the uptake of glucose and amino acids as well as several biosynthetic pathways. Insulin promotes glycogen deposition, as noted; in addition, glycolysis, fatty acid synthesis from glucose, and protein synthesis are increased, and protein degradation and triglyceride hydrolysis are decreased (Czech, 1977; Larner and Haynes, 1975; Tepperman, 1973). In fat and muscle, insulin directly stimulates the transport of glucose into the cell. Thus, increased metabolic activity could be anticipated on the basis of increased flux of nutrient to the cell, but the hormone clearly exerts a number of more specific effects on metabolism. Besides glycogen synthase, another important enzyme known to be dephosphorylated and activated following insulin action is pyruvate dehydrogenase (Denton *et al.*, 1975). This suggests a greater flux to acetyl-CoA, which can correlate with increased metabolic rate and fatty acid synthesis. In this context, a stimulation of acetyl-CoA carboxylase (Halestrap and Denton, 1974) has also been postulated. One of the well-documented effects of insulin is its antilipolytic action, although the mechanism is not clear. In general, then, the results of insulin activity are to promote the laying down of various forms of storage materials.

   Many of the gross metabolic consequences of glucagon or β-adrenergic stimulation by epinephrine are similar and can be discussed together. This is a reflection of the fact that increased cyclic AMP concentration is thought to mediate at least some of the effects of glucagon and epinephrine (as a β-adrenergic agonist). Note that many other results of catecholamine action cannot be grouped together with glucagon (see Innes and Nickerson, 1975). In contrast with insulin, both agents may be considered essentially catabolic, promoting generally the mobilization of reserve materials, and, in particular, causing the formation of glucose or glucose phosphate either from glycogen or via gluconeogenesis (see Tepperman, 1973). The production of glucose from lactate or pyruvate is stimulated by either epinephrine or glucagon. One probable site of this control in liver is pyruvate kinase, whose inactivation would favor the shunting of phosphoenolopyruvate toward gluconeogenesis rather than pyruvate formation. This enzyme undergoes phosphorylation (which inactivates) catalyzed by cyclic AMP-dependent protein kinase (Engström, 1978). This could provide a mechanistic link between hormone action and metabolic control. Another potential hormonal control point is the entry of pyruvate into the mitochondrion and its subsequent carboxylation to oxalacetic acid (Garrison and Haynes, 1975; Adam and Haynes, 1969).

Epinephrine inhibits the synthesis of fatty acids but stimulates the hydrolysis of triglycerides by hormone-sensitive lipase and the release of fatty acids (Tepperman, 1973). The lipase activation, which occurs also in response to glucagon action, is thought to be mediated by cyclic AMP-dependent protein kinase (Steinberg *et al.,* 1975). Since there is evidence for the cyclic AMP-dependent phosphorylation of acetyl-CoA carboxylase (Carlson and Kim, 1973; Seubert and Hamm, 1975; Hardie and Cohen, 1978), this could be a mechanism for the hormonal inhibition of *de novo* fatty acid synthesis.

In fact, all the substrates for cyclic AMP-dependent protein kinase *in vitro* (Table IV) may be considered potential sites for regulation by hormones that modify the cyclic AMP concentration, although this is not yet clearly established in all cases. Obviously, such a multiplicity of target enzymes could underlie a potent hormonal control of metabolism, with a more extensive coordination resulting from the broad specificity of cyclic AMP-dependent protein kinase than was discussed earlier in relation to coordinating glycogen metabolism. One may wonder also whether other protein kinases or other covalent modification mechanisms are involved in this sort of control.

A number of actions of glucagon resemble those of β-adrenergic stimulation and oppose the effects of insulin. An important difference between glucagon and catecholamines is tissue specificity which is not emphasized in this brief discussion. A major target of glucagon action is the liver where glucose-6-P, produced from glycogenolysis or gluconeogenesis, can be hydrolyzed to glucose by glucose-6-Pase for delivery into the bloodstream. On the other hand, epinephrine action on muscle, which lacks glucose-6-Pase, will produce sugar phosphate to serve primarily as a source of metabolic energy for muscular activity.

In summary, while it is not possible to present such elegant mechanistic detail as for glycogen metabolism, it is evident that hormones exert a complex and integrative effect on metabolism as a whole. In many ways, the stimuli of catecholamines, insulin, glucagon, and other agents not discussed are sensed and coordinated at various points of the metabolic process. This area will undoubtedly undergo intense investigation in the coming years.

V. CONCLUSION

The subject of this chapter, as should now be apparent, pervades much of biochemistry, and as a consequence I could not cover in detail all the regulatory systems that have been studied. I have thus tried to provide references to articles that can serve as starting points for study in depth of the topics treated here. I have, however, presented what I consider the main conceptual principles relevant to understanding enzyme regulation, and together with the examples cited, extrapolation to systems not covered should pose no fundamental problem.

In tone, I have stressed the functional aspects of enzymes and mechanism only in terms of serving that functional goal. In this context, particular emphasis has been placed on the role of specific molecular interactions, not only for the chemical interconversions of metabolism but also for their regulation. Molecular interactions, whether protein–effector, protein–protein, hormone–receptor, and so on, clearly lie at the heart of biochemical communication. In some instances, it has been possible to explore in some detail the relation between enzymic properties and metabolic function.

I would conclude, however, with a word of caution. Accepting a few premises concerning the functionality implied by evolutionary design, the concept of regulation, in broad terms, is rather simple to appreciate. Understanding and interpreting regulation in detail, however, is difficult by comparison with the clean, chemical aspects of biochemistry; the gratuitous inclusion of the term "regulation" in the titles of many papers bears witness to this difficulty. One fundamental limitation, however, is our own capacity to recognize regulatory features. The mechanized and automated background of present Western society finds one expression in the conceptual and verbal tools that can be applied to the analysis of metabolic regulation. This chapter, for example, abounds with "feedbacks" and "sensitivities" and so on. How many features essential to the coherent behavior of cells and organisms are simply beyond our powers of conceptualization or expression? Before I digress into troubled philosophical waters, my point is just that even the best-studied examples of enzyme regulation cited above must be viewed less as rigid, incontrovertible facts than as examples of our present conceptual approach to regulation, a logical framework within which this aspect of cellular chemistry may be discussed. This may seem unsatisfactory but it also represents one of the greatest challenges in biochemistry: reconciling the bare chemical properties of cellular components with their evolved roles in the life of the cell.

ACKNOWLEDGMENTS

REFERENCES

Atkinson, D. E. (1977). "Cellular Energy Metabolism and its Regulation." Academic Press, New York.

Atkinson, D. E., and Walton, G. M. (1965). Kinetics of regulatory enzymes. *Escherichia coli* phosphofructokinase. *J. Biol. Chem.* **240**, 757–763.

Atkinson, D. E., Roach, P. J., and Schwedes, J. S. (1975). Metabolite concentrations and concentration ratios in metabolic regulation. *Adv. Enzyme Regul.* **13**, 393–411.

Avruch, J. A., Leone, G. R., and Martin, D. B. (1976). Identification and subcellular distribution of adipocyte peptides and phosphopeptides. *J. Biol. Chem.* **251**, 1505–1511.

Barnes, L. D., Kuehn, G. D., and Atkinson, D. E. (1971). Yeast diphosphopyridine nucleotide specific isocitrate dehydrogenase. Purification and some properties. *Biochemistry* **10**, 3939–3944.

Barnes, L. D., McGuire, J. J., and Atkinson, D. E. (1972). Yeast diphosphopyridine nucleotide specific isocitrate dehydrogenase—Regulation of activity and unidirectional catalysis. *Biochemistry* **11**, 4322–4329.

Bell, G. I., Valenzuela, P., and Rutter, W. J. (1976). Phosphorylation of yeast RNA polymerases. *Nature (London)* **261**, 429–431.

Berridge, M. J. (1975). The interaction of cyclic nucleotides and calcium in the control of cellular activity. *Adv. Cyclic Nucleotide Res.* **6**, 1–98.

Bloch, K., and Vance, D. (1977). Control mechanisms in the synthesis of saturated fatty acids. *Annu. Rev. Biochem.* **46**, 263–298.

Bloxham, D. P., and Lardy, H. A. (1973). Phosphofructokinase. *In* "The Enzymes" (P. D. Boyer, ed.), 3rd ed., Vol. 8, pp. 240–278. Academic Press, New York.

Bourgeois, S., and Pfahl, M. (1976). Repressors. *Adv. Protein Chem.* **30**, 1–99.

Brand, I. A., and Söling, H. D. (1975). Activation and inactivation of rat liver phosphofructokinase by phosphorylation–dephosphorylation. *FEBS Lett.* **57**, 163–168.

Brand, I. A., Müller, M. K., Unger, C., and Söling, H. D. (1976). *In vivo* and *in vitro* interconversions of active and inactive forms of phosphofructokinase in liver. *FEBS Lett.* **68**, 271–274.

Brandt, H., Killilea, S. D., and Lee, E. Y. C. (1974). Activation of phosphorylase phosphatase by a novel procedure: Evidence for a regulatory mechanism involving the release of a catalytic subunit from enzyme inhibitor complex(es) of higher molecular weight. *Biochem. Biophys. Res. Commun.* **61**, 598–604.

Braun, T., and Birnbaumer, L. (1975). Hormone-sensitive adenyl cyclase systems: Properties and function. *Compr. Biochem.* **25**, 65–106.

Brenchley, J. E., and Williams, L. S. (1975). Transfer RNA involvement in the regulation of enzyme synthesis. *Annu. Rev. Microbiol.* **29**, 251–274.

Brostrom, C. O., Hunkeler, F. L., and Krebs, E. G. (1971). The regulation of skeletal muscle phosphorylase kinase by Ca^{2+}. *J. Biol. Chem.* **246**, 1961–1967.

Burchell, A., Cohen, P. T. W., and Cohen, P. (1976). Distribution of isoenzymes of the glycogenolytic cascade in different types of muscle fibre. *FEBS Lett.* **67**, 17–21.

Calhoun, D. H., and Hatfield, G. W. (1975). Autoregulation of gene expression. *Annu. Rev. Microbiol.* **29**, 275–299.

Cannon, W. B. (1926). Some general features of endocrine influence on metabolism. *Am. J. Med. Sci.* **171**, 1–20.

Cannon, W. B. (1929). Organization for physiological homeostasis. *Physiol. Rev.* **9**, 399–431.

Cantoni, G. L. (1975). Biological methylation: Selected aspects. *Annu. Rev. Biochem.* **44**, 435–451.

Carlson, C. A., and Kim, K. H. (1973). Regulation of hepatic acetyl coenzyme A carboxylase by phosphorylation and dephosphorylation. *J. Biol. Chem.* **248**, 378–380.

Carpenter, G., and Sells, B. H. (1975). Regulation of the lactose operon in *Escherichia coli* by cAMP. *Int. Rev. Cytol.* **41**, 29–58.

Cashel, M. (1969). The control of ribonucleic acid synthesis in *Escherichia coli*. IV. Relevance of unusual phosphorylated compounds from amino acid-starved stringent strains. *J. Biol. Chem.* **244**, 3133–3141.

Cashel, M. (1975). Regulation of bacterial ppGpp and pppGpp. *Annu. Rev. Microbiol.* **29**, 301–318.

Cassel, D., and Pfeuffer, T. (1978). Mechanism of cholera toxin action: Covalent modification of the guanyl nucleotide-binding protein of the adenylate cyclase system. *Proc. Natl. Acad. Sci. U.S.A.* **75**, 2669–2673.

Cassel, D., and Selinger, Z. (1976). Catecholamine-stimulated GTPase activity in turkey erythrocyte membranes. *Biochim. Biophys. Acta* **452**, 538–551.

Cassel, D., and Selinger, Z. (1977). Mechanism of adenylate cyclase activation by cholera toxin: Inhibition of GTP hydrolysis at the regulatory site. *Proc. Natl. Acad. Sci. U.S.A.* **74**, 3307–3311.

Chamberlin, M. J. (1974a). The selectivity of transcription. *Annu. Rev. Biochem.* **43**, 303–325.

Chamberlin, M. J. (1974b). Bacterial DNA-dependent RNA polymerases. *In* "The Enzymes" (P. D. Boyer, ed.), 3rd ed., Vol. 10, pp. 333–374. Academic Press, New York.

Chapman, A. G., and Atkinson, D. E. (1977). Adenine nucleotide concentrations and turnover rates. Their correlation with biological activity in bacteria and yeast. *Adv. Microb. Physiol.* **15**, 253–306.

Cherrington, A. D., Assimacopoulos, F. D., Harper, S. C., Corbin, J. D., Park, C. R., and Exton, J. H. (1976). Studies on the α-adrenergic activation of hepatic glucose output. II. Investigation of the roles of adenosine 3′ : 5′-monophosphate and adenosine 3′ : 5′-monophosphate-dependent protein kinase in the actions of phenylephrine in isolated hepatocytes. *J. Biol. Chem.* **251**, 5209–5218.

Clark, M. G., Bloxham, D. P., Holland, P. C., and Lardy, H. A. (1973). Estimation of the fructose diphosphatase-phosphofructokinase substrate cycle in the flight muscle of *Bombus affinis. Biochem. J.* **134**, 589–597.

Cleland, W. W. (1970). Steady state kinetics. *In* "The Enzymes" (P. D. Boyer, ed.), 3rd ed., Vol. 2, pp. 1–65. Academic Press, New York.

Cohen, P. (1973). The subunit structure of rabbit skeletal muscle phosphorylase kinase, and the molecular basis for its activation reactions. *Eur. J. Biochem.* **34**, 1–14.

Cohen, P., and Atoniw, J. F. (1973). The control of phosphorylase kinase phosphatase by "second site phosphorylation"; a new form of enzyme regulation. *FEBS Lett.* **34**, 43–47.

Cohen, P., Rylatt, D. B., and Nimmo, G. A. (1977). The hormonal control of glycogen metabolism: The amino acid sequence at the phosphorylation site of protein phosphatase inhibitor-1. *FEBS Lett.* **76**, 182–186.

Cohen, P., Burchell, A., Foulkes, J. G., Cohen, P. T. W., Vanaman, T. C., and Nairn, A. C. (1978). Identification of the Ca^{2+}-dependent modulator protein as the fourth subunit of rabbit skeletal muscle phosphorylase kinase. *FEBS Lett.* **92**, 287–293.

Colowick, S. P. (1973). The hexokinases. *In* "The Enzymes" (P. D. Boyer, ed.), 3rd ed., Vol. 9, pp. 1–48. Academic Press, New York.

Cuatrecasas, P. (1974). Membrane receptors. *Annu. Rev. Biochem.* **43**, 169–214.

Curnow, R. T., and Larner, J. (1979). Hormonal and metabolic control of phosphoprotein phosphatase. *In* "The Biochemical Actions of Hormones" (G. Litwack, ed.), Vol. 6, pp. 77–119. Academic Press, New York.

Czech, M. P. (1977). Molecular basis of insulin action. *Annu. Rev. Biochem.* **46**, 359–384.

Dalziel, K. (1962). Physical significance of Michaelis constants. *Nature (London)* **196**, 1203–1205.

Davie, E. W., and Fujikawa, K. (1975). Basic mechanisms in blood coagulation. *Annu. Rev. Biochem.* **44**, 799–829.

Davie, E. W., and Kirby, E. P. (1973). Molecular mechanisms in blood coagulation. *Curr. Top. Cell. Regul.* **7**, 51–86.

Davson, H. (1951). "A Textbook of General Physiology." Churchill, London.

Dawes, E. A., and Senior, P. J. (1973). The role and regulation of energy reserve polymers in micro-organisms. *Adv. Microb. Physiol.* **10**, 136–266.

Denton, R. M., Randle, P. J., Bridges, B. J., Cooper, R. H., Kerbey, A. L., Pask, H. T., Severson, D. L., Stansbie, D., and Whitehouse, S. (1975). Regulation of mammalian pyruvate dehydrogenase. *Mol. Cell. Biochem.* **9**, 27–53.

DePaoli-Roach, A. A., Roach, P. J., and Larner, J. (1979). Rabbit skeletal muscle phosphorylase kinase. Comparison of glycogen synthase and phosphorylase as substrates. *J. Biol. Chem.* **254**, 4212–4219.

Ebashi, S. (1974). Regulatory mechanism of muscle contraction with special reference to the Ca-troponin-tropomyosin system. *Essays Biochem.* **10**, 1–36.

Ebner, K. E. (1973). Lactose synthetase. *In* "The Enzymes" (P. D. Boyer, ed.), 3rd ed., Vol. 9, pp. 363–377. Academic Press, New York.

Englesberg, E., and Wilcox, G. (1974). Regulation: Positive control. *Annu. Rev. Genet.* **8**, 219–242.

Engström, L. (1978). The regulation of liver pyruvate kinase by phosphorylation-dephosphorylation. *Curr. Top. Cell. Regul.* **13**, 29–51.

Epstein, W., and Beckwith, J. R. (1968). Regulation of gene expression. *Annu. Rev. Biochem.* **37**, 411–436.

Feldman, K. Zeisel, H., and Helmreich, E. (1972). Interactions between native and chemically modified subunits of matrix-bound glycogen phosphorylase. *Proc. Natl. Acad. Sci. U.S.A.* **69**, 2278–2282.

Fischer, E. H., Pocker, A., and Saari, J. C. (1970). The structure, function and control of glycogen phosphorylase. *Essays Biochem.* **6**, 23–68.

Fischer, E. H., Heilmeyer, L. M. G., Jr., and Haschke, R. H. (1971). Phosphorylase and the control of glycogen degradation. *Curr. Top. Cell. Regul.* **4**, 221–251.

Fletterick, R. J., and Madsen, N. B. (1977). X-rays reveal phosphorylase architecture. *Trends Biochem. Sci.* **2**, 145–148.

Fletterick, R. J., Sygusch, J., Murray, N., Madsen, N. B., and Johnson, L. N. (1976). Low-resolution structure of the glycogen phosphorylase *a* monomer and comparison with phosphorylase *b*. *J. Mol. Biol.* **103**, 1–13.

Fothergill, J. E., and Anderson, W. H. K. (1978). A molecular approach to the complement system. *Curr. Top. Cell. Regul.* **13**, 259–311.

Frieden, C. (1971). Protein-protein interactions and enzymatic activity. *Annu. Rev. Biochem.* **40**, 653–696.

Friedman, D. L., and Larner, J. (1965). Studies on uridine diphosphate glucose: α-1,4-glucan-α-4-glucosyltransferase. VIII. Catalysis of the phosphorylation of muscle phosphorylase and transferase by separate enzymes. *Biochemistry* **4**, 2261–2264.

Gaertner, F. H. (1978). Unique catalytic properties of enzyme clusters. *Trends Biochem. Sci.* **3**, 63–65.

Gallant, J., Irr, J., and Cashel, M. (1971). The mechanism of amino acid control of guanylate and adenylate biosynthesis. *J. Biol. Chem.* **246**, 5812–5816.

Gallwitz, D. (1970). Enzymatic acetylation of HeLa cell histones in isolated nuclei *in vitro*. *Hoppe-Seyler's Z. Physiol. Chem.* **351**, 1050–1053.

Garrison, J. C. (1978). The effects of glucagon, catecholamines, and the calcium ionophore A23187 on the phosphorylation of rat hepatocyte cytosolic proteins. *J. Biol. Chem.* **253**, 7091–7100.

Garrison, J. C., and Haynes, R. C., Jr. (1975). The hormonal control of gluconeogenesis by regulation of mitochondrial pyruvate carboxylation in isolated rat liver cells. *J. Biol. Chem.* **250**, 2769–2777.

Gennis, R. B. (1977). Protein–lipid interactions. *Annu. Rev. Biophys. Bioeng.* **6**, 195–238.

Gerhart, J. C. (1970). A discussion of the regulatory properties of aspartate transcarbamylase from *Escherichia coli*. *Curr. Top. Cell. Regul.* **2**, 276–325.

Gill, D. M. (1977). Mechanism of action of cholera toxin. *Adv. Cyclic Nucleotide Res.* **8**, 85–118.

Gill, D. M., and Meren, R. (1978). ADP-ribosylation of membrane proteins catalyzed by cholera toxin: Basis of the activation of adenylate cyclase. *Proc. Natl. Acad. Sci., U.S.A.* **75**, 3050–3054.

Gillette, J. R., Davis, D. C., and Sasame, H. A. (1972). Cytochrome *P-450* and its role in drug metabolism. *Annu. Rev. Pharmacol.* **12**, 57–84.

Gilman, A. G., Sternweis, P. C., Howlett, A. C., and Ross, E. M. (1979). Biochemical and genetic resolution of components of the S49 lymphoma adenylate cyclase system. *Cold Spring Harbor Conf. Cell Proliferation* **6**, 299–315.

Goff, C. G. (1974). Chemical structure of a modification of the *Escherichia coli* ribonucleic acid polymerase α polypeptides induced by bacteriophage T₄ infection. *J. Biol. Chem.* **249**, 6181–6190.

Goldberg, A. L., and Dice, J. F. (1974). Intracellular protein degradation in mammalian and bacterial cells. *Annu. Rev. Biochem.* **43**, 835–869.

Goldberg, A. L., and St. John, A. C. (1976). Intracellular protein degradation in mammalian and bacterial cells. Part 2. *Annu. Rev. Biochem.* **45**, 747–803.

Goldberg, N. D., and Haddox, M. K. (1977). Cyclic GMP metabolism and involvement in biological regulation. *Annu. Rev. Biochem.* **46**, 823–896.

Goldberger, R. F., Deeley, R. G., and Mullinix, K. P. (1976). Regulation of gene expression in prokaryotic organisms. *Adv. Genet.* **18**, 1–67.

Gratecos, D., Detwiler, T., and Fischer, E. G. (1974). Purification and properties of rabbit muscle phosphorylase phosphatase. *In* "Metabolic Interconversions of Enzymes 1973" (E. H. Fischer, E. G. Krebs, H. Neurath, and E. R. Stadtman, eds.), pp. 43–52. Springer-Verlag, Berlin and New York.

Griffin, C. C., and Brand, L. (1968). Kinetic implications of enzyme–effector complexes. *Arch. Biochem. Biophys.* **126**, 856–863.

Gumaa, K. A., McLean, P., and Greenbaum, A. L. (1971). Compartmentation in relation to metabolic control in liver. *Essays Biochem.* **7**, 39–86.

Gutfreund, H. (1972). "Enzymes: Physical Principles." Wiley, New York.

Gutfreund, H. (1975). Kinetic analysis of the properties and reactions of enzymes. *Prog. Biophys. Mol. Biol.* **29**, 161–195.

Haber, E., and Wrenn, S. (1976). Problems in identification of the beta-adrenergic receptor. *Physiol. Rev.* **56**, 317–338.

Haga, T., Haga, K., and Gilman, A. G. (1977). Hydrodynamic properties of the β-adrenergic receptor and adenylate cyclase from wild type and variant S49 lymphoma cells. *J. Biol. Chem.* **252**, 5776–5782.

Halestrap, A. P., and Denton, R. M. (1974). Hormonal regulation of adipose-tissue acetyl-coenzyme A carboxylase by changes in the polymeric state of the enzyme. The role of long-chain fatty acyl-coenzyme A thioesters and citrate. *Biochem. J.* **142**, 365–377.

Hammes, G. G., and Schimmel, P. R. (1970). Rapid reactions and transient states. *In* "The Enzymes" (P. D. Boyer, ed.), 3rd ed., Vol. 2, pp. 67–114. Academic Press, New York.

Hanson, R. S., Peterson, J. A., and Yousten, A. A. (1970). Unique biochemical events in bacterial sporulation. *Annu. Rev. Microbiol.* **24**, 53–90.

Hardie, D. G., and Cohen, P. (1978). The regulation of fatty acid biosynthesis. Simple procedure for the purification of acetyl CoA carboxylase from lactating rabbit mammary gland, and its phosphorylation by endogenous cyclic AMP-dependent and independent protein kinase activities. *FEBS Lett.* **91**, 1–7.

Harris, J. I., and Waters, M. (1976). Glyceraldehyde-3-phosphate dehydrogenase. *In* "The Enzymes" (P. D. Boyer, ed.), 3rd ed., Vol. 13, pp. 1–49. Academic Press, New York.

Haselkorn, R., and Rothman-Denes, L. B. (1973). Protein synthesis. *Annu. Rev. Biochem.* **42**, 397–438.

Hayaishi, O., and Ueda, K. (1977). Poly(ADP-ribose) and ADP-ribosylation of proteins. *Annu. Rev. Biochem.* **46**, 95–116.

Hayakawa, T., Perkins, J. P., and Krebs, E. G. (1973). Studies on the subunit structure of rabbit skeletal muscle phosphorylase kinase. *Biochemistry* **12**, 574–580.

Helmreich, E. J., Zenner, H. P., Pfeuffer, T., and Cori, C. F. (1976). Signal transfer from hormone receptor to adenylate cyclase. *Curr. Top. Cell. Regul.* **10**, 41–87.

Hill, R. L., and Brew, K. (1975). Lactose synthetase. *Adv. Enzymol.* **43**, 411–490.

Hochstadt-Ozer, J., and Cashel, M. (1972). The regulation of purine utilization in bacteria. V. Inhibition of purine phosphoribosyl-transferase activities and purine uptake in isolated membrane vesicles by guanosine tetraphosphate. *J. Biol. Chem.* **247**, 7067–7072.

Hofer, H. W., and Fürst, M. (1976). Isolation of a phosphorylated form of phosphofructokinase from skeletal muscle. *FEBS Lett.* **62**, 118–122.

Holzer, H., and Duntze, W. (1971). Metabolic regulation by chemical modification of enzymes. *Annu. Rev. Biochem.* **40**, 345–374.

Holzer, H., Betz, H., and Ebner, E. (1975). Intracellular proteinases in microorganisms. *Curr. Top. Cell. Regul.* **9**, 103–156.

Huang, F. L., and Glinsmann, W. H. (1975). Inactivation of rabbit muscle phosphorylase phosphatase by cyclic AMP-dependent protein kinase. *Proc. Natl. Acad. Sci. U.S.A.* **72**, 3004–3008.

Huang, F. L., and Glinsmann, W. H. (1976). A second heat-stable protein inhibitor of phosphorylase phosphatase from rabbit muscle. *FEBS Lett.* **62**, 326–329.

Hutson, N. J., Brumley, F. T., Assimacopoulos, F. D., Harper, S. C., and Exton, J. H. (1976). Studies on the α-adrenergic activation of hepatic glucose output. I. Studies on the α-adrenergic activation of phosphorylase and gluconeogenesis and inactivation of glycogen synthase in isolated rat liver parenchymal cells. *J. Biol. Chem.* **251**, 5200–5208.

Ingebritsen, T. S., and Gibson, D. M. (1980). Reversible phosphorylation of hydroxymethylglutaryl CoA reductase. *In* "Molecular Aspects of Cellular Regulation" (P. Cohen, ed.). Elsevier/North Holland, Amsterdam. (in press).

Innes, I. R., and Nickerson, M. (1975). Norepinephrine, epinephrine, and the sympathomimetic amines. *In* "The Pharmacological Basis of Therapeutics" (L. S. Goodman and A. Gilman, eds.), 5th ed., pp. 477–513. Macmillan, New York.

Jacobson, G. R., and Stark, G. R. (1973). Aspartate transcarbamylases. *In* "The Enzymes" (P. D. Boyer, ed.), 3rd ed., Vol. 9, pp. 226–308. Academic Press, New York.

Jencks, W. P. (1969). "Catalysis in Chemistry and Enzymology." McGraw-Hill, New York.

Johnson, G. L., Kaslow, H. R., and Bourne, H. R. (1978). Reconstitution of cholera toxin-activated adenylate cyclase. *Proc. Natl. Acad. Sci. U.S.A.* **75**, 3113–3117.

Johnson, L. N., Weber, I. T., Wild, D. L., Wilson, K. S., and Yeates, D. G. R. (1978). The crystal structure of glycogen phosphorylase *b*. In "Regulatory Mechanisms of Carbohydrate Metabolism" (V. Esmann, ed.), pp. 185–194. Pergamon, Oxford.

Jovin, T. M. (1976). Recognition mechanisms of DNA-specific enzymes. *Annu. Rev. Biochem.* **45**, 889–920.

Katunuma, N. (1975). Regulation of intracellular enzyme levels by limited proteolysis. *Ergeb. Physiol., Biol. Chem. Exp. Pharmakol.* **72**, 83–104.

Katunuma, N. (1977). New intracellular proteases and their role in intracellular enzyme degradation. *Trends Biochem. Sci.* **2**, 122–125.

Katz, J., and Rognstad, R. (1976). Futile cycles in the metabolism of glucose. *Curr. Top. Cell. Regul.* **10**, 237–289.

Kayne, F. J. (1973). Pyruvate kinase. In "The Enzymes" (P. D. Boyer, ed.), 3rd ed., Vol. 8, pp. 353–382. Academic Press, New York.

Kempner, E. S. (1975). Properties of organized enzymatic pathways. *Sub-Cell. Biochem.* **4**, 213–221.

Kingdon, H. S., Shapiro, B. M., and Stadtman, E. R. (1967). Regulation of glutamine synthetase. VIII. ATP: Glutamine synthetase adenylyltransferase, an enzyme that catalyzes alterations in the regulatory properties of glutamine synthetase. *Proc. Natl. Acad. Sci. U.S.A.* **58**, 1703–1710.

Kirschner, K., and Bisswanger, H. (1976). Multifunctional proteins. *Annu. Rev. Biochem.* **45**, 143–166.

Klotz, I. M., Darnall, D. W., and Langerman, N. R. (1975). Quaternary structure of proteins. In "The Proteins" (H. Neurath, ed.), 3rd ed., Vol. 1, pp. 293–411. Academic Press, New York.

Koshland, D. E. (1970). The molecular basis for enzyme regulation. In "The Enzymes" (P. D. Boyer, ed.), 3rd ed., Vol. 1, pp. 342–396. Academic Press, New York.

Koshland, D. E., and Levitzki, A. (1974). CTP synthetase and related enzymes. In "The Enzymes" (P. D. Boyer, ed.), 3rd ed., Vol. 10, pp. 539–559. Academic Press, New York.

Koshland, D. E., Némethy, G., and Filmer, D. (1966). Comparison of experimental binding data and theoretical models in proteins containing subunits. *Biochemistry* **5**, 365–385.

Krebs, E. G., Love, D. S., Bratvold, G. E., Trayer, K. A., Meyer, W. L., and Fischer, E. H. (1964). Purification and properties of rabbit skeletal muscle phosphorylase *b* kinase. *Biochemistry* **3**, 1022–1033.

Krebs, H. A. (1954). Excursion into the borderland of biochemistry and philosophy. *Bull. Johns Hopkins Hosp.* **95**, 45–51.

Krebs, H. A. (1969). The role of equilibria in the regulation of metabolism. *Curr. Top. Cell. Regul.* **1**, 45–55.

Kretsinger, R. H. (1976). Calcium-binding proteins. *Annu. Rev. Biochem.* **45**, 239–266.

Kuehn, G. D., Barnes, L. D., and Atkinson, D. E. (1971). Yeast diphosphopyridine nucleotide specific isocitrate dehydrogenase. Binding of ligands. *Biochemistry* **10**, 3945–3951.

Laidler, K. (1965). "Chemical Kinetics." McGraw-Hill, New York.

Langan, T. A. (1973). Protein kinases and protein kinase substrates. *Adv. Cyclic Nucleotide Res.* **3**, 99–153.

Larner, J. (1971). "Intermediary Metabolism and its Regulation." Prentice-Hall, Englewood Cliffs, New Jersey.

Larner, J. (1977). "Cyclic Nucleotide Metabolism" Current Concepts Monogr. Ser. Upjohn Company, Kalamazoo, Michigan.

Larner, J., and Haynes, R. C., Jr. (1975). Insulin and oral hypoglycemic drugs: Glucagon. *In* "The Pharmacological Basis of Therapeutics" (L. S. Goodman and A. Gilman, eds.), 5th ed., pp. 1507–1533. Macmillan, New York.

Larner, J., and Villar-Palasi, C. (1971). Glycogen synthase and its control. *Curr. Top. Cell. Regul.* **3**, 195–236.

Larner, J., Lawrence, J. C., Walkenbach, R. J., Roach, P. J., Hazen, R. J., and Huang, L. C. (1978). Insulin control of glycogen synthesis. *Adv. Cyclic Nucleotide Res.* **9**, 425–439.

Lazdunski, M. (1972). Flip-flop mechanisms and half-site enzymes. *Curr. Top. Cell. Regul.* **6**, 267–310.

Lee, K. L., and Nickol, J. M. (1976). Phosphorylation of tyrosine aminotransferase *in vivo*. *J. Biol. Chem.* **249**, 6024–6026.

Lefkowitz, R. J., Mukherjee, C., Limbird, L. E., Caron, M. G., Williams, L. T., Alexander, R. W., Mickey, J. V., and Tate, R. (1976). Regulation of adenylate cyclase coupled beta-adrenergic receptors. *Recent Prog. Horm. Res.* **32**, 597–632.

LeJohn, H. B. (1968). Unidirectional inhibition of glutamate dehydrogenase by metabolites. *J. Biol. Chem.* **243**, 5126–5131.

Levitzki, A., and Koshland, D. E. (1976). The role of negative cooperativity and half-of-the-sites reactivity in enzyme regulation. *Curr. Top. Cell. Regul.* **10**, 1–40.

Limbird, L. E., and Lefkowitz, R. J. (1977). Resolution of β-adrenergic receptor binding and adenylate cyclase activity by gel exclusion chromatography. *J. Biol. Chem.* **252**, 799–802.

Lodish, H. F. (1976). Translational control of protein synthesis. *Annu. Rev. Biochem.* **45**, 39–72.

Lucas-Lenard, J., and Beres, L. (1974). Protein synthesis—Peptide chain elongation. *In* "The Enzymes" (P. D. Boyer, ed.), 3rd ed., Vol. 10, pp. 53–86. Academic Press, New York.

Lueking, D. R., and Goldfine, H. (1975). The involvement of guanosine 5'-diphosphate-3'-diphosphate in the regulation of phospholipid biosynthesis in *Escherichia coli*. *J. Biol. Chem.* **250**, 4911–4917.

Lund, E., and Kjeldgaard, N. O. (1972). Metabolism of guanosine tetraphosphate in *Escherichia coli*. *Eur. J. Biochem.* **28**, 316–326.

MacQuarrie, R. A., and Bernhard, S. A. (1971). Subunit conformation and catalytic function in rabbit-muscle glyceraldehyde-3-phosphate dehydrogenase. *J. Mol. Biol.* **55**, 181–192.

Magasanik, B. (1962). Biosynthesis of purine and prymidine nucleotides. *In* "The Bacteria" (I. C. Gunsalus and R. Y. Stanier, eds.), Vol. 3, pp. 295–334. Academic Press, New York.

Magasanik, B. (1976). Classical and post-classical modes of regulation of the synthesis of degradative bacterial enzymes. *Prog. Nucleic Acid Res. Mol. Biol.* **17**, 99–115.

Maguire, M. E., Ross, E. M., and Gilman, A. G. (1977). β-adrenergic receptor: Ligand binding properties and the interaction with adenylyl cyclase. *Adv. Cyclic Nucleotide Res.* **8**, 1–83.

Mahler, H. R., and Cordes, E. H. (1966). "Biological Chemistry." Harper, New York.

Maley, F., and Maley, G. F. (1972). The regulatory influence of allosteric effectors on deoxycytidylate deaminases. *Curr. Top. Cell. Regul.* **5**, 178–228.

Mansour, T. E. (1972). Phosphofructokinase. *Curr. Top. Cell. Regul.* **5**, 1–46.

Masters, C. J. (1977). Metabolic control and the microenvironment. *Curr. Top. Cell. Regul.* **12**, 75–105.

Mathews, B. W., and Bernhard, S. A. (1973). Structure and symmetry of oligomeric enzymes. *Annu. Rev. Biophys. Bioeng.* **2**, 257–317.

Mayr, E. (1961). Cause and effect in biology. *Science* **134**, 1501–1506.

Meyer, F., Heilmeyer, L. M. G., Jr., Haschke, R. H., and Fischer, E. H. (1970). Control of phosphorylase activity in a muscle glycogen particle. I. Isolation and characterization of the protein–glycogen complex. *J. Biol. Chem.* **245**, 6642–6648.

Mildvan, A. S. (1970). Metals in enzyme catalysis. *In* "The Enzymes" (P. D. Boyer, ed.), 3rd ed., Vol. 2, pp. 445–536. Academic Press, New York.

Miller, A. L., and Atkinson, D. E. (1972). Response of yeast pyruvate carboxylase to the adenylate energy charge and other regulatory parameters. *Arch. Biochem. Biophys.* **152**, 531–538.

Miller, D. L., and Weissbach, H. (1970). Studies on the purification and properties of factor Tu from *E. coli*. *Arch. Biochem. Biophys.* **141**, 26–37.

Milstein, S., Abita, J. P., Chang, N., and Kaufman, S. (1976). *In vitro* activation of rat liver phenylalanine hydroxylase by phosphorylation. *J. Biol. Chem.* **251**, 5310–5314.

Monod, J., Wyman, J., and Changeux, J. P. (1965). On the nature of allosteric transitions: A plausible model. *J. Mol. Biol.* **12**, 88–118.

Morgan, H. E., and Parmeggiani, A. (1964). Regulation of glycogenolysis in muscle. III. Control of muscle glycogen phosphorylase. *J. Biol. Chem.* **239**, 2440–2445.

Morrison, J. F., and Heyde, E. (1972). Enzymic phosphoryl group transfer. *Annu. Rev. Biochem.* **41**, 29–54.

Moss, J., and Vaughan, M. (1977). Mechanism of action of choleragen. Evidence for ADP-ribosyltransferase activity with arginine as an acceptor. *J. Biol. Chem.* **252**, 2455–2457.

Mowbray, J., and Moses, V. (1976). The tentative identification in *Escherichia coli* of a multienzyme complex with glycolytic activity. *Eur. J. Biochem.* **66**, 25–36.

Müller-Eberhard, H. J. (1975). Complement. *Annu. Rev. Biochem.* **44**, 697–724.

Müller-Hill, B. (1975). *Lac* repressor and *lac* operator. *Prog. Biophys. Mol. Biol.* **30**, 227–252.

Newell, P. C. (1971). The development of the cellular slime mould *Dictyostelium discoideum*: A model system for the study of cellular differentiation. *Essays Biochem.* **7**, 87–126.

Newsholme, E. A., and Crabtree, B. (1976). Substrate cycles in metabolic regulation and in heat generation. *Biochem. Soc. Symp.* **41**, 61–110.

Newsholme, E. A., and Crabtree, B. (1978). Substrate cycles in the control of energy metabolism in the intact animal. *In* "Regulatory Mechanisms of Carbohydrate Metabolism" (V. Esmann, ed.), pp. 285–295. Pergamon, Oxford.

Newsholme, E. A., and Start, C. (1973). "Regulation in Metabolism." Wiley, New York.

Nimmo, H. G., and Cohen, P. (1977). Hormonal control of protein phosphorylation. *Adv. Cyclic Nucleotide Res.* **8**, 145–266.

Novikoff, A. B., and Holtzman, E. (1970). "Cells and Organelles." Holt, New York.

Ottaway, J. H., and Mowbray, J. (1977). The role of compartmentation in the control of glycolysis. *Curr. Top. Cell. Regul.* **12**, 107–208.

Ozawa, E., Hosoi, K., and Ebashi, S. (1967). Reversible stimulation of muscle phosphorylase *b* kinase by low concentrations of calcium ions. *J. Biochem. (Tokyo)* **61**, 531–533.

Paik, W. K., and Kim, S. (1975). Protein methylation: Chemical, enzymological and biological significance. *Adv. Enzymol.* **42**, 227–286.

Pappenheimer, A. M., Jr. (1977). Diphtheria toxin. *Annu. Rev. Biochem.* **46**, 69–94.

Pettit, F. H., Pelley, J. W., and Reed, L. J. (1975). Regulation of pyruvate dehydrogenase kinase and phosphatase by acetyl-CoA/CoA and NADH/NAD ratios. *Biochem. Biophys. Res. Commun.* **65**, 575–582.

Pfeuffer, T. (1977). GTP-binding proteins in membranes and the control of adenylate cyclase activity. *J. Biol. Chem.* **252**, 7224–7234.

Pierard, A. (1966). Control of the activity of *Escherichia coli* carbamoyl phosphate synthetase by antagonistic allosteric effectors. *Science* **154**, 1572–1573.

Pine, M. J. (1972). Turnover of intracellular proteins. *Annu. Rev. Microbiol.* **26**, 103–126.

Piras, R., Rothman, L. B., and Cabib, E. (1968). Regulation of muscle glycogen synthetase by metabolites. Differential effects on the I and D forms. *Biochemistry* **7**, 56–66.

Pittendrigh, C. S. (1958). Adaptation, natural selection and behavior. *In* "Behavior and Evolution" (A. Roe and G. G. Simpson, eds.), pp. 390–416. Yale Univ. Press, New Haven, Connecticut.

Polakis, S. E., Guchhait, R. B., and Lane, M. D. (1973). Stringent control of fatty acid synthesis in *Escherichia coli*. Possible regulation of acetyl coenzyme A carboxylase by ppGpp. *J. Biol. Chem.* **248**, 7957–7966.

Preiss, J. (1973). Adenosine diphosphoryl glucose pyrophosphorylase. *In* "The Enzymes" (P. D. Boyer, ed.), 3rd ed., Vol. 8, pp. 73–119. Academic Press, New York.

Qureshi, A. A., Jenik, R. A., Kim, M., Lornitzo, F. A., and Porter, J. W. (1975). Separation of two active forms (Holo-*a* and Holo-*b*) of pigeon liver fatty acid synthetase and their interconversion by phosphorylation and dephosphorylation. *Biochem. Biophys. Res. Commun.* **66**, 344–351.

Rabin, B. R. (1967). Cooperative effects in enzyme catalysis: A possible kinetic model based on substrate-induced conformation isomerization. *Biochem. J.* **102**, 22c.

Rall, T. W., and Sutherland, E. W. (1958). Formation of a cyclic adenine ribonucleotide by tissue particles. *J. Biol. Chem.* **232**, 1065–1076.

Ramaiah, A. (1974). Pasteur effect and phosphofructokinase. *Curr. Top. Cell. Regul.* **8**, 298–345.

Ramos, F., Stalon, V., Pierard, A., and Wiame, J. M. (1967). The specialization of the two ornithine carbamoyltransferases of pseudomonas. *Biochim. Biophys. Acta* **139**, 98–106.

Reed, L. J. (1974). Multienzyme complexes. *Acc. Chem. Res.* **7**, 40–46.

Reed, L. J., and Cox, D. J. (1970). Multienzyme complexes. *In* "The Enzymes" (P. D. Boyer, ed.), 3rd ed., Vol. 1, pp. 213–240. Academic Press, New York.

Reed, L. J., Pettit, F. H., Roche, T. E., and Butterworth, P. J. (1973). Regulation of the mammalian pyruvate dehydrogenase complex by phosphorylation and dephosphorylation. *In* "Protein Phosphorylation in Control Mechanisms" (F. Huijing and E. Y. C. Lee, eds.), pp. 83–97. Academic Press, New York.

Reed, L. J., Pettit, F. H., Roche, T. E., Pelley, J. W., and Butterworth, P. J. (1976). Structure and regulation of the mammalian pyruvate dehydrogenase complex. *In* "Metabolic Interconversion of Enzymes 1975" (S. Shaltiel, ed.), pp. 121–124. Springer-Verlag, Berlin and New York.

Riou, J. P., Claus, T. H., Flockhart, D. A., Corbin, J. D., and Pilkis, S. J. (1977). *In vivo* and *in vitro* phosphorylation of rat liver fructose-1,6-biphosphatase. *Proc. Natl. Acad. Sci. U.S.A.* **74**, 4615–4619.

Roach, P. J. (1976). Intracellular concentration in discussions of metabolism. *Bull. Mol. Biol. Med.* **1**, 81–91.

Roach, P. J. (1977). Functional significance of enzyme cascade systems. *Trends Biochem. Sci.* **2**, 87–90.

Roach, P. J., and Larner, J. (1976). Rabbit skeletal muscle glycogen synthase. II. Enzyme phosphorylation state and effector concentrations as interacting control parameters. *J. Biol. Chem.* **251**, 1920–1925.

Roach, P. J., and Larner, J. (1977). Covalent phosphorylation in the regulation of glycogen synthase. *Mol. Cell. Biochem.* **15**, 179–200.

Roach, P. J., Takeda, Y., and Larner, J. (1976). Rabbit skeletal muscle glycogen synthase. I. Relationship between phosphorylation state and kinetic properties. *J. Biol. Chem.* **251**, 1913–1919.

Roach, P. J., Rosell-Perez, M., and Larner, J. (1977). Muscle glycogen synthase *in vivo* state. Effects of insulin administration on the chemical and kinetic properties of the purified enzyme. *FEBS Lett.* **80**, 95–98.

Roach, P. J., DePaoli-Roach, A. A., and Larner, J. (1978). Ca^{2+}-stimulated phosphorylation of muscle glycogen synthase by phosphorylase *b* kinase. *J. Cyclic Nucleotide Res.* **4**, 245–257.

Rowe, P. B., McCairns, E., Madsen, G., Sauer, D., and Elliott, H. (1978). *De Novo* purine synthesis in avian liver. Co-purification of the enzymes and properties of the pathway. *J. Biol. Chem.* **253**, 7711–7721.

Rubin, C. S., and Rosen, O. M. (1975). Protein phosphorylation. *Annu. Rev. Biochem.* **44**, 831–887.

Savageau, M. A. (1976). "Biochemical Systems Analysis." Addison-Wesley, Reading, Massachusetts.

Schimke, R. T. (1973). Control of enzyme levels in mammalian tissues. *Adv. Enzymol.* **37**, 135–187.

Schimke, R. T. (1975). On the properties and mechanism of protein turnover. *In* "Intracellular Protein Turnover" (R. T. Schimke and N. Katunuma, eds.), pp. 173–186. Academic Press, New York.

Scrutton, M. C., and Young, M. R. (1972). Pyruvate carboxylase. *In* "The Enzymes" (P. D. Boyer, ed.), 3rd ed., Vol. 6, pp. 1–35. Academic Press, New York.

Segal, H. L. (1976). Mechanism and regulation of protein turnover in animal cells. *Curr. Top. Cell. Regul.* **11**, 183–201.

Seubert, W., and Hamm, H. H. (1975). Inactivation and ATP-dependent reactivation of tyrosine aminotransferase *in vitro* by membrane bound enzymes from rat liver and kidney cortex. *Biochem. Biophys. Res. Commun.* **65**, 1–7.

Shapiro, B. M., and Stadtman, E. R. (1968). The novel phosphodiester residue of adenylylated glutamine synthetase from *Escherichia coli*. *J. Biol. Chem.* **243**, 3769–3771.

Sly, W. S., and Stadtman, E. R. (1963). Formate metabolism. II. Enzymatic synthesis of formylphosphate and formyl coenzyme A in *Clostridium cylindrosporum*. *J. Biol. Chem.* **238**, 2639–2647.

Sols, A., and Marco, R. (1970). Concentrations of metabolites and binding sites. Implications in metabolic regulation. *Curr. Top. Cell. Regul.* **2**, 227–273.

Srere, P. A., and Mosbach, K. (1974). Metabolic compartmentation: Symbiotic, organellar, multienzymic and microenvironmental. *Annu. Rev. Microbiol.* **28**, 61–83.

Stadtman, E. R. (1970). Mechanisms of enzyme regulation in metabolism. *In* "The Enzymes" (P. D. Boyer, ed.), 3rd ed., Vol. 1, pp. 397–459. Academic Press, New York.

Stadtman, E. R., and Chock, P. B. (1978). Interconvertible enzyme cascades in metabolic regulation. *Curr. Top. Cell Regul.* **13**, 53–95.

Stadtman, E. R., and Ginsburg, A. (1974). The glutamine synthetase of *Escherichia coli:* Structure and control. *In* "The Enzymes" (P. D. Boyer, ed.), 3rd ed., Vol. 10, pp. 755–807. Academic Press, New York.

Stark, G. R. (1977). Multifunctional proteins: One gene–more than one enzyme. *Trends Biochem. Sci.* **2**, 64–66.

Steinberg, D., Mayer, S. E., Khoo, J. C., Miller, E. A., Miller, R. E., Fredholm, B., and Eichner, R. (1975). Hormonal regulation of lipase, phosphorylase, and glycogen synthase in adipose tissue. *Adv. Cyclic Nucleotide Res.* **5**, 549–568.

Sweeney, J. R., and Fisher, J. B. (1968). An alternative to allosterism and cooperativity in the interpretation of enzyme kinetic data. *Biochemistry* **7**, 561–565.

Tepperman, J. (1973). "Metabolic and Endocrine Physiology." Yearbook Publ., Chicago, Illinois.

Tomkins, G. M. (1975). The metabolic code. *Science* **189**, 760–763.

Trevelyan, W. E. (1958). Synthesis and degradation of cellular carbohydrates by yeasts. *In* "The Chemistry and Biology of Yeasts" (A. H. Cook, ed.), pp. 369–436. Academic Press, New York.

Truffa-Bachi, P. (1973). Microbial aspartokinases. *In* "The Enzymes" (P. D. Boyer, ed.), 3rd ed., Vol. 8, pp. 509–553. Academic Press, New York.

Truffa-Bachi, P., and Cohen, G. N. (1973). Amino acid metabolism. *Annu. Rev. Biochem.* **42**, 113–134.

Utter, M. F., and Kolenbrander, H. M. (1972). Formation of oxaloacetate by CO_2 fixation on phosphoenol pyruvate. *In* "The Enzymes" (P. D. Boyer, ed.), 3rd ed., Vol. 6, pp. 117–168. Academic Press, New York.

Utter, M. F., Barden, R. E., and Taylor, B. L. (1975). Pyruvate carboxylase: An evaluation of the relationships between structure and mechanism and between structure and catalytic activity. *Adv. Enzymol.* **42**, 1–72.

Veloso, D., Guynn, R. W., Oskarsson, M., and Veech, R. L. (1973). The concentrations of free and bound magnesium in rat tissues. Relative constancy of free Mg^{2+} concentrations. *J. Biol. Chem.* **248**, 4811–4819.

Villar-Palasi, C., and Larner, J. (1960). Insulin-mediated effect on the activity of UDPG-glycogen transglucosylase of muscle. *Biochim. Biophys. Acta* **39**, 171–173.

Volpe, J. J., and Vagelos, P. R. (1973). Saturated fatty acid biosynthesis and its regulation. *Annu. Rev. Biochem.* **42**, 21–60.

Volpe, J. J., and Vagelos, P. R. (1976). Mechanisms and regulation of biosynthesis of saturated fatty acids. *Physiol. Rev.* **56**, 339–417.

von Hippel, P. H., and McGhee, J. D. (1972). DNA–protein interactions. *Annu. Rev. Biochem.* **41**, 231–300.

Vyas, S., and Maas, W. K. (1963). Feedback inhibition of acetylglutamate synthetase by arginine in *Escherichia coli*. *Arch. Biochem. Biophys.* **100**, 542–546.

Walker, D. G. (1966). The nature and function of hexokinases in animal tissues. *Essays Biochem.* **2**, 33–67.

Walsh, D. A., and Krebs, E. G. (1973). Protein kinases. *In* "The Enzymes" (P. D. Boyer, ed.), 3rd ed., Vol. 8, pp. 555–581. Academic Press, New York.

Wang, J. H. (1977). Calcium-regulated protein modulator in cyclic nucleotide systems. *In* "Cyclic Nucleotides: Mechanisms of Action" (H. Cramer and J. Schultz, eds.), pp. 37–56. Wiley, New York.

Weber, G., and Anderson, S. R. (1965). Multiplicity of binding. Range of validity and practical test of Adair's equation. *Biochemistry* **4**, 1942–1947.

Weinhouse, S. (1976). Regulation of glucokinase in liver. *Curr. Top. Cell. Regul.* **11**, 1–50.

Wood, H. G. (1977). Some reactions in which inorganic pyrophosphate replaces ATP and serves as a source of energy. *Fed. Proc., Fed. Am. Soc. Exp. Biol.* **36**, 2197–2205.

Wyngaarden, J. B. (1972). Glutamine phosphoribosylpyrophosphate amidotransferase. *Curr. Top. Cell. Regul.* **5**, 135–176.

Yeaman, S. J., and Cohen, P. (1975). The hormonal control of activity of skeletal muscle phosphorylase kinase: Phosphorylation of the enzyme at two sites *in vivo* in response to adrenalin. *Eur. J. Biochem.* **51,** 93–104.

Zillig, W., Mailhammer, R., Skorko, R., and Rohrer, H. (1977). Covalent modification of DNA-dependent RNA-polymerase as a means for transcriptional control. *Curr. Top. Cell. Regul.* **12,** 263–271.

7

The Movement of Material between Nucleus and Cytoplasm

Philip L. Paine and Samuel B. Horowitz

I. INTRODUCTION

Completion of mitosis in eukaryotes is marked by the spontaneous phase separation of protoplasm into nucleus and cytoplasm. These new entities, although superficially distinct, are in an intimate and extensive interrelationship. Cytoplasmic sources of nucleotides and enzymes provide the materials and tools for nuclear replication and transcription, as well as signals that determine which and to what extent genetic informa-

299

tion will be expressed. The nucleus, through its orderly and controlled output of ribonucleic acids, supplies the cytoplasm with templates and tools used to construct the cell.

Essential aspects of this nucleocytoplasmic commerce are the transportation and distribution functions. Their analysis, we can be sure, will form an important part of our ultimate understanding of the cell economy. However, present knowledge of these functions is primitive—mostly anecdotal rather than systematic. This unfortunate situation is traceable directly to the small size of cells, the crudeness of analytic techniques, the absolute dependence of transport systems on the spatial integrity of their parts, and the inexorability of diffusion. The diffusion tends to alter *in vivo* distributions when cells are modified in preparation for study.

Because of these difficulties and the consequent lack of reliable data, an adequate theoretical framework or model for nucleocytoplasmic movements and distributions, based upon physicochemical principles, has not been developed. Yet the physical sciences dictate certain principles which must be applied to the study of transport and equilibria in multiphase systems. Application of these principles to the movement of materials between the nucleus and cytoplasm is certain to provide insight into cell submicroscopic structure and regulatory mechanisms; ignoring the principles will result in data that are ambiguous at best and at worst uninterpretable.

Physicochemical description of the behavior of a material in a two-phase system has a dual nature, entailing both equilibrium and kinetic processes. Equilibrium is the state in which the concentrations of the material in the two phases do not change with time. Kinetics is the time rate at which a new equilibrium is approached, following a perturbation or change from a previous equilibrium. The relationship between equilibrium and kinetics is of fundamental importance. To understand the mechanisms responsible for the movement of a material between two phases, measurements of both the kinetics of movement and the equilibrium distribution are essential.

If diffusion is the only kinetic process involved, a two-phase system eventually reaches true thermodynamic equilibrium, in which the electrochemical potential of a given material is equal in the two compartments. Application of this principle reveals that many intracellular equilibrium distributions, characterized by unequal concentrations in the nucleus and cytoplasm, are determined by differing physicochemical interactions of materials with the nuclear and cytoplasmic phases.* The

* We have largely restricted this discussion to studies of intact cell systems, with the few exceptions being indicated. This permits us to avoid the always tenuous assumption that

permeability of the nuclear envelope influences the kinetics of approach to equilibria, especially for larger solutes, but it does not affect the positions of the equilibria themselves.

In dealing specifically with living systems, another consideration is also important. When active, energy-requiring kinetic processes are involved, a quasi-equilibrium or steady state is attained, characterized by constant but unequal electrochemical potentials. We discuss energy-dependent nucleocytoplasmic transport below. It should be noted here that, while there is reason to believe that energy-dependent processes are required for nucleocytoplasmic movement of large macromolecular particulates, the involvement of active mechanisms has not been clearly established for any nucleocytoplasmic transport process.

II. DETERMINANTS OF NUCLEOCYTOPLASMIC MOVEMENTS

A. Phase Properties

Endogenous molecules which move between nucleus and cytoplasm are subject to a variety of interactions with components of these phases and with the interposed nuclear envelope. Usually the interactions are so complex as to defy unambiguous interpretation. For this reason, it is helpful to consider experiments in which simple, well-characterized materials were introduced into living cells and their subsequent movement and distribution studied as guides to the parameters likely to be important.

Figure 1 shows autoradiographic grain density profiles through a cytoplasmic site at which an [³H]inulin solution was injected (Horowitz and Moore, 1974). Such profiles typically show a persistent peak at the injection site and a gradient extending from the site. Their analysis discloses three important facts. First, injected aqueous solutions do not mix locally with the cytoplasm as they would if cytoplasm were a simple aqueous solution of diffusible macromolecules, but persist as discrete volumes for 1 hr or more. Second, movement of solutes from the injection site follows diffusion kinetics, and diffusion coefficients in cytoplasm can be calculated from the profiles (solid line, Fig. 1). Third, solutes remain more concen-

---

isolated organelles and cytoplasm retain their *in vivo* kinetic and equilibrium properties. However, intact systems seldom permit one to distinguish contributions of the ground cytoplasm from those of cytoplasmic organelles. In consequence, cytoplasm, as used here, refers to a complex material consisting of ground cytoplasm and its inclusions. In discussing the equilibria of solutes between nucleus and cytoplasm, we adhere to a thermodynamic treatment which allows one to relate concentrations of different phases without implying that the phases are homogeneous.

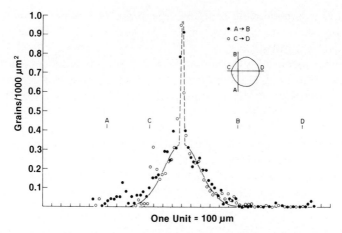

Fig. 1. Autoradiographic grain density analysis of a cytoplasmic diffusion gradient. About 10^{-6} ml of [^{3}H]inulin (molecular weight ~5500) solution was injected into the cytoplasm of the full grown oocyte of *Rana pipiens*, and 16 min allowed for the evolution of a transcellular diffusion gradient. The cell was frozen to the temperature of liquid nitrogen, and kept frozen during subsequent sectioning and autoradiography. Illustrated are two autoradiographic grain density profiles taken perpendicular to each other through the injection site. The vertical bars (A–D) mark the cell boundaries as indicated in the inset. The injection site is at the juncture of the transects A → B and C → D. The solid line is a theoretical gradient for a cytoplasmic diffusion coefficient of 2×10^{-7} cm^2 sec^{-1}. (From Horowitz and Moore, 1974, Reproduced with permission.)

trated per unit quantity of water in the persisting injection volume than in the immediately adjacent cytoplasm, indicating that they are more soluble in the water of the injection fluid than in the water of cytoplasm.

This third observation has been confirmed and extended (Horowitz and Paine, 1976) by experiments with an intracellular gelatin reference phase (RP) (Fig. 2). The fact that water solutions are only slowly miscible with cytoplasm allows an aqueous gelatin sol to be injected into a cell, displacing a localized region of cytoplasm, and to be gelled in place by cooling before it mixes with the cytoplasm. Gelatin gels, whose water has the same solvent properties as ordinary water, persist in the cytoplasm for at least 48 hr, allowing sampling of diffusible cytoplasmic solutes. Subsequent analysis of cytoplasm, nucleus, and reference phase by a variety of techniques provides information on intracellular phase properties. Table I gives the concentrations of four substances, chloride-36, sucrose, sodium, and potassium, at diffusional equilibrium between cytoplasm and an intracellular reference phase (Horowitz *et al.*, 1979). Three of these solutes, ^{36}Cl$^-$, sucrose, and K$^+$, are more concentrated in the gelatin than in the cytoplasm, while for Na$^+$ the opposite is true. These differences cannot be

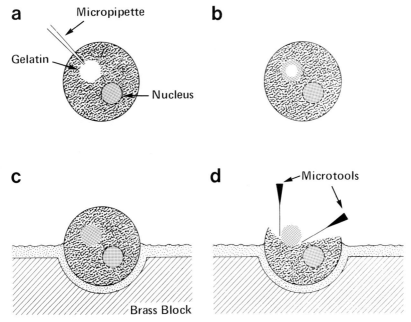

Fig. 2. The steps in the formation and isolation of an intracellular gelatin reference phase. (a) Injection of a 10–20% gelatin solution at 34°C with a glass micropipette. The injected gelatin, about 2% of the oocyte volume, is gelled in place by cooling the cell. (b) Because of its fibrous protein network, the gel excludes organelles and the structural cytoplasmic network, while the gel water comes into diffusional equilibrium with, and thus samples, the diffusible cytoplasmic solutes. (c) After equilibrium is achieved, the cell is frozen to the temperature of liquid nitrogen to prevent subsequent solute redistribution. (d) The reference phase, nucleus, and cytoplasmic samples are isolated by ultra-low temperature microdissection and separately analyzed for water and solute content. In addition, parallel equilibrium dialysis studies permit direct comparison of the properties of the gelatin RP with those of ordinary water. Hence, solute behavior in nucleus and cytoplasm can be referred to standard aqueous solutions.

accounted for by membrane or other energy-dependent transport, but reflect phase differences that distinguish cytoplasm from an ordinary water solution.

The phase properties of cytoplasm have many implications for the understanding of nucleocytoplasmic solute movements and equilibria. Aqueous regions or phases of differing composition and properties can exist in cytoplasm without necessarily being separated by membrane barriers. Mollenhauer and Morré (1978) emphasized the often overlooked existence of such cytoplasmic regions or "zones," as defined by the relative abundance of specific electron microscope-visible organelles and particles. Little is known about the origins of spatial heterogeneity in com-

TABLE I

The Equilibrium Concentrations of Solutes in Cytoplasm, an Intracellular Gelatin Reference Phase (RP), and the Nucleus in Oocytes of *Desmognathus ochrophaeus* at 5°C

	Sucrose[a]	Na^+ ($\mu Eq/ml$ H_2O)	K^+ ($\mu Eq/ml$ H_2O)	$^{36}Cl^-$ ($\mu Eq/ml$ H_2O)
Cytoplasm	0.32 ± 0.01	75.2 ± 2.7	88.6 ± 1.5	16.6 ± 1.2
Reference phase	1.00	21.0 ± 1.1	128.8 ± 2.4	46.1 ± 1.1
Nucleus	1.04 ± 0.02	20.6 ± 3.2	135.1 ± 6.8	62.8 ± 3.8

[a] Sucrose data is expressed relative to the RP concentration.

plex macromolecular systems such as cytoplasm. However, analogous phase separations are well documented in defined macromolecular mixtures, and these distinguish both solutes and particulates (Albertsson, 1971).

The segregation of nuclear material at the end of mitosis, which precedes the reformation of a complete nuclear envelope, can be viewed as the "spontaneous" creation of two phases from one that existed previously. The equilibrium distribution of a material between intracellular phases, and in particular between the nucleus and cytoplasm, can be expected to reflect the summation of the interactions between the material and the phases. Table I gives the nuclear concentrations of the four solutes studied. Each solute differs markedly in its concentration from that in cytoplasm, but more closely resembles that in the reference phase. We may hypothesize that the phase or solubility properties of the nucleus resemble gelatin gels and, therefore, are more like those of ordinary water than like those of cytoplasm.

This suggestion may be true only for certain small solutes. It ignores the potential for interaction provided by the nuclear macromolecules. The distribution of Na^+ and K^+ between isolated thymus nuclear material and saline solutions (Fig. 3) shows that in the concentration range of monovalent cations usual in cytoplasm ($<200\ \mu Eq/ml$), an approximately isomolar relation exists for Na^+ between the nuclear material and saline. This is consistent with the observed relation between RP and nuclear concentrations in Table I. However, when Na^+–saline concentration exceeds 200 $\mu Eq/ml$, isomolarity is not maintained; instead, the nuclear material accumulates excess Na^+. Similar behavior is exhibited in K^+–salines. The mechanisms involved have not been investigated. One may speculate that at high salt concentrations previously masked ionic sites become manifest as Donnan ions. This implies either the preexistence of intramacromolecu-

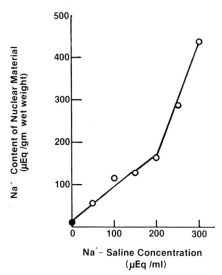

Fig. 3. Sodium content of isolated nuclear material at 37°C as a function of the Na^+ concentration of the suspending saline. (After Itoh and Schwartz, 1957, reproduced with permission.)

lar ionic (salt) linkages capable of dissociating to become sites for "free" ions or that the cations of other salts, such as the histones or calcium, are preferred at low concentrations, but replaced by Na^+ and K^+ when these are at high concentration.

B. Solute Equilibria and Kinetics

Autoradiographic grain density profiles such as that in Fig. 1 can be used to analyze the kinetics and equilibria of molecular movements into nuclei. Following solute injection, a diffusion gradient spreads concentrically from the injection site throughout the cell until thermodynamic equilibrium is reached. However, if the gradient encounters a phase whose equilibrium or kinetic properties differ from those of cytoplasm, this will be reflected by discontinuities in the profile. Figure 4 shows a series of diffusional profiles through the nuclei and adjacent cytoplasm in amphibian oocytes following the injection of size-graded tritiated dextrans into the cytoplasm (Paine *et al.*, 1975).

Figures 4a–d are for 12 Å radius dextran at 3, 5, and 16 min and 16 hr after injection. At the shorter times gradients are present in the cytoplasm and nucleus, but by 16 hr the dextran is uniformly distributed in each, although more concentrated in the nucleus. This shows that either the nuclear solvent or binding properties (equilibrium properties) differ from

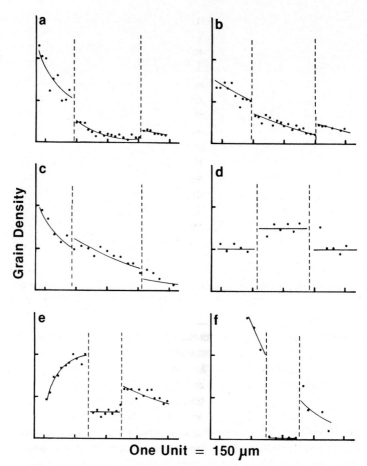

Grain Density

One Unit = 150 µm

Fig. 4. Diffusional profiles of [³H]dextrans in the nucleus and cytoplasm of injected oocytes (obtained by method described in legend of Fig. 1). The position of the nuclear envelope is indicated by the vertical dashed lines. (a–d) 12.0 Å radius dextran, diffusion times (a) 3 min, (b) 5 min, (c) 16 min, and (d) 16 hr. (e) 23.3 Å radius dextran, diffusion time 10 min.; (f) 35.5 Å radius dextran, diffusion time 42 min. (From Paine et al., 1975, reproduced with permission.)

those of cytoplasm, or that a steady state nucleocytoplasmic dextran gradient is maintained by an energy-dependent transport process. The former is the case, and the asymmetric equilibrium distribution must be taken into account if the kinetic data are to be understood.

At 3 min after injection (Fig. 4a), the 12.0 Å radius dextran gradient exhibits a sharp discontinuity at the nuclear envelope. Since the equilibrium data show that this is not due to low nuclear solubility, the discontinuity reveals that the nuclear envelope is a transport barrier. At later times, the concentration difference between cytoplasm and nucleus decreases, and at 16 min the nuclear concentration is greater than the cytoplasmic, although not as great as at equilibrium. The intranuclear gradients seen at early times (Figs. 4a–c) mirror those in the adjacent cytoplasm, demonstrating that although the nuclear envelope is a barrier sufficient to cause a discontinuity in transcellular 12 Å radius dextran diffusion gradients, it is not sufficient to isolate nuclear from cytoplasmic gradients.

Figures 4e and f are grain density profiles for larger dextrans. These gradients have bigger drops at the nuclear envelope than do 12 Å radius dextran profiles at equivalent diffusion times. The larger dextrans are uniformly distributed in the nucleus, in spite of marked gradients in the surrounding cytoplasm. These observations suggest that the nuclear envelope has a sievelike permeability to diffusional movements of macromolecules, a point that will be pursued in a later section.

III. THERMODYNAMIC EQUILIBRIUM

Every material which moves between nucleus and cytoplasm experiences a variety of interactions with components of each phase. Known interactions include relatively strong covalent bonds, weaker ionic, hydrogen, and hydrophobic bonds, and still weaker van der Waals attractions. While we know that the magnitude of these forces on any material is dependent upon the specific properties of that material, the interplay of the interactions is complex and their individual influences are seldom distinguishable. Materials respond to their net influence by accumulating in the phase in which the lowest free energy state is attained. Opposing this tendency is thermal or Brownian motion, which favors uniform distribution.

The nucleocytoplasmic equilibrium distribution of any solute, i, may be quantitatively expressed as the ratio of its molar equilibrium concentration (amounts of i per unit amounts of water) in the nucleus (n) and cytoplasm (cyt)

$$K_i = C_i^p/C_i^{cyt} \qquad (1)$$

where K_i is termed the nucleocytoplasmic equilibrium distribution ratio or the nucleocytoplasmic partition coefficient. Despite its simple formulation, K_i is seldom determined, usually because the water contents of nucleus and cytoplasm are not known. Generally, the nuclear and cytoplasmic concentrations are available only as the quantity of i per unit dry weight, total weight, or volume. Mechanistic conclusions based upon data uncorrected for differences in nuclear and cytoplasmic water are at best provisional and must be viewed with suspicion.

The tendency of a substance i to distribute between nucleus and cytoplasm in a manner which minimizes the free energy may be thought of in terms of a potential for i within each nucleus and cytoplasm. Under biological conditions of constant temperature and pressure, the appropriate potential is the chemical potential, μ_i, which is the rate of change in the Gibbs free energy of i with change in the amount of i. A more convenient related function is the absolute chemical activity, λ_i, defined by

$$\mu_i = RT \ln \lambda_i \tag{2}$$

in which T is the temperature and R the gas constant.

The difference between the chemical potential of i in a given phase, μ_i, and in a standard reference phase,* μ_i^0, is

$$\mu_i - \mu_i^0 = RT \ln \lambda_i - RT \ln \lambda_i^0 = RT \ln (\lambda_i/\lambda_i^0) = RT \ln a_i \tag{3}$$

where a_i is the ratio of the absolute activity in the given phase to the absolute activity in the reference phase, and is known as the relative activity, or simply the activity. The ratio of the activity to the concentration is the activity coefficient, γ_i

$$a_i = \gamma_i C_i \tag{4}$$

The activity coefficient is a measure of the interactions of i with other components of its phase.

When i has a net electric charge, ζ_i, we must also consider, in addition to its chemical potential, its electrical potential, U_i, in each phase. The resultant electrochemical potential, $\bar{\mu}_i$, is thus

$$\bar{\mu}_i = \mu_i^0 + RT \ln a_i + F\zeta_i U_i \tag{5}$$

where F is the Faraday constant.

At thermodynamic equilibrium, further changes in the distribution of i result in no further reduction in free energy. That is, the electrochemical potential of i is the same throughout the system. For thermodynamic equilibrium between nucleus and cytoplasm we write

$$\mu_i^0 + RT \ln a_i^n + F\zeta_i U_i^n = \mu_i^0 + RT \ln a_i^{cyt} + F\zeta_i U_i^{cyt} \tag{6}$$

Substituting in Eq. (4) and rearranging gives

$$\frac{C_i^n}{C_i^{cyt}} (= K_i) = \frac{\gamma_i^{cyt}}{\gamma_i^n} \exp\left(\frac{F\zeta_i}{RT} (U_i^{cyt} - U_i^n)\right) \tag{7}$$

---

* A defined ordinary water solution of the solute in question is often used as a standard reference phase.

From Eq. (7) we see that the ratio of the nuclear to cytoplasmic concentrations at equilibrium (that is, the nucleocytoplasmic partition coefficient) is inversely proportional to the ratio of the activity coefficients and exponentially proportional to the electrical potential between nucleus and cytoplasm. It is important to note that equilibrium does not require equal nuclear and cytoplasmic concentrations.

For solutes without a net charge, or for charged species in the absence of an electrical potential difference between nucleus and cytoplasm, Eq. (7) reduces to

$$K_i = C_i^{\text{n}}/C_i^{\text{cyt}} = \gamma_i^{\text{cyt}}/\gamma_i^{\text{n}} \tag{8}$$

In these circumstances equilibrium requires equal nuclear and cytoplasmic activities, but the nuclear and cytoplasmic concentrations will be equal only if $\gamma_i^{\text{n}} = \gamma_i^{\text{cyt}}$. When i interacts differently with nuclear and cytoplasmic components (that is, when $\gamma_i^{\text{n}} \neq \gamma_i^{\text{cyt}}$) equilibrium is characterized by unequal nuclear and cytoplasmic concentrations. The latter seems to be the rule, rather than the exception. Substances may be more concentrated in the nucleus as is the case for dextrans, $^{36}\text{Cl}^-$, sucrose, and K^+, or less concentrated as is Na^+ (Fig. 4d and Table I). Additional examples are cited below.

In seeking an explanation for differences in C_i^{n} and C_i^{cyt}, it is important to bear in mind that not only can solutes interact differently with the structural macromolecules of nucleus and cytoplasm but so can water. Hence, nuclear and cytoplasmic solute concentration differences must be thought of as reflecting differing affinities of solutes and water for the different macromolecular structures of nucleus and cytoplasm. An equivalent statement is that, in phases with different macromolecular structures, the solvent properties of water can differ. The low concentration of cytoplasmic sucrose relative to RP and nucleus, shown in Table I, is explicable in these terms. In the oocyte, cytoplasmic water is a poorer solvent for sucrose than is ordinary or nuclear water (Horowitz and Paine, 1976). Similar conclusions about cytoplasmic water appear widely in the literature. The situation for Na^+ and K^+ (Table I) is more complex, because while cytoplasmic water is a poorer solvent for these cations as well, there is also highly specific cation binding (Horowitz and Paine, 1979).

IV. KINETICS

A. Introduction

Solute and solvent movements occur between the nucleus and cytoplasm whenever distributions between the two phases are not at electrochemical equilibrium. Perturbations from equilibrium result from changes in membrane transport at the cell surface and elsewhere and from local differences in the production and consumption of substances at intracellular sources and sinks. Nonequilibrium conditions may exist for various lengths of time, depending on the relative duration of the perturbation and on the kinetics of nucleocytoplasmic redistribution. For example, transient changes in cell membrane permeability cause changes in the free ion content of cytoplasm which are reflected in the nucleus, with almost instantaneous reestablishment of nucleocytoplasmic equilibrium. On the other hand, continual cytoplasmic synthesis of a material may result in a standing nonequilibrium concentration gradient, if nucleocytoplasmic movement is slow or nuclear incorporation is rapid relative to the synthesis rate.

Nucleocytoplasmic electrochemical gradients may, in principle, be maintained by energy input. Active, energy-dependent movements, either assisting or opposing passive movements, result in steady state rather than true equilibrium distributions. Possible significances of energy-dependent nucleocytoplasmic transport mechanisms will be discussed later. For the present, it will suffice to point out that active movements do not negate the principles of thermodynamic equilibria or passive transport, but are superimposed upon them. Furthermore, active transport mechanisms make significant demands on the limited cellular energy resources. Hence, knowledge of the existence of active transport at the cell surface cannot be treated as license to postulate active transport whenever intracellular concentration asymmetries are observed.

Thermodynamic equilibria are approached with finite rates or kinetics. The process which determines the rate of passive nucleocytoplasmic movements is diffusion. The location of the nuclear envelope makes it a logical candidate for a role in controlling nucleocytoplasmic movements; indeed, the nuclear surface or cortex seems to be a significant barrier to many substances. In addition, the nucleus and cytoplasm have distinct phase properties, and the diffusional movements of materials within each of these phases are subject to modification by the interactions of the materials with the phases.

B. Intracellular Diffusion

1. Cytoplasm

Knowledge of diffusion coefficients, D, is valuable because diffusion is a potentially important determinant of cellular reaction rates, and because diffusion provides a sensitive probe of the cell's physicochemical structure. Unfortunately, diffusion coefficients are technically difficult to measure, and interpretation is often ambiguous, especially when the solute is a cellular constituent subject to metabolism or present in a mixture of physical states (i.e., bound, sequestered, and free). Table II provides examples of measured cytoplasmic values, D_c', for a series of organic solutes which appear to diffuse in the cell without strong interactions.

Points to be noted are that (a) diffusion is slower as molecular size increases, (b) the ratio of the rate of diffusion measured in cytoplasm to that in water (D_c'/D_w) is not influenced markedly by molecular weight (when <24,000), and (c) the ratio D_c'/D_w is 0.2 to 0.5. This information provides a useful rough guide to estimating cellular diffusion times when D_w is known or can be estimated (Edsall, 1953; Othmer and Thakar, 1953), as is often the case.

The mean diffusional time of a molecule in an isotropic medium is given by

$$\bar{t} = (\bar{l})^2/2D \tag{9}$$

in which $\bar{l}$ is the mean distance traveled. This allows us to deduce that a molecule of 10,000 daltons, expected to have a diffusion coefficient in water of about 10^{-6} cm²/sec (Table II), will traverse the 10 μm radius of a rather large cell in a few seconds. This is a short time compared to the time required for most experimental manipulations. On the other hand, it is a very long time relative to most biochemical and physiological reaction rates. The former observation suggests the uncertainties inherent in analyzing intracellular transport processes by conventional "wet chemistry" approaches. The latter tells us that diffusion may often be reaction limiting in cellular processes.

The ratios D_c'/D_w given in Table II may be exceptions rather than the rule. Any biochemically important compound is likely to interact with cellular elements, and consequently its diffusion will involve a series of sorption–desorption steps whose effect is to reduce the diffusion coefficient (Fenichel and Horowitz, 1969). In these circumstances, the measured coefficient, D_c', is equal to the true diffusion coefficient, D_c, divided by the ratio of total to unbound solute, R; that is, $D_c' = D_c/R$ (Horowitz *et al.*, 1970). A striking example of this "chromatographic effect" is provided by Ca^{2+}, which interacts strongly in cytoplasm, so that $R \sim 10^3–10^6$ (Duncan, 1976). Rose and Loewenstein (1975) have shown that Ca^{2+} injected into cytoplasm diffuses at immeasurably slow rates. The same phenomenon is seen in the cytoplasmic diffusion of substances capable of forming disulfide bonds with proteins (Horowitz *et al.*, 1970). Another apparent example of special interest to molecular biologists, is provided by RNA's diffusing in cytoplasm following nuclear synthesis (Lönn and Edström, 1976; Lönn, 1978). Recent work has revealed the existence of a

TABLE II

Measured Cytoplasmic Diffusion Coefficients (D_c')

Substance	Molecular weight	Measured cytoplasmic diffusion coefficients (D_c') ($\times 10^7$ cm^2/sec)	D_c'/D_w	Cell	Reference
Sorbitol	182	50	0.53	*Balanus* muscle	Caillé and Hinke (1974)
Methylene blue	320	15	0.32	*Sepia* axon	Hodgkin and Keynes (1956)
Sucrose	342	20	0.38	*Rana* oocyte	Horowitz (1972)
Eosin	648	8.0	0.20	*Sepia* axon	Hodgkin and Keynes (1956)
Dextran	3,600	3.5	0.20	*Rana* oocyte	Paine *et al.* (1975)
Inulin	5,500	3.0	0.20	*Rana* oocyte	Horowitz and Moore (1974)
Dextran	10,000	2.5	0.27	*Rana* oocyte	Paine *et al.* (1975)
Dextran	24,000	1.5	0.24	*Rana* oocyte	Paine *et al.* (1975)

proteinaceous structural framework in cytoplasm (Porter *et al.*, 1979). When the properties of the *in vivo* cytoskeleton are clarified, its interactions with solutes and water will be better understood, and the specific effects on diffusion more readily assessed.

2. Nucleus

Data on nuclear diffusion are less available than for cytoplasm. The autoradiographic results in oocytes (Fig. 4a–c) show that nuclear diffusion gradients of dextran can mirror those of the adjacent cytoplasm, indicating that nucleoplasmic diffusion coefficients, D_n, do not differ appreciably from cytoplasmic coefficients. The same conclusion can be drawn for small electrolytes on the basis of conductance measurements in *Chironomus* salivary gland (Loewenstein, 1964), and from microspectrophotometric studies of metabolites in tissue culture cells (Kohen *et al.*, 1971). As in cytoplasm, a proteinaceous skeletal matrix has been identified in the nuclei of many cells (Berezney and Coffey, 1976, 1977; Comings and Okada, 1976; Wunderlich and Herlan, 1977). A role for the matrix in intranuclear transport is at present speculative, but potentially significant (Sect. V,C).

C. Nuclear Envelope Structure

Since the initial electron microscope studies of the nuclear envelope, much attention has centered on its role in nucleocytoplasmic movements. The envelope is a permeability barrier, offering passive resistance to nucleocytoplasmic movements, and it is possibly a site of active energy-dependent transport for some materials. The envelope's structure and composition are sufficiently complex to accommodate both passive and active transport functions; but the integration of structural and functional data that could prove active transport has barely begun. Here we summarize some recent notions of the structure of the nuclear surface, emphasizing relationships of the envelope with the cytoplasm and nucleus. More detailed reviews of nuclear envelope structure are available (Franke and Scheer, 1974; Harris, 1978).

The nuclear envelope consists of two parallel membranes separated by a 150–300 Å perinuclear space (Fig. 5). At various points the outer and inner membranes join through roughly circular pores. Each pore is associated with proteinaceous elements, the whole referred to as a nuclear pore complex. Pore complex structure is common to all eukaryotic cells. However, pore complexes vary in nuclear surface density and number among cell types, and among the metabolic states and cell cycle stages of a given cell (Zerban and Werz, 1975; Schel *et al.*, 1978). The nuclear

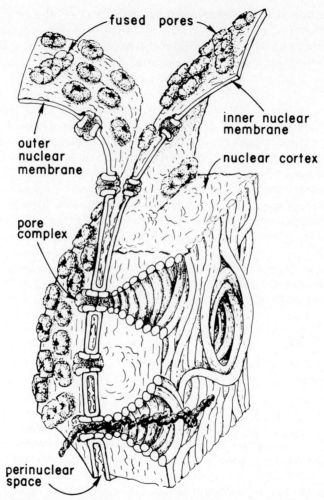

Fig. 5. Diagram of the nuclear surface, illustrating the structural relationships of the nuclear envelope membranes, the pore complexes, and the nuclear cortex (fibrous lamina). Note the hypothetical representation of RNP material in transit through the bottom pore complex. See text for full discussion. (From Schatten and Thoman, 1978, reproduced with permission.)

envelope, and in particular the pore complex, are structurally connected to the nucleus and cytoplasm. Improved ultrastructural techniques have led to replacement of earlier views of the envelope as a distinct, membranous boundary separating liquid compartments with one of a specialized interface between structured nuclear and cytoplasmic phases.

The outer membrane is continuous at points with the endoplasmic reticulum, and its cytoplasmic face often has attached ribosomes. Nuclear envelope and pore complex connections to cytoplasmic elements are numerous. Microtubules often parallel the envelope, sometimes with lateral bridges to the outer membrane, and other cytoplasmic microtubules seem to anchor terminally to the outer membrane. Additional cytoskeletal elements, the 5–7 nm actin microfilaments and the 8–12 nm intermediate filaments, have also been reported to be associated or attached to the nuclear envelope. Postulated functions include anchorage of the interphase nucleus within the cell, determination of nuclear shape, and participation in nuclear locomotion (Franke and Scheer, 1974; Osborn and Weber, 1977; Lehto et al., 1978).

The nuclear pore complex is the likely pathway for most nucleocytoplasmic movements and has been the subject of considerable ultrastructural study and model building. Many transmission and scanning electron microscopic studies exist. They use a variety of techniques, and each technique has strengths and weaknesses. Collectively they provide a coherent picture of the pore complex characteristics that seem to relate to transport. [An exhaustive review of the pore complex literature has been provided by Maul (1977).] The pore complex appears in electron micrographs as a hollow cylinder extending through the envelope and at least 200 Å into the adjacent nucleus and cytoplasm. Its wall is composed of granular and fibrillar protein structures and surrounds an electron-translucent central channel 100–150 Å in diameter. Thus, an annular appearance is presented in face-on views.

Occasionally a granule is visible in the central channel. Central granules seem to be composed of ribonucleoprotein (Maul, 1977) and may be transit forms, but they may also be images of pore complex structural elements. The pore complex has also been reported to contain ATPase activity (Yasuzumi and Tsubo, 1966). Potentially this enzyme is involved in energy-dependent translocations. However, although ATPase association with the nuclear envelope is widely acknowledged, its localization in the pore complex must be considered to be unproved (Vorbrodt, 1974; Clawson and Smuckler, 1979).

Pore complexes are connected to the adjacent nuclear cortex, most extensively to the fibrous lamina. Figure 5 is from a recent scanning electron microscope study (Schatten and Thoman, 1978). Sonication or solubilization often leaves pore complexes embedded in the fibrous lamina, even when other nuclear components and the membranous portion of the envelope are removed (Aaronson and Blobel, 1975). Gentler treatments leave behind a three-dimensional network of nuclear proteins or a nuclear skeletal infrastructure, with connections to the pore complexes (Berezney and Coffey, 1976). Analyses indicate that the nuclear skeleton is composed of a small number of proteins (Berezney and Coffey, 1976; Krohne et al., 1978; Gerace et al., 1978).

The complexity of the nuclear surface and the conservative pore complex structure seem to imply that important interactions occur between the nucleocytoplasmic interface and transported materials. But, while electron microscopy and fractionation of isolated nuclei have advanced our understanding of structure and composition, the intimate relationship of the envelope and the pore complex with the nuclear and cytoplasmic phases indicates that the central questions of envelope transport function are best approached in intact cell systems.

D. Nuclear Envelope Permeability

In theory, several paths could be involved in the movement of materials across the nuclear envelope (Stevens and André, 1969; Franke and Scheer, 1974). A route through the nuclear pore complexes is indicated by transport studies (Paine et al., 1975). Other routes can be inferred from electron micrographs, but their functional significances remain unproved.

Feldherr found that, after microinjection into amoebae, colloidal gold enters nuclei by passing through the pore complexes, and that passage is restricted to a central channel within each pore complex. By using gold particles of different diameters, Feldherr determined an upper limit for passage of 125–145 Å in *Amoeba proteus* (Feldherr, 1965), and 150–170 Å in *Chaos chaos* (Feldherr, 1966).

Comparable results emerged from a study of the movement of ferritin and fluorescein isothiocyanate (FITC)-labeled proteins in *Periplaneta americana* oocytes (Paine and Feldherr, 1972). Five hours after injection, ferritin (diameter $\sim$ 95 Å) was confined almost entirely to the cytoplasm, while smaller proteins had entered the nucleus in inverse relation to molecular size. Similar experiments using ^{125}I-labeled proteins in *Xenopus laevis* oocytes gave comparable results (Bonner, 1975a). Somatic cells (the salivary gland cells of *Chironomus thummi* larvae) behaved similarly when microinjected with FITC-protein tracers (Paine, 1975).

Quantitative determination of the permeability of the nuclear envelope was accomplished using ultra-low temperature autoradiographic analyses in *Rana* oocytes injected with [^{3}H]dextrans (Paine et al., 1975). Figure 4 demonstrates that the nuclear envelope is less permeable to larger than to smaller dextrans, and that envelope permeability limits the rate of nuclear entry. Quantitative data is provided in Fig. 6 for the three [^{3}H]dextrans investigated. The nuclear entry rates are markedly different; thermodynamic equilibrium is reached for 12.0 and 23.3 Å radius dextrans in about 0.5 and 15 hr, respectively, with an equilibrium distribution ratio,

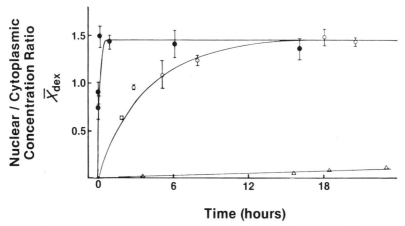

Time (hours)

Fig. 6. The time course of nuclear envelope permeation for [³H]dextrans of different radii: 12.0 Å (●), 23.3 Å (○), and 35.5 Å (△). (From Paine *et al.*, 1975, reproduced with permission).

K_{dex} of 1.45. The 35.5 Å dextran enters much more slowly, and at 23 hr is only at 10% of equilibrium.

The nuclear entry kinetics of all three [³H]dextrans are described by first-order exponential equations of the form

$$1 - (\bar{X}_{dex}/K_{dex}) = \exp[-k_n t] \tag{10}$$

in which $\bar{X}_{dex}/K_{dex}$ is the fraction of equilibrium achieved at time t, and k_n is the rate constant of nuclear envelope permeation (Fig. 7). The experimentally determined rate constants, k_n, are given in Table III with nuclear membrane permeabilities, P_n. Although the 35.5 Å [³H]dextran differs in molecular radius from the 12.0 by a factor of only 2.9, its rate of permeation differs by a factor of more than 2500. This sieving effect almost certainly arises from the porous structure of the nuclear envelope.

The functional radius of these pores can be calculated from the values of k_n, the radii of dextran molecules, and the number and length of the patent pores determined by electron microscopy. The pore radius was found to be 45 Å (Paine *et al.*, 1975), a result consistent with much of the nuclear permeation data available.

The physiological significance of the calculated pore radius can be appreciated from Fig. 8, in which the half-time, $t_{1/2}$, required to achieve nucleocytoplasmic equilibrium is plotted as a function of solute radius, for (a) the oocyte and (b) a hypothetical smaller cell in which nuclear radius is assumed to be 5 μm. Small changes in solute radius strongly influence the rate of envelope permeation. In principle, this relation provides the cell with methods of controlling the nucleocytoplasmic movement of solutes

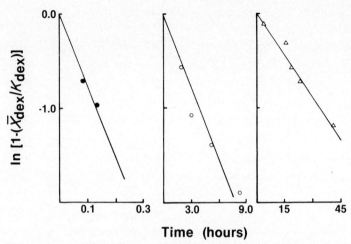

Time (hours)

Fig. 7. First-order exponential kinetics of nuclear envelope permeation illustrated by plots of ln $[1-(\bar{X}_{dex}/K_{dex})]$ versus time for dextrans with molecular radii of 12.0 Å (●, left), 23.3 Å (○, center), and 35.5 Å (△, right). Note the different time scales. The rate constants, k_n, given by the slopes of the lines are in Table III. (From Paine *et al.*, 1975, reproduced with permission.)

either by changing the dimensions of a solute (e.g., enzymatically), or varying the dimensions of the pore.

It can be shown that a 10 Å change in pore radius in the oocyte would effect a two- to threefold change in $t_{1/2}$ of ions and metabolites of about 5 Å radius, a tenfold change in the permeation of small proteins (radius ~18 Å), and in excess of a 1000-fold change for proteins larger than 31 Å. Whether the pore radius actually is subject to modulation is an unanswered question.

TABLE III

Oocyte Nuclear Envelope Permeation Constants for [³H]Dextrans[a]

Molecular weight	Molecular radius (Å)	Rate constant (k_n) (sec^{-1})	Permeability (P_n) (cm sec^{-1})
3,600	12.0	2.1×10^{-3}	1.6×10^{-5}
10,000	23.3	7.2×10^{-5}	5.4×10^{-7}
24,000	35.5	8.3×10^{-7}	6.2×10^{-9}

[a] From Paine *et al.* (1975)

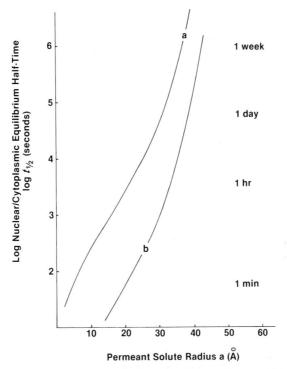

Fig. 8. Logarithm of the half-time, $t_{1/2}$ required to achieve nucleocytoplasmic equilibrium of solutes (radius a) by diffusional movement through 45 Å radius pores. (a) For the amphibian oocyte. (b) For a cell with nuclear radius of 5 μm. Discussion in text. (From Paine *et al.*, 1975, reproduced with permission.)

V. NUCLEOCYTOPLASMIC INTERRELATIONS

A. Ions and Small Molecules

The transport and equilibria of ions and small molecules between nucleus and cytoplasm are important. Metabolites and coenzyme movement is necessary because of the restriction of reaction sites to one or the other phase. For example, the nucleus is supplied by the cytoplasm with nucleotide precursors for DNA and RNA synthesis. Similarly, nicotinamide and ATP move into the nucleus for NAD synthesis, and in turn, NAD moves to the cytoplasm, from whence, after phosphorylation, it returns to the nucleus (Siebert and Humphrey, 1965; Siebert, 1978).

Nucleocytoplasmic ion distributions appear to be significant in the control of nuclear function. Ion activities affect structural transitions in chromatin (Jacobs *et al.*, 1976; Li *et al.*, 1977) and activation of specific genes in polytene chromosomes (Lezzi and Gilbert, 1970; Kroeger and Müller, 1973). Ionic changes are associated with events in mitosis (Cameron *et al.*, 1977), meiosis (O'Connor *et al.*, 1977), DNA synthesis (Orr *et al.*, 1972; McDonald *et al.*, 1972), hormonal stimulation (Kroeger *et al.*, 1973), and malignant transformation (Cone, 1974; Bader, 1976; Spaggiare *et al.*, 1976).

1. Kinetics

Several lines of evidence indicate that ions and small molecules move quickly between nucleus and cytoplasm (for recent reviews, see Harris, 1978; Siebert, 1978). Radioactive tracer studies detect no nuclear envelope hindrance to movements of water and ions (Horowitz and Fenichel, 1970; Paine, 1975). Microinjected sucrose (Horowitz, 1972), α-aminoisobutyric acid (Frank and Horowitz, 1975), and fluorescein (Paine, 1975) move between nucleus and cytoplasm at rates too great to be measured by routine methods. Higher resolution microfluorometry reveals that glycolytic intermediates experience only millisecond delays in nucleocytoplasmic transit (Kohen *et al.*, 1971). These observations are all consistent with the determination (Section IV,D) of ~45 Å radius openings in the nuclear pore complexes.

One result appears to be inconsistent with the rapid nucleocytoplasmic movement of ions. Loewenstein and co-workers (Loewenstein and Kanno, 1963; Ito and Loewenstein, 1965) and Palmer and Civan (1977), using intracellular microelectrodes, measured nuclear envelope electrical resistances R_t, of 0.5–2.0 Ω cm^2 in *Chironomus* salivary gland cells. Since, in amphibian oocytes, R_t was reported to be less than 0.001 Ω cm^2 (Loewenstein, 1964), these results have been interpreted to mean that the salivary gland cell nuclear envelope is a significant barrier to ion movement. In macromolecular flux studies, on the other hand, salivary gland cell nuclear envelope permeabilities resemble those of the oocyte (Paine, 1975). Hence, a discrepancy may appear to exist between permeability measured using macromolecular flux data and electrical resistance data. However, the discrepancy is only apparent—the result of qualitative comparison of data from two different techniques. A quantitative analysis makes this clear.

If the pore complex has a patent channel of radius r = 45 Å (Paine *et al.*, 1975), length d = 1500–3000 Å (Paine *et al.*, 1975; Schatten and Thoman, 1978), and specific resistivity ρ similar to that of cytoplasm or nucleoplasm, 100 Ω cm (Loewenstein *et al.*, 1966), the resistance of each pore complex, R_p, will be

$$R_\mathrm{p} = \frac{\rho[1 + 1.64(r/d)]d}{\pi r^2(1 - a/r)^6} = 3.25 \text{ to } 6.50 \times 10^9 \; \Omega \qquad (11)$$

where a, the hydrated radius of the current carrying ions, is taken as 2.0 Å. Since the density of patent nuclear pores, N, is about $5.5 \times 10^9/\mathrm{cm}^2$ (Paine, 1975; Schatten and Thoman, 1978), the calculated envelope resistance is

$$R_\mathrm{t} = R_\mathrm{p}/N = 0.6 \text{ to } 1.2 \; \Omega \; \mathrm{cm}^2 \qquad (12)$$

Thus, the electrical measurements and macromolecular kinetic data are in good agreement.*

The easy movement of small molecules between cytoplasm and nucleus does not rule out transport as rate-limiting in nuclear or cytoplasmic reactions. The nuclear envelope certainly provides some resistance, in excess of ordinary diffusion, to solutes. Kohen *et al.* (1971) estimated that for glycolytic metabolites, moving 15 μm from cytoplasm into nucleus, passage across the nuclear envelope, ~3,000 Å or 2% of the total distance, accounts for 35 msec, or 70% of the total time required. However, even if the envelope were absent, the rate-limiting role of diffusion itself could not be ignored. Many biochemical and biophysical reactions are so rapid that *in vivo* diffusion of substrate from distant sources and products to distant sinks must be limiting. However, the analytical problems in working with intact cells are formidable and the identification of most of these lies in the future.

2. Equilibria

Nucleocytoplasmic equilibrium distributions of ions and small molecules are as difficult to measure as their kinetics. The rapidity of diffusion, and the fragility of biological structures, mean that redistributions of materials from *in vivo* locations accompany all but the most painstaking procedures. Cryogenic methods which reduce diffusional artifacts have been helpful in this regard.

The data in Table I, which show nucleocytoplasmic small solute asymmetries for four solutes, were obtained using ultra-low temperature microdissection (ULTM). The technique (Fig. 2) involves rapidly quenching cells to liquid nitrogen temperatures and their subsequent microdissection at low temperatures to isolate cytoplasm and nucleus while preventing artifactual solute movements (Horowitz *et al.*, 1979). The technique is

---

* The absence of a measurable R_t in amphibian oocytes remains unexplained. It is likely due to a technical difficulty such as failure of the oocyte nuclear envelope to seal around the impaling electrode.

restricted to large cells, but is a general technique in the sense that it may be used in the analysis of any metabolite.

Another recent methodological development is electron probe X-ray microanalysis of frozen hydrate sections (Gupta et al., 1977). After rapid freezing to liquid nitrogen temperatures, sections are cut at −80° to −130°C and maintained at −170°C during subsequent analysis of secondary X-ray emissions in a spectrometer-equipped electron microscope. The spatial resolution is potentially adequate for the determination of nucleocytoplasmic differences in the smallest cells. However, it is a method of elemental analysis, and consequently limited in its discrimination of solutes, though of immense potential value in studies of inorganic ions.

Progress in microprobe analysis has been impressive, and the major unresolved problem is the accurate determination of local water concentrations. Nevertheless, plausible estimates of concentrations on a water basis appear possible. Even when this is not the case, the ratio of the concentrations of ions amongst cellular locations should be accurate. Figure 9 shows a microprobe analysis of epithelial cells in the Malpighian tubules of the hemipteran *Rhodnius prolix* (Gupta et al., 1976). Cytoplasmic differences in Na^+, K^+, and Cl^- on a wet weight basis are observed. Whether these translate into concentration differences on a water basis is not clear. However, the ratios of ions also differ, indicating that whatever the local water concentrations, cytoplasm varies regionally, either in ion activities and/or the concentration of specific ion binding sites. Similarly, nuclear ionic concentrations and ratios differ from those in the adjacent cytoplasm.

In view of the high permeability of the nuclear envelope to small molecules, active transport is an energetically implausible explanation for nucleocytoplasmic ion asymmetries. In the absence of active mechanisms, unequal equilibrium concentrations in nucleus and cytoplasm must result

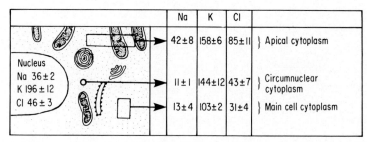

Fig. 9. Diagrammatic representation of the local concentrations of Na^+, K^+, and Cl^- in the nucleus and cytoplasm of *Rhodnius* Malpighian tubule epithelial cells as determined by electron microprobe analysis. Data are expressed in microequivalents per gram wet weight. (From Gupta et al., 1976, reproduced with permission.)

from differences in binding and/or solubility in the two phases, and be reflected in $\gamma_i^n \neq \gamma_i^{cyt}$. The differing interactions of ions with the nuclear and cytoplasmic phases *in vivo* have been approached by two methods, the intracellular reference phase (RP) technique and ion-sensitive microelectrodes.

As described in Section II,A, an intracellular gelatin reference phase can be introduced into the cytoplasm of amphibian oocytes. The RP equilibrates with and samples the diffusible solutes of the cytoplasm (Horowitz *et al.*, 1979). The oocytes are frozen and the nucleus, cytoplasm, and RP separated by ultra-low temperature microdissection, as shown in Fig. 2. Table I shows the equilibrium concentrations of Na^+, K^+, $^{36}Cl^-$, and sucrose in each of these phases. Nuclear concentrations of K^+ and Na^+ are seen to be higher and lower, respectively, than the cytoplasmic concentrations and nuclear concentrations closely resemble those in the RP.

In the oocyte, there is no electrical potential difference between the RP and cytoplasm or between the nucleus and cytoplasm. (This is deduced from the fact that neutral molecules, cations, and anions all show parallel RP–cytoplasm and nucleus–cytoplasm ratios). Hence, the chemical activities of solutes of RP, nucleus, and cytoplasm are the same

$$a_i^{RP}(= \gamma_i^{RP}C_i^{RP}) = a_i^n(= \gamma_i^n C_i^n) = a_i^{cyt}(= \gamma_i^{cyt}C_i^{cyt}) \tag{13}$$

Reference phase activity coefficients for Na^+ and K^+, γ_i^{RP}, are similar to those in ordinary saline solutions, γ_i^0, approximately 0.77. This fact allows the calculation, using the solute concentration of each phase (Table I) and Eq. (13), of the activities and the nuclear and cytoplasmic activity coefficients for Na^+ and K^+. Thus, $a_{Na} = 16.2$ $\mu Eq/ml$ and $a_K = 99.2$ $\mu Eq/ml$. The nuclear activity coefficients are $\gamma_{Na}^n = 0.78$ and $\gamma_K^n = 0.73$, indistinguishable from those of ordinary salines. Cytoplasmic coefficients are $\gamma_{Na}^{cyt} = 0.21$ and $\gamma_K^{cyt} = 1.12$, much different from each other and from those of ordinary aqueous solution and the nucleus. We conclude that in the oocyte the difference observed in the nucleocytoplasmic distributions of Na^+ and K^+ arise from specialized interactions in the cytoplasm, rather than in the nucleus or the nuclear envelope.

With one exception, the oocyte is the only cell in which data on nucleus and cytoplasmic Na^+ and K^+ activities and concentrations are available, and our knowledge of the mechanisms responsible for the nucleocytoplasmic distribution of these important cations rests disproportionately on these cells. The exception is the salivary gland epithelial cell of the fourth instar larvae of the midge, *Chironomus*. Cytoplasmic and nuclear chemical activities in these cells in *Chironomus attenuatus* were determined by Palmer and Civan (1977), using ion-sensitive microelectrodes. Concentrations were determined by Kroeger *et al.* (1973) in *Chironomus thummi* by separating nucleus and cytoplasm under oil at ambient temperatures. (The relevance of the species difference cannot be assessed, and is assumed here to be unimportant).

In the gland cell cytoplasm $a_{Na}^{cyt} = 12$, $a_K^{cyt} = 103$, $C_{Na}^{cyt} = 47$, and $C_K^{cyt} = 102$ μEq/ml. Hence, $\gamma_{Na}^{cyt} = 0.26$ and $\gamma_K^{cyt} = 1.01$, values similar to those in oocyte cytoplasm and different from those of ordinary aqueous solutions. For the nucleus, Palmer and Civan (1977) found $a_{Na}^n = 12$ and $a_K^n = 111$ μEq/ml, essentially the same activities as they measured for the cytoplasm, supporting the correctness of Eq. (13) and its underlying assumptions.* Kroeger *et al.* (1973) found $C_{Na}^n = 31$ and $C_K^n = 115$ μEq/ml, which, combined with the activities, correspond to $\gamma_{Na}^n = 0.38$ and $\gamma_K^n = 0.97$. Unlike the oocyte nucleus, the nucleus of *Chironomus* appears to behave, in its ionic properties, much like cytoplasm and different from ordinary aqueous solutions.

The circumstance that, in amphibian and dipteran salivary epithelial cytoplasm, γ_{Na}^{cyt} is lower and γ_K^n is higher than in ordinary salines is common in other cells (Lev and Armstrong, 1975). The mechanisms responsible for these differences are understood in oocytes (Horowitz and Paine, 1979). It must suffice here to explain that in oocyte cytoplasm Na$^+$, K$^+$, and water each exist in both free and bound forms, and that the observed activity coefficients arise from the differing quantities of each form.

The question of the states of Na$^+$, K$^+$, and water in the nucleus *vis à vis* those in the cytoplasm is important. It reflects on the physicochemical nature of the two phases and the mechanisms of nuclear control. The oocyte results suggest differences between nucleus and cytoplasm, while the salivary gland results suggest similarities. This is disconcerting, but it is the extent of our knowledge. Reliable determinations of nuclear and cytoplasmic concentrations and activities in other cells are needed.

B. Proteins

1. Kinetics

That materials of cytoplasmic origin control nuclear functions has been known for almost a century (Gurdon and Woodland, 1968). As regulators of gene activity, proteins are particularly important. Unfortunately, because of problems of isolation and characterization, few studies of the

---

* Palmer and Civan (1977) also measured the electrical potential difference between nuclear and cytoplasm and found it to be insignificant, under conditions when a 2.0 Ω cm² nuclear envelope resistance assured envelope intactness.

transport behavior of characterized, endogenous proteins exist. The effective 45 Å radius of the pore complex would be expected to permit diffusive passage of proteins of lesser radii. This has been confirmed with exogenous protein whose movements show that the nuclear envelope is a molecular sieve (Section IV,D).

Most endogenous cellular polypeptides have molecular weights of 20,000 to 90,000, with a modal value of about 60,000 (Kiehn and Holland, 1970). Many of these are aspherical and/or multimeric. However, if, for present purposes, we assume all have globular, monomer form, these weights may be estimated as corresponding to molecular radii of 20–42 Å, with a modal radius of 32 Å. Such dimensions would permit nucleocytoplasmic passage by passive diffusion through the nuclear envelope pores. Labeling studies have shown that nuclear concentrations of newly synthesized proteins approach cytoplasmic concentrations within minutes or seconds. This is true for newly synthesized proteins considered collectively (Smith and Ecker, 1970; Wu and Warner, 1971), histones (Robbins and Borun, 1967), and ribosomal proteins (Warner, 1974). However, such data cannot resolve mechanisms because they describe heterogeneous groups and because equilibrium distributions are unknown.

Feldherr (1975) examined the transport kinetics of newly synthesized *Xenopus laevis* oocyte proteins as a function of their size. When corrected for nuclear and cytoplasmic water, his data show that polypeptides larger than 94,000 daltons are more concentrated in the nucleus than in cytoplasm within 1 hr of supplying labeled amino acids. Figure 8 shows that the half-time for nucleocytoplasmic diffusional equilibrium of comparable size molecules (radius > 42 Å) in similar oocytes is more than a day. Diffusion through 45 Å radius pores can account for the observed nuclear uptake only if the equilibrium distribution ratio of the proteins K_{prot} is on the order of 10–50. Asymmetries of this magnitude have been identified for some endogenous proteins.

2. Equilibria

Proteins display a wide range of nucleocytoplasmic equilibrium distribution ratios. This is demonstrated by studies of newly synthesized polypeptides in *Xenopus laevis* oocytes (Bonner, 1975b; DeRobertis et al., 1978). Specific proteins ("nuclear" proteins) are more concentrated in the nucleus than in the cytoplasm (e.g., N1, N2, N3, and N4), others ("cytoplasmic" proteins) are more concentrated in the cytoplasm (e.g., tubulin), and still others ("both" proteins) distribute equally among the two phases (e.g., actin) (Fig. 10). Wassarman et al. (1979) identified a germinal vesicle (nucleus) associated protein, GVAP, in the mouse oocyte. GVAP has a molecular weight of 28,000 as determined by SDS acrylamide gel elec-

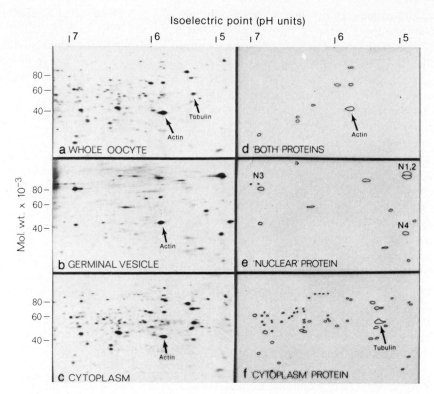

Fig. 10. Fluorography of SDS-isoelectric focusing els of proteins labeled by incubating *Xenopus laevis* oocytes in ^{14}C-labeled amino acids. Proteins extracted from (a) intact oocytes, (b) isolated nuclei (germinal vesicles), and (c) isolated cytoplasm. (d)–(f) are tracings of the three identified classes: (d) proteins present in both nucleus and cytoplasm, (e) proteins more concentrated in the nucleus, and (f) proteins more concentrated in the cytoplasm. (From DeRobertis *et al.*, 1978, reproduced with permission.)

trophoresis, and appears at least 1000-fold more concentrated in the nucleus than in the cytoplasm.

Characteristic in vivo distributions are reestablished when proteins are isolated and reintroduced into host cells. Several examples of this homing behavior are known. Experiments in *Amoeba proteus* demonstrate a class of proteins, 50–100 times more concentrated in nuclei than in cytoplasm, which redistribute in a similar ratio between host cytoplasm and nucleus in transplant experiments (Legname and Goldstein, 1972). These proteins are not associated with recognizable intranuclear structures such as nucleoli or chromatin (Goldstein, 1974).

Homing behavior is also displayed by proteins which do exhibit specific associations. Kroeger *et al.* (1963) transplanted ^{3}H-protein labeled chromosomes from salivary gland cells into the cytoplasm of unlabeled cells. Autoradiography showed donor and host chromatin to be equally labeled within a few hours. Appropriate controls permitted the conclusion that the redistribution of intact chromosomal proteins was being observed. Recently, ^{125}I-labeled nonhistone chromosomal proteins from rat liver were loaded into erythrocyte ghosts and introduced into Ehrlich ascites cells by cell fusion; they accumulated preferentially in the host nuclear chromatin (Yamaizumi *et al.*, 1978). Homing has also been observed with a defined chromosomal protein, HMG 1, after introduction into bovine fibroblasts and HeLa cells (Rechsteiner and Kuehl, 1979). HMG 1 nuclear accumulation is reversible, since the protein redistributes to other nuclei when the original host cell is fused with unlabeled cells. Finally, in heterokaryons produced by cell fusion, nucleus-specific (nucleolar, nucleoplasmic, and nuclear envelope) antigens distribute between the nuclei, assuming in each the site-specific distribution characteristic of the cell of origin (Carlsson *et al.*, 1973, 1974).

Homing experiments have been performed in the *Xenopus* oocytes in conjunction with chromatographic separation (DeRobertis *et al.*, 1978). Soluble, isotopically labeled proteins were isolated from nuclei of donor oocytes (Fig. 11a) and injected into the cytoplasm of host oocytes. Autoradiography shows that the proteins subsequently concentrate in the host nucleus, but no intranuclear structure-specific localization is observed. Donor and host cells were separated into nucleus and cytoplasm, and analyzed on two-dimensional SDS-isoelectric focusing gels. Note that N1 and N2, localized predominantly in the nucleus in donor cells (Fig. 10e), accumulate (Fig. 11b and c) in the host nucleus (to > 120 times the cytoplasmic concentration as determined by densitometry). Actin, on the other hand, which is equally concentrated in the nucleus and cytoplasm of donor cells (Fig. 10b and c), also is equally distributed in host cells (Fig. 11b and c).

What mechanisms create protein concentration differences between nucleus and cytoplasm? Of course many proteins are integral components of nuclear structures. Other proteins, more reversibly associated with nucleus or cytoplasm, may modulate cellular activities through their distributions between the two phases. If, for these proteins, the concentration gradients correspond to activity gradients, the presence of active transport mechanisms is implied. There is at present no evidence for such mechanisms, but they cannot be ruled out.

In the absence of chemical activity gradients, nucleocytoplasmic concentration differences must be attributed to differences in the sorption and solubility characteristics of the two phases. Evidence for intranuclear sorption is provided by site-specific nuclear accumulations of reinjected

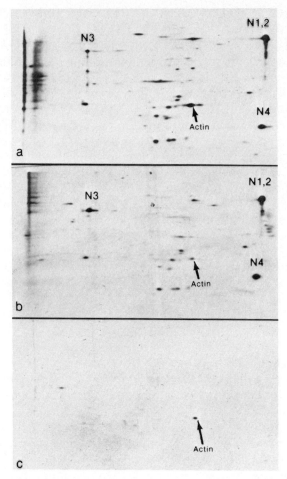

Fig. 11. Two-dimensional gels of ^{35}S-labeled *Xenopus laevis* oocyte proteins reinjected into host oocytes. Donor nuclei were isolated and aqueous extracted, yielding the proteins in (a). After reinjection into host oocytes, these proteins localized in the nucleus (b) and cytoplasm (c) in distributions quantitatively similar to those found in the donor cells (cf. Fig. 10). (Modified from DeRobertis *et al.*, 1978, reproduced with permission.)

chromosomal proteins (Yamaizumi *et al.*, 1978) and nucleolar, nucleo-plasmic, and nuclear envelope antigens (Carlsson *et al.*, 1974). Sorption is also consistent with the nuclear saturability observed in protein reinjection studies (Bonner, 1975a; Paine and Tluczek, 1978).

Evidence for solubility differences is less direct, provided by analogy between the behavior of small solutes such as sucrose and those proteins which distribute without apparent association with specific nuclear or cytoplasmic structures. The demonstration that nuclear accumulation of steroid hormone receptor obeys Henry's law (i.e., is linearly related to cytoplasmic concentration) must also be viewed as support for solubility-determined distributions (Gannon et al., 1976). In addition, a role for solubility is circumstantially supported by examples of asymmetric protein distributions between dissimilar aqueous macromolecular phases in nonliving systems (Albertsson, 1971).

In summary, the evidence is consistent with a model in which intracellular proteins move between nucleus and cytoplasm by diffusion through the pore complexes, with rates determined by passive molecular-size sieving. It is not necessary to postulate active transport, since nucleocytoplasmic activity gradients are not known to exist. Differences between nuclear and cytoplasmic protein concentrations appear adequately explained as arising from differential solubility and differential sorption to elements in the nuclear and cytoplasmic phases. Sorbed forms presumably are in mass action equilibria with diffusible forms, but distinction and quantitation of sorbed and diffusive species in vivo has been impossible. The recently introduced intracellular reference phase method (Horowitz et al., 1979) offers a tool, applicable to very large cells, for this purpose.

C. Ribonucleic Acids and Ribonucleoproteins

The nucleus is the source of RNA templates and tools for cytoplasmic protein synthesis, and the nucleocytoplasmic movement of messenger (m), ribosomal (r), and transfer (t) RNA's are obligatory steps in gene expression. In addition, low molecular weight RNA species of unknown function move in both directions between nucleus and cytoplasm.

Technical difficulties encountered in RNA transport studies are legion, resulting from the characteristics of RNA metabolism and function. Eukaryotic RNA's, with the possible exception of 5 S rRNA, are not transcribed as the final functional units, but as large polynucleotides which are associated with proteins, and subsequently enzymatically cleaved, trimmed, and modified to provide functional ribonucleoprotein (RNP) units. The complexity of this process obscures the identity of the intermediates and undermines efforts to systematically analyze intracellular RNA transport and distribution. Only limited conclusions about RNA transport are possible, based mostly on knowledge of the permeability of the nuclear envelope obtained with other materials and consideration of the likely sizes of RNP transit forms.

1. Low Molecular Weight RNA's

Some low molecular weight (LMW) RNA's ($\approx$4–10 S, 65–200 nucleotides) are found in nucleus and cytoplasm (Goldstein, 1976; Zieve and Penman, 1976; Benecke and Penman, 1977). Goldstein and Ko (1974,

1975) have shown in amoeba that some of these species move in both directions between the two phases, some distribute uniformly, and others accumulate in the nucleus to >250 times the cytoplasmic concentrations, behaving in transplant experiments as nuclear concentrating proteins. Transient increases in LMW RNA association with chromatin at specific mitotic stages have suggested to some that they regulate gene activities (Goldstein *et al.*, 1977; Goldstein and Ko, 1978). Others have found LMW RNA's associated with nuclear and cytoplasmic "skeletal" elements and suggest that they serve structural functions (Herman *et al.*, 1976).

The composition of the LMW RNA forms that move between nucleus and cytoplasm is unknown. We might speculate that because of their size they may resemble proteins in their kinetic and equilibrium properties. However, there is evidence that LMW RNA's are associated *in vivo* with proteins (Enger and Walters, 1970; Rein, 1971). If they exist as RNP structures larger than the openings in the nuclear pore complexes, low molecular weight RNA's will be more appropriately viewed as ribosomal and messenger RNP's discussed below.

2. Ribonucleoprotein Particulates

Ribosomal and messenger RNA's are packaged into ribonucleoprotein (RNP) particulates. Mature ribosomes contain 60 S and 40 S RNP sub-units. A 45 S pre-rRNA is transcribed, assembled with proteins, and cleaved to smaller RNP's, primarily within the nucleolus. Mature ribosomes are not found in the nucleus, but electron microscopy reveals numerous nucleolar RNP fibrils and granules thought to be processing intermediates. Ribosomal RNA's entering the cytoplasm are first detected as free ribosome subunits, and it is widely assumed that rRNP's cross the nuclear envelope as particulates at least as big as these subunits: 2.1×10^6 daltons, diameter ~230 Å for the 60 S subunit, and 0.8×10^6 daltons, ~230 Å $\times$ 120–140 Å for the 40 S subunit (Van Holde and Hill, 1974).

Messenger RNA's are also transcribed as parts of larger (heterogeneous, hn) RNA's, whose processing involves complexing to proteins and subsequent cleavage, trimming, and modification. Nuclear fractionation yields 30–400 S hnRNP particulates, which can be transformed into 30 S "monoparticles" by mild ribonuclease treatment (Preobrazhensky and Spirin, 1978). The 30 S particulates are 200 × 80 Å in the electron microscope (Samarina *et al.*, 1967). Among the nuclear structures identifiable with the electron microscope, the perichromatin fibrils and the 400–500 Å diameter perichromatin granules are the most likely to be the hnRNP (Puvion and Bernhard, 1975).

Understanding the transport of rRNP and mRNP is likely to require considerations over and above those applicable to smaller solutes. The sizes of these particulates approach or exceed the mean distance between the elements of the intracellular macromolecular matrices in which they exist. Under these circumstances, the inertia of the matrix alone will limit diffusional movements. Indeed, diffusional processes may contribute negligibly to the transportation of RNP particulates. Several investigators have hypothesized that RNP's are closely associated with the skeletal matrix of the nucleus, and that the matrix may play roles in RNP processing and in the intranuclear movement of RNP's toward the nuclear pore complexes (Wunderlich *et al.*, 1976; Herlan *et al.*, 1979). Diffusion is certainly negligible for RNP movement through the nuclear pore complexes, since the forms of rRNP and mRNP which transit from nucleus to cytoplasm are larger than the 90 Å patent diameter of the pore complex. These particulates must undergo extensive configurational changes in order to penetrate the pore complex.

Energy input may be required at one or more steps during the processing and translocation of RNP's, not just to propel movement against activity gradients, but also to power molecular rearrangements in particulate–pore complex interactions. Coupling of energy input to interactions of RNP's with the nuclear matrix and pore complexes could, in principle, provide bases for distinguishing between those RNA sequences which are transported to the cytoplasm and those which are not (Lewin, 1975).

To date, work on the energy dependence of nucleocytoplasmic transport has relied almost entirely on isolated nuclear systems. The relevance of such systems to living cells is suspect, since structural elements of the nucleus, nuclear envelope, and cytoplasm are modified by nuclear isolation. Nevertheless, such studies appear to implicate ATP in the export of RNP's from nuclei. A broad specificity nuclear ATPase has been identified, and its hydrolytic activity correlated with RNP release from isolated nuclei (Agutter, *et al.*, 1976; Clawson, *et al.*, 1978). The localization of nuclear ATPase is unclear. The proposition that it is specific to the nuclear pore complexes (Yasuzumi and Tsubo, 1966) has proven technically difficult to support (see Clawson and Smuckler, 1979); it may be generally distributed among several types of intranuclear RNP-containing structures (Vorbrodt and Maul, 1980). Furthermore, it is uncertain whether the role of ATP is to provide energy or to perform a more passive function (Ishikawa *et al.*, 1978).

One experimental system has allowed the application of ultrastructural, microdissection, and biochemical methods to the nucleocytoplasmic movement of messengerlike RNP. Large chromosomal puffs, Balbiani rings 1 and 2, of the salivary gland cells of *Chironomus tentans* are the sites of transcription of two distinct 75 S RNA's. These 75 S species appear

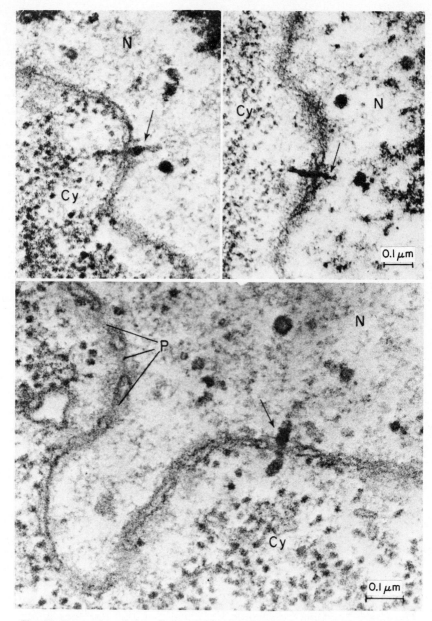

Fig. 12. Penetration of the salivary gland cell nuclear pore complex by electron dense materials (arrows), thought to be transit forms of the Balbiani ring granules. Similar images suggesting that configurational changes are involved in RNP transit are observed in many cell types. N, nucleus; Cy, cytoplasm; P, nuclear pores. (From Stevens and Swift, 1966, reproduced with permission.)

also in the cytoplasm and have been identified as messenger RNA's on the basis of their poly(A) contents and their presence in cytoplasmic polysomes (Daneholt et al., 1976). Electron micrographs of the Balbiani rings reveal nascent RNP fibrils condensing into granular structures. The largest granules are 400–500 Å in diameter and are observed throughout the nucleus. Some granules are associated with nuclear pore complexes, where they exhibit configurational changes (Fig. 12) consistent with a narrowing down to rodlike or fibrillar structures traversing the pore complex (Stevens and Swift, 1966; Monneron and Bernard, 1969; Paine et al., 1975). The granules are observed also in the cytoplasm in a gradient of concentration decreasing with distance from the nuclear envelope (Daneholt et al., 1976). Lönn (1978) has analyzed concentric cytoplasmic zones isolated by nonaqueous microdissection and found similar cytoplasmic gradients of labeled Balbiani ring 75 S RNA. Furthermore, pretreatment with cycloheximide, which prevents translation of mRNA, results in a more uniform distribution of newly synthesized 75 S RNA in the cytoplasm, consistent with the idea that the normal gradient is due to association of 75 S RNA's with the endoplasmic reticulum soon after they exit from the nucleus. The salivary gland cell system appears to provide a powerful tool, still in the early stages of development, for analyzing many of the outstanding problems in RNA transport.

ACKNOWLEDGMENTS

This work was supported by National Institutes of Health grants GM 19548, GM 26734, and CA 17456, and by an institutional grant to the Michigan Cancer Foundation from the United Foundation of Greater Detroit.

REFERENCES

Aaronson, R. P., and Blobel, G. (1975). Isolation of nuclear pore complexes in association with a lamina. Proc. Natl. Acad. Sci. U.S.A. 72, 1007–1011.

Agutter, P. S., McArdle, H. J., and McCaldin, B. (1976). Evidence for involvement of nuclear envelope nucleoside triphosphatase in nucleocytoplasmic translocation of ribonucleoprotein. Nature (London) 263, 165–167.

Albertsson, P.-Å. (1971). "Partition of Cell Particles and Macromolecules." Wiley (Interscience), New York.

Bader, J. P. (1976). Sodium: A regulator of glucose uptake in virus-transformed and nontransformed cells. J. Cell. Physiol. 89, 677–682.

Benecke, B.-J., and Penman, S. (1977). A new class of small nuclear RNA molecules synthesized by a type I RNA polymerase in HeLa cells. Cell 12, 939–946.

Berezney, R., and Coffey, D. S. (1976). The nuclear protein matrix: Isolation, structure, and functions. Adv. Enzyme Regul. 14, 63–100.

Berezney, R., and Coffey, D. S. (1977). Nuclear matrix: Isolation and characterization of a framework structure from rat liver nuclei. *J. Cell Biol.* **73**, 616–637.

Bonner, W. M. (1975a). Protein migration into nuclei. I. Frog oocyte nuclei *in vivo* accumulate microinjected histones, allow entry to small proteins, and exclude large proteins. *J. Cell Biol.* **64**, 421–430.

Bonner, W. M. (1975b). Protein migration into nuclei. II. Frog oocyte nuclei accumulate a class of microinjected oocyte nuclear proteins and exclude a class of microinjected oocyte cytoplasmic proteins. *J. Cell Biol.* **64**, 431–437.

Caillé, J. P., and Hinke, J. A. M. (1974). The volume available to diffusion in the muscle fiber. *Can. J. Physiol. Pharmacol.* **52**, 814–828.

Cameron, I. L., Sparks, R. L., Horn, K. L., and Smith, N. R. (1977). Concentration of elements in mitotic chromatin as measured by X-ray microanalysis. *J. Cell Biol.* **73**, 193–199.

Carlsson, S.-A., Moore, G. P. M., and Ringertz, N. R. (1973). Nucleo-cytoplasmic protein migration during the activation of chick erythrocyte nuclei in heterokaryons. *Exp. Cell Res.* **76**, 234–241.

Carlsson, S.-A., Ringertz, N. R., and Savage, R. E. (1974). Intracellular antigen migration in interspecific myoblast heterokaryons. *Exp. Cell Res.* **84**, 255–266.

Clawson, G., and Smuckler, E. (1979). Reciprocity between nuclear envelope NTPase activity and RNA transport. *J. Cell Biol.* **83**, 416a.

Clawson, G. A., Koplitz, M., Castler-Schechter, B., and Smuckler, E. A. (1978). Energy utilization and RNA transport: Their interdependence. *Biochemistry* **17**, 3747–3752.

Comings, D. E. and Okada, T. A. (1976). Nuclear proteins. III. The fibrillar nature of the nuclear matrix. *Exp. Cell Res.* **103**, 341–360.

Cone, C. D., Jr. (1974). The role of the surface electrical transmembrane potential in normal and malignant mitogenesis. *Ann. N.Y. Acad. Sci.* **238**, 420–435.

Daneholt, B., Case, S. T., Hyde, J., Nelson, L., and Wieslander, L. (1976). Production and fate of Balbiani ring products. *Prog. Nucleic. Acid Res. Mol. Biol.* **19**, 319–334.

DeRobertis, E. M., Longthorne, R. F., and Gurdon, J. B. (1978). Intracellular migration of nuclear proteins in *Xenopus* oocytes. *Nature (London)* **272**, 254–256.

Duncan, C. J., ed. (1976). "Calcium in Biological Systems." Cambridge Univ. Press, London and New York.

Edsall, J. T. (1953). The size, shape, and hydration of protein molecules. *In* "The Proteins" (H. Neurath and K. Bailey, eds.), 1st ed., pp. 549–726. Academic Press, New York.

Enger, M. D., and Walters, R. A. (1970). Isolation of low molecular weight, methylated ribonucleic acids from 10 S to 30 S particles of Chinese hamster cell fractions. *Biochemistry* **9**, 3551–3562.

Feldherr, C. M. (1965). The effect of the electron-opaque pore material on exchanges through the nuclear annuli. *J. Cell Biol.* **25**, 43–51.

Feldherr, C. M. (1966). Nucleocytoplasmic exchanges during cell division. *J. Cell Biol.* **31**, 199–203.

Feldherr, C. M. (1975). The uptake of endogenous proteins by oocyte nuclei. *Exp. Cell Res.* **93**, 411–419.

Fenichel, I. R., and Horowitz, S. B. (1969). Intracellular transport. *In* "Biological Membranes" (R. M. Dowben, ed.), pp. 177–221. Little, Brown, Boston, Massachusetts.

Frank, M., and Horowitz, S. B. (1975). Nucleocytoplasmic transport and distribution of an amino acid, in situ. *J. Cell Sci.* **19**, 127–139.

Franke, W. W., and Scheer, U. (1974). Structures and functions of the nuclear envelope. *In* "The Cell Nucleus" (H. Busch, ed.), Vol. 3, pp. 220–348. Academic Press, New York.

Gannon, F., Katzenellenbogen, B., Stancel, G., and Gorski, J. (1976). Estrogen receptor movement to the nucleus: Discussion of a cytoplasmic exclusion hypothesis. *In* "The

Molecular Biology of Hormone Action'' (J. Papaconstantinou, ed.), pp. 137–150. Academic Press, New York.

Gerace, L., Blum, A., and Blobel, G. (1978). Immunochemical localization of the major polypeptides of the nuclear pore complex–lamina fraction. *J. Cell Biol.* **79**, 546–566.

Goldstein, L. (1974). Movement of molecules between nucleus and cytoplasm. In "The Cell Nucleus" (H. Busch, ed.), Vol. 1, pp. 387–438. Academic Press, New York.

Goldstein, L. (1976). Role for small nuclear RNAs in programming chromosomal information? *Nature (London)* **261**, 519–521.

Goldstein, L., and Ko, C. (1974). Electrophoretic characterization of shuttling and nonshuttling small nuclear RNAs. *Cell* **2**, 259–269.

Goldstein, L., and Ko, C. (1975). The characteristics of shuttling RNAs confirmed. *Exp. Cell Res.* **96**, 297–302.

Goldstein, L., and Ko, C. (1978). Identification of the small nuclear RNAs associated with the mitotic chromosomes of *Amoeba proteus. Chromosoma* **68**, 319–325.

Goldstein, L., Wise, G. E., and Ko, C. (1977). Small nuclear RNA localization during mitosis: an electron microscope study. *J. Cell Biol.* **73**, 322–331.

Gupta, B. L., Hall, T. A., Maddrell, S. H. P., and Moreton, R. B. (1976). Distribution of ions in a fluid-transporting epithelium determined by electron-probe X-ray microanalysis. *Nature (London)* **264**, 284–287.

Gupta, B. L., Hall, T. A., and Moreton, R. B. (1977). Electron probe X-ray microanalysis. *In* "Transport of Ions and Water in Animals" (B. L. Gupta, R. B. Moreton, J. L. Oschman, and B. J. Wall, eds.), pp. 83–144. Academic Press, New York.

Gurdon, J. B., and Woodland, H. R. (1968). The cytoplasmic control of nuclear activity in animal development. *Biol. Rev. Cambridge Philos. Soc.* **43**, 233–267.

Harris, J. R. (1978). The biochemistry and ultrastructure of the nuclear envelope. *Biochim. Biophys. Acta* **515**, 55–104.

Herlan, G., Eckert, W. A., Kaffenberger, W., and Wunderlich, F. (1979). Isolation and characterization of an RNA-containing nuclear matrix from *Tetrahymena* macronuclei. *Biochemistry* **18**, 1782–1788.

Herman, R., Zieve, G., Williams, J., Lenk, R., and Penman, S. (1976). Cellular skeletons and RNA messages. *Prog. Nucleic. Acid Res. Mol. Biol.* **19**, 379–402.

Hodgkin, A. L., and Keynes, R. D. (1956). Experiments on the injection of substances into squid giant axons by means of a microsyringe. *J. Physiol.* **131**, 592–616.

Horowitz, S. B. (1972). The permeability of the amphibian oocyte nucleus, *in situ. J. Cell Biol.* **54**, 609–625.

Horowitz, S. B., and Fenichel, I. R. (1970). Analysis of sodium transport in the amphibian oocyte by extractive and radioautoradiographic techniques. *J. Cell Biol.* **47**, 120–131.

Horowitz, S. B., and Moore, L. C. (1974). The nuclear permeability, intracellular distribution, and diffusion of inulin in the amphibian oocyte. *J. Cell Biol.* **60**, 405–415.

Horowitz, S. B., and Paine, P. L. (1976). Cytoplasmic exclusion as a basis for asymmetric nucleocytoplasmic solute distributions. *Nature (London)* **260**, 151–153.

Horowitz, S. B., and Paine, P. L. (1979). Reference phase analysis of free and bound intracellular solutes. II. Isothermal and isotopic studies of cytoplasmic sodium, potassium and water. *Biophysical J.* **25**, 45–62.

Horowitz, S. B., Fenichel, I. R., Hoffman, B., Kollmann, G., and Shapiro, B. (1970). The intracellular transport and distribution of cysteamine phosphate derivatives. *Biophysical J.* **10**, 994–1010.

Horowitz, S. B., Paine, P. L., Tluczek, L., and Reynhout, J. K. (1979). Reference phase analysis of free and bound intracellular solutes. I. Sodium and potassium in amphibian oocytes. *Biophysical J.* **25**, 33–44.

Ito, S., and Loewenstein, W. R. (1965). Permeability of a nuclear membrane: Changes during normal development and changes induced by growth hormone. *Science* **150**, 909–910.

Itoh, S., and Schwartz, I. L. (1957). Sodium and potassium distribution in isolated thymus nuclei. *Am. J. Physiol.* **188**, 490–498.

Jacobs, G. A., Smith, J. A., Watt, R. A., and Barry, J. M. (1976). Ion binding and chromatin condensation. *Biochim. Biophys. Acta* **442**, 109–115.

Johnson, J. D., Douvas, A. S., and Bonner, J. (1974). Chromosomal proteins. *Int. Rev. Cytol., Suppl.* **4**, 273–362.

Kiehn, E. D., and Holland, J. J. (1970). Size distribution of polypeptide chains in cells. *Nature (London)* **226**, 544–545.

Kohen, E., Siebert, G., and Kohen, C. (1971). Transfer of metabolites across the nuclear membrane: A microfluorometric study. *Hoppe-Seyler's Z. Physiol. Chem.* **352**, 927–937.

Kroeger, H., and Müller, G. (1973). Control of puffing activity in three chromosomal segments of explanted salivary gland cells of *Chironomus thummi* by variation in extracellular Na^+, K^+, and Mg^{++}. *Exp. Cell Res.* **82**, 89–94.

Kroeger, H., Jacob, J., and Sirlin, J. L. (1963). The movement of nuclear proteins from the cytoplasm to the nucleus in salivary cells. *Exp. Cell Res.* **31**, 416–423.

Kroeger, H., Trösch, W., and Müller, G. (1973). Changes in nuclear electrolytes of *Chironomus thummi* salivary gland cells during development. *Exp. Cell Res.* **80**, 329–339.

Krohne, G., Franke, W. W., and Scheer, U. (1978). The major polypeptides of the nuclear pore complex. *Exp. Cell Res.* **116**, 85–102.

Legname, C., and Goldstein, L. (1972). Proteins in nucleocytoplasmic interactions. *Exp. Cell Res.* **75**, 111–121.

Lehto, V.-P., Virtanen, I., and Kurki, P. (1978). Intermediate filaments anchor and nuclei in nuclear monolayers of cultured human fibroblasts. *Nature (London)* **272**, 175–177.

Lev, A. A., and Armstrong, W. McD. (1975). Ionic activities in cells. *Curr. Top. Membr. Trans.* **6**, 59–123.

Lewin, B. (1975). Units of transcription and translation: Sequence components of heterogeneous nuclear RNA and messenger RNA. *Cell* **4**, 77–93.

Lezzi, M., and Gilbert, L. I. (1970). Differential effects of K^+ and Na^+ on specific bands of isolated polytene chromosomes of *Chironomus tentans*. *J. Cell Sci.* **6**, 615–628.

Li, H. J., Hu, A. W., Maciewicz, R. A., Cohen, P., Santella, R. M., and Chang, C. (1977). Structural transition in chromatin induced by ions in solution. *Nucleic Acids Res.* **4**, 3839–3854.

Loewenstein, W. R. (1964). Permeability of the nuclear membrane as determined with electrical methods. *Protoplasmatologia* **2**, 26–34.

Loewenstein, W. R., and Kanno, Y. (1963). Some electrical properties of a nuclear membrane examined with a microelectrode. *J. Gen. Physiol.* **46**, 1123–1140.

Loewenstein, W. R., Kanno, Y., and Ito, S. (1966). Permeability of nuclear membranes. *Ann. N.Y. Acad. Sci.* **137**, 708–716.

Lönn, U. (1978). Delayed flow-through cytoplasm of newly synthesized Balbiani ring 75 S RNA. *Cell* **13**, 727–733.

Lönn, U., and Edström, J-E. (1976). Mobility restriction *in vivo* of the heavy ribosomal subunit in a secretory cell. *J. Cell Biol.* **70**, 573–580.

McDonald, T. F., Sachs, H. G., Orr, C. W., and Ebert, J. D. (1972). External potassium and baby hamster kidney cells: intracellular ions, ATP, growth, DNA synthesis, and membrane potential. *Dev. Biol.* **28**, 290–303.

Maul, G. G. (1977). The nuclear and the cytoplasmic pore complex: structure, dynamics, distribution, and evolution. *Int. Rev. Cytol., Suppl.* **6**, 75–186.

Mollenhauer, H. H., and Morré, D. J. (1978). Structural compartmentation of the cytosol: zones of exclusion, zones of adhesion, cytoskeletal and intercisternal elements. In "Subcellular Biochemistry" (D. B. Roodyn, ed.), pp. 327–359. Plenum, New York.
Monneron, A., and Bernhard, W. (1969). Fine structural organization of the interphase nucleus in some mammalian cells. J. Ultrastruct. Res. 27, 266–288.
O'Connor, C. M., Robinson, K. R., and Smith, L. D. (1977). Calcium, potassium, and sodium exchange in full-grown and maturing Xenopus laevis oocytes. Dev. Biol. 61, 28–40.
Orr, C. W., Yoshikawa-Fukada, M., and Ebert, J. D. (1972). Potassium: effect on DNA synthesis and multiplication of baby-hamster kidney cells. Proc. Natl. Acad. Sci. U.S.A. 69, 243–247.
Osborn, M., and Weber, K. (1977). The detergent-resistant cytoskeleton of tissue culture cells includes the nucleus and the microfilament bundles. Exp. Cell Res. 106, 339–349.
Othmer, D. F., and Thakar, M. S. (1953). Correlating diffusion coefficients in liquids. Ind. Eng. Chem. 45, 589–593.
Paine, P. L. (1975). Nucleocytoplasmic movement of fluorescent tracers microinjected into living salivary gland cells. J. Cell Biol. 66, 652–657.
Paine, P. L., and Feldherr, C. M. (1972). Nucleocytoplasmic exchange of macromolecules. Exp. Cell Res. 74, 81–98.
Paine, P. L., and Tluczek, L. J. M. (1978). Nuclear accumulation of proteins. J. Cell Biol. 79, 244a.
Paine, P. L., Moore, L. C., and Horowitz, S. B. (1975). Nuclear envelope permeability. Nature (London) 254, 109–114.
Palmer, L. G., and Civan, M. M. (1977). Distribution of Na+, K+ and Cl− between nucleus and cytoplasm in Chironomus salivary gland cells. J. Membr. Biol. 33, 41–61.
Porter, Keith R., Byers, H. R., and Ellisman, M. H. (1979). The cytoskeleton. In "The Neurosciences" (F. O. Schmitt and F. G. Worden, eds.) pp. 703–722. MIT Press, Cambridge, Massachusetts.
Preobrazhensky, A. A., and Spirin, A. S. (1978). Informosomes and their protein components: the present stage of knowledge. Prog. Nucleic Acid Res. Mol. Biol. 21, 2–38.
Puvion, E., and Bernhard, W. (1975). Ribonucleoprotein components in liver cell nuclei as visualized by cryoultramicrotomy. J. Cell Biol. 67, 200–214.
Rechsteiner, M. and Kuehl, L. (1979). Microinjection of the nonhistone chromosomal protein HMGI into bovine fibroblasts and HeLa cells. Cell 16, 901–908.
Rein, A. (1971). The small molecular weight monodisperse nuclear RNAs in mitotic cells. Biochim. Biophys. Acta. 232, 306–313.
Robbins, E. and Borun, T. W. (1967). The cytoplasmic synthesis of histones in HeLa cells and its temporal relationship to DNA replication. Proc. Natl. Acad. Sci. U.S.A. 57, 409–416.
Rose, B., and Loewenstein, W. R. (1975). Calcium ion distribution in cytoplasm visualized by aequorin: Diffusion in cytosol restricted by energized sequestering. Science 190, 1204–1206.
Samarina, O. P., Krichevskaya, A. A., Molnar, J., Bruskov, V. I., and Georgiev, G. P. (1967). Nuclear ribonucleoproteins containing messenger RNA (isolation and some properties). Mol. Biol. 1, 129–141.
Schatten, G., and Thoman, M. (1978). Nuclear surface complex. J. Cell Biol. 77, 517–535.
Schel, J. H. N., Steenbergen, L. C. A., Bekers, A. G. M., and Wanka, F. (1978). Change in the nuclear pore frequency during the nuclear cycle of Physarum polycephalum. J. Cell Sci. 34, 225–232.
Siebert, G. (1978). The limited contribution of the nuclear envelope to metabolic compartmentation. Biochem. Soc. Trans. 6, 5–9.

Siebert, G., and Humphrey, G. B. (1965). Enzymology of the nucleus. *Adv. Enzymol. Relat. Subj. Biochem.* **27,** 239–288.

Smith, L. D., and Ecker, R. E. (1970). Regulatory processes in the maturation and early cleavage of amphibian eggs. *Dev. Biol.* **5,** 1–38.

Smith, E. L., DeLange, R. J., and Bonner, J. (1970). Chemistry and biology of the histones. *Physiol. Rev.* **50,** 159–170.

Spaggiare, S., Wallach, M. J., and Tupper, J. T. (1976). Potassium transport in normal and transformed mouse 3T3 cells. *J. Cell Physiol.* **89,** 403–416.

Stein, G. S., Spelsberg, T. C., and Kleinsmith, L. J. (1974). Nonhistone chromosomal proteins and gene regulation. *Science* **183,** 817–824.

Stevens, B. J., and André, J. (1969). The nuclear envelope. *In* "Handbook of Molecular Cytology" (A. Lima-de-Faria, ed.), pp. 837–871. North-Holland Publ., Amsterdam.

Stevens, B. J., and Swift, H. (1966). RNA transport from nucleus to cytoplasm in *Chironomus* salivary glands. *J. Cell Biol.* **31,** 55–77.

VanHolde, K. E., and Hill, W. E. (1974). General physical properties of ribosomes. *In* "Ribosomes" (M. Nomura, A. Tissières, and P. Lengyel, eds.), pp. 53–92. Cold Spring Harbor Lab., Cold Spring Harbor, New York.

Vorbrodt, A. (1974). Cytochemistry of nuclear enzymes. *In* "The Cell Nucleus" (H. Busch, ed.), Vol. 3, pp. 309–344. Academic Press, New York.

Vorbrodt, A., and Maul, G. G. (1980). Cytochemical studies on the relation of nucleoside triphosphatase activity to ribonucleoproteins in isolated rat liver nuclei. *J. Histochem. Cytochem.* **28,** 27–35.

Warner, J. R. (1974). The assembly of ribosomes in eukaryotes. *In* "Ribosomes" (M. Nomura, A. Tissières, and P. Lengyel, eds.), pp. 461–488. Cold Spring Harbor Lab., Cold Spring Harbor, New York.

Wassarman, P. M., Schultz, R. M., and Letourneau, G. E. (1979). Protein synthesis during meiotic maturation of mouse oocytes in meiotic maturation of mouse oocytes *in vitro:* Synthesis and phosphorylation of a protein localized in the germinal Vesicle. *Dev. Biol.* **69,** 94–107.

Wu, R. S., and Warner, J. R. (1971). Cytoplasmic synthesis of nuclear proteins. *J. Cell Biol.* **51,** 643–652.

Wunderlich, F. and Herlan, G. (1977). A reversibly contractile nuclear matrix: Its isolation structure, and composition. *J. Cell Biol.* **73,** 271–278.

Wunderlich, F., Berezney, R., and Kleinig, H. (1976). The nuclear envelope: An interdisciplinary analysis of its structure, composition, and functions. *In* "Biological Membranes" (D. Chapman and D. F. H. Wallach, eds.), pp. 241–333. Academic Press, New York.

Yamaizumi, M., Uchida, T., Okada, Y., Furusawa, M., and Mitsui, H. (1978). Rapid transfer of non-histone chromosomal proteins to the nucleus of living cells. *Nature (London)* **273,** 782–784.

Yasuzumi, G., and Tsubo, I. (1966). The fine structure of nuclei as revealed by electron microscopy. III. Adenosine triphosphatase activity in the pores of nuclear envelope of mouse choroid plexus epithelial cells. *Exp. Cell Res.* **43,** 281–292.

Zerban, H., and Werz, G. (1975). Changes in frequency and total number of nuclear pores in the life cycle of *Acetabularia. Exp. Cell Res.* **93,** 472–476.

Zieve, G., and Penman, S. (1976). Small RNA Species of the HeLa cell: metabolism and subcellular localization. *Cell* **8,** 19–31.

8

Regulation of Location of Intracellular Proteins

Aldons J. Lusis and Richard T. Swank

339

I. INTRODUCTION AND OVERVIEW

Ultrastructural and biochemical observations have shown that eukaryotic cells are subdivided by membranes into a variety of compartments with specific functions and constituents (Fig. 1). There is extensive ordering within the compartments, as many proteins and other macromolecules are present in complexes or are associated with subcellular structures. In this chapter we consider the processes responsible for localizing proteins to particular intracellular sites. Among the questions addressed are the following: Where in the cell are proteins synthesized? How do proteins move between membrane-bounded compartments? What are the mechanisms leading to the segregation of populations of proteins? What is the role of covalent modification in protein localization? How are proteins inserted into membranes and assembled into complexes?

In physical terms, we assume that localization consists of a combination of self-assembly and catalytic interactions. These are discussed in some detail in Chapters 5 and 9 of this volume. In biological terms, we under-

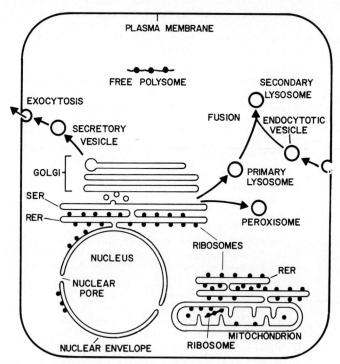

Fig. 1. Diagramatic representation of cell organelles and sites of protein synthesis. RER, rough endoplasmic reticulum; SER, smooth endoplasmic reticulum.

stand localization as a sequence of processes beginning with the synthesis of messenger RNA (mRNA), followed by synthesis of protein at the proper intracellular site, transport by passive diffusion or specific biological mechanisms, and, finally, insertion or assembly into a particular organelle or complex. The segregation of populations of proteins occurs both at the time of synthesis and during subsequent transport, and the information directing localization is encoded entirely within the mRNA and polypeptide sequences. The expression of this information involves recognition by the processing apparatus of structural features present in populations of proteins. The clearest example to date of such a recognition feature is the "signal sequence" of secretory proteins, which directs polysomes to receptors in the rough endoplasmic reticulum (RER), resulting in the "vectorial discharge" of secretory proteins into the lumen of the RER. Other populations of proteins which are segregated into organelles, such as nuclear proteins, lysosomal and peroxisomal enzymes, and various membrane proteins, must also contain specific "traffic-directing" structural features (Table I). Frequently, transport of proteins between intracellular compartments is accompanied by covalent modification, and it is likely that modifications are directly involved in localization and assembly. In eukaryotes, protein synthesis occurs not only in the cytoplasm but also in

TABLE I

Processes Involved in the Segregation of Populations of Proteins

---

A. Sites of synthesis
   1. Cytoplasm
     (a) Free ribosomes (release into cell sap)
     (b) Ribosomes bound to RER membrane (release into ER lumen, ER membrane and possibly cell sap)
   2. Mitochondria (and chloroplasts)
     (a) Free ribosomes (release into matrix)
     (b) Ribosomes bound to matrical side of inner membrane (release into membrane)
B. Posttranslational transport
   1. From cell sap to
     (a) Nuclei (involving passage through nuclear pores)
     (b) Membranes (involving insertion into lipid bilayer or attachment at cytoplasmic face of membrane)
     (c) Various membrane-bounded organelles (involving transfer across membranes)
   2. From RER membrane to other membranes (involving lateral flow in plane of membrane and membrane fission-fusion)
   3. From RER luminal space to other organelles (involving direct and intermittent membrane contacts)
   4. From extracellular space to lysosomes (involving endocytotic uptake and fusion to primary lysosomes)
C. Assembly into complexes (involving specific associations with other macromolecules)

---

certain semiautonomous organelles such as mitochondria and chloroplasts. The proteins synthesized in these organelles are derived from organellar DNA and are probably not exported to other compartments.

Since organelles are bounded by membranes, and since membranes form diffusion barriers to proteins, the interaction of proteins with membranes is of major importance in protein localization. The basic structure of biological membranes is a phospholipid bilayer in which proteins are embedded. A widely accepted view of membranes is that they are essentially fluid in nature and that the protein and lipid constituents are able to move laterally in the plane of the membrane unless somehow restricted. On the other hand, proteins are generally unable to "flip-flop" or reverse orientations with respect to the plane of the membrane. The functional and topological interrelatedness of the intracellular network of membranes is becoming apparent. Membranes of different organelles often form direct contacts or intermittent contracts by means of fission and fusion. From this, one concludes that membranes are very dynamic structures, frequently in a state of flux. We are beginning to understand how proteins bind to, recognize, and cross particular membranes. Proteins interact not only with the hydrophobic interior of the lipid bilayer of membranes but also form specific associations at the membrane surfaces. Special biological mechanisms are required for transporting proteins across membranes, and presently there is evidence for two such mechanisms, one obligately coupled to protein translation and the other involving posttranslational transfer. Also, membrane fission and fusion provide means for transporting proteins between certain membrane-bounded compartments. Membranes are apparently not assembled *de novo* but rather by insertion of lipid and protein into preexisting membranes. In certain cases, the insertion of proteins into membrane requires special mechanisms, while in other cases insertion may occur spontaneously.

Several strategies have been employed in studying protein localization and the related problems of organelle assembly. Microscopy, cytochemistry, and cell fractionation have yielded information relating to the topological arrangement of intracellular proteins and the relationships between organelles. Pulse labeling with radioactive precursors of proteins has made it possible to examine the synthesis and intracellular transport pathways of proteins. In recent years, several reconstitution systems have been developed, and these have been particularly useful in analyzing the molecular mechanisms involved in the segregation of populations of proteins and the transport of proteins across membranes. Finally, genetic approaches, involving the identification and characterization of mutants altered in protein localization, are being used increasingly. Such mutants are particularly valuable since they allow us to test the function of individual components of the localization apparatus in a living cell. One suspects that in order to obtain satisfying answers to the many fundamental questions remaining to be answered, a combination of approaches will be required.

The aims of this chapter are to (1) discuss the general mechanisms involved in protein localization, (2) summarize the available evidence concerning the localization pathways for different classes of proteins, and (3) describe in some detail certain systems which have been particularly informative in terms of localization mechanisms.

II. GENERAL MECHANISMS

Protein localization consists of a combination of several processes, including synthesis, transport, self-assembly, and modification. These processes are interrelated, occurring together or alternately in a defined sequence. Let us consider each in turn.

A. Sites of Synthesis

In most eukaryotic cells, proteins are synthesized on at least two distinct populations of ribosomes in the cytoplasm, free ribosomes and ribosomes bound to the cytoplasmic face of the membranes of the rough endoplasmic reticulum (RER). In addition, certain proteins are synthesized within mitochondria and chloroplasts, semiautonomous organelles which contain their own self-replicating DNA and the associated apparatus for RNA and protein synthesis (Fig. 1). It is now clear that free and bound ribosomes have different functional roles and that they synthesize different classes of proteins (Table I). In the case of secretory proteins, synthesis on ribosomes bound to RER allows proteins to be "vectorially discharged" across the membrane into the lumen (or cisternal space), effectively segregating them from proteins synthesized on free polysomes, which are released into the cytoplasm (Blobel and Dobberstein, 1975a,b; Palade, 1975). Accumulating evidence indicates that certain membrane proteins are inserted into membrane by incomplete vectorial discharge (Rothman and Lenard, 1977; Katz et al., 1977a). Vectorial translation is also utilized by prokaryotes where, again, two populations of ribosomes are found, free ribosomes and ribosomes bound to the cytoplasmic surface of the cell membrane (e.g., Smith et al., 1977; Sekizawa et al., 1977; Randall et al., 1978).

Experiments with reconstitution systems have shown that the recognition features directing synthesis on membrane-bound ribosomes, as opposed to free ribosomes, are encoded within the messenger RNA and are not determined by the protein synthesizing system (Blobel and Dobber-

stein, 1975a) (Section V,A, p. 371). In general, proteins which are synthesized on membrane-bound ribosomes contain characteristic amino-
terminal regions, hydrophobic in nature, and the "signal hypothesis"
proposes that these sequences, present in the nascent chains, function in
binding polysomes to receptors in the membrane (Blobel and Sabatini,
1971; Blobel and Dobberstein, 1975a). In the case of most secretory proteins, these sequences are proteolytically clipped from the remainder of
the protein before the completion of the translation. In addition, mRNA
may itself associate in a specific manner with receptors on the membrane
surface (reviewed by Sabatini and Kreibich, 1976; Shore and Tata, 1977c).

B. Intracellular Transport

While some proteins are synthesized at the site of their final localization, clearly many others are not. These latter proteins undergo transport,
either passive "physical" transport by means of diffusion or "biological"
transport by means of various cellular processes, some of which require
an input of energy. For globular proteins of average size, diffusion within
the cell appears to be a relatively rapid process, on the order of minutes or
less over cellular dimensions. The rates of "biological" transport, including the movement of proteins across membranes, probably vary considerably. These rates have been estimated for some proteins using pulse
labeling with radioactive protein precursors and are generally on the order
of minutes to hours (e.g., Sections IV,A and E). Neurons are an especially
interesting system for studies of intracellular transport, and in extended
axons there appear to be distinct classes of proteins differing in their rates
of transport (from 1 mm/day up to 400 mm/day) and in their destinations
(Lorenz and Willard, 1978; reviewed in Wilson, 1978). In considering
intracellular transport of proteins, membranes are of particular importance, since in the absence of special biological mechanisms they form
barriers that are impermeable to proteins (Section III).

While localization is dependent upon synthesis at the proper intracellular site, segregation of various protein populations must also occur after
translation (Table I). For example, most cytoplasmic proteins, certain
peroxisomal enzymes, and certain membrane proteins are synthesized on
free ribosomes in the cytoplasm. This suggests that proteins which
undergo biological transport contain "traffic-directing" structural features
which are recognized by the transport apparatus.

C. Self-Assembly

Weak, noncovalent forces (including ionic bonding, hydrophobic bonding, hydrogen bonding, and van der Waals interactions) are responsible

for the folding of polypeptide chains and for the association of individual polypeptide chains with each other and with other cellular constituents (see Chapter 9, this volume). While these interactions can be relatively easily disrupted (for example, by changes in temperature), they are sufficiently stable to maintain cellular form and substructure and exhibit remarkable specificity. In recent years there have been several striking demonstrations of self-assembly *in vitro,* including the renaturation of individual polypeptides, the polymerization of large protein complexes (such as microtubules), the reconstitution of functional ribosomes from their separated constituents (over 50 proteins and 3 RNA species), and the insertion of proteins into membranes. Mechanisms involved in the assembly of certain viruses, such as tobacco mosaic virus, have been elaborated in great molecular detail (e.g., Butler and Klug, 1978). In several systems assembly of cerain complexes requires the involvement of polypeptides not incorporated into the final or "mature" structures. This happens in the assembly of ribosomes (Section IV,C) and nucleosomes (Section IV,D) and the polymerization of microtubules (Vallee and Borisy, 1978). In other cases, assembly may be obligately coupled to covalent modification (Section II,D).

In addition to being largely responsible for the ordering of proteins within subcellular compartments, specific noncovalent interactions with previously localized constituents allow the selective accumulation of proteins in compartments or on surfaces. A well-known example is the binding of peptide hormones and extracellular proteins to receptors on the cell surface. Another is the accumulation of DNA-binding proteins in the nucleus as opposed to the cytoplasm, assuming they are able to move freely through the pores in the nuclear envelope (Table I).

D. Role of Covalent Modification

Covalent modifications, such as proteolytic cleavage of the polypeptide chain and the addition of conjugant groups, are often associated with the translocation of proteins and their assembly into complex cellular structures (Uy and Wold, 1977; Chapter 5, this volume). In functional terms, modification may play a role in the following aspects of enzyme localization:

1. "Traffic-Directing" Labels

Various conjugant groups, as well as portions of the polypeptide chain, probably provide "traffic-directing" labels for classes of proteins. Strong evidence for this comes from the role of carbohydrate in the cellular uptake of serum proteins in mammals. Ashwell, Morell, and co-workers have shown that the clearance of orosomucoid and certain other glycoproteins from the circulation is mediated by a hepatocyte cell surface receptor which specifically recognizes terminal galactose residues (Ashwell and Morell, 1974). Recently, this receptor was found to be present in the membranes of several subcellular compartments as well as the plasma membrane, although the function of the internal receptor is not known (Pricer and Ashwell, 1976). The endocytotic uptake of lysosomal acid hydrolases by Kupffer cells and fibroblasts is mediated by a different set of receptors, specific for mannosyl or phosphomannosyl residues (Hickman et al., 1974; Stahl et al., 1976; Kaplan et al., 1977).

Whether glycosylation plays a role in the translocation of proteins within the cell is less certain. Mutants with altered glycosylation would be particularly useful in examining this question, but as yet few such mutants have been reported. Waring (1978) showed that several mouse myeloma variant lines defective in the secretion of immunoglobulins (IgA) synthesize both the heavy and light chains but are unable to glycosylate heavy chains, suggesting that at least partial glycosylation is required for secretion of some immunoglobulin species. Defective glycosylation and secretion of α-antitrypsin occur concurrently in the inherited disease α-antitrypsin deficiency Pi-ZZ, but since the primary defect is apparently an alteration in the structural gene for the protein, it is premature to draw conclusions concerning the function of glycosylation in the secretion of the protein (Yoshida et al., 1976; Hercz et al., 1978). That glycosylation is not necessarily required for export of proteins from the cell is indicated by the fact that some secreted proteins lack carbohydrate. Some recent studies have employed the antibiotic tunicamycin, which specifically inhibits core glycosylation of proteins, to examine the role of carbohydrate in transport processes. While tunicamycin did not inhibit the export of a glycoprotein of chick embryo fibroblasts to the cell surface (Olden et al., 1978) nor the secretion of transferrin, very low density lipoprotein, or serum albumin (Struck et al., 1978), it did inhibit the incorporation of certain Sindbis and vesicular stomatitis virus proteins into plasma membranes of the host cells (Leavitt et al., 1977). However, such studies must be interpreted with caution, particularly since tunicamycin results in some inhibition of protein synthesis.

2. Transfer between Compartments

Modifications that alter the structural characteristics of protein can result in a relocation of that protein among the compartments of the cell. This could involve the addition or removal of labels recognized by specific receptors or simply an alteration of the general properties of the protein (e.g., its charge, size, or solubility) which in turn could affect its interactions with other cellular constituents, such as membranes, nucleic acids, or other proteins. Such a mechanism could function in the various processes whereby proteins appear to be "passed" from one cellular compartment to another and, in particular, provide a means of allowing certain proteins (such as diptheria toxin, see Section III,D) to cross membranes. For example, a protein could be concentrated in an organelle (or a membrane surface) by diffusion, if at that site it was modified in such a way that transfer out of the organelle (or dissociation from the surface) was no longer possible. One such modification would be the removal of a recognition feature (e.g., part of the peptide chain or a carbohydrate moiety) required for receptor-mediated transfer across membranes.

Although no case is understood in detail, several examples of modification that appear to play a role in protein relocation are known. These include the insertion of cytoplasmically synthesized subunits of ATPase and of glutamate dehydrogenase into mitochondria (Section IV,E) and the insertion of the small subunit of ribulose-1,5-bisphosphatase carboxylase into chloroplasts (Section IV,F). Each of these proteins is synthesized on cytoplasmic ribosomes as precursors which are converted into the mature subunits during or after transport into the organelles. An example of a modification that alters the binding of a protein to lipid is the activation of porcine pancreatic phospholipase A_2. The enzyme is secreted as an inactive zymogen unable to hydrolyze substrates present in a lipid–water interface and is activated by proteolytic cleavage, exposing a hydrophobic region which allows the enzyme to penetrate micellar surfaces (van DamMieras *et al.*, 1975; Slotboom and de Haas, 1975). Conceivably, a protein could become localized to a particular membrane (containing the appropriate protease) by a similar mechanism. Finally, the addition and removal of carbohydrate moieties recognized by glycoprotein receptors (discussed above) could well be involved in regulating the transfer of glycoproteins within cells as well as between cells.

3. Coupled Modification–Assembly

The construction of certain cellular structures may require a coupling of modification and self-assembly. The best evidence for this comes from studies on the assembly of certain viruses, such as bacteriophage T4 (e.g., Laemmli, 1970; King and Laemmli, 1971).

III. ROLE OF MEMBRANES: PROTEIN–MEMBRANE INTERACTIONS

With the exception of certain macromolecular complexes, such as ribosomes, all organelles are bounded by membranes or composed of membranes, and, in the absence of special biological mechanisms, proteins are unable to cross membranes. The nucleus is a special case since its envelope contains pores through which certain proteins are able to pass freely. Thus, membranes assume a central role in enzyme localization. For the purposes of this review we are concerned primarily with how proteins bind to, recognize, and cross particular membranes.

A. Structural Aspects

Historically, membrane proteins have been divided into two types—integral proteins, which require the use of detergents to be removed from membrane, and peripheral proteins, which are released by relatively mild treatments such as high ionic strength. It now appears that there is a topological basis for this classification, as integral membrane proteins are generally embedded in the nonpolar region of the lipid bilayer, while peripheral proteins are attached only at the surface of the bilayer or to other membrane proteins (Fig. 2) (reviewed by Singer, 1974).

Integral membrane proteins are directly bound to lipid by means of hydrophobic regions of the polypeptide chain, with more hydrophilic regions protruding from the membrane. They may span the membrane (in some cases probably more than once) or they may be anchored to the membrane by means of a small hydrophobic domain, as in the case of the

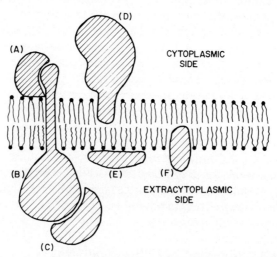

Fig. 2. Topology of membrane proteins. Integral membrane proteins (B, D, F) interact with interior of lipid bilayer, while peripheral membrane proteins (A, C, E) bind to integral membrane proteins or the lipid surface. Integral membrane proteins which have a considerable mass exposed on the extracytoplasmic side (B, F) are termed "ectoproteins," and those which have the bulk of their mass on the cytoplasmic side (D) are "endoproteins."

amphipathic protein cytochrome b_5 (see below). The topological arrangement of a protein in the lipid bilayer is well understood in only a few cases; probably the best understood is the major glycoprotein of erythrocyte membranes (Section IV,A,5).

Peripheral membrane proteins appear to be bound to membrane primarily by means of specific, noncovalent complexing with integral membrane proteins. These interactions are often relatively weak, and thus it has often been difficult to rule out the possibility that binding results from nonspecific adsorption. However, this mechanism of membrane binding has now been clearly demonstrated for a number of proteins, including β-glucuronidase (Section VI,A), components of the oxidative phosphorylation system in mitochondria (Section IV,E), and proteins of the red blood cell membrane (Section IV,A,5). Structurally, noncovalent complexing with integral membrane proteins provides a means of binding very polar molecules to membrane and provides a level of membrane specificity.

Receptors are a special class of membrane proteins which "anchor" other proteins to membrane. Generally, receptors are considered to be located on the outside surface of the plasma membrane and to function in the binding and uptake of peptide hormones and specific extracellular proteins. It is becoming apparent, however, that receptors are present in other cellular membranes as well and that they function in the transfer of proteins between intracellular compartments. For example, there is good evidence that certain integral membrane proteins of the RER function as ribosome-binding sites (Kreibich et al., 1978a,b), and the uptake of certain cytoplasmically derived polypeptides by mitochondria and chloroplasts likely involves membrane receptors (Sections IV,E and IV,F).

Two key concepts in understanding the role of membranes in protein localization are membrane fluidity and membrane asymmetry. According to the fluid mosaic model of membrane structure (Singer and Nicholson, 1972; Singer, 1974), now supported by a large body of evidence, membranes are essentially a solution of proteins embedded in a fluid lipid matrix. Thus, the protein and lipid constituents are able to move (or flow) laterally in the plane of the membrane, although under certain conditions the mobility may be restricted. Membrane fluidity is of central importance in considering processes such as membrane biogenesis, the transport of proteins across membranes, the interaction of proteins within the membranes, and membrane fission and fusion (e.g., Singer, 1975). Although membrane proteins are capable of lateral mobility, they are generally unable to reverse their orientations within the plane of the membrane. Such an inversion would require that the hydrophilic regions of the protein, exposed at the membrane surface, move through the nonpolar interior of the lipid bilayer, and this is thermodynamically implausible. A number of experimental approaches agree in indicating that biological membranes have an asymmetric distribution of their protein components, and, as discussed below, this appears to result from asymmetric insertion of proteins into the lipid bilayer. Thus, membrane asymmetry is of interest in understanding the function and topology of membranes and provides clues as to how proteins are inserted into membranes (Bretscher, 1973; Singer, 1974; Rothman and Lenard, 1977).

B. Transport of Proteins across Membranes

There is now evidence for two distinct biological mechanisms functioning in the transport of proteins across membranes. The first and best understood of these involves "vectorial translation," in which the synthesis of a protein and its transport are coupled. We will not discuss this mechanism in detail here, since it has been extensively reviewed (for example, Sabatini and Kreibich, 1976; Shore and Tata, 1977c) and is covered elsewhere in this volume (chapters 1, 2, 10). Briefly it involves the binding of a hydrophobic amino-terminal sequence of the nascent chain (the "signal sequence") to receptors on the membrane, and the vectorial discharge of the polypeptide through a channel in the membrane as synthesis continues.

The second, and still poorly understood, mechanism for transport of proteins across membranes involves binding of the completed protein to membrane receptors, followed by direct transport across the membrane. The first evidence for such a process came from studies of the transport of certain bacterial toxins into cells (Section III,D). Evidence for transfer of completed protein across intracellular membranes comes from studies of the transport of cytoplasmically synthesized polypeptides into mitochondria and chloroplasts (Sections IV,E and F). As yet, we know little of the molecular details involved, but the proteins presumably first bind to specific membrane receptors and then move across the membrane by means of polar channels. It is notable that in each case the transport is accompanied by covalent modification. In the case of the mitochondrial and chloroplast proteins this appears to involve proteolytic removal of a terminal peptide containing 20–60 amino acids (Dobberstein et al., 1977; Cashmore et al., 1978; Highfield and Ellis, 1978; Maccecchini et al., 1979). However, a number of quite different mechanisms may be responsible for transmembrane movement of proteins in the absence of protein synthesis.

Finally, it should be noted that the problem of crossing two membranes (or any even number of membranes), rather than a single membrane, can be solved by membrane fusion, a well-recognized biological phenomenon.

Thus, for example, proteins present in intracellular membrane vesicles are secreted into the extracellular space by fusion of the vesicle with the plasma membrane (Fig. 3). It seems likely that protein or lipid constituents of different membranes function in directing membrane fusion (Section IV,A).

C. Insertion of Proteins into Membranes

In the cell, membranes are probably not assembled *de novo* from pools of lipid and protein; rather, they appear to be made by insertion of lipid and protein into preexisting membrane, at rates sufficient to account for growth and degradation (Dallner *et al.*, 1966a,b). Peripheral membrane proteins can attach to membrane spontaneously (and specifically), by

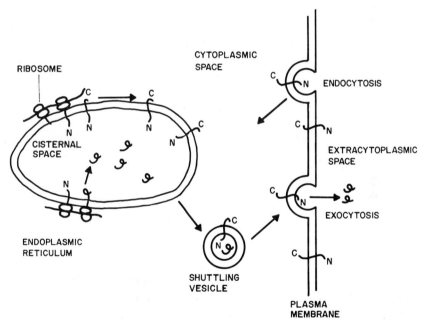

Fig. 3. Transport of membrane-bound and soluble polypeptides by fission–fusion. Different compartments of the cisternal continuum can communicate by direct contacts or by shuttling visicles, involving fission and fusion of membranes. Thus, topologically, the cisternal space, the interior of a shuttling vesicle, and the extracellular space can be considered equivalent. Similarly, the asymmetry of the membrane is conserved; for example, a protein inserted with its carboxyl terminus (C) on the cytoplasmic side of the membrane and its amino terminus (N) on the extracytoplasmic side will retain this orientation during transport and after integration into the plasma membrane.

means of the same interactions that bind subunits of an enzyme together. However, the problem of inserting a protein into the phospholipid bilayer, as in the case of integral membrane proteins, is rather similar to the problem of transporting a protein across a membrane. That is, the displacement of the polar residues of a protein across the hydrophobic interior of the lipid bilayer has a high-energy requirement. Although certain amphipathic proteins, which bind to lipid by means of a hydrophobic "tail," may become inserted spontaneously, it seems that special biological mechanisms are required for inserting transmembrane proteins containing hydrophilic groups exposed on the extracytoplasmic side of the membrane.

Evidence is now accumulating that many such transmembrane proteins are inserted into membrane by means of "vectorial translation" (e.g., Katz *et al.*, 1977a). Presumably, the synthesis of these proteins resembles that for secretory proteins except that they are not discharged completely on the other side of the membrane but remain embedded in the lipid bilayer. According to this hypothesis, these proteins should be synthesized on ribosomes bound to the RER in eukaryotes (and plasma membrane in prokaryotes), and they should have an asymmetric orientation, with the carboxyl-terminal portion of the polypeptide facing toward the cytoplasmic face of the membrane. This is precisely what has been observed for a number of membrane proteins in a variety of cell types (Bretscher, 1973; Steck, 1974; Sabatini and Kreibich, 1976; Rothman and Lenard, 1977). Furthermore, *in vitro* reconstitution experiments support this scheme, as the insertion of certain proteins into membrane appears to be obligatorily coupled to protein translation (Section IV,A,6).

Returning to the question of whether certain proteins become spontaneously inserted into membrane, it should be noted that isolated cytochrome b_5 binds directly to ER membranes, suggesting that insertion into membrane of amphipathic proteins may be feasible in the absence of simultaneous protein synthesis (Strittmatter *et al.*, 1972; Spatz and Strittmatter, 1973) (Section IV,A,4). Wickner *et al.* (1978) have shown that phage M13 coat protein is synthesized *in vitro* as a water-soluble complex which can insert itself spontaneously and in an integral fashion into either cell membranes or synthetic lipid vesicles, and on this basis they have proposed that proteins can become asymmetrically inserted into membranes by a process of refolding upon encountering a lipid bilayer. A study of the interaction of proteins with detergents supports such a "refolding" mechanism (Clarke, 1977). Finally, covalent modification could participate in the posttranslational insertion of proteins into the lipid bilayer (Section II,D,2).

D. A Model System of Protein Transport across Membranes: Cellular Uptake of Bacterial Toxins

Studies of the mechanism of action of certain toxins and glycoprotein hormones indicate the existence of "localization peptides" which serve to bind associated proteins to particular receptors on plasma membranes or to transport these proteins across the plasma membrane into subcellular compartments.

One example is diptheria toxin, a protein of molecular weight 62,000 specified by plasmid genes of *Corynebacterium diptheriae* (reviewed in Pappenheimer, 1977, 1978). It can be split by mild proteolysis into two fragments, A (molecular weight 22,000) and B (molecular weight 40,000), which remain linked by disulfide bridges. Topological studies of the intact toxin with specific antibodies suggest that most of A is buried within B. When intact toxin is added to susceptible cells there is a 30–60 min lag during which binding and internalization occur before protein synthesis is completely inhibited by ADP-ribosylation of elongation factor 2 (EF-2) of the translation apparatus. The A fragment alone is capable of inactivation of EF-2 in cell-free extracts but has no effect on whole cells. Comparisons of protein synthesis inhibition in cell-free extracts versus intact cells and using intact toxin, isolated A or B fragments, or toxins with mutant A or B subunits have led to a general scheme in which (a) the A subunit is responsible for inhibition of protein synthesis and (b) the intact toxin is necessary to transport A to the inside of the cell. Further, the B subunit is thought to be necessary for binding to the plasma membrane, since isolated B subunits or mutant toxins with normal B and abnormal A subunits compete with the binding of intact toxin to plasma membrane.

The mechanism of binding of B to the plasma membrane and of the transfer of catalytically active A to the cell interior is incompletely defined. About 4000 molecules are specifically bound per cell per hour with a K_d of approximately $10^{-8} M$. However, a single toxin molecule may suffice to kill the cell. Fragment B binds large quantities of the detergent Triton X-100, and the experiments of Boquet (1977) suggest that at least a portion of binding of the intact toxin to cells involves the hydrophobic region of subunit B. On the other hand, Draper *et al.* (1978) have argued that the plasma membrane receptor may be an oligosaccharide, since they find toxin binding is inhibited by certain complex oligosaccharides or by pretreatment of cells with lectins, such as concanavalin A or wheat germ agglutinin. The transfer of A across the plasma membrane apparently accounts for much of the 30–60 min lag time between the binding of receptor and the inhibition of protein synthesis. The molecular mechanism by which A transverses the plasma membrane is uncertain, but it has been proposed that the hydrophobic B region becomes inserted into the lipid bilayer, thereby providing a polar channel for the transfer of the hydrophilic A region to the cytoplasmic face of the membrane. Here, reduction of the disulfide bridge and proteolytic "nicking" presumably occur, causing a release of the A fragment into the cytoplasm (Boquet, 1977).

Several additional toxins and several glycoprotein hormones, including cholera toxin, exotoxin A, the plant toxins abrin and ricin, and the toxin of *Bacillus thuringiensis* (Van Heyningen, 1977), act by the same general strategy. That is, one part is catalytically inactive but is required for transport of the active part to the inside of the cell. In the case of cholera toxin interaction with cell membranes also occurs via a B subunit, and GM 1 ganglioside appears to be the most likely receptor (Van Heyningen, 1977). The glycoprotein hormones, thyrotropin, human chorionic gonadotropin, leutinizing hormone, and follicle-stimulating hormone contain α and β subunits. The β subunit differs among the hormones and is presumably responsible for tissue specificity, while the α subunit is similar in the different hormones and acts to stimulate adenylate cyclase. Specific ganglioside–thyrotropin interactions have been demonstrated (Mullin *et al.*, 1976a). Although these ganglioside interactions differ in specificity from that reported for cholera toxin, the binding of [^{125}I]thyrotropin to thyroid plasma membranes is inhibited 40% by unlabeled cholera toxin (Mullin *et al.*, 1976b). A final suggestion that these toxins and glycoprotein hormones may utilize related mechanisms to transverse plasma membranes is the finding that the B chain of cholera toxin and the β subunits of thyrotropin, leutinizing hormone, human chorionic gonadotropin, and follicle-stimulating hormone have regions of sequence homology (Ledley *et al.*, 1976).

IV. ORGANELLE BIOGENESIS AND LOCALIZATION PATHWAYS

In this section the pathways for incorporation of proteins into particular compartments and complexes are reviewed. Because of their topological relatedness, organelles comprising the internal system of membranes (with the exception of mitochondria and chloroplasts) have been considered together under the heading "cisternal continuum." A detailed description of the structure and function of cellular organelles and their relationships with one another is beyond the scope of this chapter, and we consider here only certain aspects relating to the localization of their protein constituents.

A. Cisternal Continuum

1. General Aspects

The internal system of cellular membranes is now recognized to be interdependent in terms of biogenesis and function. The organelles comprising this system, collectively termed the cisternal continuum, include the rough and smooth ER, nuclear envelope, Golgi apparatus, plasma membrane, lysosomes, peroxisomes, and various cytoplasmic vesicles (Fig. 1). While these organelles differ in composition and function, they form either direct contacts or intermittent contacts by means of shuttling vesicles. They are thus said to "communicate" with each other. The semiautonomous organelles, mitochondria and choloroplasts, are not generally considered to be part of this system, although there is evidence that their outer membranes form contacts with the ER (see below).

Our understanding of the nature of the contacts between the internal system of membranes is still very incomplete and is derived largely from cytochemical and electron microscopic studies. Primary lysosomes and peroxisomes may be derived by budding from specialized regions of the ER or Golgi apparatus (deDuve, 1969; Poole, 1969; Novikoff, 1973, 1976). The Golgi complex communicates with the ER either directly or by means of shuttling vesicles (Morré et al., 1974; Palade, 1975). Secondary lysosomes are derived by fusion of primary lysosomes and various endocytotic vesicles. The plasma membrane is probably assembled from vesicle precursors originally derived from the ER (e.g., Doyle et al., 1978).

In terms of protein constituents and functions there are clear differences between these organelles. The ER is a major site of protein and lipid synthesis and its membranes contain a number of proteins not present in significant levels in other organelles (Palade and Siekevitz, 1956; DePierre and Dallner, 1975). Lysosomes contain a variety of hydrolases characterized by acid pH optima (deDuve and Wattiaux, 1966), and peroxisomes contain a characteristic set of enzymes (e.g., uricase and catalase) with diverse functions (deDuve and Baudhuin, 1966; Poole, 1969; Novikoff and Holtzman, 1976). The Golgi complex appears to play a central role in organelle biogenesis. Along with the ER, it functions in the biosynthesis of glycoproteins and is involved in the translocation (and possibly sorting) of membrane proteins, secretory proteins, and lysosomal enzymes (Cook, 1973; Morré, 1977).

The mechanisms that operate in fusion of membranes, and in "budding" or endocytosis, are unclear (Jacques, 1969; Lawson et al., 1977). However, they are specific for particular membranes or membrane regions. Electron microscopic studies of mucus secretion in Tetrahymena, involving the fusion of secretory vesicles with the plasma membrane, show that fusion occurs only in specific regions of the plasma membrane characterized by a ring of intramembranous particles (Satir *et al.*, 1973; Satir, 1974). Thus, fusion may be mediated by specific protein–protein or protein–lipid interactions, although it could occur spontaneously as well. There is evidence that endocytosis also occurs at distinct sites on the plasma membrane; for example, the receptors for low density lipoprotein are concentrated in regions termed "coated pits" (section VI,B).

Since the individual organelles of the cisternal continuum "communicate" by direct contacts or by shuttling vesicles (involving fission and fusion of membranes), their interior spaces are in a sense topologically equivalent. Thus, the cisternal space of the ER, the interior of a lysosome or a shuttling vesicle, and the extracellular space can be considered spatially continuous (Fig. 3) (deDuve, 1969). Since the protein components of biological membranes are unable to invert their orientations with respect to the plane of the membrane (although they exhibit lateral mobility), a conservation of membrane asymmetry exists among the membranes of the cisternal continuum. For example, a protein which is inserted into membrane with its carboxyl terminus facing the cell sap will maintain this orientation as it "flows" laterally along the membrane or is shuttled by means of vesicles to separate membranous compartments (Fig. 3) (Rothman and Lenard, 1977). Once a protein is inserted into membrane or transferred to the interior of a membranous compartment, the problem of transport can be solved by direct or intermittent "communication" between the different compartments. In order to account for the very different compositions of the various compartments, specific sorting reactions must direct the transport of the proteins. Such sorting could occur during insertion into the membrane system or during transport between different membrane compartments.

Since protein synthesis occurs in the cell sap, and since membranes are normally impermeable to proteins, special mechanisms are required in the biogenesis of the compartments of the cisternal continuum. Clues concerning the nature of these mechanisms were first provided by studies of the cellular secretion of proteins (see Chapter 10 of this volume). Briefly, secretory proteins are synthesized on polysomes bound to the cytoplasmic face of the RER membrane and "vectorially" transported across the membrane to the luminal space during synthesis (Redman *et al.*, 1966; Palade, 1975). These proteins are synthesized as precursors containing an amino-terminal "signal sequence" which functions in binding the growing nascent chain to receptors in the membrane and which is generally removed before the completion of translation (Blobel and Dobberstein, 1975a,b). The pathway for transport of secretory proteins involves migration from the ER to the Golgi apparatus, where the proteins are packaged into vesicles which in turn "condense" by fusion to form larger vacuoles. The extrusion of the proteins from the cell involves fusion of the vacuoles with the plasma membrane, releasing the vacuolar contents (Palade, 1975). The mechanisms elucidated in studies of secretory proteins have been useful in considering models for localization of other types of proteins of the cisternal continuum.

2. Membrane Proteins

In examining the localization of membrane proteins, it,has been useful to group them according to their topological arrangement in the membrane. Two types of integral membrane proteins have been distinguished: ectoproteins, which have considerable mass exposed on the extracytoplasmic side of the membrane, and endoproteins, which have the bulk of their mass on the cytoplasmic side of the membrane (Fig. 2). Likewise, peripheral membrane proteins are located either on one face of the membrane or the other (Sabatini and Kreibich, 1976; Rothman and Lenard, 1977).

Since protein synthesis occurs in the cytoplasm, ectoproteins and peripheral membrane proteins located on the extracytoplasmic face of the membrane must at least partially cross the membrane. While little information is available concerning the localization of extracytoplasmic peripheral membrane proteins, accumulating evidence suggests that many ectoproteins are synthesized primarily on polysomes bound to ER and that they are inserted into membrane by a "vectorial translation" process similar to that observed for secretory proteins. First, *in vivo* labeling studies of total integral membrane proteins and of a few specific ectoproteins (such as the membrane glycoproteins of Sindbis virus and vesicular stomatitis virus) indicate that synthesis occurs largely on RER (Dallner *et al.*, 1966a,b; Omura and Kuriyama, 1971; Palade, 1975; Wirth *et al.*, 1977). Second, transmembrane ectoproteins are generally observed to be oriented such that the amino terminus projects from the extracytoplasmic face of the membrane and the carboxyl terminus from the cytoplasmic face, consistent with insertion by "vectorial translation" (Bretscher, 1973; Rothman and Lenard, 1977). Finally, with the development of *in vitro* systems capable of "vectorial translation," it has been possible to demonstrate that the insertion of the glycoprotein of vesicular stomatitis virus into membrane is obligately coupled to translation (Section IV,A,6). As in the case of secretory proteins, an amino-terminal "signal sequence" probably functions in directing synthesis of ectoproteins on membrane-bound polysomes. Presumably, ectoproteins contain structural features which result in only incomplete discharge across the membrane. There is little information on the mechanisms of transport of ectoproteins from RER to other organelles; however, since their asymmetric orientations are retained, transport is thought to involve lateral diffusion in the plane of the membrane and shuttling vesicles. Since there are distinct differences in the constituents of membranes of different organelles (DePierre and Ernster, 1977), sorting mechanisms would probably be required. Ectoproteins of the red blood cell membrane interact with cytoskeletal structures located at the cytoplasmic face of the membrane, and it is possible that the cytoskeleton functions as a sorting device (Rothman and Lenard, 1977).

The mechanisms involved in the insertion of endoproteins into membrane are less clear. Perhaps they become inserted after the completion of synthesis; for example, cytochrome b_5, the endoprotein about which the most information is available, can insert spontaneously into phospholipid vesicles (Section IV,A,4). If endoproteins are synthesized on free polysomes and subsequently inserted, then mechanisms would be required to ensure specific delivery to the target membrane. This could be achieved by membrane receptors which facilitate insertion. On the other hand, some evidence suggests that certain endoproteins, including cytochrome b_5 and cytochrome b_5 reductase, are synthesized on a special class of ribosomes "loosely bound" to RER membranes (Rothman and Lenard, 1977; Harano and Omura, 1978). These ribosomes are released by 0.6 M KCl and are probably not anchored to membrane by the nascent chain, although a "signal sequence" could be involved in binding polysomes to receptors in the RER membrane. Also, since cytochrome b_5 is anchored to membrane by a carboxyl-terminal sequence, insertion presumably does not occur by "vectorial translation."

Peripheral membrane proteins located on the cytoplasmic face of the membrane do not require special transport mechanisms and could become localized by specific complexing with membrane constituents exposed on the cytoplasmic face of individual organelles. For example, proteins located on the cytoplasmic face of the erythrocyte plasma membrane are synthesized on free polysomes (Bretscher, 1973; Lodish, 1973; Lodish and Small, 1975).

3. Lysosomal and Peroxisomal Enzymes

Cytochemical observations have indicated that lysosomes and peroxisomes are derived by "budding" from specialized regions of ER or Golgi apparatus (deDuve, 1969; Novikoff, 1976). Thus, components of the membranes of these organelles could be synthesized and inserted into membrane elsewhere (presumably the RER) and transported by direct and indirect membrane contacts. Similarly, lysosomal and peroxisomal enzymes located within the organelles could be "vectorially discharged" into the ER cisternal space (as in the case of secretory proteins) and subsequently transferred selectively into budding lysosomes or peroxisomes. Since many lysosomal and certain peroxisomal enzymes are glycosylated, an attractive feature of such a hypothesis is that it accounts for the glycosylation of the enzymes, which is thought to occur primarily on the extracytoplasmic side of the ER and Golgi apparatus (Waechter and Lennarz, 1976; Struck and Lennarz, 1977).

However, recently it has been shown that two peroxisomal enzymes, catalase and uricase, are localized by an alternate pathway. In some earlier studies, utilizing *in vivo* incorporation of a radioactive precursor into rat liver catalase, it was concluded that catalase synthesis occurs primarily on the ER (Higashi and Peters, 1963a,b). But more recent studies, indicating that a considerable fraction of the newly synthesized catalase is present in the cell sap, dispute this conclusion (Lazarow and deDuve, 1973a,b). The latter study suggests that catalase is incorporated into peroxisomes between 8 min and 1 hr after synthesis. Recently, translation *in vitro* by free and membrane-bound polysomes from rat liver showed that both catalase and uricase were synthesized almost entirely by free ribosomes (Goldman and Blobel, 1978). This is strong evidence that at least for these two enzymes, transport to peroxisomes occurs by a posttranslational process rather than "vectorial discharge" into the ER luminal space. It is not yet known how the enzymes are transferred from the cytoplasmic to the extracytoplasmic side of the membrane, although presumably specific membrane receptors would be required to achieve insertion into a target organelle. In other studies where evidence has been obtained for posttranslation transport of proteins across membranes, the proteins were modified during transport (Sections IV,E and F); however, there is no evidence for such precursors to catalase and uricase (Robbi and Lazarow, 1978; Goldman and Blobel, 1978).

As yet, little is known about the mechanisms for localization of lysosomal enzymes, but the available information suggests that the ER and Golgi complex may have a role in their transport (Novikoff, 1973). When rat liver β-glucuronidase was pulse labeled *in vivo* with radioactive precursors, the specific radioactivity of the enzyme rose first in microsomes and thereafter in lysosomes, suggesting that the enzyme is transported from ER to lysosomes. During transport, the enzyme is covalently modified (the chemical nature of the modifications is uncertain), and Golgi fractions contained a form of the enzyme resembling that in lysosomes, suggesting that transport occurs via the Golgi complex (Tsuji *et al.,* 1977; Tsuji and Kato, 1977). Time-course studies of the induction of β-glucuronidase in mouse kidney after androgen treatment also implicated the Golgi apparatus in the translocation of the enzyme (e.g., Lin, 1975).

However, such *in vivo* studies of β-glucuronidase in mouse and rat are complicated by cell heterogeneity, the dual intracellular localization of the enzyme in lysosomes and microsomes, and the problem of amino acid reincorporation (e.g., Lusis and Paigen, 1977a; Section VI,A).

It is well known that lysosomal enzymes can be transferred between cells, and Neufeld and colleagues have proposed that in certain cell types enzymes are inserted into lysosomes by a secretion–recapture mechanism (Neufeld *et al.*, 1977). According to this proposal, lysosomal enzymes are secreted from the cell (perhaps by a pathway resembling that for secretory proteins) and subsequently recaptured by receptor-mediated endocytosis. The recent findings of cell surface receptors functioning in the specific uptake of lysosomal enzymes (Section II,D,1) support such a scheme. Furthermore, the hypothesis provides an explanation for certain human diseases in which several lysosomal enzymes are deficient in certain cell types (Neufeld *et al.*, 1975).

4. A Model System of ER Assembly: Cytochrome b_5

Cytochrome b_5 is localized in liver ER by an amphipathic mechanism (Strittmatter *et al.*, 1972; Robinson and Tanford, 1975; Ozols and Gerard, 1977). That is, this enzyme consists of a hydrophobic carboxyl-terminal domain which anchors the molecule in the lipid bilayer and a hydrophilic amino-terminal catalytic domain which remains in the aqueous phase at the cytoplasmic surface of the membrane. Such a topological arrangement is ideally suited to the function of an enzyme that is a vital link in a membrane-bound electron transport system that includes, in order, NADH, the flavoprotein cytochrome b_5 reductase, cytochrome b_5, and the nonheme iron protein fatty acid desaturase. It appears, in fact, that all three protein components of this electron transport chain may be bound to ER by amphipathic mechanisms (Rogers and Strittmatter, 1974). The susceptibility of these proteins to proteases and to reduction by external NADH indicate that they are bound to the cytoplasmic face of the ER membranes.

When extracted from microsomes by detergents, cytochrome b_5 is a monomer of molecular weight about 16,700, which rapidly aggregates when detergent is removed. The purified protein binds to and can be disaggregated by a wide variety of detergents and membrane lipids. This binding is not a high affinity type but rather involves physical incorporation of the hydrophobic domain together with any amphiphilic ligands in a common micelle. In reconstitution studies a 10- to 20-fold excess of the enzyme can be bound to microsomal membranes, where it displays normal electron transport kinetics. In such binding studies cytochrome b_5 prefers ER membranes to lysosomal, Golgi, and inner mitochondrial membrane (e.g., Remacle, 1978). The hydrophobic and hydrophilic domains of cytochrome b_5 can be separated by proteolytic cleavage, a useful step in examining domain structure and function. The hydrophilic domain obtained by protease treatment has a molecular weight of about 11,000 and has negligible affinity for detergents and membranes. The hydrophobic carboxyl-terminal domain, on the other hand, exhibits strong affinity for detergents and lipids, as does the intact molecule.

In vivo, cytochrome b_5 does not show an absolute specificity in subcellular localization, and significant amounts of it are present in outer mitochondrial and peroxisomal membranes in addition to ER (Fowler *et al.,* 1976). This is in agreement with the idea that it is capable of binding nonspecifically to membrane phospholipids. Despite the fact that cytochrome b_5 is present in several subcellular compartments, it is not uniformly distributed among intracellular membranes, and, therefore, an as yet unresolved mechanism for specific subcellular localization must exist. It is possible that its synthesis occurs on free polysomes, and the preexisting distribution of proteins and lipids determines the binding of newly synthesized cytochrome b_5 to particular membranes (Bretscher, 1973). Similarly, its insertion into specific membranes could be facilitated *in vivo* by membrane-bound modifying enzymes (Rothman and Lenard, 1977). Alternatively, cytochrome b_5 localization might be coupled to translation in the sense that its synthesis could occur on ribosomes bound to ER membranes (Omura and Kuriyama, 1971), and its presence in adjacent membranes could result from diffusion. It has been reported, in fact, that cytochrome b_5 is synthesized on a class of ribosomes bound to ER by mechanisms not involving nascent chains (Harano and Omura, 1978).

5. A Model System of Plasma Membrane Topology and Assembly: The Red Blood Cell Membrane

The organization of proteins in membranes has been particularly well characterized in red blood cell membrane. Experimentally, this system offers a variety of useful features, and a combination of approaches has yielded a coherent picture of the toplogy of the major protein constituents of the membrane. Several major integral membrane proteins, including some glycoproteins, have been characterized. In each case the N terminus is oriented toward the cell surface, while the C terminus is embedded in the membrane or, in the case of transmembrane proteins, extends to the inside of the lipid bilayer. Among the glycoproteins, the carbohydrate moieties are externally disposed and linked to polypeptide backbones through either O-glycosidic (to serine or threonine) or N-glycosidic (to asparagine) bonds. About 40% of the total membrane protein is peripheral, and nearly all of that is located on the cytoplasmic side of the lipid bilayer. Many of these peripheral membrane proteins appear to form complexes with the C-terminal portions of the integral membrane proteins, and some, such as spectrin, may also interact with the lipid bilayer (Steck, 1974; Marchesi et al., 1976).

Detailed structural investigations have been carried out on glycophorin A, the major sialoglycoprotein of red cell membrane. Glycophorin A consists of a single polypeptide chain of 131 amino acids, that spans the red cell membrane. Carbohydrate, accounting for about 60% of the mass of the protein, is attached at 16 separate sites along the polypeptide chain, all within the externally disposed N-terminal half of the chain. An extremely hydrophobic sequence of about 20 amino acids, near the middle of the chain, extends through the lipid bilayer, while the C-terminal third of the molecule is probably exposed on the cytoplasmic side of the membrane. Several lines of evidence suggest that glycophorin A may interact with other integral membrane proteins (forming so-called "intramembranous particles"), as well as with peripheral membrane proteins. In this respect, the N- and C-terminal portions of the molecule are quite hydrophilic and capable of interaction with hydrophilic peripheral proteins (Marchesi et al., 1972, 1976; Shotten et al., 1978).

The fact that all of the major ectoproteins present in the red cell membrane are asymmetrically oriented such that the N-terminal regions face the extracytoplasmic space suggests that they may be inserted by "vectorial translation" on ribosomes bound to ER membrane (Rothman and Lenard, 1977). Subsequent transfer to the plasma membrane would presumably occur by shuttling vesicles. On the other hand, the peripheral membrane proteins present on the cytoplasmic face are synthesized on free polysomes (Lodish, 1973; Lodish and Small, 1975).

6. A Model System of Plasma Membrane Assembly: Vesicular Stomatitis Virus

Vesicular stomatitis virus consists of a ribonucleoprotein core containing three proteins, L, N, and NS. Surrounding the core is a lipoprotein envelope containing proteins G and M (Fig. 4). Since the envelope can be viewed as a lipid bilayer derived from the plasma membrane, the biogenesis and processing of the envelope proteins is a valuable model system for plasma membrane assembly (Atkinson, 1978; Katz et al., 1977a,b). Also the limited coding capacity of the virus mandates that it uses cellular factors for membrane assembly.

M protein is nonglycosylated. Several independent experimental methods have shown that it is a matrix or membrane protein embedded on the cytosolic side of the plasma membrane (Fig. 4). Newly synthesized M protein is released into the cytosolic compartment and rapidly migrates to

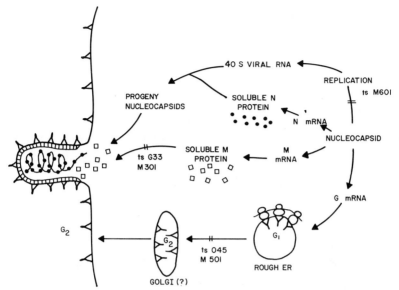

Fig. 4. Model for assembly of vesicular stomatitis virus. (Redrawn from Knipe *et al.*, 1977c, reproduced with permission.)

the plasma membrane of the host cell with a transit time of about 2 min. In HeLa cells it appears to be added by unknown mechanisms directly to the cytosolic surface of the plasma membrane without prior attachment to other intracellular membranes, and it randomly mixes with preexisting M protein (Atkinson, 1978). The pulse labeling and subcellular fractionation studies of Knipe *et al.* (1977b) suggest that in CHO cells very little of the M protein is in the plasma membrane per se but is very quickly incorporated into budding virus. It is possible that the M protein may be incorporated into budding intermediates which are nucleating sites for binding nascent nucleocapsids.

Glycoprotein G is accessible to external labeling in infected cells and forms the spikes of the envelope of the completed virion. Unlike the M protein, it is synthesized exclusively by membrane polysomes and is never found in the soluble fraction. Its mechanism of synthesis and processing has been studied in intact infected cells (Atkinson, 1978; Knipe *et al.*, 1977b) and in *in vitro* protein synthesizing systems reconstituted with viral specific mRNA and ER membrane fractions (Toneguzzo and Ghosh, 1978; Rothman and Lodish, 1977).

Synthesis and insertion of G into membranes *in vitro* can be achieved during cell-free translation with viral mRNA and stripped pancreatic microsomes. The inserted protein spans the lipid bilayer in the same orientation as observed *in vivo* and it is glycosylated. Since the ability to cross the membrane is rapidly lost if protein synthesis is allowed to continue beyond about 80 residues before adding microsomal membranes, the insertion must be tightly coupled to translation. The mechanism of insertion may be via an N-terminal signal peptide such as detected in secretory proteins by Blobel and Dobberstein (1975a). Synthesis of the G protein differs from that of secretory proteins, however, since it is not completely extruded into the lumen of the endoplasmic reticulum. Rather, a carboxyl-terminal sequence of molecular weight about 3000 remains accessible to exogenous proteases. Two glycoprotein intermediates, probably containing one or two of the two oligosaccharide core chains of G, are separable on sodium dodecyl sulfate polyacrylamide gels. The chains appear to be added at rather precise points (40 and 70%, respectively) in the completion of the nascent polypeptide (Rothman and Lodish, 1977).

In infected cells, glycosylation of G likewise occurs very rapidly, probably on polysome-bound nascent chains. All sugars are externally disposed on the mature virion, consistent with glycosylation occurring within the lumen of the ER. Noncore sugars, such as fucose (Atkinson, 1978) and sialic acid (Knipe *et al.,* 1977c), are added at much later times during transit to the plasma membrane. For example, in HeLa cells (Atkinson, 1978) the total time of transit of G protein to the plasma membrane is about 75 min, and fucose is added about 20 min before arrival at the plasma membrane. The kinetics of processing and assembly of the G protein are, therefore, very different from that of the envelope M protein.

The detailed molecular mechanisms of the construction of the viral envelope remain to be elucidated, but Knipe *et al.* (1977a) have utilized temperature sensitive vesicular stomatitis virus mutants (Fig. 4) in combination with subcellular fractionation techniques to examine the effects of defects in specific viral proteins on viral maturation. They have noted that some aspects of morphogenesis are relatively independent of other viral proteins. For example, mutations in the envelope M protein (ts M301) or capsid N protein (ts M601), which cause increased rates of degradation of these proteins, do not greatly affect the migration of the G protein to the cell surface. Other mutations indicate an interdependence of viral proteins in morphogenesis. For example, mutations which prevent the addition of sialic acid to G protein (ts M501 or ts 045) cause G to accumulate in ER, thus preventing its normal migration to the cell surface, and also cause decreased membrane binding of the M and N proteins. Moreover, mutations in the capsid N protein (ts M601) decreased binding of M protein to membranes.

B. Cell Sap

Many of the soluble proteins in the cell sap are synthesized on free polysomes (Redman, 1969; Ganoza and Williams, 1969; Hicks *et al.*, 1969; Tagaki *et al.*, 1970; Andrews and Tata, 1971; Rolleston, 1974) and their localization is determined by the fact that they simply cannot leave this compartment. However, some proteins present in the cell sap, such as rat liver serine dehydratase, are apparently produced on ribosomes bound to the ER (McLaughlin and Pitot, 1976). While it is possible that certain proteins synthesized on RER are discharged on the cytoplasmic face of the membrane (discussed in Sabatini and Kreibich, 1976), nearly all proteins synthesized on rat liver RER *in vitro* are transferred across or embedded deep into the membrane (Shore and Harris, 1977).

Certain proteins of the cell sap are known to form large complexes (e.g., fatty acid synthetase multienzyme complexes), and evidence is accumulating that many proteins generally regarded as soluble are in fact present in such complexes (e.g., Ouadi and Keleti, 1978). Furthermore, in the cell there appear to be interactions (often easily disrupted) between certain soluble proteins and various subcellular structures (Masters, 1978). Microtubules and microfilaments and their associated proteins are examples of components of the cell sap which show a strikingly regular topology and which are responsive to the physiological state of the cell (Lazarides and Weber, 1974; Heggeness *et al.*, 1977). Mollenhauer and Morré (1978) have suggested that there is a type of compartmentalization of the cell sap intermediate between that of membranes and soluble multienzyme complexes. For example, electron microscopic studies have consistently revealed "zones of exclusion" (differentiated regions of the cell sap in which ribosomes, glycogen and organelles are scarce or absent) surrounding the Golgi apparatus. How such regions are maintained is not known, but components such as microtubules or microfilaments may be involved. Thus, the proteins of the cell sap, and "soluble" proteins in general, may exhibit a greater degree of order than previously recognized.

C. Ribosomes

Bacterial ribosomes consist of two subunits, a 30 S subunit (containing over 20 different proteins and 1 RNA molecule) and 50 S subunit (containing over 30 proteins and 2 RNA molecules). Despite this complexity, a combination of physical, immunological, and genetic approaches has yielded remarkably detailed information concerning the topological and functional relationships of these components (reviewed by Wittmann, 1977). In terms of ribosome assembly, two approaches have been particularly informative: (1) *in vitro* reconstitution and (2) the utilization of mutants affecting the structure and processing of ribosomal components. Both bacterial subunits have now been reconstituted *in vitro* from their separated RNA and protein constituents, indicating that the information directing ribosome assembly is inherent in the constituents themselves. Several conditional mutants for ribosome assembly have been identified, and all appear to affect either a ribosomal protein or an enzyme which modifies ribosomal components (Jaskunas *et al.*, 1974). By examining the effects of such mutations, and by perturbing *in vitro* reconstitution (for example, by omitting various components), the approximate sequence of events occurring during assembly has been determined (reviewed by Nomura and Held, 1974). Some ribosomal proteins function only in assembly and are not required for the protein synthesizing functions of the mature ribosome (Spillmann and Nierhaus, 1978).

Eukaryotic ribosomes are somewhat larger than bacterial ribosomes, consisting of a 60 S subunit and a 40 S subunit which together contain about 70 distinct proteins and 3 molecules of ribosomal RNA (rRNA). Assembly of ribosomes in eukaryotes occurs in large part in the nucleolus, the site of synthesis of 18 S and 28 S rRNA. Ribosomal proteins are synthesized on cytoplasmic polysomes and rapidly transported through the nuclear envelope to the nucleolus, where they associate with newly transcribed "immature" rRNA. Some evidence suggests that during the initial stages of ribosome assembly, the RNA–protein interactions are readily reversible, allowing proteins to move between RNA molecules, and that the association of the individual components is cooperative. Subsequently, the rRNA undergoes modification, including cleavage and methylation, and the resulting ribonucleoprotein particles (or preribosomes) are transported to the cytoplasm. These preribosomes contain some proteins not present in mature cytoplasmic ribosomes, but their function is still obscure. In addition, some proteins probably become incorporated into ribosomes at late stages of formation, perhaps in the cytoplasmic compartment (Wool and Stoffler, 1974; Warner, 1974; Auger-Buendia and Longuet, 1978). Since a few ribosomal proteins are apparently glycosylated (Yoshida, 1978), they may be synthesized on membrane-bound polysomes.

D. Nuclei

The nucleus is bounded by a double membrane structure termed the "nuclear envelope," which appears to have direct contacts with the ER (Franke, 1974). Ribosomes are often associated with the cytoplasmic face of the envelope, but their function is uncertain. Attached at the surface of the inner envelope membrane is a protein network, termed the lamina, which may function in disassembly of the envelope during cell division (Gerace *et al.*, 1978). Scattered across the surface of the envelope are pores (at which the outer and inner membranes are fused) that presumably serve as a route of exchange between the nuclear and cytoplasmic compartments (e.g., Novikoff and Holtzmann, 1976). Many chemically diverse materials can enter the nucleus, providing they do not exceed a critical size (90–125 Å). Also, mechanical damage to the nuclear envelope has no apparent quantitative or qualitative effects on the uptake of endogenous polypeptides into nuclei (Feldherr and Pomerantz, 1978). The mechanisms responsible for the selective accumulation of proteins in the nucleus are unknown, but specific recognition features appear to be involved. In this respect, experiments involving microinjection of nuclear proteins into frog oocytes have been particularly informative (Section V,B). A more detailed discussion of transport processes between the nucleus and cytoplasm is presented in Chapter 7 of this volume.

While some evidence suggests that limited protein synthesis may occur in nuclei (discussed in Goidl, 1978; Allen, 1978), the nuclear proteins which have been examined in detail have been found to be synthesized in the cytoplasm. For example, histones are synthesized in the cytoplasm on free polysomes and, possibly, a population of polysomes loosely associated with membrane (Bloch and Brack, 1964; Zauderer *et al.*, 1973).

A mechanism for the assembly of one important nuclear component, the nucleosome, is suggested by the experiments of Laskey *et al.* (1978). They found that in a cell-free system derived from frog eggs the ordered interaction of histones with DNA to form nucleosomes apparently requires an acidic protein factor. Normally, mixtures of histones and DNA form a nonspecific precipitate at physiological pH, but, in the presence of the protein factor, typical nucleosomes containing all four histone core proteins and supercoiled DNA were formed. It is speculated that the interaction of acidic protein and histones partially neutralizes histone positive charges, preventing nonspecific histone–DNA interaction.

E. Mitochondria

This organelle is enclosed by two membranes separated from one another by a narrow space. The inner membrane is highly folded and encloses the mitochondrial matrix (Fig. 1). Mitochondria contain their own self-replicating DNA, as well as an apparatus for protein synthesis, and several mitochondrial proteins are known to be synthesized entirely in mitochondria. However, a variety of biochemical and genetic studies have established that the bulk of mitochondrial proteins are coded by nuclear

DNA (Gibor and Granick, 1964; Haldar *et al.*, 1966; Schatz and Mason, 1974).

Only fragmentary information is available concerning how proteins coded by nuclear DNA are incorporated into mitochondria. Since mitochondria are bounded by an outer and inner membrane, it would appear that the cytoplasmically derived matrix proteins would have to cross two membranes. Most of these proteins are synthesized on cytoplasmic ribosomes and little, if any, mRNA of nuclear origin is transported into mitochondria for translation (Schatz and Mason, 1974). One proposal suggests that proteins are synthesized on cytoplasmic ribosomes bound to the outer mitochondrial membrane and transported across the membrane by means of "vectorial translation" (Kellems and Butow, 1974; Kellems *et al.*, 1975). However, this possibility has not been verified for any mitochondrial proteins. Other studies suggest that the principal site of synthesis of these proteins is a differentiated portion of the ER (Kadenbach, 1966, 1970; Godinot and Lardy, 1973; Shore and Tata, 1977a,b,c; Kawajiri *et al.*, 1977a,b). Presumably, such proteins could subsequently be delivered to mitochondria in association with membrane vesicles, as suggested by cases where the outer mitochondrial membrane is seen to be continuous with membranes of the ER (Franke and Kartenbeck, 1971; Morré *et al.*, 1971). It is quite possible that not all cytoplasmically derived mitochondrial proteins utilize the same processing pathways.

The topology and assembly of the enzymes of the mitochondrial oxidative phosphorylation system, bound to the matrix side of the inner mitochondrial membrane, have been extensively examined (Schatz and Mason, 1974; Hatefi, 1976; Tzagaloff, 1976; DePierre and Ernster, 1977). The best understood of these are ATPase and cytochrome oxidase, which are present as complexes containing both mitochondrially and cytoplasmically derived subunits. Mitochondrial ATPase can be solubilized, using detergents, as a complex of ten tightly associated polypeptides. Four of these polypeptides are very hydrophobic and appear to be integral membrane proteins synthesized in mitochondria and coded for by mitochondrial DNA; others are more hydrophilic and are synthesized on cytoplasmic ribosomes. The synthesis and transport of the three largest cytoplasmically derived polypeptides of yeast ATPase have recently been examined (Maccecchini *et al.*, 1979). Both *in vivo* and *in vitro,* each is synthesized as a precursor 2000–6000 daltons larger than the mature subunit, and the precursors are taken up directly when incubated with isolated mitochondria. Associated with the uptake is a conversion of each to the mature subunit, but mature subunits are not taken up. Thus, the results suggest that the transport of these three proteins into mitochondria is not achieved by "vectorial translation," but appears to involve their interaction with specific membrane receptors. Unidirectional transport may be achieved by the irreversible conversion of the precursors within the mitochondrial matrix (Maccecchini *et al.,* 1979).

Cytochrome oxidase has several features in common with ATPase. It is a complex containing four hydrophilic polypeptides synthesized on cytoplasmic ribosomes and three hydrophobic polypeptides (which are probably integral membrane proteins) synthesized on mitochondrial ribosomes. A variety of experiments have suggested that the hydrophilic polypeptides of these complexes are anchored in the inner mitochondrial membrane by specific noncovalent associations with the hydrophobic ones. For example, in cytoplasmic petite mutants of yeast unable to produce the hydrophobic mitochondrially derived polypeptides of cytochrome oxidase, three of the four hydrophilic peptides showed only loose association with mitochondrial membrane while the fourth was absent altogether (Schatz and Mason, 1974). In both rat liver and yeast, the four cytoplasmically derived subunits are apparently synthesized as a single precursor protein of molecular weight about 50,000, which is subsequently cleaved to yield the mature subunits (Poyton and Kavanagh, 1976; Ries *et al.,* 1978). After a 1 hr pulse with radioactive leucine, the labeled rat liver precursor was associated entirely with the microsomal fraction (Ries *et al.,* 1978).

Rat liver cytochrome *c,* which is tightly complexed with the cytochrome oxidase complex, was reported to be synthesized primarily on ribosomes bound to ER membrane, followed by transfer to the inner mitochondrial membrane (Kadenbach, 1966, 1970; Gonzalez-Cadavid *et al.,* 1971). Results from other laboratories, however, suggest that rat liver microsomal cytochrome *c* is probably not a biosynthetic precursor of mitochondrial cytochrome *c* (Robbi *et al.,* 1978). In *Neurospora,* newly synthesized cytochrome *c* was detected first in a postribosomal soluble fraction and was rapidly incorporated into the mitochondrial fraction (Korb and Neupert, 1978).

One of the few soluble proteins of the mitochondrial matrix whose synthesis and processing has been examined is rat liver glutamate dehydrogenase (GDH). The synthesis of this protein is thought to occur on membrane-bound polysomes of a particular fraction of the rough endoplasmic reticulum, and the GDH polypeptide is apparently discharged on the cytoplasmic surface of the membrane. Subsequently, the enzyme becomes inserted into the mitochondrial matrix, with a translocation time of several hours. Some evidence suggests that GDH may undergo posttranslational processing closely tied with its transport into mitochondria. A comparison of GDH from mitochondria with the precursor from ER revealed differences in charge, whereas in studies of enzyme rebinding to

ER membrane, the mitochondrial form had lower affinity than the ER precursor. Another mitochondrial matrix enzyme probably synthesized on rough ER, malate dehydrogenase, was found to be concentrated in similar regions of the ER, and in rebinding studies it competed with GDH for membrane binding sites (Godinot and Lardy, 1973; Kawajiri, 1977a,b).

In general, a major function of mitochondrial protein synthesis may be in the synthesis of integral membrane proteins of the inner membrane, and some evidence suggests that insertion into membrane occurs during the translation process on ribosomes bound to the matrix side of inner mitochondrial membrane. On the other hand, the soluble proteins of the mitochondrial matrix, the peripheral proteins of the inner mitochondrial membrane, and the proteins of the outer mitochondrial membrane are apparently synthesized primarily on cytoplasmic ribosomes (Schatz and Mason, 1974).

F. Chloroplasts

Chloroplasts, like mitochondria, are bounded by two separate membranes and possess self-replicating DNA and an apparatus for protein synthesis. In addition, they contain an internal system of membranes functioning in photosynthesis. Many of the chloroplast proteins are coded by nuclear DNA and only become incorporated into chloroplasts after synthesis on cytoplasmic ribosomes, but little is known of the details of their biosynthesis and translocation (Gibor and Granick, 1964; Smillie and Scott, 1969).

In terms of topology, some chloroplast membrane proteins resemble the ATPase and cytochrome oxidase complexes of the inner mitochondrial membrane (e.g., DePierre and Ernster, 1977). One such complex, that functions in photophosphorylation, apparently consists of a cluster of several polypeptides bound to membrane by specific noncovalent complexing with a set of polypeptides which are embedded in membrane (e.g., Younis et al., 1977).

Some recent studies have suggested an interesting pathway for the synthesis of ribulose-1,5-bisphosphatase carboxylase, a major protein of chloroplast stroma (Dobberstein et al., 1977; Cashmore et al., 1978; Highfield and Ellis, 1978). The enzyme is present as a complex containing two different subunits: a smaller subunit made on cytoplasmic ribosomes and a larger one made within the chloroplast. The smaller subunit is synthesized on free cytoplasmic ribosomes as a 20,000 molecular weight precursor to the mature 14,000 molecular weight subunit, the former being inserted into chloroplasts and cleaved by a specific endopeptidase within the chloroplast to produce the latter. There is the intriguing possibility that the precursor functions in transport of the subunit across the chloroplast envelope membranes, possibly in a manner analogous to the transport of diptheria toxin and the cytoplasmically derived subunits of mitochondrial ATPase (discussed above). When incubated with isolated chloroplasts the precursor is selectively taken up, judging from its resistance to protease digestion (Highfield and Ellis, 1978). Significantly, "vectorial translation" is not involved in the transport of this protein across chloroplast membranes, and the absence of ribosomes from the outer chloroplast membrane suggests that vectorial discharge is probably not involved in the incorporation of most other cytoplasmically derived chloroplast proteins either. However, certain chloroplast proteins might first be inserted into other membrane vesicles which then fuse with chloroplasts.

V. RECONSTITUTION SYSTEMS

In this section we discuss two types of reconstitution experiments that have been particularly informative in terms of the mechanisms of protein localization, the first utilizing cell-free protein synthesis in the presence of membranes and, the second, microinjection of mRNA or protein into frog oocytes. A variety of other *in vitro* approaches currently being explored will undoubtedly also be useful in examining questions pertaining to the transport and segregation of proteins. For example, artificial phospholipid vesicles (liposomes) are proving useful for studies of membrane fusion (Papahadjopoulos *et al.*, 1974; Blumenthal *et al.*, 1977) and membrane–protein interactions (Section IV,A,4) (Eytan and Broza, 1978). Also studies involving *in vitro* glycosylation and cleavage of proteins should be helpful in answering questions relating to the role of modification in protein localization (Pless and Lennarz, 1977; Jackson and Blobel, 1977).

A. *In Vitro* Protein Synthesis, Segregation, and Insertion into Membrane

In recent years cell-free protein-synthesizing systems have been valuable in examining mechanisms for the segregation and transport of proteins, in part because they allow the primary translation product to be characterized and the role of any modifications to be assessed. For example, such systems provided evidence that secretory proteins are synthesized as precursors which undergo proteolytic cleavage during vectorial discharge into the lumen of the ER (Swan *et al.*, 1972; Mach *et al.*, 1973; Blobel and Dobberstein, 1975a). Although there may be exceptions (Palmiter *et al.*, 1978; Lingappa *et al.*, 1978) this conclusion has subsequently been confirmed for a variety of secretory proteins (reviewed in Shore and Tata, 1977c) and has been extended to prokaryotic as well as eukaryotic cells (e.g., Smith *et al.*, 1977; Randall *et al.*, 1978). Also, this type of analysis has been applied to other classes of proteins, including cytoplasmically synthesized mitochondrial and chloroplast proteins (Sections IV,E and F). Recently, cell-free systems incorporating membrane preparations have been shown to be capable of asymmetric insertion into membrane and glycosylation of a transmembrane protein, the glycoprotein of vesicular stomatitis virus (Katz *et al.*, 1977a; Toneguzzo and Ghosh, 1978). Significantly, insertion was coupled to translation, suggesting a "vectorial translation" mechanism.

It seems likely that such *in vitro* translation systems will provide further insights into mechanisms involved in the segregation of different classes of proteins. For example, several studies utilizing *in vitro* translation suggest that topologically and functionally distinct populations of polysomes may be bound to ER membranes (Shore and Tata, 1977a,b,c; Kawajiri *et al.*, 1977a,b; Harano and Omura, 1978).

B. Frog Oocytes

From the time frog oocyte microinjections began (Gurdon *et al.*, 1971; Lane *et al.*, 1971) the method has proved useful for testing ideas of how information for directing protein localizations is expressed. Frog oocytes possess several features useful for examining factors regulating the intracellular location of enzymes. First, they are very large and durable, allowing simple microinjection of solutions. Second, they have a relatively low rate of endogenous protein synthesis, yet are capable of translating injected mRNA's very efficiently and accurately (Gurdon *et al.*, 1971; Lane, 1976). Third, they contain an apparatus which packages and processes proteins (Berns *et al.*, 1972; Laskey *et al.*, 1972; Berridge and Lane, 1976; Ghysdael *et al.*, 1977; Zehavi-Willner and Lane, 1977).

The injection of isolated mRNA's coding for cytoplasmic proteins, such as globin, direct the synthesis of cytoplasmic products, while the injection of mRNA's coding for secretory proteins, such as albumin, vitellogenin, and milk proteins, result in the synthesis of proteins enclosed by membrane vesicles. This suggests that the information required for locating the sites at which the proteins will be synthesized and how they will be packaged is encoded within the mRNA, and that the origin of the mRNA is unimportant. Moreover, since the injection of albumin proteins into oocytes did not result in their insertion into vesicles, membrane transfer of secretory proteins apparently is coupled to translation (Zehavi-Willner and Lane, 1977). These results are in accord with the conclusions of Blobel and Dobberstein (1975a,b) based on *in vitro* translation of isolated mRNA in the presence of appropriate membrane preparations (see Section V,A). An interesting conclusion that emerges from these studies is that the signals for localization present in proteins or mRNA show little specificity with respect to cell type and species. For example, messenger RNA's isolated from viruses, arthropods, or mammals were translated, and the resulting proteins were often faithfully processed.

Experiments of Gurdon (1970), Bonner (1975), and DeRobertis *et al.* (1978) suggest that nuclear proteins have recognition features that allow their selective accumulation in nuclei. An important facet of these experiments is that they should enable further dissection of these putatative recognition features. These workers microinjected into *Xenopus* oocyte cytoplasm preparations of radiolabeled proteins obtained from purified oocyte nuclei. After 1 day's incubation, the majority of injected nuclear proteins were found by two-dimensional polyacrylamide gel analysis to be stable in oocytes and were recovered in higher concentration in reisolated nuclei than in the cytoplasm. A few injected nuclear proteins normally found in both nucleus and cytoplasm (e.g., actin), were also present in reisolated cytoplasm. When radiolabeled cytoplasmic proteins were injected, most of the proteins were not incorporated into the nucleus, but proteins normally found in both nucleus and cytoplasm were partitioned between the two compartments. While the transport mechanisms are unknown, autoradiography of sectioned oocytes showed that most of the injected nuclear proteins were distributed uniformly in the nucleus, indicating that proteins do not accumulate by binding to chromosomes or the nuclear envelope. Both high and low molecular weight and acidic and basic proteins migrated into the nucleus. The putative signals are not all species specific, since injected calf thymus histones also accumulate in frog nuclei.

VI. GENETIC SYSTEMS

The isolation and characterization of mutants altered in enzyme localization is proving to be a powerful tool in identifying the individual steps involved in directing an enzyme to its proper intracellular site. In addition to the systems discussed below, genetic variants have been useful in examining the assembly of mitochondria (Section IV,F), ribosomes (Section IV,C), and plasma membrane (Section IV,A,6).

A. Mouse Glucuronidase

Mouse glucuronidase has been especially useful in genetic studies of enzyme regulation and processing (reviewed by Lusis and Paigen, 1977a; Swank *et al.*, 1978). Although acid hydrolases such as glucuronidase are typically located in lysosomes, mouse glucuronidase is unusual in that a significant proportion of the enzyme is also present in microsomes of some tissues. In livers of most strains of mice, for example, nearly half of the enzyme is microsomal. The enzyme in microsomes is firmly bound to membrane, while that in lysosomes is soluble and is released upon rupture of the lysosomal membrane (Paigen, 1961). Glucuronidase is not unique in being localized at multiple intracellular sites; cytochrome b_5 is present in a variety of intracellular membranes (Section IV,A,4), actin is present in both the cell sap and the nucleus (Section V,B), and a significant amount of another mouse acid hydrolase, β-galactosidase, is microsomal in certain tissues (Lusis *et al.*, 1977).

Biochemical studies have suggested a model for the dual localization of glucuronidase (Fig. 5). The enzyme at both intracellular sites is derived from the same structural gene, *Gus,* located on chromosome 5, as shown by mutations of this gene that simultaneously alter the properties of both enzymes. Six different forms of the enzyme, resulting from processing of the glucuronidase polypeptide have been identified. Lysosomes contain a single form, L, consisting of a tetramer of identical subunits, each of molecular weight about 70,000. On the other hand, the microsomal en-

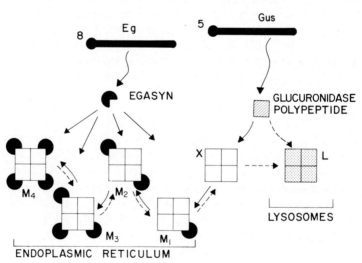

Fig. 5. Model for intracellular localization of mouse glucuronidase. (From Lusis and Paigen, 1977a, reproduced with permission.)

zyme extracted with the detergent Triton X-100 occurs as a series of higher molecular weight complexes, termed M_1, M_2, M_3, and M_4. These consist of a glucuronidase tetrameric core, known as X, complexed with 1 to 4 molecules of an accessory protein, egasyn. The X and L glucuronidase tetramers differ by covalent modification, with X being slightly larger and less negatively charged. The accessory protein egasyn is a glycoprotein of molecular weight about 64,000 (Swank and Paigen, 1973; Tomino and Paigen, 1975; Lusis et al., 1976).

The discovery of a mutation affecting glucuronidase localization was instrumental in the elucidation of these structural relationships, and in clarifying the function of egasyn. Mice of certain strains lack microsomal glucuronidase almost entirely, but retain about normal levels of the lysosomal enzyme (Ganschow and Paigen, 1967). The defect results from mutation at a single genetic locus, Eg, on chromosome 8, unlinked to the glucuronidase structural gene (Karl and Chapman, 1974). Initial characterization revealed that the mutant lacked M form complexes (Swank and Paigen, 1973), and subsequently the block in the assembly of the M forms was shown to be an inability to produce egasyn (Lusis et al., 1976).

Several observations support the conclusion that egasyn stabilizes the membrane binding of glucuronidase, functioning as a membrane anchor. Egasyn is localized entirely in the microsomal fraction of liver and, judging from conditions required to extract it, appears to be an integral membrane protein (Lusis et al., 1976). Moreover, the levels of egasyn and microsomal glucuronidase are strongly correlated among tissues and during development. Tissues such as kidney and liver are rich in both egasyn and microsomal glucuronidase, while tissues such as brain and spleen lack egasyn entirely and contain almost exclusively lysosomal glucuronidase (Owerbach and Lusis, 1976; Lusis and Paigen, 1977b). In controlling the distribution of glucuronidase among cell types, egasyn serves a regulatory as well as a structural role.

Using strains of mice carrying different allelic forms of glucuronidase, it has also been possible to examine directly the importance of enzyme structure in determining enzyme localization. Strains of mice carrying one allelic form of glucuronidase (with altered heat stability but otherwise quite similar to the other known allelic forms) were found to have significantly different proportions of enzyme in the two main intracellular sites (lysosomes and the membranes of the ER) of liver. That the altered localization resulted from changes in enzyme structure, rather than other strain differences, was shown by the fact that in genetic crosses the altered localization pattern cosegregated with the glucuronidase structural gene, Gus. In addition differences in the cellular concentration of glucuronidase, as well as its structure, can have pronounced effects on the intracellular distribution of the enzyme (Paigen and Ganschow, 1965; Ganschow and Paigen, 1968; Lusis and Paigen, 1977a).

Finally, a series of mutants which affect the secretion of glucuronidase and other lysosomal enzymes from kidney proximal tubule cells have been identified (Brandt *et al.*, 1975; Meisler, 1978; Novak and Swank, 1979). With respect to enzyme localization, a particularly interesting feature of these mutants is that they all have altered coat color patterns, a condition that led to their initial identification. This suggests that certain processing steps are shared in the biogenesis of lysosomes and melanosomes (organelles structurally similar to lysosomes that contain the pigment producing apparatus). Such mutants should be useful in clarifying relationships between organelles and their associated enzymes.

B. Low Density Lipoprotein Receptor "Mislocation Mutant"

Among the various forms of human hypercholesterolemia, there is an interesting "mislocation mutation" of the fibroblast plasma membrane receptor for low density lipoprotein. Normally cultured fibroblasts adsorb serum low density lipoprotein by a specific high affinity receptor localized in coated pits on the surface membrane (Fig. 6). The low density lipoprotein together with bound cholesterol esters is subsequently rapidly transported by endocytic vesicles to lysosomes where the cholesteral esters and some protein components are degraded. The resulting free cholesterol serves in membrane biosynthesis (Anderson *et al.*, 1977).

Kinetic data suggest that there is rapid recycling of receptors back to

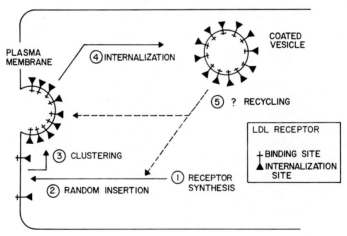

Fig. 6. Model for uptake of low density lipoprotein (LDL). (Redrawn from Anderson *et al.*, 1977, with permission.)

the cell surface where they migrate laterally to coated pits. It is thought that the clustering of low density lipoprotein receptors into coated pits is not dependent on the binding of external low density lipoprotein, as occurs in plasma membrane "capping" phenomena. The majority of receptors are found in coated pits even after prior fixation of cells with formaldehyde.

The receptor mislocation mutant is able to bind the normal amount of low density lipoprotein with high affinity but does not internalize it (Brown and Goldstein, 1976). As judged by electron microscopy of low density lipoprotein covalently coupled to ferritin, receptors in the mutant are scattered randomly over the fibroblast cell surface. In contrast, in normal cells 50–80% of receptors are localized in coated pits. Low density lipoprotein binds normally to the randomly scattered receptors in the mutant and is internalized only very slowly.

The size, configuration, and distribution of coated pits in fibroblasts of the mislocation mutant are normal. Furthermore, the mislocation mutant does not complement genetically with another class of receptor-negative mutants (Goldstein et al., 1977). This suggests that the receptor-negative mutants arise from defects in the structural gene for low density lipoprotein receptor, and that the receptor has two domains, one required for binding low density lipoprotein and another necessary for proper insertion into coated pits. The mutant defects have not yet been biochemically characterized; nevertheless, the data are consistent with a model (Fig. 6) in which the receptor is a transmembrane protein whose external domain contains the receptor binding site, while the cytoplasmic portion has a domain necessary for segregation into coated pits.

C. Effect of Gene Fusion and "Signal Sequence" Mutations on Enzyme Localization

The advanced genetic methodology available in prokaryotes has enabled studies of enzyme localization mechanisms not yet possible in eukaryotic systems.

Silhavy et al. (1976, 1977) have described an elegant genetic method by which hybrid proteins were formed by fusing portions of E. coli genes which normally code for membrane proteins with the gene coding for the soluble protein β-galactosidase. Because of the nature of the genetic fusion the amino terminus of the hybrid protein is likely derived from the membrane protein and the carboxyl terminus from β-galactosidase, which retains its catalytic activity. Fusion strains with varying amounts of membrane protein in the amino terminus were constructed, and the subsequent subcellular localization of the hybrid proteins monitored by β-galactosidase assay.

In four fusion strains containing a hybrid protein derived from an inner membrane protein for maltose transport and β-galactosidase, 80–90% of cellular β-galactosidase activity was found in the inner membrane. When the β-galactosidase gene was fused by similar techniques to genes coding for soluble proteins, no alteration in the localization of β-galactosidase occurred. It is likely that the β-galactosidase fused to the membrane protein faced the cytoplasm in these strains, since introduction of a mutation in the lactose transport system abolished the ability to grow on lactose. An externally oriented β-galactosidase would have cleaved extracellular lactose to metabolically useful products.

Two fusion strains in which the outer membrane receptor for bacteriophage adsorption was fused to β-galactosidase were particularly interesting. In one strain only a small portion (about 10% of the length) of the membrane protein was fused to β-galactosidase while in a second strain a larger proportion (probably greater than 20%) was fused. β-Galactosidase retained its normal soluble location in the strain with a small portion of the membrane protein; however, 50% of β-galactosidase activity was localized in the outer membrane in the strain with a larger portion of the amino terminus of the membrane protein. Both the ability of this latter strain to hydrolyze external lactose (the λ receptor also plays a role in sugar transport) and fluorescent antibody studies indicate that β-galactosidase was on the external face of the outer membrane.

Emr et al. (1978) have partially characterized two mutations selected in strains in which the gene for bacteriophage λ adsorption was fused to the gene for cytoplasmic β-galactosidase. In the parent fusion strain, the fused protein was located predominantly in the outer membrane, while in two mutant strains the fused protein was shifted in location to the cytoplasm. It is likely that both mutations lie very early in the receptor gene, in a gene region corresponding to the amino terminus of the fused protein. One is apparently a small deletion, and the other, a revertible point mutation. It is likely that the mutations are in the portion of the λ receptor gene which codes for the signal sequence such that protein can no longer be exported to the outer membrane.

Lin et al. (1978) have isolated and characterized a mutant in E. coli with an alteration in the signal sequence of an outer membrane lipoprotein. The protein is normally synthesized as a precursor containing 20 amino acids not present in the mature protein at the amino terminus and is converted into the mature polypeptide during transport to the outer membrane. An amino acid substitution at position 14 of the precursor (an aspartate residue being substituted for a glycyl residue) affected several aspects of the processing of the protein. Only about half of the mutant protein was assembled into the outer membrane, and the normal covalent modifications, including cleavage of the signal sequence, the covalent coupling of diglyceride, and the joining of the lipoprotein to the murein sacculus, did not occur. Since a significant fraction of the mutant protein was incorporated into the outer membrane, the removal of the signal sequence is not obligatorily coupled to the translocation of the protein. A likely possibility is that the mutation affects the attachment of the nascent chain of the precursor to the cytoplasmic membrane (Lin *et al.,* 1978).

These experiments indicate that the information for localization of *E. coli* membrane proteins is present in specific regions of the structural gene for the protein and that this information can direct even a large soluble enzyme such as β-galactosidase to specific subcellular sites. Further characterization of these types of mutations may enable determination of the minimal structure required for normal intracellular transport and membrane insertion. Genetic selection techniques available in such systems should enable further dissection of mechanisms of protein export and localization.

VII. SUMMARY

1. The cells of higher organisms are characterized by a complex network of membranes, consisting of protein and lipid, that subdivide the cell into a variety of compartments with specific functions and constituents. Ultrastructural and biochemical observations indicate that many of the membranes form close associations, being either continuous or in contact by means of shuttling vesicles.

2. Synthesis at the proper intracellular site is an important part of correct protein localization and is responsible for topologically segregating certain classes of proteins. Cytoplasmic protein synthesis occurs on at least two distinct populations of ribosomes—free ribosomes in the cell sap and ribosomes bound to the cytoplasmic face of the ER. The information determining the site of synthesis of a protein is encoded within the mRNA, and the properties of the newly emerging nascent transcript determine, at least in part, on which population of ribosomes translation occurs.

3. After translation, many proteins must be transported to particular sites in the cell. Since some proteins are synthesized at the same site yet are transported to different sites, the processing apparatus must contain sorting mechanisms which recognize different classes of proteins. The nature of these sorting mechanisms is unknown, but the origin of this recognition can only reside in the amino acid sequence of the proteins.

4. In eukaryotic cells, protein synthesis occurs in mitochondria and chloroplasts in addition to the cytoplasm. These proteins are coded by organellar DNA, and we have no reason to suppose that the proteins synthesized in these organelles are transported to other cellular locations. However, many proteins coded by nuclear DNA are transported into these organelles after synthesis in the cytoplasm.

5. It is becoming apparent that covalent modification of proteins serves an important role in protein localization. There is evidence that conjugant groups, such as carbohydrates, function as traffic-directing labels and that proteolytic cleavage is involved in transfer of proteins between membrane-bound compartments.

6. In the absence of special biological mechanisms, membranes are impermeable to proteins, and the problem of protein localization is in large part one of inserting proteins into particular membrane-bound compartments. Two distinct mechanisms have been observed for the transfer of proteins across membranes, one coupled to translation ("vectorial discharge") and the other posttranslational. In addition, fission and fusion of membranes makes possible movement of proteins between certain membrane-bound organelles.

7. Membranes are apparently assembled by insertion of protein and lipid into preexisting membrane. The insertion of certain integral membrane proteins into membrane is coupled to translation, although some integral and peripheral membrane proteins are inserted by posttranslational processes. Once inserted into membrane, proteins do not "flip-flop" with respect to the plane of the membrane, although they do exhibit lateral movement in the plane of the membrane. Membrane proteins are transported between different cellular membranes by means of lateral migration and membrane fission–fusion. During transport asymmetric distribution of proteins between the two faces of the membrane is maintained.

8. Reconstitution systems and genetic approaches are proving particularly useful in examining mechanisms of protein localization. Reconstitution systems which more closely approximate the true architecture of the cell are being developed. Genetic studies summarized here, including those of the low density lipoprotein receptor mutant, the glucuronidase structural gene, and the signal sequence of various prokaryotic proteins, have demonstrated the importance of information in the amino acid sequence in maintaining fidelity of protein subcellular localization. The importance of genes regulating posttranslational processing of proteins is exemplified by studies on the anchor protein, egasyn, which binds glucuronidase to subcellular membranes. It is expected that the number of identified genes affecting protein self-assembly and catalytic interactions

will expand as advanced methods of genetic engineering and mutant selection become available. Genetic approaches should provide information difficult to obtain by purely microscopic or cell fractionation methodologies.

ACKNOWLEDGMENTS

We are grateful to Drs. Reneé LeBoeuf and Franklin Berger for their criticism and Maccecchini *et al.* (1979) for making their manuscript available to us prior to publication.

REFERENCES

Allen, W. R. (1978). Does protein synthesis occur within the nucleus? *Trends Biochem. Sci.* 3, N225–228.

Anderson, R. G. W., Goldstein, J. L., and Brown, M. S. (1977). A mutation that impairs the ability of lipoprotein receptors to localize in coated pits on the cell surface of human fibroblasts. *Nature (London)* 270, 695–699.

Andrews, T. M., and Tata, J. R. (1971). Protein synthesis by membrane-bound and free ribosomes by secretory and nonsecretory tissues. *Biochem. J.* 121, 683–694.

Ashwell, G., and Morell, A. G. (1974). The role of surface carbohydrates in the hepatic recognition and transport of circulating glycoproteins. *Adv. Enzymol.* 41, 99–128.

Atkinson, P. H. (1978). Glycoprotein and protein precursors to plasma membranes in vesicular stomatitis virus infected HeLa cells. *J. Supramol. Struct.* 8, 89–109.

Auger-Buendia, M., and Longuet, M. (1978). Characterization of proteins from nucleolar preribosomes of mouse leukemia cells by two-dimensional polyacrylamide gel electrophoresis. *Eur. J. Biochem.* 85, 105–114.

Berns, A. J. M., van Kraaikamp, M., Bleomendal, H., and Lane, C. D. (1972). Calf crystallin synthesis in frog cells: The translation of lens-cell 14 S RNA in oocytes. *Proc. Natl. Acad. Sci. U.S.A.* 69, 1606–1609.

Berridge, M. V., and Lane, C. D. (1976). Translation of *Xenopus* liver messenger RNA in *Xenopus* oocytes: Vitellogenin synthesis and conversion to yolk platelet proteins. *Cell* 8, 283–297.

Blobel, G., and Dobberstein, B. (1975a). Transfer of proteins across membranes. I. Presence of proteolytically processed and unprocessed nascent immunoglobulin light chains on membrane-bound ribosomes of murine myeloma. *J. Cell Biol.* 67, 835–851.

Blobel, G., and Dobberstein, B. (1975b). Transfer of proteins across membranes. II. Reconstitution of functional rough microsomes from heterologous components. *J. Cell Biol.* 67, 852–862.

Blobel, G., and Sabatini, D. D. (1971). Ribosome–membrane interaction in eukaryotic cells. *Biomembranes* 2, 193–195.

Bloch, D. P., and Brack, S. D. (1964). Evidence for the cytoplasmic synthesis of nuclear histone during spermiogenesis in the grssshopper *Chortophaga viridifascists* (de Geer). *J. Cell Biol.* 22, 327–340.

Blumenthal, R., Weinstein, J. N., Sharrow, S. O., and Henkart, P. (1977). Liposome–lymphocyte interaction: Saturable sites for transfer and intracellular release of liposome contents. *Proc. Natl. Acad. Sci. U.S.A.* 74, 5603–5607.

Bonner, W. M. (1975). Frog oocyte nuclei accumulate a class of microinjected oocyte nuclear proteins and exclude a class of microinjected oocyte cytoplasmic proteins. *J. Cell Biol.* **64,** 431–437.

Boquet, P. (1977). Transport of diptheria toxin fragment A across mammalian cell membranes. *Biochem. Biophys. Res. Commun.* **75,** 696–702.

Brandt, E. J., Elliott, R. W., and Swank, R. T. (1975). Defective lysosomal enzyme secretion in kidneys of Chediak-Higashi (beige) mice. *J. Cell Biol.* **67,** 774–788.

Bretscher, M. S. (1973). Membrane structure: Some general principles. *Science* **181,** 622–629.

Brown, M. S., and Goldstein, J. L. (1976). Analysis of a mutant strain of human fibroblasts with a defect in the internalization of receptor-bound low density lipoprotein. *Cell* **9,** 663–674.

Butler, P. J. G., and Klug, A. (1978). The assembly of a virus. *Sci. Am.* **239,** 62–69.

Cashmore, A. R., Broadhurst, M. K., and Gray, R. E. (1978). Cell-free synthesis of leaf protein: Identification of an apparent precursor of the small subunit of ribulose-1,5-biphosphate carboxylase. *Proc. Natl. Acad. Sci. U.S.A.* **75,** 655–659.

Clarke, S. (1977). The interaction of Triton X-100 with soluble proteins: Possible implications for the transport of proteins across membranes. *Biochem. Biophys. Res. Commun.* **79,** 46–52.

Cook, G. M. W. (1973). The Golgi apparatus: Form and function. *In* "Lysosomes in Biology and Pathology" (J. T. Dingle, ed.), pp. 237–277. North-Holland Publ., Amsterdam.

Dallner, G., Siekevitz, P., and Palade, G. E. (1966a). Biogenesis of endoplasmic reticulum membranes. I. Structural and chemical differentiation in developing rat hepatocyte. *J. Cell Biol.* **30,** 73–96.

Dallner, G., Siekevitz, P., and Palade, G. E. (1966b). Biogenesis of endoplasmic reticulum membranes. II. Synthesis of constitutive microsomal enzymes in developing rat hepatocyte. *J. Cell Biol.* **30,** 97–117.

deDuve, C. (1969). The lysosome in retrospect. *In* "Lysosomes in Biology and Pathology" (J. T. Dingle and H. B. Fell, eds.), pp. 3–40. North-Holland Publ., Amsterdam.

deDuve, C., and Baudhuin, P. (1966). Peroxisomes (microbodies and related particles). *Physiol. Rev.* **46,** 323–358.

deDuve, C., and Wattiaux, R. (1966). Functions of lysosomes. *Annu. Rev. Biochem.* **28,** 435–492.

DePierre, J. W., and Dallner, G. (1975). Structural aspects of the membrane of the endoplasmic reticulum. *Biochim. Biophys. Acta* **415,** 411–472.

DePierre, J. W., and Ernster, L. (1977). Enzyme topology of intracellular membranes. *Annu. Rev. Biochem.* **46,** 201–202.

DeRobertis, E. M., Longthorne, R. F., and Gurdon, J. B. (1978). Intracellular migration of nuclear proteins in *Xenopus* oocytes. *Nature (London)* **272,** 254–256.

Dobberstein, B., Blobel, G., and Chau, N. (1977). *In vitro* synthesis and processing of a putative precursor for the small subunit of ribulose-1,5-bisphosphate carboxylase of *Chlamydomaras reinhardtii*. *Proc. Natl. Acad. Sci. U.S.A.* **74,** 1082–1085.

Doyle, D., Baumann, H., England, B., Friedman, E., Hou, E., and Tweto, J. (1978). Biogenesis of plasma membrane glycoproteins in hepatoma tissue culture cells. *J. Biol. Chem.* **253,** 965–973.

Draper, R. K., Chin, D., and Simon, M. I. (1978). Diptheria toxin has the properties of a lectin. *Proc. Natl. Acad. Sci. U.S.A.* **75,** 261–265.

Emr, S. D., Schwartz, M., and Silhavy, T. J. (1978). Mutations altering the cellular localization of the phage receptor, an *Escherichia coli* outer membrane protein. *Proc. Natl. Acad. Sci. U.S.A.* **75,** 5802–5806.

Eytan, G. D., and Broza, R. (1978). Role of charge and fluidity in the incorporation of cytochrome oxidase into liposomes. *J. Biol. Chem.* **253**, 3196–3202.

Feldherr, C. M., and Pomerantz, J. (1978). Mechanism for the selection of nuclear polypeptides in *Xenopus* oocytes. *J. Cell Biol.* **78**, 168–175.

Fowler, S., Remacle, J., Trouvet, A., Beaufay, H., Berthet, J., Wibo, M., and Hauser, P. (1976). Immunological localization of cytochrome b_5 by electron microscopy: Methodology and application to various subcellular fractions. *J. Cell Biol.* **71**, 535–550.

Franke, W. W. (1974). Nuclear envelopes. Structure and biochemistry of the nuclear envelope. *Int. Rev. Cytol.* **4**, Suppl., 71–236.

Franke, W. W., and Kartenbeck, J. (1971). Outer mitochondrial membrane continuous with endoplasmic reticulum. *Protoplasma* **73**, 35–41.

Ganoza, M. C., and Williams, C. A. (1969). *In vitro* synthesis of different categories of specific protein by membrane-bound and free ribosomes. *Proc. Natl. Acad. Sci. U.S.A.* **63**, 1370–1376.

Ganschow, R. E., and Paigen, K. (1967). Separate genes determining the structure and intracellular location of hepatic glucuronidase. *Proc. Natl. Acad. Sci. U.S.A.* **58**, 938–945.

Ganschow, R. E., and Paigen, K. (1968). Glucuronidase phenotypes in inbred mouse strains. *Genetics* **59**, 335–349.

Gerace, L., Blum, A., and Blobel, G. (1978). Immunocytochemical localization of the major polypeptides of the nuclear pore complex–lamina fraction. Interphase and mitotic distribution. *J. Cell Biol.* **79**, 546–566.

Ghysdael, J., Hubert, E., Travnicek, M., Bolobnesi, D. P., Burney, A., Cleuter, Y., Huez, G., Kettmann, R., Marbaix, G., Portetelle, D., and Ghantrenne, H. (1977). Frog oocytes synthesize and completely process the precursor polypeptide to virion structural proteins after microinjection of avian myeloblastosis virus RNA. *Proc. Natl. Acad. Sci. U.S.A.* **74**, 3230–3234.

Gibor, A., and Granick, S. (1964). Plastids and mitochondria: Inheritable systems. *Science* **145**, 890–897.

Godinot, C., and Lardy, H. A. (1973). Biosynthesis of glutamate dehydrogenase in rat liver. Demonstration of its microsomal localization and hypothetical mechanism of transfer to mitochondria. *Biochemistry* **12**, 2051–2060.

Goidl, J. A. (1978). Does protein synthesis occur within the nucleus? *Trends Biochem. Sci.* **3**, N225–N228.

Goldman, B. M., and Blobel, G. (1978). Biogenesis of peroxisomes: Intracellular site of synthesis of catalase and uricase. *Proc. Natl. Acad. Sci. U.S.A.* **75**, 5066–5070.

Goldstein, J. L., Brown, M. S., and Stone, N. J. (1977). Genetics of the LDL receptor: Evidence that the mutations affecting binding and internalization are allelic. *Cell* **12**, 629–641.

Gonzalez-Cadavid, N. F., Ortega, J. P., and Gonzalez, M. (1971). The cell-free synthesis of cytochrome *c* by a microsomal fraction from rat liver. *Biochim. J.* **124**, 685–694.

Gurdon, J. B. (1970). Nuclear transplantation and control of gene activity in animal development. *Proc. R. Soc. London, Ser. B* **176**, 303–314.

Gurdon, J. B., Lane, C. D., Woodland, H. R., and Marbaix, G. (1971). Use of frog oocytes for the study of messenger RNA and its translation in living cells. *Nature (London)* **233**, 177–182.

Haldar, D., Freeman, K., and Work, T. S. (1966). Biogenesis of mitochondria. *Nature (London)* **211**, 9–12.

Harano, T., and Omura, T. (1978). Biogenesis of endoplasmic reticulum membrane in rat liver cells. III. Biosynthesis of NADPH cytochrome b_5 reductase and cytochrome b_5 by loosely bound ribosomes. *J. Biochem. (Tokyo)* **84**, 213–233.

Hatefi, Y. (1976). The enzymes and enzyme complexes of the mitochondrial oxidative phosphorylation system. *In* "The Enzymes of Biological Membranes" (A. Martonosi, ed.), Vol. 4, pp. 3–41. Plenum, New York.

Heggeness, M. H., Wang, K., and Singer, S. J. (1977). Intracellular distributions of mechanochemical proteins in cultured fibroblasts. *Proc. Natl. Acad. Sci. U.S.A.* **74**, 3883–3887.

Hercz, A., Katona, E., Cutz, E., Wilson, J. R., and Barton, M. (1978). α-Antitrypsin: The presence of excess mannose in the Z variant isolated from liver. *Science* **201**, 1229–1232.

Hickman, S., Shapiro, L. J., and Neufeld, E. F. (1974). A recognition marker required for uptake of a lysosomal enzyme by cultured fibroblasts. *Biochem. Biophys. Res. Commun.* **57**, 55–61.

Hicks, S. J., Drysdale, J. W., and Munro, H. N. (1969). Preferential synthesis of ferritin and albumin by different populations of liver polysomes. *Science* **164**, 584–585.

Higashi, T., and Peters, T. (1963a). Studies on rat liver catalase. I. Combined immunochemical and enzymatic determination of catalase in liver cell fractions. *J. Biol. Chem.* **238**, 3945–3951.

Higashi, T., and Peters, T. (1963b). Studies of rat liver catalase. II. Incorporation of ^{14}C-leucine into catalase of liver cell fractions *in vivo*. *J. Biol. Chem.* **238**, 3952–3954.

Highfield, P. E., and Ellis, R. J. (1978). Synthesis and transport of the small subunit of chloroplast ribulose bisphosphate carboxylase. *Nature (London)* **271**, 420–424.

Jackson, R. C., and Blobel, G. (1977). Post-translational cleavage of presecretory proteins with an extract of rough microsomes from dog pancreas containing signal peptidase activity. *Proc. Natl. Acad. Sci. U.S.A.* **74**, 5598–5602.

Jacques, P. J. (1969). Endocytosis. *In* "Lysosomes in Biology and Pathology" (J. T. Dingle and H. Fell, eds.), Vol. 2, pp. 395–420. North-Holland Publ., Amsterdam.

Jaskunas, S. R., Nomura, M., and Davies, J. (1974). Genetics of bacterial ribosomes. *In* "Ribosomes" (M. Nomura, A. Tissières, and P. Lengyel, eds.), pp. 333–366. Cold Spring Harbor Lab., Cold Spring Harbor, New York.

Kadenbach, B. (1966). Synthesis of mitochondrial proteins: Demonstration of transfer of proteins from microsomes to mitochondria. *Biochim. Biophys. Acta* **134**, 430–442.

Kadenbach, B. (1970). Biosynthesis of cytochrome *c*. The sites of synthesis of apoprotein and holoenzyme. *Eur. J. Biochem.* **12**, 392–398.

Kaplan, A., Achord, D. T., and Sly, W. S. (1977). Phosphohexosyl receptors on human fibroblasts. *Proc. Natl. Acad. Sci. U.S.A.* **74**, 2026–2030.

Karl, T. R., and Chapman, V. M. (1974). Linkage and expression of the *Eg* locus controlling inclusion of β-glucuronidase into microsomes. *Biochem. Genet.* **11**, 367–372.

Katz, F., Rothman, J. E., Lingappa, V. R., Blobel, G., and Lodish, H. (1977a). Membrane assembly *in vitro*: Synthesis, glycosylation, and assymetric insertion of a transmembrane protein. *Proc. Natl. Acad. Sci. U.S.A.* **74**, 3278–3282.

Katz, F. N., Rothman, J. E., Knipe, D. M., and Lodish, H. F. (1977b). Membrane assembly: Synthesis and intracellular processing of the vesicular stomatitis viral glycoprotein. *J. Supramol. Struct.* **7**, 353–370.

Kawajiri, K., Harano, T., and Omura, T. (1977a). Biogenesis of the mitochondrial matrix enzyme, glutamate dehydrogenase, in rat liver cells. I. Subcellular localization, biosynthesis and intracellular translocation of glutamate dehydrogenase. *J. Biochem. (Tokyo)* **82**, 1403–1416.

Kawajiri, K., Harano, T., and Omura, T. (1977b). Biogenesis of mitochondrial matrix en-

zyme, glutamate dehydrogenase, in rat liver cells. II. Significance of binding of glutamate dehydrogenase to microsomal membrane. *J. Biochem. (Tokyo)* **82**, 1417–1423.

Kellems, R. E., and Butow, R. A. (1974). Cytoplasmic type 80 S ribosomes associated with
yeast mitochondria. III. Changes in the amount of bound ribosomes in response to
changes in metabolic state. *J. Biol. Chem.* **249**, 3304–3310.

Kellems, R. E., Allison, U. F., and Butow, R. A. (1975). Cytoplasmic type 80 S ribosomes
associated with yeast mitochondria. IV. Attachment of ribosomes to the outer membrane of isolated mitochondria. *J. Cell Biol.* **65**, 1–14.

King, J., and Laemmli, U. K. (1971). Polypeptides of tail fibers of bacteriophage T4. *J. Mol.
Biol.* **62**, 465–471.

Knipe, D. M., Lodish, H. F., and Baltimore, D. (1977a). Localization of two cellular forms
of the vesicular stomatitis viral glycoprotein. *J. Virol.* **21**, 1121–1127.

Knipe, D. M., Baltimore, D., and Lodish, M. F. (1977b). Separate pathways of maturation of
the major structural proteins of vesicular stomatitis virus. *J. Virol.* **21**, 1128–1139.

Knipe, D. M., Baltimore, D., and Lodish, M. F. (1977c). Maturation of viral proteins in cells
infected with temperature-sensitive mutants of vesicular stomatitis virus. *J. Virol.* **21**,
1149–1158.

Korb, H., and Neupert, W. (1978). Biogenesis of cytochrome c in *Neurospora crassa*. *Eur.
J. Biochem.* **91**, 609–620.

Kreibich, G., Ulrich, B. L., and Sabatini, D. D. (1978a). Proteins of rough microsomal
membranes related to ribosome binding. I. Identification of ribophorins I and II, membrane proteins characteristic of rough microsomes. *J. Cell Biol.* **77**, 464–487.

Kreibich, G., Freienstein, C. M., Peregra, B. W., Ulrich, B. L., and Sabatini, D. D. (1978b).
Proteins of rough microsomal membranes related to ribosome binding. II. Crosslinking
of bound ribosomes to specific membrane proteins at the binding sites. *J. Cell Biol.* **77**,
488–506.

Laemmli, U. K. (1970). Cleavage of structural proteins during the assembly of the head of
bacteriophage T4. *Nature (London)* **227**, 442–444.

Lane, C. (1976). Rabbit hemoglobin from frog eggs. *Sci. Am.* **235**, 60–71.

Lane, C. D., Marbaix, G., and Gurdon, J. B. (1971). Rabbit haemoglobin synthesis in frog
cells: The translation of reticulocyte 9 S RNA in frog oocytes. *J. Mol. Biol.* **61**, 73–91.

Laskey, R. A., Gurdon, J. B., and Crawford, L. V. (1972). Translation of encephalomyocarditis viral RNA in oocytes of *Xenopus laevis*. *Proc. Natl. Acad. Sci. U.S.A.* **69**,
3665–3669.

Laskey, R. A., Hinde, B. M., Mills, A. A., and Finch, J. T. (1978). Nucleosomes are
assembled by an acidic protein which binds histones and transfers them to DNA.
Nature (London) **275**, 416–420.

Lawson, D., Raff, M. C., Gomperts, B., Fewtrell, C., and Gilula, N. B. (1977). Molecular
events during membrane fusion. A study of exocytosis in rat peritoneal mast cells. *J.
Cell Biol.* **72**, 242–259.

Lazarides, E., and Weber, K. (1974). Actin antibody: The specific visualization of actin
filaments in non-muscle cells. *Proc. Natl. Acad. Sci. U.S.A.* **71**, 2268–2272.

Lazarow, P. B., and deDuve, C. (1973a). The synthesis and turnover of rat liver peroxisomes. IV. Biochemical pathway of catalase synthesis. *J. Cell Biol.* **59**, 491–506.

Lazarow, P. B., and deDuve, C. (1973b). The synthesis and turnover of rat liver peroxisomes. V. Intracellular pathway of catalase synthesis. *J. Cell Biol.* **59**, 507–524.

Leavitt, R., Schlesinger, S., and Kornfeld, S. (1977). Impaired intracellular migration and
altered solubility of nonglycosylated glycoproteins of vesicular stomatitis virus and
Sindbis virus. *J. Biol. Chem.* **252**, 9018–9023.

Ledley, F. D., Mullin, B. R., Lee, G., Aloj, S. M., Fishman, P. H., Hunt, L. T., Dayhoff, M.

D., and Kohn, L. D. (1976). Sequence similarity between cholera toxin and glycoprotein hormones: Implications for structure activity relationship and mechanism of action. *Biochem. Biophys. Res. Commun.* **69**, 852–859.

Lin, G.-W. (1975). Multiple forms of β-glucuronidase: Molecular nature, transformation, and subcellular translocation. *In* "Isozymes" (C. L. Markert, ed.), Vol. 1, pp. 637–651. Academic Press, New York.

Lin, J. J. C., Kanazawa, H., Ozols, J., and Wu, H. C. (1978). An *Escherichia coli* mutant with an amino acid alteration within the signal sequence of outer membrane prolipoprotein. *Proc. Natl. Acad. Sci. U.S.A.* **75**, 4891–4895.

Lingappa, V. R., Shields, D., Woo, S. L. C., and Blobel, G. (1978). Nascent chicken ovalbumin contains the functional equivalent of a signal sequence. *J. Cell Biol.* **79**, 567–572.

Lodish, H. F. (1973). Biosynthesis of reticulocyte membrane proteins by membrane-free polyribosomes. *Proc. Natl. Acad. Sci. U.S.A.* **70**, 1526–1530.

Lodish, H. F., and Small, B. (1975). Membrane proteins synthesized by rabbit reticulocytes. *J. Cell Biol.* **65**, 51–64.

Lorenz, T., and Willard, M. (1978). Subcellular fractionation of intra-axonally transported polypeptides in the rabbit visual system. *Proc. Natl. Acad. Sci. U.S.A.* **75**, 505–509.

Lusis, A. J., and Paigen, K. (1977a). Mechanisms involved in the intracellular localization of mouse glucuronidase. *In* "Isozymes" (M. Rattazzi, J. G. Scandalios, and G. S. Whitt, eds.), Vol. 2, pp. 63–106. Alan R. Liss, Inc., New York.

Lusis, A. J., and Paigen, K. (1977b). Relationships between levels of membrane-bound glucuronidase and the associated protein egasyn in mouse tissues. *J. Cell Biol.* **73**, 728–735.

Lusis, A. J., Tomino, S., and Paigen, K. (1976). Isolation, characterization and radioimmunoassay of murine egasyn, a protein stabilizing glucuronidase membrane binding. *J. Biol. Chem.* **251**, 7753–7760.

Lusis, A. J., Breen, G. A. M., and Paigen, K. (1977). Nongenetic heterogeneity of mouse β-galactosidase. *J. Biol. Chem.* **252**, 4613–4618.

Maccecchini, M.-L., Rudin, Y., Blobel, G., and Schatz, G. (1979). Import of proteins into mitochondria: Precursor forms of the extramitochondrially-made F_1-ATPase subunits in yeast. *Proc. Natl. Acad. Sci. U.S.A.* **76**, 343–347.

Mach, B., Faust, C., and Vinsall, P. (1973). Purification of 14 S messenger RNA of immunoglobulin light chain that codes for a possible light chain precursor. *Proc. Natl. Acad. Sci. U.S.A.* **70**, 451–455.

McLaughlin, C. A., and Pitot, H. C. (1976). Hormonal and nutritional effects on the binding of ^{125}I-labelled anti-serine dehydratase Fab to rat tissue polysomes. *Biochemistry* **15**, 3550–3556.

Marchesi, V. T., Tillack, T. W., Jackson, P. L., Segrest, J. P., and Scott, R. E. (1972). Chemical characterization and surface orientation of the major glycoprotein of human erythrocyte membrane. *Proc. Natl. Acad. Sci. U.S.A.* **69**, 1445–1449.

Marchesi, V. T., Furthmayr, H., and Tomita, M. (1976). The red cell membrane. *Annu. Rev. Biochem.* **45**, 667–698.

Masters, C. J. (1978). Interactions between soluble enzymes and subcellular structure. *Trends Biochem. Sci.* **3**, 206–208.

Meisler, M. H. (1978). Synthesis and secretion of kidney β-galactosidase in mutant *le/le* mice. *J. Biol. Chem.* **253**, 3129–3134.

Mollenhauer, H. H., and Morré, D. J. (1978). Structural compartmentation of the cytosol: Zones of exclusion, zones of adhesion, cytoskeletal and intercisternal elements. *Sub-Cell. Biochem.* **5**, 327–357.

Morré, D. J. (1977). The Golgi apparatus and membrane biogenesis. *In* "The Synthesis, Assembly and Turnover of Cell Surface Compartments" (G. Poste and G. L. Nicolson, eds.), pp. 1–83. Elsevier, Amsterdam.

Morré, D. J., Merrit, W. D., and Lembi, C. (1971). Connections between mitochondria and endoplasmic reticulum in rat liver and onion stem. *Protoplasma* **73**, 43–49.

Morré, D. J., Keanan, T. W., and Huang, C. M. (1974). Membrane growth and differentiation: Origin of Golgi apparatus membranes from endoplasmic reticulum. *Adv. Cytopharmacol.* **2**, 107–125.

Mullin, B. R., Fishman, P. H., Lee, G., Aloj, S. M., Ledley, F. D., Winard, R. J., Kohn, L. D., and Brady, R. O. (1976a). Thyrotropin-ganglioside interactions and their relationship to the structure and function of thyrotropin receptors. *Proc. Natl. Acad. Sci. U.S.A.* **73**, 842–846.

Mullin, B. R., Aloj, S. M., Fishman, P. H., Lee, G., Kohn, L. D., and Brady, R. O. (1976b). Cholera toxin interactions with thyrotropin receptors on thyroid plasma membranes. *Proc. Natl. Acad. Sci. U.S.A.* **73**, 1679–1683.

Neufeld, E. F., Lim, T. W., and Shapiro, L. J. (1975). Inherited disorders of lysosomal metabolism. *Annu. Rev. Biochem.* **44**, 357–376.

Neufeld, E. F., Sando, G. N., Garvin, A. J., and Rome, L. H. (1977). The transport of lysosomal enzymes. *J. Supramol. Struct.* **6**, 95–101.

Nomura, M., and Held, W. A. (1974). Reconstitution of ribosomes: Studies of ribosome structure, function and assembly. *In* "Ribosomes" (M. Nomura, A. Tissiéres, and P. Lengyel, eds.), pp. 193–223. Cold Spring Harbor Lab., Cold Spring Harbor, New York.

Novak, E., and Swank, R. T. (1979). Lysosomal dysfunctions associated with mutations at mouse pigment genes. *Genetics* **92**, 189–204.

Novikoff, A. B. (1973). Lysosomes: A personal account. *In* "Lysosomes and Storage Diseases" (H. G. Hers and F. Van Hoof, eds.), pp. 1–41. Academic Press, New York.

Novikoff, A. B. (1976). The endoplasmic reticulum: A cytochemists view. *Proc. Natl. Acad. Sci. U.S.A.* **73**, 2781–2787.

Novikoff, A. B., and Holtzman, E. (1976). "Cell and Organelles." Holt, New York.

Olden, K., Pratt, R. M., and Yamada, M. (1978). Role of carbohydrates in protein secretion and turnover: Effects of tunicamycin on the major cell surface glycoprotein of chick embryo fibroblasts. *Cell* **13**, 461–473.

Omura, T., and Kuriyama, Y. (1971). Role of rough and smooth microsomes in the biosynthesis of microsomal membranes. *J. Biochem. (Tokyo)* **69**, 651–658.

Ouadi, J., and Keleti, T. (1978). Kinetic evidence for interaction between aldolase and D-glyceraldehyde-3-phosphate dehydrogenase. *Eur. J. Biochem.* **85**, 157–161.

Owerbach, D., and Lusis, A. J. (1976). Phenobarbital induction of egasyn: Availability of egasyn *in vivo* determines glucuronidase binding to membrane. *Biochem. Biophys. Res. Commun.* **69**, 628–634.

Ozols, J., and Gerard, C. (1977). Primary structure of the membranes segment of cytochrome b_5. *Proc. Natl. Acad. Sci. U.S.A.* **74**, 3725–3729.

Paigen, K. (1961). The effect of mutation on the intracellular location of β-glucuronidase. *Exp. Cell Res.* **25**, 286–301.

Paigen, K., and Ganschow, R. (1965). Genetic factors in enzyme realization. *Brookhaven Symp. Biol.* **18**, 99–114.

Palade, G. (1975). Intracellular aspects of the process of protein synthesis. *Science* **189**, 347–358.

Palade, G. E., and Siekevitz, P. (1956). Liver microsomes. An integrated morphological and biochemical study. *J. Biophys. Biochem. Cytol.* **2**, 171–201.

Palmiter, R. D., Gagnon, J., and Walsh, K. A. (1978). Ovalbumin: A secreted protein without a transient hydrophobic leader sequence. *Proc. Natl. Acad. Sci. U.S.A.* **75**, 94–98.

Papahadjopoulos, D., Poste, G., Schaeffer, B. E., and Vail, W. J. (1974). Membrane fusion and molecular segregation in phospholipid vesicles. *Biochim. Biophys. Acta* **352**, 10–28.

Pappenheimer, A. M., Jr. (1977). Diphtheria toxin. *Annu. Rev. Biochem.* **46**, 69–94.

Pappenheimer, A. M., Jr. (1978). Diptheria: Molecular biology of an infectious process. *Trends Biochem. Sci.* **3**, N220–N224.

Pless, D. P., and Lennarz, W. L. (1977). Enzymatic conversion of proteins to glycoproteins. *Proc. Natl. Acad. Sci. U.S.A.* **74**, 134–138.

Poole, B. (1969). The biogenesis and turnover of rat liver peroxisomes. *Ann. N.Y. Acad. Sci.* **168**, 229–243.

Poyton, R. O., and Kavanagh, J. (1976). Regulation of mitochondrial protein synthesis by cytoplasmic proteins. *Proc. Natl. Acad. Sci. U.S.A.* **73**, 3947–3951.

Pricer, W. E., Jr., and Ashwell, G. (1976). Subcellular distribution of mammalian hepatic binding protein specific for asialoglycoproteins. *J. Biol. Chem.* **251**, 7539–7544.

Randall, L. L., Hardy, S. J. S., and Josefsson, L. (1978). Precursors of three exported proteins in *Escherichia coli. Proc. Natl. Acad. Sci. U.S.A.* **75**, 1209–1212.

Redman, C. M. (1969). Biosynthesis of serum proteins and ferritin by free and attached ribosomes of rat liver. *J. Biol. Chem.* **244**, 4308–4315.

Redman, C. M., Siekevitz, G. E., and Palade, G. F. (1966). Synthesis and transfer of amylase in pigeon pancreatic microsomes. *J. Biol. Chem.* **241**, 1150–1158.

Remacle, J. (1978). Binding of cytochrome b_5 to membranes of isolated subcellular organelles from rat liver. *J. Cell Biol.* **79**, 291–313.

Ries, G., Hundt, E., and Kadenbach, B. (1978). Immunoprecipitation of a cytoplasmic precursor of rat liver cytochrome oxidase. *Eur. J. Biochem.* **91**, 179–191.

Robbi, M., and Lazarow, P. B. (1978). Synthesis of catalase in two cell-free protein synthesizing systems and in rat liver. *Proc. Natl. Acad. Sci. U.S.A.* **75**, 4344–4348.

Robbi, M., Berthet, J., and Beaufay, H. (1978). The biosynthesis of cytochrome *c. Eur. J. Biochem.* **84**, 341–346.

Robinson, N. C., and Tanford, C. (1975). The binding of deoxycholate, Triton X-100, sodium dodecyl sulfate and phosphatidylcholine vesicles to cytochrome b_5 . *Biochemistry* **14**, 369–378.

Rogers, M. J., and Strittmatter, P. (1974). The binding of reduced nicotinamide adenine dinucleotide-cytochrome b_5 reductase to hepatic microsomes. *J. Biol. Chem.* **249**, 5565–5569.

Rolleston, F. S. (1974). Membrane-bound and free ribosomes. *Sub-Cell. Biochem.* **3**, 91–117.

Rothman, J. E., and Lenard, J. (1977). Membrane assymetry. *Science* **195**, 743–753.

Rothman, J. E., and Lodish, H. F. (1977). Synchronized transmembrane insertion and glycosylation of a nascent membrane protein. *Nature (London)* **269**, 775–780.

Sabatini, D. D., and Kreibich, G. (1976). Functional specialization of membrane-bound ribosomes in eukaryotic cells. *In* "The Enzymes of Biological Membranes" (A. Martonosi, ed.), Vol. 2, pp. 531–579. Plenum, New York.

Satir, B. (1974). Ultrastructural aspects of membrane fusion. *J. Supramol. Struct.* **2**, 529–537.

Satir, B., Schooley, C., and Satir, P. (1973). Membrane fusion in a model system. Mucocyst secretion in *Tetrahymena. J. Cell Biol.* **56**, 153–176.

Schatz, G., and Mason, T. L. (1974). The biosynthesis of mitochondrial proteins. *Annu. Rev. Biochem.* **43**, 51–87.

Sekizawa, J., Inouye, S., Halegova, S., and Inouye, M. (1977). Precursors of major outer membrane proteins of *Escherichia coli. Biochem. Biophys. Res. Commun.* **77**, 1126–1133.

Shore, G. C., and Harris, R. (1977). Fate of polypeptides synthesized on rough microsomal vesicles in a messenger-dependent rabbit reticulocyte system. *J. Cell Biol.* **74**, 315–321.

Shore, G. C., and Tata, J. R. (1977a). Two fractions of rough endoplasmic reticulum from rat liver. I. Recovery of rapidly sedimenting endoplasmic reticulum in association with mitochondria. *J. Cell Biol.* **72**, 714–725.

Shore, G. C., and Tata, J. R. (1977b). Two fractions of rough endoplasmic reticulum from rat liver. II. Cytoplasmic messenger RNA's which code for albumin and mitochondrial proteins are distributed differently between the two fractions. *J. Cell Biol.* **72**, 726–743.

Shore, G. C., and Tata, J. R. (1977c). Function of polyrobosome–membrane interactions in protein synthesis. *Biochim. Biophys. Acta* **472**, 197–236.

Shotten, D., Thompson, K., Wofsy, L., and Branton, D. (1978). Appearance and distribution of surface proteins of the human erythrocyte membrane. An electron microscope and immunochemical labeling study. *J. Cell Biol.* **76**, 512–531.

Silhavy, T. J., Casadaban, M. J., Shuman, H. A., and Beckwith, J. A. (1976). Conversion of β-galactosidase to a membrane-bound state by gene fusion. *Proc. Natl. Acad. Sci. U.S.A.* **73**, 3423–3427.

Silhavy, T. J., Shuman, H. A., Beckwith, J. A., and Schwartz, M. (1977). Use of gene fusions to study outer membrane protein localization in *Escherichia coli. Proc. Natl. Acad. Sci. U.S.A.* **74**, 5411, 5415.

Singer, S. J. (1974). The molecular organization of membranes. *Annu. Rev. Biochem.* **43**, 805–833.

Singer, S. J. (1975). Membrane fluidity and cellular function. *In* "Control Mechanisms in Development" (R. A. Meints and E. Davies, eds.), pp. 181–192. Plenum, New York.

Singer, S. J., and Nicholson, G. L. (1972). The fluid mosaic model of the structure of cell membranes. *Science* **175**, 720–731.

Slotboom, A. J., and de Haas, G. H. (1975). Specific transformations at the N-terminal region of phospholipase A_2. *Biochemistry* **14**, 5394–5399.

Smillie, R. M., and Scott, W. S. (1969). Organelle biosynthesis: The chloroplast. *Prog. Mol. Subcell. Biol.* **1**, 136–202.

Smith, W. P., Tai, P. C., Thompson, R. C., and Davis, B. D. (1977). Extracellular labeling of nascent polypeptides traversing the membrane of *Escherichia coli. Proc. Natl. Acad. Sci. U.S.A.* **74**, 2830–2834.

Spatz, L., and Strittmatter, P. (1973). A form of reduced micotinamide adenine dinucleotide-cytochrome b_5 reductase containing both the catalytic site and an additional hydrophobic membrane-binding segment. *J. Biol. Chem.* **248**, 793–799.

Spillmann, S., and Nierhaus, K. H. (1978). The ribosomal protein L24 of *Escherichia coli* is an assembly protein. *J. Biol. Chem.* **253**, 7047–7050.

Stahl, P., Schlesinger, P. H., Rodman, J. S., and Doebber, T. (1976). Recognition of lysosomal glycosidases *in vivo* inhibited by modified glycoproteins. *Nature (London)* **264**, 86–88.

Steck, T. L. (1974). The organization of proteins in the human red blood cell membrane: A review. *J. Cell Biol.* **74**, 1–19.

Strittmatter, P., Rogers, M. J., and Spatz, L. (1972). The binding of cytochrome b_5 to liver microsomes. *J. Biol. Chem.* **247**, 7188–7194.

Strittmatter, P., Spatz, L., Corcoran, D., Rogers, M. J., Setlow, B., and Redline, R. (1974). Purification and properties of rat liver microsomal stearyl coenzyme A desaturase. *Proc. Natl. Acad. Sci. U.S.A.* **71**, 4565–4569.

Struck, D. K., and Lennarz, W. J. (1977). Evidence for the participation of saccharide-lipids in the synthesis of the oligosaccharide chain of ovalbumin. *J. Biol. Chem.* **252,** 1007–1013.

Struck, D. K., Sivta, P. B., Lane, M. D., and Lennarz, W. J. (1978). Effect of tunicamycin on the secretion of serum proteins by primary cultures of rat and chick hepatocytes. Studies of transferrin, very low density lipoprotein, and serum albumin. *J. Biol. Chem.* **253,** 5332–5337.

Swan, D., Aviv, H., and Leder, P. (1972). Purification and properties of biologically active messenger RNA for a myeloma light chain. *Proc. Natl. Acad. Sci. U.S.A.* **69,** 1967–1971.

Swank, R. T., and Paigen, K. (1973). Biochemical and genetic evidence for a macromolecular β-glucuronidase complex in microsomal membranes. *J. Mol. Biol.* **77,** 371–389.

Swank, R. T., Paigen, K., Davey, R., Chapman, V. M., Labarca, C., Watson, G., Ganschow, R., Brandt, E. J., and Novak, E. (1978). Genetic regulation of mammalian glucuronidase. *Recent Prog. Horm. Res.* **34,** pp. 401–436.

Takagi, M., Tanaka, T., and Ogata, K. (1970). Functional differences in protein synthesis between free and bound polysomes of rat liver. *Biochim. Biophys. Acta* **217,** 148–158.

Tomino, S., and Paigen, K. (1975). Egasyn, a protein complexed with microsomal glucuronidase. *J. Biol. Chem.* **250,** 1146–1148.

Toneguzzo, F., and Ghosh, H. P. (1978). *In vitro* synthesis of vesicular stomatitis virus membrane glycoprotein and insertion into membranes. *Proc. Natl. Acad. Sci. U.S.A.* **75,** 715–719.

Tsuji, H., and Kato, K. (1977). The synthesis of rat liver lysosomes. II. Intracellular transport of β-glucuronidase. *J. Biochem. (Tokyo)* **82,** 637–644.

Tsuji, H., Hattori, N., Yamamoto, J., and Kato, K. (1977). The synthesis of rat liver lysosomes. I. Comparison of microsomal, Golgi, and lysosomal β-glucuronidase. *J. Biochem. (Tokyo)* **82,** 619–636.

Tzagoloff, A. (1976). The adenosine triphosphatase complex of mitochondria. *In* "The Enzymes of Biological Membranes" (A. Martonosi, ed.), Vol. 4, pp. 103–142. Plenum, New York.

Uy, R., and Wold, F. (1977). Posttranslation covalent modifications of proteins. *Science* **198,** 890–896.

Vallee, R. B., and Borisy, G. G. (1978). The non-tubulin component of microtubule protein oligomers. *J. Biol. Chem.* **253,** 2834–2845.

van Dam-Mieras, M. C. E., Slotboom, A. J., Pieterson, W. A., and de Haas, G. H. (1975). The interaction of phospholipase A_2 with micellar interfaces. The role of the N-terminal region. *Biochemistry* **14,** 5387–5393.

Van Heyningen, S. (1977). Cholera toxin. *Biol. Rev. Cambridge Philos. Soc.* **52,** 509–549.

Waechter, C. J., and Lennarz, W. J. (1976). The role of polyprenol-linked sugars in glycoprotein synthesis. *Annu. Rev. Biochem.* **45,** 95–112.

Waring, G. L. (1978). Nonsecreting myeloma variants with heavy-chain carbohydrate deficiencies. *Biochem. Genet.* **16,** 69–78.

Warner, J. R. (1974). The assembly of ribosomes in eukaryotes. *In* "Ribosomes" (M. Nomura, A. Tissières, and P. Lengyel, eds.), pp. 461–488. Cold Spring Harbor Lab., Cold Spring Harbor, New York.

Wickner, W., Mandel, G., Zwizinski, C., Bates, M., and Killick, T. (1978). Synthesis of phage M13 coat protein and its assembly into membranes *in vitro. Proc. Natl. Acad. Sci. U.S.A.* **75,** 1754–1758.

Wilson, D. L. (1978). The building of neurons: From gene regulation to protein destination. *Trends Biochem. Sci.* **3,** 230–232.

Wirth, D. F., Katz, F., Small, B., and Lodish, H. F. (1977). How a Sindbis virus mRNA directs the synthesis of one soluble protein and two integral membrane glycoproteins. *Cell* **10**, 253–263.

Wittmann, H. G. (1977). Structure and function of *Escherichia coli* ribosomes. *Fed. Proc., Fed. Am. Soc. Exp. Biol.* **36**, 2075–2080.

Wool, I. G., and Stoffler, G. (1974). Structure and function of eukaryotic ribosomes. *In* "Ribosomes" (M. Nomura, A. Tissières, and P. Lengyel, eds.), pp. 417–460. Cold Spring Harbor Lab., Cold Spring Harbor, New York.

Yoshida, A., Lieberman, J., Gaidulis, L., and Ewing, C. (1976). Molecular abnormality of human alpha-antitrypsin variant (Pi-ZZ) associated with plasma activity deficiency. *Proc. Natl. Acad. Sci. U.S.A.* **73**, 1324–1328.

Yoshida, K. (1978). The presence of ribosomal glycoproteins. Agglutination of free and membrane-bound ribosomes from wheat germ by concanavalin A. *J. Biochem. (Tokyo)* **83**, 1609–1614.

Younis, H. M., Winget, G. D., and Racker, E. (1977). Requirement of the δ subunit of chloroplast coupling factor 1 for photophosphorylation. *J. Biol. Chem.* **252**, 1814–1818.

Zauderer, M., Liberto, P., and Baglioni, C. (1973). Distribution of histone mRNA among free and membrane-associated polyribosomes of a mouse myeloma cell line. *J. Mol. Biol.* **79**, 577–586.

Zehavi-Willner, T., and Lane, C. (1977). Subcellular compartmentation of albumin and globin made in oocytes under the direction of injected mRNA. *Cell* **11**, 683–693.

9

The Biogenesis of Supramolecular Structures

Richard G. W. Anderson

> *The molecular and the organismic are but two different vantage points from which to look at a living system, neither of them granting a monopoly to insight.* (Weiss, 1967)

I. INTRODUCTION

The supramolecular structures produced by cells are diverse; yet all are composed of single or multiple units of the four basic macromolecules: protein, lipid, carbohydrate, and nucleic acid. Some of these structures, such as oligomeric enzymes and microtubules, are primarily composed of

393

repeating units of a single type of macromolecule, whereas structures such as membranes are a mosaic of lipid, carbohydrate, and protein. The rules that govern the assembly of individual structures depend on the type of subunit macromolecule(s), the number of these macromolecules, and the geometry of the structure to be formed. Furthermore, once the supramolecular structure is formed, it may function as an individual unit (either inside or outside of the cell), or it may become the subunit of an intracellular organelle. In aggregate, these supramolecular structures constitute the form and function of cells and tissues.

A study of the assembly of a supramolecular structure must begin with a consideration of morphology. Besides developing an appreciation that protein–protein interactions lead to structures with quite a different morphology than those that involve lipid–lipid interactions, the morphology of the structure also is related to its function. Primarily, we have come to depend on microscopic techniques to assess the form of individual structures. Our understanding of how macromolecules interact within these structures can be refined through the use of various biophysical techniques; however, such electron microscopic techniques as negative staining, thin sectioning, or freeze-fracturing allow one to delineate the intermediates in the assembly of a structure as well as to appreciate its variation in form. Microscopy also tells us that some structures always have the same shape but that others are able to take on different shapes and configurations brought about by local environmental conditions within and around the cell. Finally, morphological studies give one an intuitive feeling that some supramolecular structures are organizationally more complex than others (e.g., T4 bacteriophage versus microtubules).

Form, shape, and geometry, all nicely revealed through the microscope, are not by themselves sufficient for understanding assembly. The determination of function is of equal importance, and microscopic techniques are of limited use for examining function. The competence to function is manifest in many ways, and only when present can it be said that the supramolecular structure is assembled.

With techniques in hand to determine form and function, many investigators have turned their efforts to the study of organizational hierarchies in cells and tissues. The assembly of viruses, ribosomes, chromosomes, and membranes, to mention a few, have been scrutinized over the past two decades. Considering the diversity in molecular composition, morphology, and function, it would be surprising if all these structures were generated by a common assembly process. Yet, as numerous reviews (Crane, 1950; Caspar and Klug, 1962; Kushner, 1969; Holtzman, 1974; Perham, 1975; Bouck and Brown, 1976) and textbooks (Lehninger, 1975; Dyson, 1978) indicate, many consider supramolecular morphogenesis to be due to a self-assembly process where all of the information for assembly is contained within the three-dimensional form of the subunit molecules. In this case, information refers to the expression of the cell genome.

However, various studies on the assembly of bacteriophages (see Kellenberger, 1972; Wood, 1973, 1974) have clearly established that these structures are not assembled by a simple, self-assembly process. For these structures, Wood (1974) has suggested three kinds of molecules, not found in the assembled structure, that might participate in assembly; (a) covalent bond-forming or -breaking enzymes, (b) proteins that promote the formation of hydrophobic interactions, and (c) template or measuring devices. If these viruses do not self-assemble, could it be that the synthesis of subunit molecules is not sufficient for the assembly of other supramolecular structures? Recently I addressed this question (Anderson, 1977) and found that in terms of the expression of genetic information, the assembly of most structures requires more than that contained within the subunit molecules. In this analysis, I concluded that it was useful to assign structures to one of four basic types of assembly reactions. It is these four types of assembly reactions that I wish to characterize in more detail in this review.

The assembly of a supramolecular structure is directly related to the expression of genetic information. This information, which initially is contained primarily within the nucleic acids of the cell, is present within cells in the form of the various macromolecules that are produced by standard synthetic pathways. These macromolecules can be divided into two categories: (a) structural macromolecules found within the supramolecular structure and (b) nonstructural macromolecules not incorporated into the structure but required for assembly. The relative participation of either structural or nonstructural macromolecules in assembly appears to depend on both the complexity of the structure to be assembled (e.g., how many different types of structural molecules) and the constraints imposed by the intracellular environment (e.g., the temporal arrangement of subunit synthesis). One way to organize the enormous amount of data on the assembly of supramolecular structures is to arrange the structures into categories based upon the role of structural and nonstructural molecules in assembly (see Fig. 1). According to this scheme structures can be assigned to four different types of assembly reactions: (1) structures that require only structural molecules for assembly; (2) structures that require nonstructural molecules to modify structural molecules either before or during the assembly process; (3) structures that require nonstructural molecules to facilitate, either directly or indirectly, the association of structural molecules during the assembly step; and (4) structures that require nonstructural molecules to modify the assembled structure to

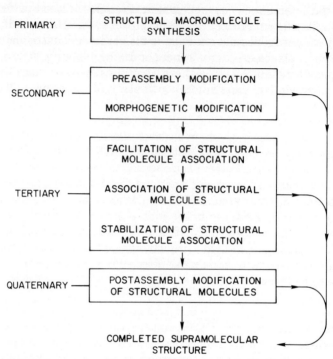

Fig. 1. A flow diagram that illustrates the pathway for supramolecular assembly. The designation of the assembly reaction as primary, secondary, tertiary, or quaternary depends on which of the various events (e.g., preassembly modifications, association of structured molecules, etc.) is required for morphogenesis. Except for a primary reaction, all of the subsequent levels of assembly require the activity of nonstructured molecules. See text for further discussion.

make it functionally competent within the cell. For the purposes of future discussion, these four types of assembly reactions are referred to as primary, secondary, tertiary, and quaternary, respectively. Each structure is assigned to the highest-order reaction even though lower-order reactions may take place during assembly.

When a structure assembles in a cellular or extracellular environment, the process fundamentally depends on the ability of the structural molecules to maintain a stable association. For structures that assemble in a primary reaction, such stable associations arise directly from subunit–subunit interactions. But for secondary and tertiary reactions, nonstructural molecules are required either to facilitate the stable interaction of structural molecules or to prevent association until the proper time in the assembly sequence. In the quaternary reaction, molecules specifically associate to form a morphologically definable structure, but for various reasons the interactions are not specific enough to create a functional structure. To understand better the proposed system for classifying the assembly of supramolecular structures, it is necessary first to consider the types of molecular interactions that occur during assembly. The emphasis will be on elucidating how stable associations between molecules are formed. Following this, the four types of assembly reactions will be illustrated.

II. MOLECULAR INTERACTIONS DURING ASSEMBLY

The form and function of any supramolecular structure depend on the intermolecular bonding patterns that hold the subunits together. Furthermore, the supramolecular structure must exist in a stable free energy state relative to the component macromolecules. Thus, the bonding interactions of macromolecules within a structure and the thermodynamic parameters governing equilibrium processes must be considered in analyzing the assembly of biological structures.

A. Thermodynamic Considerations

The assembly of any supramolecular structure requires that subunit molecules associate, usually by relatively weak interactions, to form the structure. Because the assembled structure is in a lower free energy state than the subunits, the reaction favors the assembly step. This free energy difference includes changes in both enthalpy and entropy during assembly. For many structures, principally those held together by ionic and hydrophobic interactions, the assembly reaction is favored because of an increase in entropy (Tanford, 1973, 1978; Lauffer, 1975). It may seem incongruous that order be created by an increase in disorder, but remembering that thermodynamics considers the components, the structure, and the environment as an integrated system, it is easy to appreciate that the formation of such structures as tobacco mosaic viruses, microtubules, and actin filaments are entropy driven structure-forming processes because ". . . although the formation of the structure involves a decrease in entropy, this decrease is overbalanced by an increase in entropy coming from the release of bound solvent molecules" (Lauffer, 1975). The thermodynamics of assembly emphasizes the dependence of these reactions on the microenvironment within the cell and, in particular, on the solvent properties of that environment.

Although all such reactions must obey the laws of thermodynamics and in one sense are driven by the free energy difference between reactants and products, often such reactions will not occur. This may be either because high-energy barriers must be overcome before assembly can proceed or because the initial association of subunits is so weak that the structure is not stable relative to the kinetic energy of motion. Activation energies can be minimized through enzymelike activity. On the other hand, the stabilization of weakly interacting subunits favoring assembly may involve the binding of the subunits to an intermediate molecule or group of molecules that promotes the association of subunits by concentrating them and allowing them to interact in a favorable environment. Such nucleation reactions (Sonneborn, 1970; Oosawa and Higashi, 1967) may result in the nucleating molecules being incorporated into the structure (homonucleation) or being discarded after assembly (heteronucleation). Therefore, even though a particular assembly reaction may be energetically favorable, catalytic or nucleating molecules may be required.

B. Molecular Interactions

The energy balance between the subunits, the environment, and the assembled product is only one aspect of supramolecular assembly. To conceptualize how molecules can become arranged into complex geometrical arrays, it is necessary to know how these molecules associate and how they remain associated in the final product.

It has long been recognized that for the most part, supramolecular structures are held together by weak bonds with energies in the range of 0.5 to 6 kcal/mole. These bonds fall into four categories: (1) electrostatic forces, (2) van der Waals' or London dispersion forces, (3) hydrogen bonds, and (4) hydrophobic interactions. The nature of these bonds has been discussed in several treatises (Davidson, 1967; Némethy, 1975; Watson, 1976). Each of these bonds may play a different role in an assembly reaction; on the other hand, for a given structure one particular type of bond may be dominant in holding it together (e.g., hydrophobic interactions predominate in tobacco mosaic virus and membrane structures, but electrostatic bonds are of primary importance in ribosomes).

Much of the research that has been done on supramolecular structures, such as viruses, membranes, ribosomes, etc., has been concerned with the bond interactions within the fully formed structure. Of equal importance, but more difficult to study, are those properties of the subunits that account for their ability to associate specifically during an assembly reaction. Which type of weak interactions have the required specificity to determine the initial recognition events? As pointed out by Némethy (1975) and Hopfinger (1977), dispersive and repulsive forces, as well as hydrophobic interactions, do not seem to have that specificity. On the other hand, whereas ionic and hydrogen bonds may have the required specificity, the rates of assembly of various structures seem to be too rapid to involve the point by point interaction that is required for the formation of a hydrogen or ionic bond. Hopfinger (1977) has suggested that for many structures, hydrophobic interactions determine the initial aggregation events, followed by the formation of specific ionic and hydrogen bonds. The key event in subunit stabilization, then, would be the secondary formation of electrostatic interactions.

But ionic and hydrogen bond formation may be deterministic for only certain types of structures (e.g., those involving protein–protein interactions or protein–nucleic acid interactions). Structures such as membranes, which depend on lipid–lipid interactions or lipid–protein interactions, may rely on other mechanisms to determine specificity during assembly. Therefore, it is necessary to consider the type of molecules that are interacting during the assembly of a particular structure to ascertain how specificity in the association of subunits is achieved.

1. Protein–Protein Interactions

Fundamentally, protein–protein recognition events during assembly depend on the subunits having the proper shape or conformation. The mechanism by which linear strands of amino acids fold to form specific and unique three-dimensional structures has been reviewed by several authors (Wetlaufer and Ristow, 1973; Anfinsen, 1973; Scheraga, 1974; Anfinsen and Scheraga, 1975). Although only a relatively few proteins have been studied, some general concepts about the mechanism of protein folding have emerged. A folded protein is in a lower energy state relative to the unfolded species; however, it may not be in the lowest possible energy state. Rather, there are kinetically favorable pathways for the folding of the protein, and the pathway of folding rather than the absolute difference in energy states is deterministic. The initial events in the folding process involve short-range interactions (predominantly hydrogen bond formation) between amino acids within the backbone of the molecule. This results in the formation of α-helical or β-sheet regions that act as a nucleus for the subsequent folding of the molecule. Although these nucleation steps may largely determine the pathway of folding, the long-range interactions between side chains (e.g., hydrophobic interactions) contribute to the overall stability of the folded state of the polypeptide.

Even though the folding process is generally considered to be a spontaneous event that depends on interactions within the molecule, the fact that the formation of the proper structure depends on the selection of the proper pathway of folding suggests that environmental molecules interacting with the folding chain may influence the final structure of the protein. For example, some types of membrane proteins translated into an aqueous environment would fold differently than if translated directly into the lipid environment of the membrane (see Section III,D). Similarly, proteins that fold as proproteins, with amino acid sequences not found in the biologically active protein, apparently require these amino acid sequences solely for proper folding to occur.

Following the proper folding of a protein, in some cases multiple copies of the molecule will associate to form the quaternary structure of a functional protein (reviewed by Klotz *et al.*, 1970). Proteins with quaternary structures fall into two general classes: (1) those, such as glutamate dehydrogenase (Josephs *et al.*, 1972), composed of only one type of subunit and (2) those, represented by aspartate transcarbamylase (Schachman, 1972; Bothwell and Schachman, 1974) or respiratory proteins (Antonini and Chiancone, 1977; McLachlan *et al.*, 1972), made up of multiple copies of two or more different polypeptide subunits. Regardless of the type of protein, the formation of quaternary structure involves stable protein–protein associations similar to those required for the assembly of more complex, protein-containing, supramolecular structures.

Several kinds of data suggest the following model for protein–protein interaction during assembly of quaternary protein structure. Because the subunit polypeptides are in an aqueous environment, the ionic components of the molecule will strongly interact with water as well as various counterions. There are three classes of bound water (Fig. 2): (a) type I water, which is referred to as bulk water, is freely exchangeable with the environment; (b) type II water is bound more tightly but can still exchange; and (c) type III water, which is very tightly bound (Richards, 1977). Since water is always associated with proteins, the first interactions that occur between two closely opposed molecules involve the displacement of water. Thus, when molecules that have some type of conformational complimentarity collide, the displaced water leads to entropy changes that stabilize the complex. However, these types of interactions probably occur between a variety of polypeptide molecules within a cell, but do not lead to the formation of quaternary structures. If it is assumed that the initial hydrophobic interactions stabilize the complex long enough for more specific hydrogen and ionic bonds to form (Hopfinger, 1977), then it is possible to understand how specificity in association can be achieved. In fact, X-ray diffraction studies of assembled proteins (Table I) indicate that hydrogen bonds and ion pairs are key bonding patterns between subunits (Liljas and Rossmann, 1974; Matthews and Bernhard, 1973). Furthermore, studies on substrate binding to chymotrypsin (Segal *et al.*, 1971), the binding of proteins to subtilisin BPN (Robertus *et al.*,

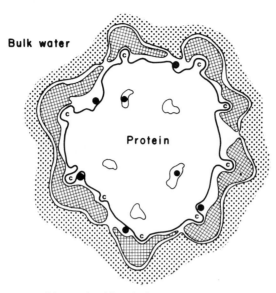

c   Charged side chains
●   Rigidly bound molecules
▦   Bound water
⸬   Transition water

Fig. 2. Protein–water relationships, as illustrated in a cross-sectional, schematic drawing. (From Hopfinger, 1977, "Intermolecular Interactions and Biomolecular Organization." Reprinted with permission of John Wiley & Sons, Inc.)

1972), and trypsin–trypsin inhibitor complexes (Janin and Chothia, 1976; Chothia and Janin, 1975; Wodak and Janin, 1978; Rühlmann *et al.,* 1973) indicate that these interactions involve the temporary formations of short anti-parallel β structures, implying that hydrogen bonds provide the specificity and stability for subunit association. However, maximal subunit interactions, as reflected by the symmetry of association of subunits and the van der Waals' interactions, are important contributions to stabilizing the quaternary structure of the protein.

2. Protein–Nucleic Acid Interactions

These are an important class of macromolecular interactions involved in the assembly of such supramolecular structures as chromosomes, ribosomes, and viruses. How do these two classes of molecules interact with specificity during assembly and how are these interactions stabilized within the assembled structure? Chromatin and ribosomes illustrate how they interact.

Chromatin consists of a linear DNA molecule with regions coated with

TABLE I

Types of Interpolypeptide Bonds Contributing to
Organization of Representative Proteins[a,b]

Protein	van der Waal's contacts	H bonds	Ion pairs
α-Chymotrypsin	500	15	1
Concanavalin A	316	30	6
Hemoglobin			
Oxy	190	6	
Deoxy	167	6	4
Insulin	210	10	1

[a] From Liljas and Rossmann, 1974, reproduced with permission.

[b] These data are based on X-ray diffraction studies of intact protein. It is proposed that the hydrogen bonds and ion pairs are crucial for proper polypeptide interaction, but that van der Waal's contacts contribute to stabilizing the proper three-dimensional structure.

histone clusters alternating with regions relatively free of histones (Hopfinger, 1977; Pardon and Richards, 1973; Kornberg, 1974). Current models suggest that histones [which can be divided into three classes: lysine-rich histones, moderately lysine-rich histones, and arginine-rich histones (reviewed in Pardon and Richards, 1973)] bind to DNA chains in an α-helical conformation by ionic interactions between the basic amino acids of the histone and the acidic phosphate groups of the DNA. Within the histone clusters, interhistone association via β-pleated sheets may also take place. Interhistone associations may induce secondary structural changes in the chromatin (see Fig. 3).

Studies on the structure of chromatin suggest that ionic interactions between DNA and histone, as well as hydrogen bonding between adjacent histones, must be important aspects of histone–DNA interaction during assembly. What is unclear is why only certain segments of DNA bind histones. There is some evidence (Clark and Felsenfeld, 1972; Combard and Vendrely, 1970; Sponar and Sormová, 1972) that histones may interact with specific nucleotide sequences in the DNA. However, Pardon and Richards (1973) suggest that base specificity is not involved. Recently it has been proposed that the formation of hydrogen bonds (at least two) may be an essential aspect of the histone–DNA interaction event (Seeman et al., 1976).

Recently, Laskey et al. (1977) have described a physiological system for

A - binding

B - binding

DNA₁

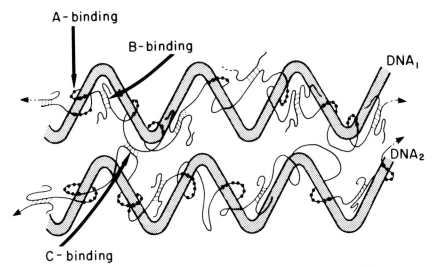

DNA₂

C - binding

Fig. 3. A model of histone–DNA and histone–histone interaction in condensed regions of chromatin. A-binding regions are where histone interacts with DNA. B- and C-binding regions represent interhistone binding either within one segment of DNA or between different DNA segments, respectively. (From Hopfinger, 1977, from "Intermolecular Interactions and Biomolecular Organization." Reprinted by permission of John Wiley & Sons, Inc.)

studying chromatin assembly. This system utilizes the supernatant fraction from *Xenopus laevis* eggs to assemble chromatin from purified SV40 virus DNA. They conclude that in addition to DNA, histones, and nicking-closing enzyme, an undefined thermolabile component is required for assembly at physiologic ionic strength. This kind of study suggests that chromatin assembly may in part require nonstructural molecules.

As with chromatin, a major aspect of ribosome assembly is the specific interaction of ribosomal RNA's with sets of ribosomal proteins (see Fig. 4 for the structure of a ribosome). As recently reviewed by Ebel *et al.* (1977), both hydrogen bonds and ionic interactions must be of primary importance in this process. Yet it is still unclear whether the ribosomal proteins interact only with certain nucleotide sequences. The initial association between RNA and proteins may be relatively nonspecific, and specificity may be achieved through secondary interactions between the different types of ribosomal proteins.

For both protein–protein interactions and protein–nucleic acid interactions, specificity is achieved through hydrogen bond formation and ionic interactions. In addition, protein–protein recognition is heavily dependent on the tertiary structure of the subunits. Properly folded sequences of amino acids create complimentary conformations that allow the fitting

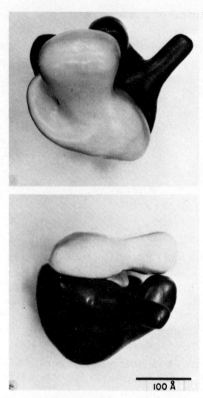

Fig. 4. Two views of a model of the 70 S *E. coli* ribosome. Unlike many supramolecular structures, the ribosome does not have geometric symmetry. The diverse functional activity of a ribosome may not be compatible with a symmetrically organized structure. (From Lake, 1976, reprinted with permission.)

together of the subunits. One would predict that similar complimentarity must exist between proteins and nucleic acids for recognition to take place; yet, little is known about what aspects of nucleic acid structure contribute to complimentarity. One possibility is that the secondary structure depends on the sequence of nucleotides within the nucleic acid.

3. Lipid–Lipid Interactions

Anyone familiar with cell functions has to be impressed by the ubiquitous presence of membrane containing structures. Membranes, which are specialized arrangements of lipids and proteins, are unique in their ability to compartmentalize regions of a cell and to direct communications vectorially between compartments. Membranes are among the most common supramolecular structures, and the mechanism of their assembly deserves attention. A great deal of research has been devoted to studying the assembly of these structures both *in vitro* and *in vivo*.

The ability of lipids to organize into characteristic membrane bilayers depends on their amphipathic properties, that is, on the presence of a hydrophobic hydrocarbon tail and a hydrophilic ionic head. All membranes are composed of both glycerolipids and sphingolipids. In addition, eukaryotic membranes also contain sterols. Membrane bilayers are easily assembled *in vitro* from either pure species of glycerolipids or from a mixture of several types of glycerolipids (reviewed in Chapman and Cornell, 1977). As pointed out by Tanford (1973, 1978), the presence of two fatty acid chains on the hydrophobic portion of the glycerolipids is responsible for the lipid bilayer organization.

The results of *in vitro* experiments show that the formation of bilayers is dependent on the concentration of lipid in the aqueous medium. Once the lipid concentration reaches a critical concentration, characteristic for that species of lipid [e.g., the critical micelle concentration for phospholipids, the major type of glycerolipids in cell membranes, is $\sim 10^{-9}\ M$ (Tanford, 1973, 1978)], the hydrophobic hydrocarbon chains associate so as to allow maximum interaction of the ionic head portion with the aqueous environment. Thus, the assembly of membrane lipid bilayers is primarily due to hydrophobic interactions.

The relatively weak hydrophobic interactions are uniquely suited for maintaining membrane structure because they allow for a great deal of flexibility and deformability (Tanford, 1973, 1978). On the other hand, as discussed before, hydrophobic interactions do not seem to have the specificity to account for the unique lipid compositions found in cell membranes. The lipid composition within various compartments of the cell seems to be specific for that compartment (Korn, 1966). In addition, the distribution of lipids is asymmetric across the bilayer so that the composition of lipids on each side of the bilayers of a given membrane is different (Bergelson and Barsukov, 1977; Cullis *et al.*, 1977; Bretscher, 1974). An important question is how does a membrane acquire a unique lipid composition during assembly?

It is known from *in vitro* studies that when diacylphospholipids assemble, they form either unilamellar or multilamellar vesicles and Israelachvili (1977) and Israelachvili *et al.* (1977) have proposed that the distribution of lipid in the membrane is a function of geometric packing. The hydrocarbon chain length, its relative saturation or unsaturation, and the size of the head group of the phospholipid can influence how the lipids are arranged in the membrane. Thus, when mixtures of lipids are assembled into lamellar vesicles, one finds that they distribute asymmetrically across the bilayer. It is conceivable that the arrangement of lipids in biological membranes, in particular the asymmetric arrangement, is due in part to geometric packing.

Even though geometric packing may contribute to the initial distribution of lipids, other mechanisms must be involved in determining the characteristic lipid organization found in biological membranes. In part, this requires the activity of nonstructural molecules and will be discussed further in the section on quaternary assembly reactions, where membrane assembly is considered in more detail.

4. Lipid–Protein Interactions

Since biological membranes also contain proteins [up to 75% in mitochondrial inner membranes (reviewed in Hackenbrock, 1977)], lipid–protein interactions are important in assembly. The important lipid–protein interactions involve the intrinsic membrane proteins (reviewed in Singer, 1971, 1974, 1977; Singer and Nicolson, 1972). These are amphipathic molecules embedded in the lipid bilayer and stabilized in this configuration by hydrophobic interactions. Just as with lipids, membrane proteins are asymmetrically distributed across the membrane, and each type of membrane in a cell has a unique protein composition (Rothman and Lenard, 1977; Steck, 1974).

Under the proper conditions, isolated intrinsic membrane proteins can be intercalated into lipid bilayers *in vitro* (Eytan *et al.*, 1977). The incorporation of proteins into membranes depends on hydrophobic interactions, and the lipid–protein interactions are much the same as lipid–lipid interactions (reviewed by Gennis and Jonas, 1977; Segrest and Jackson, 1974; Lenaz, 1977). But, as with lipid–lipid interactions, the specificity in protein composition and distribution must be determined by factors other than hydrophobic interactions.

The situation for lipid–lipid interactions and lipid–protein interactions is quite different than that for protein–protein or protein–nucleic acid interactions. The bilayer morphology of membranes is relatively invariant from compartment to compartment, even though the types of lipids and proteins contained in the different membranes is not the same. Hydrophobic interactions adequately explain how the characteristic membrane morphology can spontaneously form. In fact, such reactions are so spontaneous that the problem for the cell is how to achieve the specificity in molecular composition and distribution that is required for the proper function of this supramolecular structure. The conceptual problem is not in understanding the generation of form, as it is with the protein- or nucleic acid-containing structure, but rather the generation of function.

III. THE ASSEMBLY REACTIONS

The assembly of a supramolecular structure necessarily is a multifactorial event. The first event must be the synthesis of the subunit macromolecules. This is followed by the coming together of the subunit molecules and the recognition events that lead to a morphologically and functionally unique structure. Finally, once formed, the structure may need to be modified either to stabilize the subunit interactions or to produce a functionally competent structure. Considered below are various aspects of the intracellular strategy for achieving these objectives during supramolecular assembly.

A. Primary Assembly Reaction

This assembly reaction proceeds spontaneously once the proper number of structural molecules are synthesized through standard pathways. The structural macromolecules, then, contain all the genetic information required for assembly. A primary assembly reaction is synonymous with and corresponds to a self-assembly reaction (Caspar and Klug, 1962; Kushner, 1969; Perham, 1975; Holtzman, 1974; Bouck and Brown, 1976). The assembly of rod-shaped viruses (such as tobacco mosaic virus) and microtubules are suitable examples of structures that form by a primary assembly reaction.

1. Tobacco Mosaic Virus

In 1955, Fraenkel-Conrat and Williams demonstrated that tobacco mosaic virus (TMV) could be dissociated into its subunits and then reassembled to form a functional virus, an achievement that fostered a great deal of investigation into the way proteins and nucleic acids interact to form a supramolecular structure (see reviews by Butler and Durham, 1977; Butler, 1976; Lauffer, 1975). Because of the impact of this system on current thinking about supramolecular assembly, it is important to relate TMV assembly to the four types of assembly reactions that lead to the biogenesis of cellular structures.

The assembly of TMV is an entropy-driven reaction (Lauffer, 1975) that involves the association of a single species of protein (A protein, 17,500 dalton) with a single species of single-stranded RNA (2.1×10^6 dalton). In the fully formed structure, the protein is helically arranged around the RNA molecule such that each protein subunit interacts with three nucleotides of the RNA. The virus has a discrete length of 300 nm, an inner diameter of 4 nm (reviewed by Caspar, 1963), and the RNA molecule determines the length of the virion.

According to Butler and Durham (1977), for assembly to take place, multiple A protein subunits first arrange into a disk of 34 subunits that then associates with the RNA molecule to initiate assembly. This type of nucleation (homonucleation) seems to be a common aspect of rod-shaped virus assembly and has been demonstrated in both tobacco rattle virus (Morris and Semancik, 1973) and barley stripe mosaic virus (Atabekov *et al.*, 1968). Once nucleation has occurred, the elongation of the virus proceeds by the addition of protein subunits. Butler and Klug (1971) proposed that elongation proceeds by the stepwise addition of the performed disks to the growing rods; however, Ohno and Okada (Okada and Ohno, 1972; Okada, 1975; Okada *et al.*, 1975; Ohno *et al.*, 1977a) have suggested that elongation of the rods proceeds by the addition of individual A protein subunits. Both laboratories agree that the nucleation step requires the preformed disks (Ohno *et al.*, 1977a), and, as discussed by Butler and Durham (1977), maybe both mechanisms are operating within the cell, since it is difficult to conceive of the 34 subunit disk being transformed into a helical arrangement of A protein subunits without some breakage and rejoining of the subunits. Studies by Ohno *et al.* (1975) indicate that cucumber green mottle mosaic virus, another rod-shaped virus, cannot elongate by the addition of disk aggregates.

Assuming TMV is formed by a primary assembly reaction [it may be that A protein must undergo posttranslational modification before it is competent for assembly (Roberts *et al.*, 1974)], how do the preformed disks of A protein specifically interact with the viral RNA? Several approaches have been used to assess the specificity of this interaction *in vivo* and *in vitro*. Atabekov *et al.*, (1975) reported that when plants were doubly infected by two different strains of TMV, there was little or no mixed particle production within the cells; however, some mixed particle formation was detected when mixed viruses were reassembled *in vitro*. Goodman and Ross (1974) also found that doubly infected cells did not produce mixed particles, but Sarkar (1969) found that mixed virions could be produced in doubly infected cells as did Otsuki and Takebe (1978) using isolated tobacco mesophyll protoplasts doubly infected with a common strain, OM and tomato strain, T, of TMV. More recent work from Atabekov's laboratory (Taliansky *et al.*, 1977) suggests that the formation of mixed particles *in vivo* is dependent on the type of TMV strains used to doubly infect the cells. In other words, some strains cross-react, but others do not. Presumably, these differences reflect the specificity of interaction between the different strains of A protein and their homologous

RNA. Since host RNA is rarely incorporated into forming virions *in vivo*, the recognition process must be fairly specific.

Recent work has been directed at determining the site at which disk protein interacts with RNA during nucleation (Ohno *et al.*, 1977b; Zimmern and Wilson, 1976; Zimmern, 1976). Perham and Wilson (1978) have shown that the portion of the RNA involved in nucleation corresponds to a region where strong protein–nucleotide interaction takes place. By mapping RNA from virus fragments, Zimmern and Wilson (1976) located the nucleotide sequence at which initial protein–nucleic acid interaction takes place. This initiation site is 1000 nucleotides from the 3'-end of the RNA and has an average length of approximately 250 nucleotides. Zimmern (1976) was not able to find a highly repetitive iteration of any one trinucleotide in this region, but it may contain more complicated repeats; therefore, the exact mechanism for recognition within the RNA molecule is still not clear. Even less is known about A protein properties necessary for assembly; however, Hubert *et al.* (1976) have found that the aspartic acid residue at position 88 in the protein is crucial for TMV assembly. Future X-ray diffraction and nucleotide sequencing studies can be expected to reveal exactly how the nucleotides and coat protein interact during nucleation.

Regardless of the basic mechanism of recognition, recent evidence (Butler *et al.*, 1977; Lebeurier *et al.*, 1977; Ohno *et al.*, 1977b) suggests that once nucleation occurs, assembly proceeds by the bidirectional addition of coat protein. Based on electron microscopy studies, Butler *et al.* (1977) have shown that during assembly " . . . the uncoated RNA is folded back along the growing rod, probably down the central hole" (diagrammed in Fig. 5).

2. Microtubules

The ubiquitous 24-nm diameter microtubules found within the cytoplasm of all eukaryotic cells, as well as in centrioles, flagella, and cilia, to a first approximation are formed by a primary assembly reaction. As first demonstrated by Weisenberg (1972) and Borisy and Olmsted (1972), the principal structural protein of this supramolecular structure, tubulin, can be extracted from brain and assembled *in vitro* into microtubules at 37°C in the presence of GTP, Mg^{2+}, and low levels of Ca^{2+}. Tubulin from nonneural cells also assemble *in vitro* (Doenges *et al.*, 1977). The assembly process is unidirectional (Olmsted *et al.*, 1974; Margolis and Wilson, 1978) in that the addition of subunits preferentially occurs at one end of the growing tubule.

The initial studies suggested that only tubulin is required for assembly;

Growth

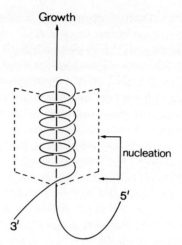

nucleation

5'

3'

Fig. 5. A diagram of the possible arrangement of the TMV RNA in a forming virus. The direction of assembly (from 3' to 5') is upward at the end opposite the two RNA tails. The initiation region on the RNA is also indicated. (Kindly provided by P. J. G. Butler, see Butler *et al.*, 1977.)

however, two microtubule-associated proteins (MAP) have now been implicated in the assembly reaction. It is possible to assemble and disassemble microtubules cyclically *in vitro*. Regardless of how many times this is done, two nontubulin proteins are always present in the assembled tubule. In 1975 Weingarten *et al.* demonstrated that one of these proteins, called tau (60,000–80,000 dalton), was required for the proper assembly *in vitro*. Murphy and Borisy (1975) and Keates and Hall (1975) reported that the high molecular weight MAP, called HMW (MW 270,000–280,000), also could stimulate microtubule assembly. Since it is possible to assemble chromatographically pure, MAP-free tubulin into microtubules (Burton and Himes, 1978; Himes *et al.*, 1976, 1977; Lee and Timasheff, 1975), the morphology of the microtubule is derived from the specific associative properties of tubulin even though, under some conditions, tubulin assembles into sheets rather than tubules (reviewed in Baker and Amos, 1978). However, the conditions for these reactions are nonphysiological, and most likely *in vivo* tau and/or HMW are required for assembly.

The involvement of accessory proteins, which are structural in the sense that they are part of the fully formed microtubule, suggests that like rod-shaped viruses, the assembly event involves a copolymerization process. Despite the extensive work establishing a role for both HMW (Marcum and Borisy, 1978a,b; Murphy *et al.*, 1977a,b; Vallee and Borisy, 1978; Borisy *et al.*, 1975; Johnson and Borisy, 1977; Scheele and Borisy, 1978) and tau (Cleveland *et al.*, 1977a,b) in the assembly of microtubules, *in vitro*

(Burns, 1978), it is not clear, whether both MAP's are necessary for the assembly reaction *in vivo*. Whereas there is good evidence that both HMW (Sheterline, 1978; Amos, 1977; Murphy and Borisy, 1975; Dentler *et al.*, 1975; Sherline and Schiavone, 1977; Klein *et al.*, 1978) and tau (Connolly *et al.*, 1977; Lockwood, 1978) are associated with fully formed microtubules *in vivo*, recent work by Fellous *et al.* (1977) and Francon *et al.* (1978) suggests that tau is the essential protein for promoting initiation and elongation *in vivo*. Furthermore, Schmitt *et al.* (1977) suggest that the HMW is not required for *in vivo* assembly. Apparently some major discrepancies between *in vivo* and *in vitro* studies need to be clarified to understand how the MAP's function within the cell.

The preceding discussion suggests that microtubule formation may be a primary assembly reaction. However, these studies do not consider the relationship of assembly to the organization of the cell. Microtubule assembly within cells is a regulated event, and this regulation may involve the activity of nonstructural molecules, an idea supported by three lines of evidence. First, several laboratories have studied the translation of tubulin mRNA in a cell-free system (Cleveland *et al.*, 1978; Bryan *et al.*, 1978; Gilmore-Hebert and Heywood, 1976; Saborio *et al.*, 1978) in order to determine whether freshly translated tubulin is competent for assembly into microtubules. Interestingly, although tubulin is composed of α and β subunits, newly synthesized α subunits can be incorporated into microtubules much more efficiently than β subunits (Cleveland *et al.*, 1978). Perhaps posttranslational phosphorylation of the β subunit is required to make it competent for assembly (Eipper, 1972). In some cells a subpopulation of tubulin is modified by the addition of a tyrosine residue onto the C-terminal end of the polypeptide (Rodriquez and Borisy, 1978; Raybin and Flavin, 1975; Thompson, 1977), a process that can be correlated with stages of brain development. These and related studies suggest that nonstructural molecules may modify tubulin subunits as a means of regulating microtubule assembly. Second, it has long been recognized that microtubules do not polymerize at random within cells, but instead polymerize in association with microtubule organizing centers (Pickett-Heaps, 1969). Microtubule organizers occur in a variety of morphological forms, including centrioles, kinetochores, and amorphous bodies (reviewed by Bouck and Brown, 1976). Exactly how the microtubule organizing centers function is not clear, even though the activity of such centers has been demonstrated both *in vitro* (Snell *et al.*, 1974; Snyder and McIntosh, 1975; Telzer *et al.*, 1975; Weisenberg and Rosenfeld, 1975) and *in vivo* (Heidemann and Kirschner, 1975; Heidemann *et al.*, 1977). Stearns and Brown (1977) were able to extract four types of proteins from isolated basal bodies, some of which may have microtubule initiating activity.

Therefore, it is possible that these organizing centers contain molecules that initiate assembly but are not incorporated into the microtubule (heteronucleation). Third, recent studies (Marcum *et al.*, 1978) have related the activity of calcium-dependent regulatory protein to the control of microtubule polymerization within cells. Calcium at certain concentrations inhibits microtubule polymerization (Nishida and Sakai, 1977) and will depolymerize microtubules *in vitro* (Weisenberg, 1972; Borisy and Olmsted, 1972). Thus, it is possible that calcium-dependent regulatory protein, which immunofluorescent studies (Welsh *et al.*, 1978) indicate is not a structural component of microtubules, can act in a specific way to sequester or release calcium and thereby control microtubule formation.

In conclusion, both microtubules and rod-shaped viruses [as well as spherical plant viruses (for reviews, see Anderson, 1977; Adolph and Butler, 1976)] are the best examples of structures formed by primary assembly reactions. This classification may prove to be incorrect if in future studies it can be unequivocally established that nonstructural molecules are essential for assembly *in vivo*.

B. Secondary Assembly Reactions

The secondary assembly reaction is the first of the four reactions in which there is a clear requirement for nonstructural molecules during the assembly step. In this reaction nonstructural molecules function to modify the structural molecules prior to or during the assembly step (see Fig. 1). The type of modifications mediated by these nonstructural molecules include: (a) the addition of functional groups to the precursor molecule (e.g., phosphorylation, acetylation, methylation or the addition of carbohydrate chains) and (b) the removal of portions of the subunit molecules (e.g., the removal of nucleotides from RNA during ribosome assembly or the cleavage of proteins during virus assembly).

As indicated in Fig. 1, the secondary assembly reaction can involve modifications of the subunit structural molecules either before or during the assembly event. It might be argued that as defined, the secondary assembly reaction is a trivial classification since posttranslational modifications seem to be widespread within cells (Uy and Wold, 1977; Gallop and Paz, 1975). The distinction is that usually these nonstructural molecules are specific gene products that only operate to modify structural proteins during assembly. These modifications are required for proper assembly to take place.

1. Virus Assembly

Posttranslational cleavage of polypeptides is an important modification in many viral assembly sequences (Hershko and Fry, 1975). Consider, for

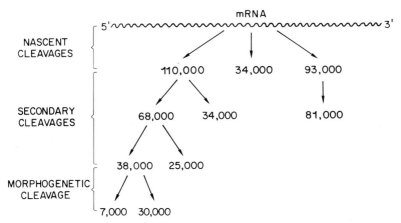

Fig. 6. Sites of posttranslational cleavage of precursor proteins during the assembly of poliovirus. For details, see text. (Drawn from Hershko and Fry, 1975, reproduced with permission.)

example, the assembly of the poliovirus. The assembly of this double-stranded RNA-containing icosahedral virus involves three different types of cleavages (diagrammed in Fig. 6). (1) The total genome of these viruses is translated as a single unit to yield a precursor polypeptide or polyprotein from which the viral precursor proteins are derived by a cleavage reaction called nascent cleavage. (2) Following the release of three proteins (110,000, 34,000, and 93,000 daltons) by the nascent cleavage reaction, a set of secondary cleavages take place that converts the 110,000 dalton protein to two proteins of 68,000 and 34,000 daltons and the 93,000 dalton protein to an 81,000 dalton protein. The 68,000 dalton protein released in the secondary cleavage is further cleaved to form a 38,000 dalton and a 25,000 dalton protein. (3) During assembly itself, the 38,000 dalton protein is cleaved to form two proteins of 7000 and 30,000 daltons. Since this reaction takes place during assembly and is required for assembly, Hershko and Fry (1975) have called it a morphogenetic cleavage. These cleavage reactions are both specific and necessary for assembly. Notice, however, that whereas most of the cleavages are designed to make the subunits competent for assembly, it is not clear how morphogenetic cleavages function during assembly.

2. Ribosome Assembly

The bacterial 30 S ribosomal subunit is another example of a supramolecular structure formed by a secondary assembly reaction. This subunit is composed of about 60% RNA and 40% protein. It contains one single-stranded 16 S RNA (0.5×10^6 dalton) and 21 proteins (average molecular weight 16,000–22,000). This structure can readily be disassembled and then reassembled *in vitro* in the following sequence (Nomura, 1973).

16 S RNA $\xrightarrow[\substack{+7 \text{ proteins} \\ 0°C}]{}$ RI particles $\xrightarrow[37°C]{}$ RI* particles $\xrightarrow[\substack{+14 \text{ proteins} \\ 0°C}]{}$ 30 S ribosome

The conversion of RI (reconstitution intermediate) particles to RI* particles is rate limiting and has an activation energy of 27 kcal/mole. Despite the successful *in vitro* reconstitution of 30 S ribosomes utilizing only structural molecules, a variety of genetic and biochemical studies have established that *in vivo* morphogenesis involves a secondary assembly reaction.

Work from several laboratories (Hecht and Woese, 1968; Adensik and Levinthal, 1969; Dahlberg and Peacock, 1971; Lowry and Dahlberg, 1971) suggests that *in vivo* a precursor 16 S RNA (P16 S RNA) 150–200 nucleotides longer than the mature 16 S RNA is involved in the initial formation of the 30 S ribosomal subunits (RNA maturation is reviewed in Hadjiolov and Nikolaev, 1976). Nine proteins, most of which correspond with the initial seven proteins that react *in vitro* (Bollen *et al.*, 1970; Homann and Nierhaus, 1971) associate with the P16 S RNA *in vivo* to form a P30 S (21 S) particle (Nierhaus *et al.*, 1973). Subsequent to the primary association of proteins and precursor RNA, short segments of nucleotides are removed from each end of the precursor RNA (Hecht and Woese, 1968; Adesnik and Levinthal, 1969; Dahlberg and Peacock, 1971; Lowry and Dahlberg, 1971), the remaining RNA is methylated (Thammana and Held, 1974; Nierhaus *et al.*, 1973), and twelve more proteins are added, converting the P30 S particle to the 30 S subunit. In this scheme, at least two groups of nonstructural macromolecules (ribonucleases and methylases) are involved in the modification of the RNA. Furthermore, Nierhaus *et al.* (1973) have shown that the P30 S particle can be converted to the 30 S subunit *in vitro* if ATP, *S*-adenosylmethionine, and S100 enzymes are in the incubation medium, which further supports a role for nonstructural molecules in the *in vivo* assembly reaction.

The 50 S bacterial ribosome subunit also has been assembled *in vitro* according to the reactions shown in Table II (Dohme and Nierhaus, 1976; Sieber and Nierhaus, 1978). Nevertheless, RNA processing, as well as posttranslational modification of ribosomal proteins, occurs during cellular assembly (Chang and Chang, 1974, 1975; Colson and Smith, 1977). Of more interest, however, is the recent evidence that the assembly of 50 S subunits requires nonstructural molecules that function in subunit maturation.

The first suggestion that nonstructural proteins are required for the maturation of 50 S subunits came from the genetic studies of Sypherd *et al.* (1974) and Bryant and Sypherd (1974). Guha *et al.* (1975) found

TABLE II

Scheme of *in Vitro* Assembly of the Large Ribosomal Subunit[a,b]

Incubation	Reaction	S value
First	$(5 S + 23 S)RNA + proteins \xrightarrow{0°C, 4 mM Mg^{2+}} RI_{50}(1)$	33
	$RI_{50}(1) \xrightarrow{44°C, 4 mM Mg^{2+}} RI_{50}*(1)$	41
Second	$RI_{50}*(1) + proteins \xrightarrow[50°C, 20 mM Mg^{2+}]{44°C, 4 mM Mg^{2+} or} RI_{50}(2)$	48
	$RI_{50}(2) \xrightarrow{50°C, 20 mM Mg^{2+}} 50 S \text{ ribosomal subunit}$	50

[a] From Sieber and Nierhaus, 1978, reproduced with permission from *Biochemistry* **17**, 3505–3511. Copyright 1978 American Chemical Society.

[b] See Dohme and Nierhaus (1976). The conversion $RI_{50}*(1) \to RI_{50}(2)$ can occur in both the first and the second incubations.

nonstructural proteins were associated with precursor ribosomal subunit particles during assembly in *Bacillus subtilis*. Such proteins may act catalytically to promote two steps in the assembly reaction (see Table II) known to have a high energy of activation: (a) the conversion of $RI_{50}(1)$ to $RI_{50}*(1)$ (activation energy 70 kcal/mole) and (b) the conversion of $RI_{50}(2)$ to 50 S (activation energy 54 kcal/mole) (Sieber and Nierhaus, 1978). Interestingly, the high activation energy step in 30 S ribosome assembly may require the activity of a maturation factor in *in vivo* (Mangiarotti *et al.*, 1975; Silengo *et al.*, 1977).

The bulk of what is known about bacterial ribosome assembly clearly places this reaction in the secondary category. Yet, the evidence on 50 S subunit assembly indicates that, as with microtubule assembly, future studies may establish that ribosome formation actually requires a higher-order assembly reaction.

C. Tertiary Assembly Reactions

The structures resulting from this class of assembly reactions require nonstructural molecules for the fitting together of structural molecules during morphogenesis. As shown in Fig. 1, these nonstructural molecules may be involved at three points in the assembly reaction: (a) they may be required to facilitate the association of subunit molecules during structure formation; (b) they may be required to join together the structural molecules or groups of structural molecules; (c) once the subunits or structural molecules have associated, in some cases they may be required to stabilize these associations to maintain the functional morphology of the structure. Most structures within this assembly class also require nonstructural molecules to modify structural molecules during assembly.

The most effective way to determine whether structures belong to this class is to utilize genetic techniques to identify and characterize the activity of the nonstructural molecules. Since the development by Edgar *et al.* (1964) and Edgar and Wood (1966) of *in vitro* complementation techniques to study conditional lethal mutations in viral assembly, a large number of studies (reviewed by Casjens and King, 1975; Cummings and Bolin, 1976; Hohn and Katsura, 1977; Hershko and Fry, 1975) have demonstrated the importance of tertiary assembly reactions in viral morphogenesis.

1. Facilitation of Structural Molecule Association

The biogenesis of T4 bacteriophage has been one of the most thoroughly studied assembly reactions (reviewed in Casjens and King, 1975; Kellenberger, 1976; Murialdo and Becker, 1978). T4 is a complex, DNA-containing virus that consists of an elongated icosahedral head and a tail segment to which is attached a base plate with tail fibers (see Fig. 7). At least eighteen gene products are required for head assembly, but only eleven have been detected in the purified head (Casjens and King, 1975). The most abundant structural protein in the head (P23) is the product of gene *23*. The assembly of the head occurs in several steps that include the formation of a prehead, the processing of prehead proteins, the removal of the core, the packaging of the DNA, and finally the maturation of the head. It has been generally assumed that, once synthesized, structural molecules are present in the cytoplasm in an uncomplexed state where they associate to form the supramolecular structure. However, for the assembly of the head of T4 bacteriophage this is not the case; the head structural molecules must first be bound to a nonstructural molecule.

A major nonstructural protein (P31) involved in T4 head assembly is

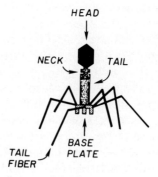

Fig. 7. A diagram of the structure of T4 bacteriophage.

produced by gene *31*. A mutation in this gene results in the aggregation of the P23 into amorphous lumps that associate with the host membrane (reviewed in Laemmli *et al.,* 1974). Recently, Castillo and Black (1978) have been able to purify P31 from infected cells and have been able to test various hypotheses about how it functions in prehead assembly (Castillo *et al.,* 1977). They concluded that P31 influences directly the solubility properties of P23 and, although they were not able to detect any direct interaction between P31 and P23 *in vitro,* they suggest that without P31, P23 molecules nonspecifically associate and become unable to participate in normal assembly. These conclusions are tentative, but it appears that the major structural protein of the phage must first associate with P31 protein to remain in a state that will allow maximum interactions among the P23 structural proteins during assembly. In this sense P31 functions to facilitate the association of the structural molecules.

Similarly, nonstructural proteins seem to be involved in facilitating the association of structural molecules during assembly of T4 tail fiber (Revel *et al.,* 1976; Bishop and Wood, 1976; Wood and Revel, 1976). Six genes are involved in the assembly of the tail fiber, four of which code for structural polypeptides. The products of the remaining two, P38 and P37, seem to be involved in catalyzing assembly either by binding transiently to the structural molecules in such a way as to facilitate assembly or by transiently associating with the structural molecules to prevent them from interacting with other cellular components.

Based on morphologic observations, several supramolecular structures appear to assemble in association with other organelles or supramolecular structures. One example is the association of early T4 head precursors with host cytoplasmic membrane during assembly (Laemmli, 1970; Simon, 1972). Another example, discussed earlier, is microtubule formation in association with microtubule organizing centers. Also, during the assembly of centrioles and basal bodies, a strict association is observed between the forming organelle and a secondary supramolecular structure (Anderson and Brenner, 1971; Kalnins and Porter, 1969; Dirksen and Crocker, 1966; Anderson and Hein, 1976). In all these situations the secondary supramolecular structure may function to facilitate assembly of structural molecules.

One of the more interesting discoveries to come from the study of virus assembly is that for many viruses, head assembly requires a nonstructural protein that acts as a scaffold or superstructure. During the morphogenesis of the bacteriophage P22, approximately 250 molecules of the gene *8* protein, termed scaffolding protein, associated with 420 molecules of the coat protein (the product of gene *5*) to form a prohead (King and Casjens, 1974; Casjens and King, 1974). The scaffolding protein, essential for efficient assembly of closed shells, may facilitate the formation of the proper bond angles between the coat protein. A unique aspect of this assembly step is that once the DNA is encapsulated, the scaffolding protein is released and the prohead is transformed into a mature head structure. The scaffolding protein is released intact and is reused to assemble another virus head (Earnshaw and Casjens, 1976; King *et al.*, 1976). Recently, King *et al.* (1978) demonstrated that the synthesis of the scaffolding protein is tightly coupled to capsid assembly. Thus, the scaffolding protein acts catalytically during the assembly of the head to facilitate structural molecule interactions.

Scaffolding proteins or scaffolding core structures seem to be involved in the assembly of head structures of a variety of viruses including: lambda phage (Hohn and Katsura, 1977); ϕX174 bacteriophage (Fujisawa and Hyashi, 1977); ϕ29 bacteriophage (Hagen *et al.*, 1976; Viñuela *et al.*, 1976; Jiménez *et al.*, 1977); T7 bacteriophage (Roeder and Sadowski, 1977); and T4 bacteriophage (van Driel and Couture, 1978a,b; Showe and Onorato, 1978; Paulson and Laemmli, 1977). In contrast to the assembly of the P22 phage, the scaffolding core structures in these phages appear to be removed by protease activity during the conversion to the final head structure and are not recycled.

How these scaffolding structures determine the form of the mature virus is not known. Four mechanisms have been proposed.

1. The scaffold functions as a *template*. The scaffolding protein assembles into a scaffold that has a characteristic shape. This scaffold has binding sites for capsid protein that are distributed over its surface. Once bound, the multiple copies of the capsid protein become arranged into the shape of the scaffold. The capsid protein at each free end eventually join together to form a mature capsid.

2. A *vernier mechanism* for determining shape or length. In this case, the scaffold consists of a number of subunits n, whereas the capsid contains a different number m of protein subunits. Assuming that m and n do not have a common denominator, during the copolymerization of the two different subunits, the shape or length is complete at that point when m and n are in register.

3. The *cummulative strain* model for shape determination. As the scaffold subunits interact with the capsid subunits, the latter undergo a slight conformational change that is transmitted to contiguous capsid subunits. As more and more subunits join together, they undergo a cummulative conformational change. Eventually, the bound capsid subunits reach a conformational state that does not permit the further addition of free capsid subunits. At this point, the structure is complete. These three models are reviewed by Cummings and Bolin (1976).

4. The *kinetic method,* by which the shape of the head is determined by the relative rates of assembly of the scaffold and the shell (Showe and Onorato, 1978). For the T4 bacteriophage, Paulson and Laemmli (1977) favor a vernier mechanism, Showe and Onorato (1978) prefer the kinetic model, and van Driel and Couture (1978a) prefer a ruler or template model.

Regardless of the precise mechanism, the involvement of scaffolding core structures in virus morphogenesis is well established, and similar mechanisms may be operating in the assembly of other supramolecular structures. Komeda *et al.* (1978) have suggested that a scaffolding protein is required for the formation of the hook-basal body complex attached to the bacterial flagellum. Likewise, it may be that the cartwheel structure found in the lumen of forming centrioles and basal bodies (see Fig. 8) functions as a scaffold (Anderson and Brenner, 1971).

2. The Joining of Subunits during Assembly

During the assembly of some viruses, nonstructural molecules may function to join structural elements together. This can involve covalent or noncovalent joining of individual structural molecules or the joining of like or dissimilar subassemblies of structural molecules.

The formation of the lambda phage head structure illustrates a structure that involves the catalytic joining of subunit molecules. The assembly of this phage head requires the products from nine phage genes and at least one host gene (reviewed by Hohn and Katsura, 1977; Sternberg, 1976; Künzler and Hohn, 1978). The assembly begins with the formation of prohead II, for which phage gene products pB, pC, pE, and pNu3 and host gene product pgroE are required. First, pC, pB, and pNu3 associate to form a scaffolding for the proper assembly of the capsid protein, pE. pNu3 acts as the scaffolding protein. The prohead I formed in this reaction, is converted, in the presence of pgroE, to prohead II in the following sequence: (1) the pNu3 leaves the prohead; (2) the pC and pE proteins are covalently joined to form a new protein, X2; and (3) pB is cleaved to form pB'. The prohead II, together with phage gene proteins pA, pD and pFI, cut and package the phage DNA. Finally, phage proteins pW and pFII attach the tail segment to the fully formed head structure.

The reaction of interest is the fusion of pC with pE during the conversion of prohead I to prohead II, as first described by Hendrix and Casjens (1974). Presumably, the pC and pNu3 act in some way to guide the proper assembly of the capsid protein (Casjens and King, 1975; Hendrix and Casjens, 1975) because without them, coat protein (pE) polymerizes into long tubular polyheads, or spiral head-related structures. However, even though all of the pC molecules appear to be joined to an equal number of

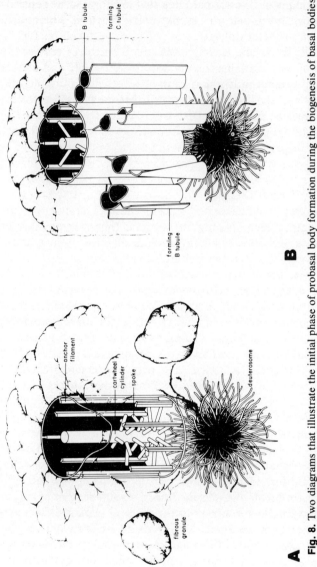

Fig. 8. Two diagrams that illustrate the initial phase of probasal body formation during the biogenesis of basal bodies in the primate oviduct. (A) The probasal body begins as an anulus of electron-dense material that is derived from fibrous granules, discrete 4 to 8 nm diameter granules found in the cytoplasm of cells making basal bodies. Within the anulus is a cartwheel structure that consists of a cartwheel cylinder and radiating spokes that connect to nine anchor filaments. The nine anchor filaments specify the location for the nine sets of three microtubules that will occupy the wall. In this sense, the cartwheel functions as a scaffold. (B) Once the cartwheel forms, nine sets of three microtubules polymerize from the fibrous granule material in juxtapositions to each of the anchor filaments. The deuterosome functions as a nucleation center during this assembly reaction. (From Anderson and Brenner, 1971, reproduced with permission.)

pE molecules during assembly, the significance of the reaction is not understood. Perhaps these kinds of reactions stabilize the assembly intermediates and ensure the proper order of subunit addition.

The catalytic joining of precursor particles that are formed by separate subassembly reactions is another important joining reaction. The attachment of tail fibers to the bacteriophage T4 is a good example of this type of reaction (reviewed in Wood and Revel, 1976). Four genes are required to produce the nonstructural proteins involved in the assembly and attachment of the tail fiber component (Bishop *et al.*, 1974). *In vitro* complementation experiments have shown that gene product 63 is necessary for the joining of the tail fiber to the phage baseplate (see Fig. 7). *In vitro* this assembly step can be stimulated up to 50-fold by the presence of gene product 63, which suggests that it acts catalytically.

The isolation and purification of the protein product (42,000 dalton) of gene *63* (Wood *et al.*, 1978) has permitted a more detailed study of assembly. Wood *et al.* (1978) conclude that the attachment event does not involve a covalent linkage. Furthermore, in contrast to earlier assumptions (Wood and Henninger, 1969), it is no longer clear that the protein acts catalytically, because large molar amounts are required for a detectable *in vitro* effect. Moreover, it cannot be ruled out that a small amount of this protein binds and remains associated with the attached fiber.

Figure 9 shows diagrammatically the assembly of the ϕX174 virus (Fujisawa and Hayashi, 1977) and illustrates, for another viral species, the two classes of tertiary assembly reactions discussed thus far. Scaffolding protein (D) is required to facilitate the association of capsid structural proteins and protein B activity is necessary to join by noncovalent bonds a 9 S particle and a 6 S particle to form a 12 S particle (Siden and Hayashi, 1974).

3. Stabilization of Subunit Interaction

In this reaction nonstructural molecules stabilize the association between assembled structural molecules. The formation of collagen fibrils *in situ* belongs to this class of reactions (reviewed by Grant and Jackson, 1976; Fessler and Fessler, 1978).

The ordered arrangement of collagen fibrils results from a complex process that begins with the synthesis of procollagen molecules. The procollagen subunits are next enzymatically modified, secreted, and converted to tropocollagen, probably through the activity of two separate enzymes. As soon as they are formed, tropocollagen molecules are organized into fibrils. Although this step presumably does not involve nonstructural molecules, other tissue components that determine such factors as place, size, and direction of fibril assembly are involved. In one

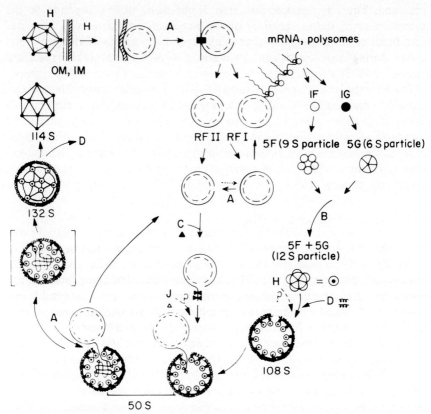

Fig. 9. A diagram of the morphogenesis of bacteriophage ϕX174. The virion (H) enters the bacterium by passing through the outer membrane (OM) and the inner membrane (IM). Once inside the cell, the DNA (represented by A) transcribes a mRNA. This mRNA in turn translates the various structural and nonstructural proteins required for viral assembly. There are several parts to this diagram and the reader is referred to Fujisawa and Hayashi (1977) for further details. Attention should be directed to two reactions: (a) Gene *B* protein acts catalytically to promote the aggregation of 5F (the 9 S particle) and 5G (the 6 S particle) to form a 12 S particle. (b) The assembly of the 108s capsomeric structure requires that 12 of the 12 S particles interact with protein D, a scaffolding protein. Once the DNA is encapsulated, the protein D leaves the structure and the 114 S particle is formed. (From Fujisawa and Hayashi, 1977, reprinted with permission.)

sense, these tissue components function as a template. Even though the fibrils form spontaneously, they have little tensile strength. The stabilization of the fibrils results from interfibril cross-links brought about by the activity of the enzyme, lysyl oxidase, which deaminates particular lysine or hydroxylysine residues to produce aldehydes that condense with other strategically located lysine, hydroxylysine, or glycosylated hydroxylysine residues. Thus, the proper function of the fibril matrix depends on the interfibril cross-linking activity of the enzyme. Transamidating enzymes also function in this way by cross-link proteins (e.g., the fibrinogen molecules during clot formation) in various types of assembly reactions (Lorand, 1977).

I have used virus assembly almost exclusively to illustrate the tertiary assembly reaction. With the development of new techniques, future studies of other assembly systems undoubtably will establish the generality of this type of assembly reaction. One must conclude at this point that the association of structural molecules often requires more than the recognition properties inherent in the ionic bonds and hydrophobic interactions mentioned earlier.

D. Quaternary Assembly Reactions

The endpoint of the three kinds of assembly reactions already described is the characteristic morphology of the structure. However, even though such structures may be morphologically complete, they may be unable to function. To achieve a functional state, the structure may have to be modified by other nonstructural molecules, a process referred to as a quaternary assembly reaction.

These postassembly modifications usually involve: (a) the addition of a structural molecule or macromolecule to the structure, (b) the removal of all or a portion of a structural molecule, or (c) the chemical modification of a structural molecule. These modifications are not detected by morphological changes, but by the appearance of the function normally associated with the structure. For example, the assembly of biological membranes is a quarternary assembly reaction.

In discussing lipid–lipid interactions and lipid–protein interactions (see Section II,B,3 and 4), I mentioned that a fundamental dichotomy exists between the apparent ease with which these molecules can spontaneously associate to form membranes *in vitro* and the unique lipid and protein composition found within different biological membranes. Since the hydrophobic interactions seem to have limited specificity, the assembly of cellular membranes must include mechanisms for modifying the spontaneously formed bilayer. Membrane assembly might take place in the following way: (a) lipids spontaneously associate to form a bilayer as they reach a critical micelle concentration within the cytoplasm of the cell (Tanford, 1978); (b) certain classes of integral membrane proteins are synthesized on one side of the bilayer and then spontaneously insert by hydrophobic interactions between the lipid and the hydrophobic region of the protein; (c) because of the large amount of energy required to position a transmembrane protein in the bilayer with polar head groups at both surfaces (Singer, 1977), special cellular machinery (i.e., nonstructural molecules or supramolecular structures) insert these proteins into the membrane; (d) to achieve the characteristic lipid composition of membranes, and also to achieve the asymmetry in lipid distribution across the membrane, specific sets of nonstructural molecules enzymatically alter the lipid composition of the membrane; (e) once the proteins and lipids are properly positioned in the membrane, nonstructural molecules add new molecules or remove portions of the existing structural molecules to modify membrane function; and (f) in some cases, nonstructural molecules translocate lipid and protein molecules in the plane of the membrane in specific compartments of the cell.

If this schema is correct, since steps (c)–(f) are postassembly reactions, the formation of functional biological membranes would belong to the quaternary class of assembly reactions. The evidence to support several aspects of this schema is discussed below, but lest one think that membranes are unique to this class of assembly reactions, the biogenesis of eukaryotic ribosomes most likely involves postassembly modifications (Kruiswijk et al., 1978a,b,c; Lastick and McConkey, 1976).

1. Lipid Composition of Membranes

There are several examples of nonstructural enzymes that alter the lipid composition of membranes. Several investigators have reported that the polar head groups of certain lipids can be methylated by a transfer reaction that utilizes methionine as a methyl donor (Moore et al., 1978; reviewed in Hirata et al., 1978; Leonard et al., 1978). One of these reactions, the methylation of phosphotidylethanolamine to form phosphatidylcholine, has been studied in some detail by Hirata et al. (1978) and Hirata and Axelrod (1978), who showed that two methyltransferases are involved in this reaction. The first catalyzes the addition of a methyl group to phosphatidylethanolamine to form phosphatidyl-N-monomethylethanolamine. This transferase activity is magnesium dependent. The second methyltransferase then adds two methyl groups to the phosphatidyl-N-monomethylethanolamine to form phosphatidylcholine. The conversion of phosphatidylethanolamine to phosphatidylcholine results in the movement of the modified lipids from the cytoplasmic surface to the cisternal surface of the membrane within the microsome fraction. Similarly, in erythrocyte membranes, methylation of phosphatidylethanolamine causes a translocation of the lipid across the membrane; in addition, this conversion results in an increase in the fluidity of the membrane (Hirata and Axelrod, 1978). Thus, these enzymes may be responsible in part for the characteristic asymmetry in phospholipid distribution seen in functional cell surface membranes (Bergelson and Bar-sukov, 1977). The activity of such enzymes may also explain how the phospholipids are asymmetrically distributed in the membrane of *Bacillus megaterium* (Rothman and Kennedy, 1977).

Lipid transport proteins may be another group of nonstructural proteins able to generate asymmetric lipid distributions in membranes (reviewed in Bergelson and Barsukov, 1977; Stern and Dales, 1974). Such proteins may recognize only certain lipid species and bind to and transfer them from one membrane compartment to another.

2. Insertion of Transmembrane Proteins

Whereas it is not difficult to understand how certain integral membrane proteins might bind and intercalate into one or the other sides of the lipid bilayer, the insertion of transmembrane proteins presents a special problem (Singer, 1977). Recent research on the synthesis and delivery of secretory proteins to exocytic vesicles has provided a working model for how transmembrane proteins are inserted into the lipid bilayer. The salient features of this model, called the signal hypothesis (reviewed in Blobel, 1977) are (a) proteins that must go through the endoplasmic reticulum (ER) membrane are synthesized with a 25–30 amino acid lead sequence (the signal sequence) that triggers the binding of the ribosome to the ER membrane and then inserts into the lipid bilayer [the binding step also involves ER membrane proteins, ribophorins, that may facilitate the insertion process (Kreibich *et al.*, 1978a,b)]; (b) during the subsequent synthesis of the protein, it vectorially moves across the bilayer until a portion of it reaches the cisternal space of the ER; (c) within the cisternae, proteases remove the signal sequence.

Support for this concept of transmembrane insertion comes from several sources. The vesicular stomatitis virus (VSV) contains an icosahedral virion core surrounded by a membrane, the major protein of which (protein G) is a glycosylated integral membrane protein asymmetrically positioned in the membrane such that its glycosylated end is exposed to the extracellular environment. The synthesis and the positioning of this protein within the membrane takes place in the ER of the infected cell (Katz *et al.*, 1977a; Atkinson *et al.*, 1976; David, 1977; Little and Huang, 1977). Recently a cell-free protein synthesis system has been utilized to investigate how the protein is inserted into the membrane (Rothman and Lodish, 1977; Dal Canto *et al.*, 1976; Katz *et al.*, 1977b; Morrison and McQuain, 1977, 1978). These studies suggest that the appearance of the G protein in the membrane is tightly coupled to its synthesis since it is never found free in the system. Furthermore, there is an absolute requirement that the ribosome bind to the ER membrane for protein insertion to take place.

Recently it has been shown that during synthesis the G protein contains a 10–15 amino acid "signal sequence" (Rothman and Lodish, 1977).

Further evidence that this is how transmembrane proteins are inserted into membranes comes from studies on the assembly of the filamentous phages, M13 (Wickner, 1975; Wickner and Killick, 1977; Marvin and Wachtel, 1975; Zwizinski and Wickner, 1977; Wickner *et al.*, 1978), FD (Chamberlain *et al.*, 1978), and F1 (Chang *et al.*, 1978). During the life cycle of the phage, the major coat protein of all of these viruses is inserted into the host's plasma membrane. The coat protein is positioned asymmetrically in the membrane and ultimately it associates with the nucleic acid during viral assembly. Although it has been possible to assemble the coat protein in an asymmetric or partially asymmetric fashion into synthetic membranes (Zwizinski and Wickner, 1977; Chamberlain *et al.*, 1978), during cellular synthesis this coat protein contains a signal sequence which is involved in the ribosome-mediated asymetric insertion of the protein into the bacterial membrane (Chang *et al.*, 1978). As with the insertion of membrane proteins into ER membranes and into bacterial membranes (reviewed recently by DiRienzo *et al.*, 1978; De Leij *et al.*, 1978), protein insertion involves the attachment of the ribosome to the membrane and the removal of the signal sequence once the protein is inserted (Chang *et al.*, 1978). Thus, ribosomes and proteases are nonstructural groups of molecules involved in the post assembly modification of membranes.

3. Modification of Membrane Structural Molecules

A final example of postassembly modification of membranes is the glycosylation of membrane proteins and lipids (Fleischer, 1977). A number of different types of lipid and protein subunits within membranes are known to have covalently attached carbohydrate residues, which, in plasma membranes, are exposed on the extracellular surface. Proteins are not inserted into membranes in a glycosylated form. The addition of the carbohydrate subunits is a complex, enzymatically mediated process that involves the transfer of lipid–carbohydrate complexes to the protein (Waechter and Lennarz, 1976). There is considerable evidence that this glycosylation step occurs in the ER and in the Golgi apparatus (reviewed in Leblond and Bennett, 1977). The glycosylation of the viral integral membrane protein for VSV (G protein) has been shown with *in vitro* protein-synthesizing systems, to take place in the microsome fraction (Rothman and Lodish, 1977; Katz *et al.*, 1977b) and to be required for the proper assembly of the virus (Gibson *et al.*, 1978; Leavitt *et al.*, 1977).

IV. CONCLUSION

This review is incomplete. Much more could be said about the various structures that were discussed, and other structures were not even mentioned. My goal, however, has been to be illustrative rather than exhaustive, and I hope the reader has developed an appreciation for the considerable diversity in the way that supramolecular structures are assembled.

In my mind, such diversity demands that the numerous structures be catalogued and catagorized to appreciate the general strategy for cell-mediated assembly. Grouping structures according to the ways that structural and nonstructural molecules interact during assembly is one of several possible formats that one could choose. However, this method allows one to relate the biogenesis of structures to the expression of genetic information. The expression of genetic information through structural molecule interactions appears to be uniform among structures. However, this is not the case for nonstructural molecules. Sometimes the latter behave like typical enzymes that modify the structural molecules, and at other times they function in quite novel ways, as best illustrated by the viral scaffolding structures. I have tried to illustrate these ideas in Fig. 10.

In part, the four different types of assembly reactions reflect a temporal arrangement in the expression of genetic information. For example, the activity of nonstructural molecules occurs primarily before assembly in a secondary reaction, during assembly in a tertiary reaction, and after initial assembly in a quaternary reaction. However, it is interesting that the type of activity nonstructural molecules engage in seems to be unique to each class. Investigations into the mechanism of these activities should be emphasized in future research. Do microtubule organizing centers function as templates, or do they simply concentrate structural molecules through some type of binding reaction? Why are morphogenetic cleavage reactions required for assembly? What is the mechanistic basis of scaffolding protein activity during phage assembly? These are just a few questions that, when answered, will allow a fuller understanding of how nonstructural molecules function.

To understand better how and where nonstructural molecules function in the assembly of structures, more studies on the *in vivo* assembly of supramolecular structures are needed. The numerous studies on viral biogenesis and ribosome assembly illustrate how the judicious use of biochemical and genetic techniques can be used to unravel the extraordinary ways that nonstructural molecules are active within cells. For structures such as membranes, centrioles, cilia, and chromatin, new ways to utilize the power of these techniques are needed to study assembly processes.

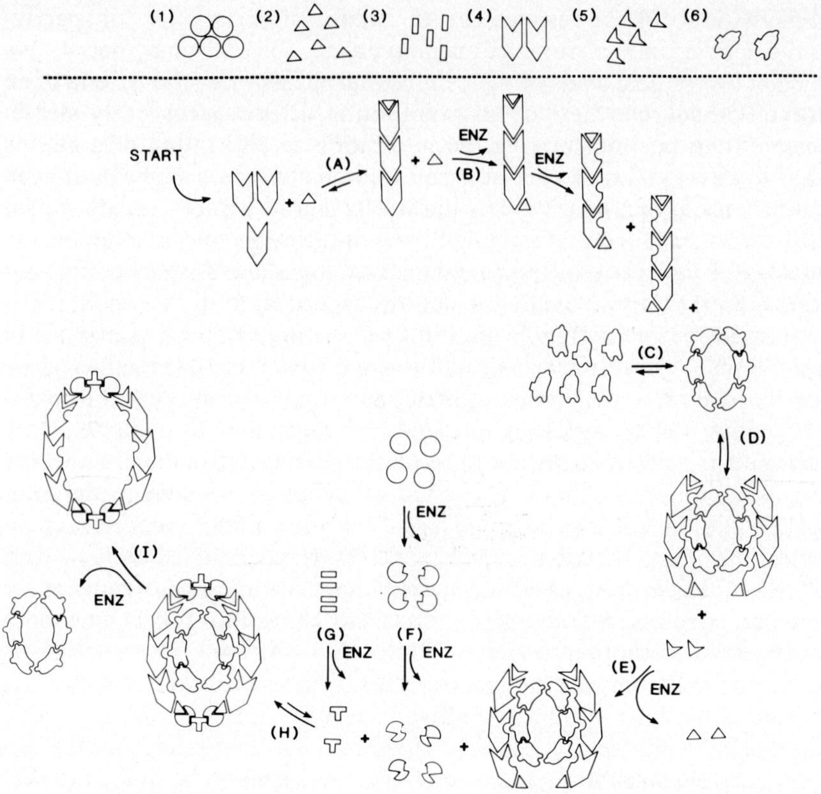

Fig. 10. Several of the assembly reactions discussed in the text are illustrated in this diagram of the assembly of a hypothetical structure. Subunits only associate during assembly if they have complimentary shapes. Six subunits are involved in the assembly reaction with (1) to (5) corresponding to structural molecules and (6) corresponding to the subunits for a scaffolding structure. In addition, several nonstructural molecules, denoted ENZ, are required at key points during assembly. There are two problems that occur during the assembly of a multisubunit structure: (a) to control subunit interactions so that they only occur at the proper time during assembly and (b) to fit together those products of subassembly reactions that ordinarily would not interact. Assume that all of the subunits are synthesized by standard synthetic pathways. Since (2) and (5) could both interact with (4) in the early phases of assembly, assume that (5) is synthesized late in the assembly reaction and is not initially available. (A) is a spontaneous reaction, and the length of the linear sequence, composed of subunit (4), is controlled by the insertion of a cap (2). (B) involves two nonstructural molecules, one of which adds one copy of (2) and the other modifies each of the (4) subunits. Reactions (A) and (B) represent a subassembly pathway leading to the formation of oligomers that are subsequently incorporated into the structure. However, these oligomers are straight chains that must be shaped to fit into the assembled structure. The shaping step requires the scaffolding structure that assembles spontaneously in reaction (C). Notice that the side group on (6) is square and therefore cannot interact with the oligomers; however, once this subunit is incorporated into the scaffolding component, a

Finally, it should be realized that the source of information for specifying the activity of structural and nonstructural molecules may not always reside in the nucleic acids of the cell. Nucleic acids meet the criteria of an information source; they are self-replicating, they are relatively stable, and they can be transferred from generation to generation of a cell or virus. If we were looking for other information sources, they would have to have these properties. Despite the arguments why nucleic acids are the information source (Watson, 1976), it is interesting that studies on the assembly of supramolecular structures have uncovered some possible exceptions to this rule. Sonneborn and his collaborators have shown (Sonneborn, 1970) that basal body replication and unit territory organization in *Paramecium* probably depends on non-nucleic acid-based information sources that are carried from generation to generation by unidentified components within the cortex of these protoza (see de Terra, 1978, for a recent review). The structural macromolecules are apparently specified for by the nucleic acids of this cell, but the temporal, the spatial, and possibly the morphological organization of these structures depends on cortical information sources. Nothing is known about the chemistry and function of these sources of information; but the establishment of their existence presents the tantalizing possibility that other non-nucleic acid sources of information operate within cells. Certainly for structure biogenesis, these possibilities must be seriously considered.

ACKNOWLEDGMENTS

I would like to thank Dr. Frederick Grinnell for critically reading the manuscript. I am especially indebted to Ms. Carolyn Mosse for the enormous amount of clerical work that was required and for the typing of the manuscript.

---

conformational change occurs in (6), producing a rounded side group that can interact with the oligomer. (D) is a spontaneous reaction that involves the binding of two oligomers to the scaffolding superstructure (this would involve both the *template* and *cummulative strain* models for scaffolding function). (E) requires a nonstructural molecule to remove (2) from the cap of the oligomers and to insert subunit (5). Subunits (1) and (3) are converted to active intermediates by the action of nonstructural molecules in reactions (G) and (F), respectively. Reaction (G) involves a covalent joining of two (3) subunits. (H) is a spontaneous reaction that joins together the two oligomers held in the proper alignment by the scaffold. Once the oligomers are joined, a nonstructural molecule removes the scaffold in reaction (I) to give the fully formed supramolecular structure. Reaction (I) may involve the degradation of the scaffolding structure. Once assembled, the projecting portion of the T-shaped subunit, produced in reaction (G), is modified (not shown) to produce the functional structure (a quaternary assembly reaction). Thus, reactions (A), (C), (D), and (H) are primary; reactions (B), (E), (F), and (G) are secondary; and reactions (D) and (I) together are tertiary.

REFERENCES

Adesnik, M., and Levinthal, C. (1969). Synthesis and maturation of ribosomal RNA in *Escherichia coli. J. Mol. Biol.* **46,** 281–303.

Adolph, K. W., and Butler, P. J. G. (1976). Assembly of a spherical plant virus. *Philos. Trans. R. Soc. London, Ser. B* **276,** 113–122.

Amos, L. A. (1977). Arrangement of high molecular weight associated proteins on purified mammalian brain microtubules. *J. Cell Biol.* **72,** 642–654.

Anderson, R. G. W. (1977). The biogenesis of cell structures and the expression of assembly information. *J. Theor. Biol.* **67,** 535–548.

Anderson, R. G. W., and Brenner, R. M. (1971). The formation of basal bodies (centrioles) in the rhesus monkey oviduct. *J. Cell Biol.* **50,** 10–34.

Anderson, R. G. W., and Hein, C. E. (1976). Estrogen dependent ciliogenesis in the chick oviduct. *Cell Tissue Res.* **171,** 459–466.

Anfinsen, C. B. (1973). Principles that govern the folding of protein chains. *Science* **181,** 223–230.

Anfinsen, C. B., and Scheraga, H. A. (1975). Experimental and theoretical aspects of protein folding. *Adv. Protein Chem.* **29,** 205–300.

Antonini, E., and Chiancone, E. (1977). Assembly of multisubunit respiratory proteins. *Annu. Rev. Biophys. Bioeng.* **6,** 239–271.

Atabekov, J. G., Novikov, V. K., Kiselev, N. A., Kaftanova, A. S., and Egorov, A. M. (1968). Stable intermediate aggregates formed by the polymerization of barley stripe mosaic virus protein. *Virology* **36,** 620–638.

Atabekova, T. I., Taliansky, M. E., and Atabekov, J. G. (1975). Specificity of protein–RNA and protein–protein interaction upon assembly of TMV *in vivo* and *in vitro. Virology* **67,** 1–13.

Atkinson, P. H., Moyer, S. A., and Summers, D. F. (1976). Assembly of vesicular stomatitis virus glycoprotein and matrix protein into HeLa cell plasma membranes. *J. Mol. Biol.* **102,** 613–631.

Baker, T. S., and Amos, L. A. (1978). Structure of the tubulin dimer in zinc-induced sheets. *J. Mol. Biol.* **123,** 89–106.

Bergelson, L. D., and Barsukov, L. I. (1977). Topological asymmetry of phospholipids in membranes. *Science* **197,** 224–230.

Bishop, R. J., and Wood, W. B. (1976). Genetic analysis of T4 tail fiber assembly. I. A gene 37 mutation that allows bypass of gene 38 function. *Virology* **72,** 244–254.

Bishop, R. J., Conley, M. P., and Wood, W. B. (1974). Assembly and attachment of bacteriophage T4 tail fibers. *J. Supramol. Struct.* **2,** 196–201.

Blobel, G. (1977). Synthesis and segregation of secretory proteins: The signal hypothesis. *In* "International Cell Biology" (B. Brinkley and K. Porter, eds.), pp. 318–325. Rockefeller Univ. Press, New York.

Bollen, A., Herzog, A., Favre, A., Thibault, J., and Gros, F. (1970). Fluorescence studies on the 30 S ribosome assembly process. *FEBS Lett.* **11,** 49–54.

Borisy, G. G., and Olmsted, J. B. (1972). Nucleated assembly of microtubules in porcine brain extracts. *Science* **177,** 1196–1197.

Borisy, G. G., Marcum, J. M., Olmsted, J. B., Murphy, D. B., and Johnson, K. A. (1975). Purification of tubulin and associated high molecular weight proteins from porcine brain and characterization of microtubule assembly *in vitro. Ann. N.Y. Acad. Sci.* **253,** 107–132.

Bothwell, M., and Schachman, H. K. (1974). Pathways of assembly of aspartate transcarbamoylase from catalytic and regulatory subunits. *Proc. Natl. Acad. Sci. U.S.A.* **71**, 3221–3225.

Bouck, G. B., and Brown, D. L. (1976). Self-assembly in development. *Annu. Rev. Plant Physiol.* **27**, 71–94.

Bretscher, M. S. (1974). Some aspects of membrane structure. *In* "Perspectives in Membrane Biology" (S. Estrada-O and C. Gitler, eds.), pp. 3–20. Academic Press, New York.

Bryan, R. N., Cutter, G. A., and Hayashi, M. (1978). Separate mRNAs code for tubulin subunits. *Nature (London)* **272**, 81–83.

Bryant, R. E., and Sypherd, P. S. (1974). Genetic analysis of cold-sensitive ribosome maturation mutants of *Escherichia coli*. *J. Bacteriol.* **117**, 1082–1092.

Burns, R. (1978). Rings, MAPs and microtubules. *Nature (London)* **273**, 709–710.

Burton, P. R., and Himes, R. H. (1978). Electron microscope studies of pH effects on assembly of tubulin free of associated proteins: Delineation of substructure by tannic acid staining. *J. Cell Biol.* **77**, 120–133.

Butler, P. J. G. (1976). Assembly of tobacco mosaic virus. *Philos. Trans. R. Soc. London, Ser. B* **276**, 151–163.

Butler, P. J. G., and Durham, A. C. H. (1977). Tobacco mosaic virus protein aggregation and the virus assembly. *Adv. Protein Chem.* **31**, 187–251.

Butler, P. J. G., and Klug, A. (1971). Assembly of the particle of tobacco mosaic virus from RNA and disks of protein. *Nature (London), New Biol.* **229**, 47–50.

Butler, P. J. G., Finch, J. T., and Zimmern, D. (1977). Configuration of tobacco mosaic virus RNA during virus assembly. *Nature (London)* **265**, 217–219.

Casjens, S., and King, J. (1974). P22 morphogenesis. 1. Catalytic scaffolding protein in capsid assembly. *J. Supramol. Struct.* **2**, 202–224.

Casjens, S., and King, J. (1975). Virus assembly. *Annu. Rev. Biochem.* **44**, 555–611.

Caspar, D. L. D. (1963). Assembly and stability of the tobacco mosaic virus particle. *Adv. Protein Chem.* **18**, 37–121.

Caspar, D. L. D., and Klug, A. (1962). Physical principles in the construction of regular viruses. *Cold Spring Harbor Symp. Quant. Biol.* **27**, 1–24.

Castillo, C. J., and Black, L. W. (1978). Purification and properties of the bacteriophage T4 gene 31 protein required for prehead assembly. *J. Biol. Chem.* **253**, 2132–2139.

Castillo, C. J., Hsiao, C.-L., Coon, P., and Black, L. W. (1977). Identification and properties of bacteriophage T4 capsid-formation gene products. *J. Mol. Biol.* **110**, 585–601.

Chamberlain, B. K., Nozaki, Y., Tanford, C., and Webster, R. (1978). Association of the major coat protein of fd bacteriophage with phospholipid vesicles. *Biochim. Biophys. Acta* **510**, 18–37.

Chang, C. N., and Chang, F. N. (1974). Methylation of ribosomal proteins *in vitro*. *Nature (London)* **251**, 731–733.

Chang, C. N., and Chang, F. N. (1975). Methylation of the ribosomal proteins in *Escherichia coli*: Nature and stoichiometry of the methylated amino acids in 50S ribosomal proteins. *Biochemistry* **14**, 468–477.

Chang, C. N., Blobel, G., and Model, P. (1978). Detection of prokaryotic signal peptidase in an *Escherichia coli* membrane fraction: Endoproteolytic cleavage of nascent f1 precoat protein. *Proc. Natl. Acad. Sci. U.S.A.* **75**, 361–365.

Chapman, D., and Cornell, B. A. (1977). Phase transitions, protein aggregation and membrane fluidity. *In* "Structure of Biological Membranes" (S. Abrahamsson and I. Pascher, eds.), pp. 85–93. Plenum, New York.

Chothia, C., and Janin, J. (1975). Principles of protein–protein recognition. *Nature (London)* **256**, 705–708.

Clark, R. J., and Felsenfeld, G. (1972). Association of arginine-rich histones with GC-rich regions of DNA in chromatin. *Nature (London), New Biol.* **240,** 226–229.

Cleveland, D. W., Hwo, S.-Y., and Kirschner, M. W. (1977a). Purification of tau, a microtubule-associated protein that induces assembly of microtubules from purified tubulin. *J. Mol. Biol.* **116,** 207–225.

Cleveland, D. W., Hwo, S.-Y., and Kirschner, M. W. (1977b). Physical and chemical properties of purified tau factor and the role of tau in microtubule assembly. *J. Mol. Biol.* **116,** 227–247.

Cleveland, D. W., Kirschner, M. W., and Cowan, N. J. (1978). Isolation of separate mRNAs for α- and β-tubulin and characterization of the corresponding *in vitro* translation products. *Cell* **15,** 1021–1031.

Colson, C., and Smith, H. O. (1977). Genetics of ribosomal protein methylation in *Escherichia coli.* I. A mutant deficient in methylation of protein L11. *Mol. Gen. Genet.* **154,** 167–173.

Combard, A., and Vendrely, R. (1970). Analytical study of the degradation of nucleohistone during calf thymus chromatin autolysis. *Biochem. J.* **118,** 875–881.

Connolly, J. A., Kalnins, V. I., Cleveland, D. W., and Kirschner, M. W. (1977). Immunoflourescent staining of cytoplasmic and spindle microtubules in mouse fibroblasts with antibody to tau protein. *Proc. Natl. Acad. Sci. U.S.A.* **74,** 2437–2440.

Crane, H. R. (1950). Principles and problems of biological growth. *Sci. Mon.* **70,** 376–389.

Cullis, P. R., De Kruijff, B., McGrath, A. E., Morgan, C. G., and Radda, G. K. (1977). Lipid asymmetry, clustering and molecular motion in biological membranes and their models. *In* "Structure of Biological Membranes" (S. Abrahamsson and I. Pascher, eds.), pp. 389–407. Plenum, New York.

Cummings, D. J., and Bolin, R. W. (1976). Head length control in T4 bacteriophage morphogenesis: Effect of canavanine on assembly. *Bacteriol. Rev.* **40,** 314–359.

Dahlberg, A. E., and Peacock, A. C. (1971). Studies of 16 and 23 S ribosomal RNA of *Escherichia coli* using composite gel electrophoresis. *J. Mol. Biol.* **55,** 61–74.

Dal Canto, M. C., Rabinowitz, S. G., and Johnson, T. C. (1976). *In vivo* assembly and maturation of vesicular stomatitis virus. *Lab. Invest.* **35,** 515–524.

David, A. E. (1977). Assembly of the vesicular stomatitus virus envelope: Transfer of viral polypeptides from polysomes to cellular membranes. *Virology* **76,** 98–108.

Davidson, N. (1967). Weak interactions and the structure of biological macromolecules. *In* "The Neurosciences: A Study Program" (G. Quarton, T. Melnechuk, and F. Schmitt, eds.), pp. 46–56. Rockefeller Univ. Press, New York.

De Leij, L., Kingma, J., and Witholt, B. (1978). Insertion of newly synthesized proteins into the outer membrane of *Escherichia coli. Biochim. Biophys. Acta* **512,** 365–376.

Dentler, W. L., Granett, S., and Rosenbaum, J. L. (1975). Ultrastructural localization of the high molecular weight proteins associated with *in vitro* assembled brain microtubules. *J. Cell Biol.* **65,** 237–241.

de Terra, N. (1978). Some regulatory interactions between cell structures at the supramolecular level. *Biol. Rev. Cambridge Philos. Soc.* **53,** 427–463.

DiRienzo, J. M., Nakamura, K., and Inouye, M. (1978). The outer membrane proteins of gram-negative bacteria: Biosynthesis, assembly, and functions. *Annu. Rev. Biochem.* **47,** 481–532.

Dirksen, E. R., and Crocker, T. T. (1966). Centriole replication in differentiating ciliated cells of mammalian respiratory epithelium. An electron microscope study. *J. Microsc. (Paris)* **5,** 629–644.

Doenges, K. H., Nagle, B. W., Uhlmann, A., and Bryan, J. (1977). *In vitro* assembly of tubulin from nonneural cells (Ehrlich ascites tumor cells). *Biochemistry* **16,** 3455–3459.

Dohme, F., and Nierhaus, K. H. (1976). Total reconstitution and assembly of 50 S subunits from *Escherichia coli* ribosomes *in vitro*. *J. Mol. Biol.* **107**, 585–599.

Dyson, R. D. (1978). "Cell Biology, A Molecular Approach." Allyn & Bacon, Boston, Massachusetts.

Earnshaw, W., and Casjens, S. (1976). Assembly of the head of bacteriophage P22: X-Ray diffraction from heads, proheads and related structures. *J. Mol. Biol.* **104**, 387–410.

Ebel, J. P., Ehresmann, B., Backendorf, C., Reinbolt, J., Tritsch, D., Ehresmann, C., and Branlant, C. (1977). Ribosomal protein–nucleic acid interaction. *In* "Gene Expression" (B. Clark, H. Klenow, and J. Zeuthen, eds.), Vol. 43, pp. 109–120. Pergamon, Oxford.

Edgar, R. S., and Wood, W. B. (1966). Morphogenesis of bacteriophage T4 in extracts of mutant-infected cells. *Proc. Natl. Acad. Sci. U.S.A.* **55**, 498–505.

Edgar, R. S., Denhardt, G. H., and Epstein, R. H. (1964). A comparative genetic study of conditional lethal mutations of bacteriophage T4D. *Genetics* **49**, 635–648.

Eipper, B. A. (1972). Rat brain microtubule protein purification and determination of covalently bound phosphate and carbohydrate. *Proc. Natl. Acad. Sci. U.S.A.* **69**, 2283–2287.

Eytan, G. D., Schatz, G., and Racker, E. (1977). Incorporation of integral membrane proteins into liposomes. *In* "Structure of Biological Membranes" (S. Abrahamsson and I. Pascher, eds.), pp. 373–387. Plenum, New York.

Fellous, A., Francon, J., Lennon, A.-M., and Nunez, J. (1977). Microtubule assembly *in vitro*. *Eur. J. Biochem.* **78**, 167–174.

Fessler, J. H., and Fessler, L. I. (1978). Biosynthesis of procollagen. *Annu. Rev. Biochem.* **47**, 129–162.

Fleischer, B. (1977). Localization of some glycolipid glysocylating enzymes in the Golgi apparatus of rat kidney. *J. Supramol. Struct.* **7**, 79–89.

Fraenkel-Conrat, H., and Williams, R. C. (1955). Reconstitution of active tobacco mosaic virus from its inactive protein and nucleic acid components. *Proc. Natl. Acad. Sci. U.S.A.* **41**, 690–698.

Francon, J., Fellous, A., Lennon, A. M., and Nunez, J. (1978). Requirement for 'factor(s)' for tubulin assembly during brain development. *Eur. J. Biochem.* **85**, 43–53.

Fujisawa, H., and Hayashi, M. (1977). Assembly of bacteriophage ϕX174: Identification of a virion capsid precursor and proposal of a model for the functions of bacteriophage gene products during morphogenesis. *J. Virol.* **24**, 303–313.

Gallop, P. M., and Paz, M. A. (1975). Posttranslational protein modifications, with special attention to collagen and elastin. *Physiol. Rev.* **55**, 418–487.

Gennis, R. B., and Jonas, A. (1977). Protein–lipid interactions. *Annu. Rev. Biophys. Bioeng.* **6**, 195–238.

Gibson, R., Leavitt, R., Kornfeld, S., and Schlesinger, S. (1978). Synthesis and infectivity of vesicular stomatitus virus containing nonglycosylated G protein. *Cell* **13**, 671–679.

Gilmore-Hebert, M. A., and Heywood, S. M. (1976). Translation of tubulin messenger ribonucleic acid. *Biochim. Biophys. Acta* **454**, 55–66.

Goodman, R. M., and Ross, A. F. (1974). Independent assembly of virions in tobacco doubly infected by potato virus X and potato virus Y or tobacco mosaic virus. *Virology* **59**, 314–318.

Grant, M. E., and Jackson, D. S. (1976). The biosynthesis of procollagen. *Essays Biochem.* **12**, 77–113.

Guha, S., Roth, H. E., and Nierhaus, K. (1975). Ribosomal proteins of *Bacillus subtilis* vegetative and sporulating cells. *Mol. Gen. Genet.* **138**, 299–307.

Hackenbrock, C. R. (1977). Molecular organization and the fluid nature of the mitochondrial energy transducing membrane. *In* "Structure of Biological Membranes" (S. Abrahamsson and I. Pascher, eds.), pp. 199–234. Plenum, New York.

Hadjiolov, A. A., and Nikolaev, N. (1976). Maturation of ribosomal ribonucleic acids and the biogenesis of ribosomes. *Prog. Biophys. Mol. Biol.* **31**, 95–144.

Hagen, E. W., Reilly, B. E., Tosi, M. E., and Anderson, D. L. (1976). Analysis of gene function of bacteriophage φ29 of *Bacillus subtilis:* Identification of cistrons essential for viral assembly. *J. Virol.* **19**, 501–517.

Hecht, N. B., and Woese, C. R. (1968). Separation of bacterial ribosomal ribonucleic acid from its macromolecular precursors by polyacrylamind gel electrophoresis. *J. Bacteriol.* **95**, 986–990.

Heidemann, S. R., and Kirschner, M. W. (1975). Aster formation in eggs of *Xenopus laevis:* Induction by isolated basal bodies. *J. Cell Biol.* **67**, 105–117.

Heidemann, S. R., Sander, G., and Kirschner, M. W. (1977). Evidence for a functional role of RNA in centrioles. *Cell* **10**, 337–350.

Hendrix, R. W., and Casjens, S. R. (1974). Protein fusion: A novel reaction in bacteriophage λ head assembly. *Proc. Natl. Acad. Sci. U.S.A.* **71**, 1451–1455.

Hendrix, R. W., and Casjens, S. R. (1975). Assembly of bacteriophage lambda heads: Protein processing and its genetic control in petit λ assembly. *J. Mol. Biol.* **91**, 187–199.

Hershko, A., and Fry, M. (1975). Post-translational cleavage of polypeptide chains: Role in assembly. *Annu. Rev. Biochem.* **44**, 775–797.

Himes, R. H., Burton, P. R., Kersey, R. N., and Pierson, G. B. (1976). Brain tubulin polymerization in the absence of "microtubule-associated proteins." *Proc. Natl. Acad. Sci. U.S.A.* **73**, 4397–4399.

Himes, R. H., Burton, P. R., and Gaito, J. M. (1977). The dimethyl sulfoxide-induced self-assembly of tubulin lacking associated proteins. *J. Biol. Chem.* **252**, 6222–6228.

Hirata, F., and Axelrod, J. (1978). Enzymatic methylation of phosphatidyl-ethanolamine increases erythrocyte membrane fluidity. *Nature (London)* **275**, 219–220.

Hirata, F., Viveros, O. H., Diliberto, E. J., Jr., and Axelrod, J. (1978). Identification and properties of two methyltransferases in conversion of phosphatidylethanolamine to phosphatidylcholine. *Proc. Natl. Acad. Sci. U.S.A.* **75**, 1718–1721.

Hohn, T., and Katsura, I. (1977). Structure and assembly of bacteriophage lambda. *Curr. Top. Microbiol. Immunol.* **78**, 69–110.

Holtzman, E. (1974). The biogenesis of organelles. *Hosp. Pract.* **9**, 75–89.

Homann, H. E., and Nierhaus, K. H. (1971). Ribosomal proteins: Protein compositions of biosynthetic precursors and artificial subparticles from ribosomal subunits in *Escherichia coli* K12. *Eur. J. Biochem.* **20**, 249–257.

Hopfinger, A. J. (1977). "Intermolecular Interactions and Biomolecular Organization." Wiley, New York.

Hubert, J. J., Bourque, D. P., and Zaitlin, M. (1976). A tobacco mosaic virus mutant with non-functional coat protein and its revertant: Relationship to the virus assembly process. *J. Mol. Biol.* **108**, 789–798.

Israelachvili, J. N. (1977). Refinement of the fluid-mosaic model of membrane structure. *Biochim. Biophys. Acta* **469**, 221–225.

Israelachvili, J. N., Mitchell, D. J., and Ninham, B. W. (1977). Theory of self-assembly of lipid bilayers and vesicles. *Biochim. Biophys. Acta* **470**, 185–201.

Janin, J., and Chothia, C. (1976). Stability and specificity of protein–protein interactions: The case of the trypsin–trypsin inhibitor complexes. *J. Mol. Biol.* **100**, 197–211.

Jiménez, F., Camacho, A., De La Torre, J., Viñuela, E., and Salas, M. (1977). Assembly of *Bacillus subtilis* phage φ29. 2. Mutants in the cistrons coding for the non-structural proteins. *Eur. J. Biochem.* **73**, 57–72.

Johnson, K. A., and Borisy, G. G. (1977). Kinetic analysis of microtubule self-assembly *in vitro*. *J. Mol. Biol.* **117**, 1–31.

Josephs, R., Eisenberg, H., and Reisler, E. (1972). Subunits to superstructures: Assembly of glutamate dehydrogenase. *In* "Protein–Protein Interactions" (R. Jaenicke and E. Helmreich, eds.), pp. 57–90. Springer-Verlag, Berlin and New York.

Kalnins, V. I., and Porter, K. R. (1969). Centriole replication during ciliogenesis in the chick tracheal epithelium. *Z. Mikrosk.-Anat. Forsch.* **100**, 1–30.

Katz, F. N., Rothman, J. E., Knipe, D. M., and Lodish, H. F. (1977a). Membrane assembly: Synthesis and intracellular processing of the vesicular stomatitis viral glycoprotein. *J. Supramol. Struct.* **7**, 353–370.

Katz, F. N., Rothman, J. E., Lingappa, V. R., Blobel, G., and Lodish, H. F. (1977b). Membrane assembly *in vitro*: Synthesis, glycosylation, and asymmetric insertion of a transmembrane protein. *Proc. Natl. Acad. Sci. U.S.A.* **74**, 3278–3282.

Keates, R. A. B., and Hall, R. H. (1975). Tubulin requires an accessory protein for self-assembly into microtubules. *Nature (London)* **257**, 418–420.

Kellenberger, E. (1972). Assembly in biological systems. *Ciba Found. Symp.* **7** (new ser.), 189–203.

Kellenberger, E. (1976). DNA viruses. *Philos. Trans. R. Soc. London, Ser. B* **276**, 3–13.

King, J., and Casjens, S. (1974). Catalytic head assembling protein in virus morphogenesis. *Nature (London)* **251**, 112–119.

King, J., Botstein, D., Casjens, S., Earnshaw, W., Harrison, S., and Lenk, E. (1976). Structure and assembly of the capsid of bacteriophage P22. *Philos. Trans. R. Soc. London, Ser. B* **276**, 37–49.

King, J., Hall, C., and Casjens, S. (1978). Control of the synthesis of phage P22 scaffolding protein is coupled to capsid assembly. *Cell* **15**, 551–560.

Klein, I., Willingham, M., and Pastan, I. (1978). A high molecular weight phosphoprotein in culture fibroblasts that associates with polymerized tubulin. *Exp. Cell Res.* **114**, 229–238.

Klotz, I. M., Langerman, N. R., and Darnall, D. W. (1970). Quaternary structure of proteins. *Annu. Rev. Biochem.* **39**, 25–62.

Komeda, Y., Silverman, M., Matsumura, P., and Simon, M. (1978). Genes for the hook-basal body proteins of the flagellar apparatus in *Escherichia coli. J. Bacteriol.* **134**, 655–667.

Korn, E. D. (1966). Structure of biological membranes. *Science* **153**, 1491–1498.

Kornberg, R. D. (1974). Chromatin structure: A repeating unit of histones and DNA. *Science* **184**, 868–870.

Kreibich, G., Ulrich, B. L., and Sabatini, D. D. (1978a). Proteins of rough microsomal membranes related to ribosome binding. I. Identification of ribophorins I and II, membrane proteins characteristic of rough microsomes. *J. Cell Biol.* **77**, 464–487.

Kreibich, G., Freienstein, C. M., Pereyra, B. N., Ulrich, B. L., and Sabatini, D. D. (1978b). Proteins of rough microsomal membranes related to ribosome binding. II. Cross-linking of bound ribosomes to specific membrane proteins exposed at the binding sites. *J. Cell Biol.* **77**, 488–506.

Kruiswijk, T., De Hey, J. T., and Planta, R. J. (1978a). Modification of yeast ribosomal proteins: Phosphorylation. *Biochem. J.* **175**, 213–219.

Kruiswijk, T., Kunst, A., Planta, R. J., and Mager, W. H. (1978b). Modification of yeast ribosomal proteins: Methylation. *Biochem. J.* **175**, 221–225.

Kruiswijk, T., Planta, R. J., and Krop, J. M. (1978c). The course of the assembly of ribosomal subunits in yeast. *Biochim. Biophys. Acta* **517**, 378–389.

Künzler, P., and Hohn, T. (1978). Stages of bacteriophage lambda head morphogenesis; physical analysis of particles in solution. *J. Mol. Biol.* **122,** 191–215.

Kushner, D. J. (1969). Self-assembly of biological structures. *Bacteriol. Rev.* **33,** 302–345.

Laemmli, U. K. (1970). Cleavage of structural proteins during the assembly of the head of bacteriophage T4. *Nature (London)* **227,** 680–685.

Laemmli, U. K., Paulson, J. R., and Hitchins, V. (1974). Maturation of the head of bacteriophage T4. V. *J. Supramol. Struct.* **2,** 276–301.

Lake, J. A. (1976). Ribosome structure determined by electron microscopy of *Escherichia coli* small subunits, large subunits and monomeric ribosomes. *J. Mol. Biol.* **105,** 131–159.

Laskey, R. A., Mills, A. D., and Morris, N. R. (1977). Assembly of SV40 chromatin in a cell-free system from *Xenopus* eggs. *Cell* **10,** 237–243.

Lastick, S. M., and McConkey, E. H. (1976). Exchange and stability of HeLa ribosomal proteins *in vivo*. *J. Biol. Chem.* **251,** 2867–2875.

Lauffer, M. A. (1975). "Entropy-Driven Processes in Biology." Springer-Verlag, Berlin and New York.

Leavitt, R., Schlesinger, S., and Kornfeld, S. (1977). Impaired intracellular migration and altered solubility of nonglycosylated glycoproteins of vesicular stomatitis virus and Sindbis virus. *J. Biol. Chem.* **252,** 9018–9023.

Lebeurier, G., Nicolaieff, A., and Richards, K. E. (1977). Inside-out model for self-assembly of tobacco mosaic virus. *Proc. Natl. Acad. Sci. U.S.A.* **74,** 149–153.

Leblond, C. P., and Bennett, G. (1977). Role of the golgi apparatus in terminal glycosylation. *In* "International Cell Biology" (B. Brinkley and K. Porter, eds.), pp. 326–336. Rockefeller Univ. Press, New York.

Lee, J. C., and Timasheff, S. N. (1975). The reconstitution of microtubules from purified calf brain tubulin. *Biochemistry* **14,** 5183–5187.

Lehninger, A. L. (1975). "Biochemistry." Worth Publ. Co., New York.

Lenaz, G. (1977). Lipid properties and lipid–protein interactions. *In* "Membrane Proteins and their Interactions with Lipids" (R. Capaldi, ed.), pp. 47–149. Dekker, New York.

Leonard, E. J., Skeel, A., Chiang, P. K., and Cantoni, G. L. (1978). The action of the adenosylhomocysteine hydrolase inhibitor, 3-deazaadenosine, on phagocytic function of mouse macrophages and human monocytes. *Biochem. Biophys. Res. Commun.* **84,** 102–109.

Liljas, A., and Rossmann, M. G. (1974). X-ray studies of protein interactions. *Annu. Rev. Biochem.* **43,** 475–507.

Little, S. P., and Huang, A. S. (1977). Synthesis and distribution of vesicular stomatitis virus-specific polypeptides in the absence of progeny production. *Virology* **81,** 37–47.

Lockwood, A. H. (1978). Tubulin assembly protein: Immunochemical and immunofluorescent studies on its function and distribution in microtubules and cultured cells. *Cell* **13,** 613–627.

Lorand, L. (1977). Biological functions of transamidating enzymes which cross-link proteins. *In* "Search and Discovery: A Tribute to Albert Szent-Györgyi" (B. Kaminer, ed.), pp. 177–193. Academic Press, New York.

Lowry, C. V., and Dahlberg, J. E. (1971). Structural differences between the 16 S ribosomal RNA of *E. coli* and its precursor. *Nature (London), New Biol.* **232,** 52–54.

McLachlan, A. D., Perutz, M. F., and Pulsinelli, P. D. (1972). Subunit interactions in haemoglobin. *In* "Protein–Protein Interactions" (R. Jaenicke and E. Helmreich, eds.), pp. 91–110. Springer-Verlag, Berlin and New York.

Mangiarotti, G., Turco, E., Perlo, C., and Altruda, F. (1975). Role of precursor 16 S RNA in assembly of *E. coli* 30 S ribosomes. *Nature (London)* **253,** 569–571.

Marcum, J. M., and Borisy, G. G. (1978a). Characterization of microtubule protein oligomers by analytical ultracentrifugation. *J. Biol. Chem.* **253**, 2825–2833.

Marcum, J. M., and Borisy, G. G. (1978b). Sedimentation velocity analyses of the effect of hydrostatic pressure on the 30 S microtubule protein oligomer. *J. Biol. Chem.* **253**, 2852–2857.

Marcum, J. M., Dedman, J. R., Brinkley, B. R., and Means, A. R. (1978). Control of microtubule assembly-disassembly by clacium-dependent regulator protein. *Proc. Natl. Acad. Sci. U.S.A.* **75**, 3771–3775.

Margolis, R. L., and Wilson, L. (1978). Opposite end assembly and disassembly of microtubules at steady state *in vitro*. *Cell* **13**, 1–8.

Marvin, D. A., and Wachtel, E. J. (1975). Structure and assembly of filamentous bacterial viruses. *Nature (London)* **253**, 19–23.

Matthews, B. W., and Bernhard, S. A. (1973). Structure and symmetry of oligomeric enzymes. *Annu. Rev. Biophys. Bioeng.* **2**, 257–317.

Moore, C., Blank, M. L., Lee, T.-C., Benjamin, B., Piantadosi, C., and Snyder, F. (1978). Membrane lipid modifications: Biosynthesis and identification of phosphatidyl-N-methyl-N-isopropylethanolamine in rat liver microsomes. *Chem. Phys. Lipids* **21**, 175–178.

Morris, T. J., and Semancik, J. S. (1973). *In vitro* protein polymerization and nucleoprotein reconstitution of tobacco rattle virus. *Virology* **53**, 215–224.

Morrison, T. G., and McQuain, C. O. (1977). Assembly of viral membranes. I. Association of vesicular stomatitis virus membrane proteins and membranes in a cell-free system. *J. Virol.* **21**, 451–458.

Morrison, T. G., and McQuain, C. O. (1978). Assembly of viral membranes: Nature of the association of vesicular stomatitis virus proteins to membranes. *J. Virol.* **26**, 115–125.

Murialdo, H., and Becker, A. (1978). Head morphogenesis of complex double-stranded deoxyribonucleic acid bacteriophages. *Microbiol. Rev.* **42**, 529–576.

Murphy, D. B., and Borisy, G. G. (1975). Association of high-molecular-weight proteins with microtubules and their role in microtubule assembly *in vitro*. *Proc. Natl. Acad. Sci. U.S.A.* **72**, 2696–2700.

Murphy, D. B., Johnson, K. A., and Borisy, G. G. (1977a). Role of tubulin-associated proteins in microtubule nucleation and elongation. *J. Mol. Biol.* **117**, 33–52.

Murphy, D. B., Vallee, R. B., and Borisy, G. G. (1977b). Identity and polymerization-stimulatory activity of the nontubulin proteins associated with microtubules. *Biochemistry* **16**, 2598–2605.

Némethy, G. (1975). Molecular interactions and allosteric effects. *In* "Subunits in Biological systems, Part C" (S. Timasheff and G. Fasman, eds.), pp. 1–90. Dekker, New York.

Nierhaus, K. H., Bordasch, K., and Homann, H. E. (1973). Ribosomal proteins. XLIII. *In vivo* assembly of *Escherichia coli* ribosomal proteins. *J. Mol. Biol.* **74**, 587–597.

Nishida, E., and Sakai, H. (1977). Calcium-sensitivity of the microtubule reassembly system. *J. Biochem. (Tokyo)* **82**, 303–306.

Nomura, M. (1973). Assembly of bacterial ribosomes. *Science* **179**, 864–873.

Ohno, T., Okada, Y., Nonomura, Y., and Inoue, H. (1975). Assembly of a rod-shaped virus. *J. Biochem. (Tokyo)* **77**, 313–319.

Ohno, T., Takahashi, M., and Okada, Y. (1977a). Assembly of tobacco mosaic virus *in vitro*: Elongation of partially reconstituted RNA. *Proc. Natl. Acad. Sci. U.S.A.* **74**, 552–555.

Ohno, T., Sumita, M., and Okada, Y. (1977b). Location of the initiation site on tobacco mosaic virus RNA involved in assembly of the virus *in vitro*. *Virology* **78**, 407–414.

Okada, Y. (1975). Mechanism of assembly of tobacco mosaic virus *in vitro*. *Adv. Biophys.* **7**, 1–41.

Okada, Y., and Ohno, T. (1972). Assembly mechanism of tobacco mosaic virus particle from its ribonucleic acid and protein. *Mol. Gen. Genet.* **114**, 205–213.

Okada, Y., Ohno, T., and Nonomura, Y. (1975). Assembly of tobacco mosaic virus *in vitro*. *J. Biochem. (Tokyo)* **77**, 1157–1163.

Olmsted, J. B., Marcum, J. M., Johnson, K. A., Allen, C., and Borisy, G. G. (1974). Microtubule assembly: Some possible regulatory mechanisms. *J. Supramol. Struct.* **2**, 429–450.

Oosawa, F., and Higashi, S. (1967). Statistical thermodynamics of polymerization and polymorphism of protein. *Prog. Theor. Biol.* **1**, 79–165.

Otsuki, Y., and Takebe, I. (1978). Production of mixedly coated particles in tobacco mesophyll protoplasts doubly infected by strains of tobacco mosaic virus. *Virology* **84**, 162–171.

Pardon, J., and Richards, B. (1973). The structure of nucleoprotein systems. *In* "Subunits in Biological Systems, Part B" (G. Fasman and S. Timasheff, eds.), pp. 1–70. Dekker, New York.

Paulson, J. R., and Laemmli, U. K. (1977). Morphogenetic core of the bacteriophage T4 head. Structure of the core in polyheads. *J. Mol. Biol.* **111**, 459–485.

Perham, R. N. (1975). Self-assembly of biological macromolecules. *Philos. Trans. R. Soc. London, Ser. B* **272**, 123–136.

Perham, R. N., and Wilson, T. M. A. (1978). The characterization of intermediates formed during the disassembly of tobacco mosaic virus at alkaline pH. *Virology* **84**, 293–302.

Pickett-Heaps, J. D. (1969). The evolution of the mitotic apparatus: An attempt at comparative ultrastructural cytology in dividing plant cells. *Cytobios* **1**, 257–280.

Raybin, D., and Flavin, M. (1975). An enzyme tyrosylating α-tubulin and its role in microtubule assembly. *Biochem. Biophys. Res. Commun.* **65**, 1088–1095.

Revel, H. R., Herrmann, R., and Bishop, R. J. (1976). Genetic analysis of T4 tail fiber assembly. *Virology* **72**, 255–265.

Richards, F. M. (1977). Areas, volumes, packing, and protein structure. *Annu. Rev. Biophys. Bioeng.* **6**, 151–176.

Roberts, B. E., Paterson, B. M., and Sperling, R. (1974). The cell-free synthesis and assembly of viral specific polypeptides into TMV particles. *Virology* **59**, 307–313.

Robertus, J. D., Kraut, J., Alden, R. A., and Birktoft, J. J. (1972). Subtilisin: A stereochemical mechanism involving transition-state stabilization. *Biochemistry* **11**, 4293–4303.

Rodriguez, J. A., and Borisy, G. G. (1978). Modification of the C-terminus of brain tubulin during development. *Biochem. Biophys. Res. Commun.* **83**, 579–586.

Roeder, G. S., and Sadowski, P. D. (1977). Bacteriophage T7 morphogenesis: Phage-related particles in cells infected with wild-type and mutant T7 phage. *Virology* **76**, 263–285.

Rothman, J. E., and Kennedy, E. P. (1977). Rapid transmembrane movement of newly synthesized phospholipids during membrane assembly. *Proc. Natl. Acad. Sci. U.S.A.* **74**, 1821–1825.

Rothman, J. E., and Lenard, J. (1977). Membrane asymmetry. *Science* **195**, 743–753.

Rothman, J. E., and Lodish, H. F. (1977). Synchronised transmembrane insertion and glycosylation of a nascent membrane protein. *Nature (London)* **269**, 775–780.

Rühlmann, A., Kukla, D., Schwager, P., Bartels, K., and Huber, R. (1973). Structure of the complex formed by bovine trypsin and bovine pancreatic trypsin inhibitor. *J. Mol. Biol.* **77**, 417–436.

Saborio, J. L., Palmer, E., and Meza, I. (1978). *In vivo* and *in vitro* synthesis of rat brain α- and β-tubulins. *Exp. Cell Res.* **114**, 365–373.

Sarkar, S. (1969). Evidence of phenotypic mixing between two strains of tobacco mosaic virus. *Mol. Gen. Genet.* **105**, 87–90.

Schachman, H. K. (1972). Structure, function and dynamics of a regulatory enzyme—aspartate transcarbamylase. *In* "Protein–Protein Interactions" (R. Jaenicke and E. Helmreich, eds.), pp. 17–56. Springer-Verlag, Berlin and New York.

Scheele, R. B., and Borisy, G. G. (1978). Electron microscopy of metal-shadowed and negatively stained microtubule protein. *J. Biol. Chem.* **253**, 2846–2851.

Scheraga, H. A. (1974). Poly(amino acids), interatomic energies, and protein folding. *In* "Peptides, Polypeptides, and Proteins" (E. Blout, F. Bovey, M. Goodman, and N. Lotan, eds.), pp. 49–70. Wiley, New York.

Schmitt, H., Josephs, R., and Reisler, E. (1977). A search for *in vivo* factors in regulation of microtubule assembly. *Nature (London)* **265**, 653–655.

Seeman, N. C., Rosenberg, J. M., and Rich, A. (1976). Sequence-specific recognition of double helical nucleic acids by proteins. *Proc. Natl. Acad. Sci. U.S.A.* **73**, 804–808.

Segal, D. M., Powers, J. C., Cohen, G. H., Davies, D. R., and Wilcox, P. E. (1971). Substrate binding site in bovine chymotrypsin A. A crystallographic study using peptide chloromethyl ketones as site-specific inhibitors. *Biochemistry* **10**, 3728–3738.

Segrest, J. P., and Jackson, R. L. (1977). Molecular properties of membrane proteins. *In* "Membrane Proteins and their Interactions with Lipids" (R. Capaldi, ed.), pp. 21–45. Dekker, New York.

Sherline, P., and Schiavone, K. (1977). Immunofluorescence localization of proteins of high molecular weight along intracellular microtubules. *Science* **198**, 1038–1040.

Sheterline, P. (1978). Localisation of the major high-molecular-weight protein on microtubules *in vitro* and in cultured cells. *Exp. Cell Res.* **115**, 460–464.

Showe, M. K., and Onorato, L. (1978). Kinetic factors and form determination of the head of bacteriophage T4. *Proc. Natl. Acad. Sci. U.S.A.* **75**, 4165–4169.

Siden, E. J., and Hayashi, M. (1974). Role of the gene *B* product in bacteriophage φX174 development. *J. Mol. Biol.* **89**, 1–16.

Sieber, G., and Nierhaus, K. H. (1978). Kinetic and thermodynamic parameters of the assembly *in vitro* of the large subunit from *Escherichia coli* ribosomes. *Biochemistry* **17**, 3505–3511.

Silengo, L., Altruda, F., Dotto, G. P., Lacquaniti, F., Perlo, C., Turco, E., Mangiarotti, G. (1977). Ribosome maturation in *E. coli*. *Ital. J. Biochem.* **26**, 133–143.

Simon, L. D. (1972). Infection of *Escherichia coli* by T2 and T4 bacteriophages as seen in the electron microscope: T4 head morphogenesis. *Proc. Natl. Acad. Sci. U.S.A.* **69**, 907–911.

Singer, S. J. (1971). The molecular organization of biological membranes. *In* "Structure and Function of Biological Membranes" (L. Rothfield, ed.), pp. 145–222. Academic Press, New York.

Singer, S. J. (1974). The molecular organization of membranes. *Annu. Rev. Biochem.* **43**, 805–833.

Singer, S. J. (1977). The fluid mosaic model of membrane structure. *In* "Structure of Biological Membranes" (S. Abrahamsson and I. Pascher, eds.), pp. 443–462. Plenum, New York.

Singer, S. J., and Nicolson, G. L. (1972). The fluid mosaic model of the structure of cell membranes. *Science* **195**, 720–731.

Snell, W. J., Dentler, W. L., Haimo, L. T., Binder, L. T., and Rosenbaum, J. L. (1974). Assembly of chick brain tubulin onto isolated basal bodies of *Chlamydomonas reinhardtii*. *Science* **185**, 357–359.

Snyder, J. A., and McIntosh, J. R. (1975). Initiation and growth of microtubules from mitotic centers in lysed mammalian cells. *J. Cell Biol.* **67**, 744–760.

Sonneborn, T. M. (1970). Gene action in development. *Proc. R. Soc. London, Ser. B* **176,** 347–366.

Sponar, J., and Sormová, Z. (1972). Complexes of histone F1 with DNA in 0.15 M NaCl. *Eur. J. Biochem.* **29,** 99–102.

Stearns, M. E., and Brown, D. L. (1977). Comparison of microtubule initiating proteins purified from basal body rootlets and brain microtubules. *J. Cell Biol.* **75,** 271a.

Steck, T. L. (1974). The organization of proteins in the human red blood cell membrane. *J. Cell Biol.* **62,** 1–19.

Stern, W., and Dales, S. (1974). Biogenesis of vaccinia: Concerning the origin of the envelope phospholipids. *Virology* **62,** 293–306.

Sternberg, N. (1976). A genetic analysis of bacteriophage λ head assembly. *Virology* **71,** 568–582.

Sypherd, P. S., Bryant, R., Dimmitt, K., and Fujisawa, T. (1974). Genetic control of ribosomal assembly. *J. Supramol. Struct.* **2,** 166–177.

Taliansky, M. E., Atabekova, T. I., and Atabekov, J. G. (1977). The formation of phenotypically mixed particles upon mixed assembly of some tobacco mosaic virus (TMV) strains. *Virology* **76,** 701–708.

Tanford, C. (1973). "The Hydrophobic Effect: Formation of Micelles and Biological Membranes." Wiley, New York.

Tanford, C. (1978). The hydrophobic effect and the organization of living matter. *Science* **200,** 1012–1018.

Telzer, B. R., Moses, M. J., and Rosenbaum, J. L. (1975). Assembly of microtubules onto kinetochores of isolated mitotic chromosomes of HeLa cells. *Proc. Natl. Acad. Sci. U.S.A.* **72,** 4023–4027.

Thammana, P., and Held, W. A. (1974). Methylation of 16S RNA during ribosome assembly *in vitro*. *Nature (London)* **251,** 682–686.

Thompson, W. C. (1977). Post-translational addition of tyrosine to alpha-tubulin *in vivo* in intact brain and in myogenic cells in culture. *FEBS Lett.* **80,** 9–13.

Uy, R., and Wold, F. (1977). Posttranslational covalent modification of proteins. *Science* **198,** 890–896.

Vallee, R. B., and Borisy, G. G. (1978). The non-tubulin component of microtubule protein oligomers. *J. Biol. Chem.* **253,** 2834–2845.

van Driel, R., and Couture, E. (1978a). Assembly of bacteriophage T4 head-related structures. II. *In vitro* assembly of prehead-like structures. *J. Mol. Biol.* **123,** 115–128.

van Driel, R., and Couture, E. (1978b). Assembly of the scaffolding core of bacteriophage T4 preheads. *J. Mol. Biol.* **123,** 713–719.

Viñuela, E., Camacho, A., Jiménez, F., Carrascosa, J. L., Ramírez, G., and Salas, M. (1976). Structure and assembly of phage φ29. *Philos. Trans. R. Soc. London, Ser. B* **276,** 29–35.

Waechter, C. J., and Lennarz, W. J. (1976). The role of polyprenol-linked sugars in glycoprotein synthesis. *Annu. Rev. Biochem.* **45,** 95–112.

Watson, J. D. (1976). "Molecular Biology of the Gene." Benjamin, Menlo Park, California.

Weingarten, M. D., Lockwood, A. H., Hwo, S.-Y., and Kirschner, M. W. (1975). A protein factor essential for microtubule assembly. *Proc. Natl. Acad. Sci. U.S.A.* **72,** 1858–1862.

Weisenberg, R. C. (1972). Microtubule formation *in vitro* in solutions containing low calcium concentrations. *Science* **177,** 1104–1105.

Weisenberg, R. C., and Rosenfeld, A. C. (1975). *In vitro* polymerization of microtubules into asters, and spindles in homogenates to surf clam eggs. *J. Cell Biol.* **64,** 146–158.

Weiss, P. (1967). 1 + 1 ≠ 2 (One plus one does not equal two). *In* "The Neurosciences: A

Study Program" (G. Quarton, T. Melnechuk, and F. Schmitt, eds.), pp. 801–821. Rockefeller Univ. Press, New York.

Welsh, M. J., Dedman, J. R., Brinkley, B. R., and Means, A. R. (1978). Calcium-dependent regulator protein: Localization in mitotic apparatus of eukaryotic cells. *Proc. Natl. Acad. Sci. U.S.A.* **75**, 1867–1871.

Wetlaufer, D. B., and Ristow, S. (1973). Acquisition of three-dimensional structure of proteins. *Annu. Rev. Biochem.* **42**, 135–158.

Wickner, W. (1975). Asymmetric orientaton of a phage coat protein in cytoplasmic membrane of *Escherichia coli. Proc. Natl. Acad. Sci. U.S.A.* **72**, 4749–4753.

Wickner, W., and Killick, T. (1977). Membrane-associated assembly of M13 phage in extracts of virus-infected *Escherichia coli. Proc. Natl. Acad. Sci. U.S.A.* **74**, 505–509.

Wickner, W., Mandel, G., Zwizinski, C., Bates, M., and Killick, T. (1978). Synthesis of phage M13 coat protein and its assembly into membranes *in vitro. Proc. Natl. Acad. Sci. U.S.A.* **75**, 1754–1758.

Wodak, S. J., and Janin, J. (1978). Computer analysis of protein–protein interaction. *J. Mol. Biol.* **124**, 323–342.

Wood, W. B. (1973). Genetic control of bacteriophage T4 morphogenesis. *In* "Genetic Mechanisms of Development" (F. Ruddle, ed.), pp. 29–46. Academic Press, New York.

Wood, W. B. (1974). Undelivered remarks for the 1974 Squaw Valley meeting on assembly mechanisms. *J. Supramol. Struct.* **2**, 512–514.

Wood, W. B., and Henninger, M. (1969). Attachment of tail fibers in bacteriophage T4 assembly: Some properties of the reaction *in vitro* and its genetic control. *J. Mol. Biol.* **39**, 603–618.

Wood, W. B., and Revel, H. R. (1976). The genome of bacteriophage T4. *Bacteriol. Rev.* **40**, 847–868.

Wood, W. B., Conley, M. P., Lyle, H. L., and Dickson, R. C. (1978). Attachment of tail fibers in bacteriophage T4 assembly. *J. Biol. Chem.* **253**, 2437–2445.

Zimmern, D. (1976). The region of tobacco mosaic virus RNA involved in the nucleation of assembly. *Philos. Trans. R. Soc. London, Ser. B* **276**, 189–204.

Zimmern, D., and Wilson, T. M. A. (1976). Location of the origin for viral reassembly on tobacco mosaic virus RNA and its relation to stable fragment. *FEBS Lett.* **71**, 294–298.

Zwizinski, C., and Wickner, W. (1977). Studies of asymmetric membrane assembly. *Biochim. Biophys. Acta* **471**, 169–176.

10

Protein Secretion and Transport

C. M. Redman and D. Banerjee

443

I. INTRODUCTION

A wide variety of cells produce proteins destined for extracellular export. These include all plant eukaryotes, which secrete proteins and polysaccharides for the cell wall, and many animal cells which, to a variable degree, produce proteins that are needed for various purposes outside of the cell. Among animal cells there are some differentiated cells whose special functions are to produce and secrete large amounts of proteins. Notable among these, not only because of a well-developed secretory process but also because they have been studied extensively, are the cells of the exocrine pancreas. The hepatocyte is another well characterized cell type, and much of the information given in this chapter is obtained from studies with these two kinds of cells. It can be assumed that the general basic mechanisms described for these two types of cell also apply to other eukaryotic cells, although some variations will be discussed.

II. GENERAL MORPHOLOGY OF SECRETORY CELLS

Cells engaged in the synthesis of secretory proteins are characterized by having a large proportion of their polysomes attached to the membranes of the endoplasmic reticulum (ER). In the exocrine pancreas these membranes occupy a large area of the cytoplasm and appear in stacks termed rough ER because the cisternae are studded with ribosomes attached to the cytoplasmic surface (see Fig. 1). In the hepatocyte the rough ER occupies a smaller proportion of the cytoplasm than in the exocrine pancreas, and it is often segregated in specific areas of the cytoplasm. Hepatic rough ER also appears, however, as characteristic stacks of flat ribosome-studded cisternae.

Contiguous with the rough ER are other portions of the ER which do not contain attached polysomes. In the hepatocyte this smooth ER is spread throughout a large portion of the cytoplasm, and extensive biochemical analyses have shown that the membranes of the rough and smooth ER are qualitatively very similar in protein, lipid, and enzymatic content (see DePierre and Dallner, 1975). The exocrine pancreas, which is geared to producing large amounts of secretory protein, is characterized by an amplified rough ER system with the smooth ER being limited to areas termed transitional elements of the rough ER.

Secretory cells are further compartmentalized by other membranous

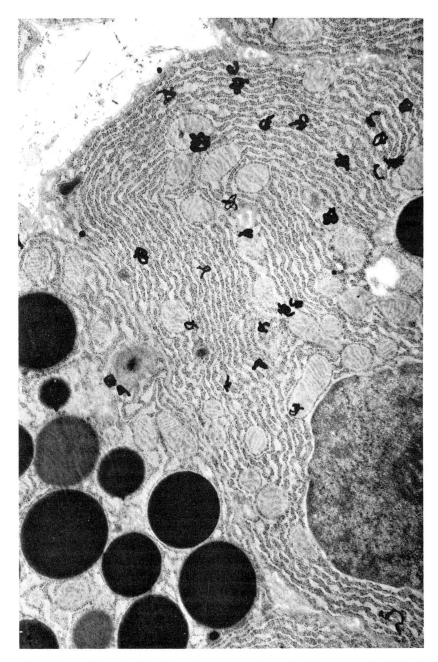

Fig. 1. Autoradiograph of a guinea pig pancreas acinar cell. Slices were pulse-labeled in *vitro* with L-[³H]leucine for 3 min. The autoradiographic grains are located over the stacks of rough endoplasmic reticulum. (From Jamieson and Palade, 1967b, with the permission of the *Journal of Cell Biology*).

elements that participate in secretion. These are the Golgi apparatus, which is comprised of flattened cisternae, GERL, a specialized acid phosphatase-containing region of the smooth ER, and secretory vesicles derived from the Golgi and GERL complex. In the exocrine pancreas these membrane-enclosed compartments, which are part of the secretion process are obviously polarized; they start from the base of the cell, which is rich in rough ER, and extend to the apex of the cell where secretory vesicles (zymogen granules) are in abundance (see Fig. 4).

From morphological and autoradiographic observations and from biochemical studies combined with cell fractionation, it has become apparent that the membranous compartments described above are involved in the secretion of proteins and that secretion involves the dynamic progression of nascent secretory proteins, within membranous compartments, from their site of synthesis on polysomes to their eventual discharge from the cell. This progression within the cell necessitates specific interactions between successive compartments, is accompanied by several posttranslational modifications of the secretory proteins while they are in transit through these compartments, and is finely regulated so that specific proteins can be segregated and transported to their correct locations.

III. METHODOLOGY

Several experimental approaches, pioneered to a large extent by George Palade and co-workers in studies on the exocrine pancreas, have been used to identify the cellular organelles involved in the secretion of proteins and to elucidate the temporal relationship between these various cell compartments. These methods include *in vivo* studies in which secretory proteins are labeled with radioactive amino acids followed by cell fractionation at various time intervals, identification of the isolated fractions by morphological and biochemical procedures, and determination of the presence and chemical nature of the nascent radioactive secretory proteins in the various fractions. Another useful method, which also involves labeling the nascent proteins *in vivo* with radioactive precursors, localizes the nascent proteins at different times in the secretory cell by electron microscope autoradiography (Figs. 1–4). These techniques have allowed investigators to outline the intracellular pathways that proteins follow from their site of synthesis on polysomes to their site of discharge from the cells and have allowed the measurement of elapsed time for these proteins to travel this route.

To obtain more detailed information regarding the intracellular progression of secretory proteins from one cellular compartment to another, sev-

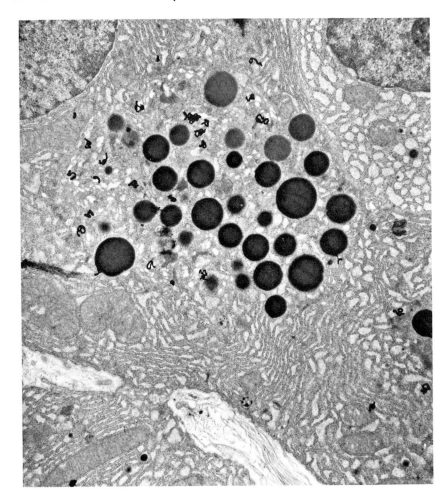

Fig. 2. Guinea pig pancreas slices were pulse-labeled for 3 min with L-[³H]leucine and then postincubated for 7 min. The autoradiographic grains in the acinar cells appear at the periphery of the Golgi complex. (From Jamieson and Palade, 1967b, with the permission of the *Journal of Cell Biology*.)

eral *in vitro* systems have been developed which limit the study to the processes that occur in well defined cellular organelles. Taken together, these experimental approaches performed on many different tissues, allow us to describe a general secretory pathway which is common to most eukaryotic cells and whose basic sequence is as follows. Secretory proteins are synthesized on polysomes attached to the membranes of the ER, the proteins are vectorially transferred to the lumen of the ER, and then in

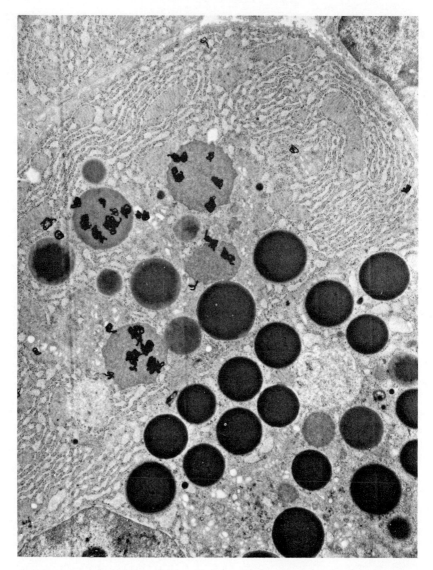

Fig. 3. A part of an acinar cell prepared as in Fig. 1 from slices postincubated for 37 min. The autoradiographic grains appear over the condensing vacuoles. The zymogen granules are unlabelled. (From Jamieson and Palade, 1967b, with the permission of the *Journal of Cell Biology*.)

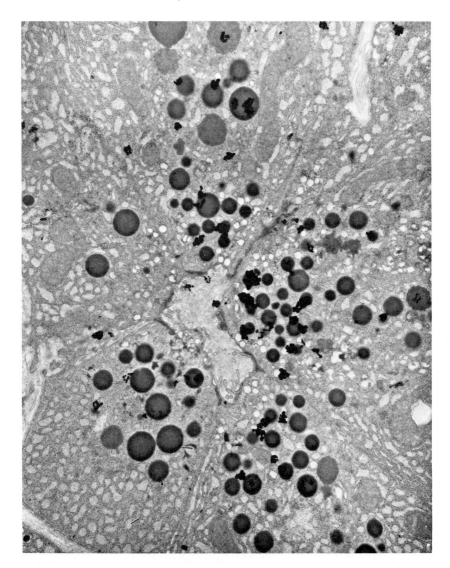

Fig. 4. Autoradiograph of several guinea pig pancreatic acinar cells prepared as in Fig. 1 and postincubated for 117 min. The autoradiographic grains are concentrated over the zymogen granules near the apex of the cells, with some residual grains left in the rough ER and some radioactive secretory material is noticed in the acinar lumen. (From Jamieson and Palade, 1967b, with permission from the *Journal of Cell Biology.*)

stepwise fashion these proteins are transferred to the smooth ER, to the Golgi apparatus, and to secretory vesicles which move to the site of discharge, fuse with the plasma membrane, and secrete the protein from the cell (see Palade, 1975).

IV. SYNTHESIS AND SEGREGATION OF SECRETORY PROTEINS

A. Site of Synthesis

Secretory proteins are synthesized on ER-bound polysomes, as was first suggested by the morphological observation that cells actively engaged in the production of secretory protein have a greater percentage of their polysomes attached to the membranes of the ER than do nonsecreting cells. More direct evidence for this came from studies of both the pancreas and liver when Siekevitz and Palade (1960) showed that chymotrypsinogen was labeled *in vivo* at 1 to 3 min after the intravenous injection of radioactive leucine into a guinea pig, and that radioactive chymotrypsinogen was preferentially located on membrane-attached polysomes. Later it was shown in rat liver that membrane-attached polysomes and free polysomes synthesize different classes of protein. Using both *in vivo* and *in vitro* techniques it was shown that the secretory protein albumin is synthesized on membrane-attached polysomes (Redman, 1968; Takagi and Ogata, 1968) and that free polysomes preferentially synthesize nonsecretory proteins (Ganoza and Williams, 1969). Some nonsecretory proteins, such as ferritin, while being predominantly synthesized by free polysomes, are also made by membrane-attached polysomes (Redman, 1969; Hicks *et al.,* 1969). These early experiments established the concept that the ER plays a major role in segregating the various proteins made by cytoplasmic polysomes and that some initial segregation is accomplished by selection of different sites of synthesis. In the last few years major advances have been made in our understanding of how secretory cells initially segregate the secretory proteins from other proteins not destined to be exported.

B. Vectorial Discharge

A central feature of the secretory process is that proteins to be exported are confined throughout most of their intracellular life to membrane compartments and do not enter the cytosol. Synthesis of the proteins, however, begins on polysomes which face the cytoplasm. Thus, the first step in the secretory process is the passage of the secretory proteins from their site of synthesis on polysomes to the cisternal space of the rough ER. This process, which involves passage of the newly made protein through the ER membrane, is commonly termed vectorial discharge and was first demonstrated in the exocrine pancreas using an *in vitro* microsomal system capable of labeling amylase with radioactive amino acids (Redman *et al.*, 1966). The segregation process occurs very early while the polypeptide chain is being translated on membrane-attached polysomes. *In vitro* studies with rat liver microsomes demonstrated that incomplete polypeptide chains have already entered into a channel which leads into the lumen of the ER, since interruption of protein synthesis by puromycin caused release of peptidyl–puromycin from the polysomes and a concomitant entrance of this aborted peptide into the lumen of the ER (Redman and Sabatini, 1966). These experiments demonstrated two salient features of the secretory process: (a) that secretory proteins are directly transferred from the membrane-attached polysomes to the vesicular cavities without first passing through the soluble cytoplasm, and (b) that the secretory polypeptides cross the ER membrane at specific ribosomal membrane junctions during the translational process.

The ER has specialized regions to facilitate the ordered transfer of secretory proteins from polysomes to the cisternae of the rough ER. In these areas of polysome–membrane interactions the polysomes are attached to the ER membrane by their large ribosomal subunits which contain the nascent polypeptide chains (Sabatini *et al.*, 1966). At this junction there exist specific membrane proteins, ribophorins, which are involved in the binding of ribosomes to the membrane (Kreibich *et al.*, 1978). Ribophorins, which are transmembrane proteins of molecular weight 61,000 to 63,000, are absent from the smooth ER and are thought to be functionally associated with both the ribosomes and the ER membrane in order to facilitate both proper alignment of the polysome to the ER membrane and subsequent vectorial discharge of the secretory protein into the cisternae of the ER.

Polysomes involved in the synthesis of secretory proteins are thought to be directed toward the proper segment of the rough ER by the amino-terminal portion of the nascent secretory proteins. In 1972 Milstein *et al.* showed that immunoglobulin light chains, when synthesized *in vitro* by a heterologous cell-free system containing mRNA from a myeloma cell, appeared as a precursor of slightly higher molecular weight than the normal immunoglobulin light chain and that this precursor was later converted into the authentic light chain product. Subsequently a number of different nascent proteins have been shown to contain extended polypeptide pieces at the amino-terminal end, and an attractive hypothesis based on these and other observations has been formulated by Blobel (1977). This hypothesis postulates that mRNA for secretory proteins contains a unique sequence of codons, localized on the 3'-side of the AUG codon, which is translated into a "signal" sequence of amino acids located at the amino terminal portion of the secretory polypeptide. After this "signal" sequence of the growing nascent polypeptide chains has emerged from the large ribosomal subunit, it allows the pertinent polysomes to attach to the ribosomal receptors on the rough ER membrane. The vectorial transfer of the newly formed secretory protein through the membrane is coupled to protein synthesis, and as the nascent protein passes into the luminal space an enzyme, signal peptidase, cleaves the signal portion of the nascent peptide. The processed secretory protein continues to grow and is released into the cisternal space of the ER after chain termination. This effectively segregates the newly made secretory protein within the cisternal space of the ER and places it in the correct location for further intracellular transport. The process is irreversible and tightly coupled to translation. The actual passage of secretory protein through the membrane is, however, not dependent on energy other than that needed for protein synthesis (Redman, 1967).

A large number of secretory proteins from different tissues has now been shown to be synthesized initially with an extended amino-terminal sequence. The length of the signal sequence varies from 10 to 30 residues, and they have in common the presence of a large proportion (usually 80 to 95%) of nonpolar amino acids. These signal sequences are usually cleaved during vectorial discharge, with the exception of ovalbumin which in the chick oviduct is synthesized and secreted without further processing (Gagnon et al., 1978). The chick oviduct, however, is capable of handling other secretory proteins, such as egg white lysozymes, in the typical manner (Palmiter et al., 1978).

As evidenced by the studies on ovalbumin, the presence of a cleavable signal sequence on nascent secretory proteins is not mandatory for vectorial discharge. Another example of a protein that can be transferred from polysomes across an intracellular membrane without processing of the signal extension is a lipoprotein from E. coli. It has been shown that an E. coli strain carrying a mutation for the structural gene of cell wall lipoprotein contains a mutant lipoprotein with a single amino acid replacement on residue 14 of the "signal" sequence. This mutation allows for transfer of the protein across the membrane but results in a failure of the "signal" sequence to be cleaved. The mutation, while not affecting the transfer of this protein across the membrane, does, however, affect its subsequent cellular destination in that this protein in the wild-type strain is located on the outer membrane of the cell, while in the mutant the lipoprotein is recovered in both inner and outer cell membranes (Lin *et al.*, 1978).

Some proteins, that contain hydrophobic amino acid residues on the amino-terminal region and are synthesized on membrane-attached polysomes, are not vectorially discharged but are inserted into the ER membrane. An example of this class of protein is hepatic cytochrome *P*-450, which is a nonexportable integral ER membrane protein. This protein from rat liver microsomes has been shown to contain N-terminal methionine and 17 hydrophobic amino acids among the first 20 amino acids at the N terminus (Haugen *et al.*, 1977). This suggests that the finished cytochrome *P*-450 retains an uncleaved hydrophobic amino terminal sequence.

Vectorial discharge of newly made secretory proteins into the lumen of the rough ER is specific, but the properties of the signal which allows the protein to cross the membrane are not presently identified. However, we can conclude that vectorial discharge involves a number of complex processes, such as protein synthesis, recognition of parts of the "signal" sequence of nascent secretory proteins with specific rough ER membrane segments, reorganization of membrane proteins and/or lipids around the ribosomal binding regions to form a channel, in most cases the removal of the "signal" sequence, proximal glycosylation, and protein folding. Some polysomes, which do not synthesize secretory protein but which produce membrane proteins, also attach to the rough ER membrane, but the nascent membrane proteins, although they may contain a hydrophobic portion at the amino-terminal end, are distinguished from secretory proteins and do not enter the cisternal space of the rough ER. This raises many questions about the role of protein-membrane recognition and the identity of the membrane proteins around the region of vectorial discharge.

C. Proteins Present in the Lumen of the ER

It is to be expected, if all secretory proteins are vectorially discharged into the cisternal space of a cell's ER, that the lumen of isolated rough ER preparations should contain a full complement of secretory proteins produced by that cell. The protein contents of isolated ER cell fractions can be obtained by physical disruption, by freezing and thawing, by sonication, or by treatment of isolated ER vesicles (rough microsomes) with sublytic concentrations of detergents that make the vesicles permeable but do not markedly disassemble the membrane (Kreibich and Sabatini, 1974). Using this latter technique it was shown that the proteins inside of hepatic rough and smooth ER are a complex mixture containing not only secretory plasma proteins and their precursors, but also some proteins not readily identifiable as secretory proteins. Since the major secretory path in the liver is toward the blood, with perhaps only a few proteins being secreted to the bile, the possibility is raised that the lumen of the ER may also contain proteins not destined for secretion. Kinetic analyses of the *in vivo* incorporation of radioactive amino acids into internal rough ER proteins showed that some of the proteins were rapidly synthesized and transfered to the smooth ER, a characteristic of secretory proteins, while other proteins were labeled more slowly, which may be a characteristic of membrane proteins (Kreibich and Sabatini, 1974). The eventual cellular destination or fate of these nonsecretory proteins found in the contents of the ER is not known. They may be peripheral membrane proteins permanently stationed at the inner surface of the ER, but skimmed off the membrane on treatment with sublytic levels of detergent, or they may be precursors of nonsecretory proteins in transit to other cellular organelles. These may include either membrane proteins fated to be inserted into the luminal surface of other organelles, or proteins destined for such organelles as lysosomes.

The finding of different types of proteins within the cisternae of the rough ER raises the intriguing question of how the ER can separate these different proteins and can direct them to their correct destination.

Other evidence, however, suggests that some of this sorting out may be done at a different level of the ER. For instance, it had been suggested that the synthesis and intracellular path of the peroxisomal proteins partly follows that of secretory proteins in that they enter the rough ER and are later diverted to peroxisomes (deDuve and Baudhuin, 1966). *In vivo* labeling studies on catalase show, however, that newly made catalase is first found in the soluble cytoplasm cell fraction and not in the rough ER, and that radioactive catalase does not enter the smooth ER prior to its deposition into peroxisomes (Redman *et al.,* 1972; Lazarow and deDuve, 1973). Since, however, in these *in vivo* studies some radioactive catalase was detected in a cell fraction which cosedimented with the rough ER prior to the entry of radioactive catalase into peroxisomes, Redman *et al.* (1972) suggested that catalase is released from polysomes directly into the cytoplasm and not into the lumen of the rough ER, but later would be packaged into peroxisomes on the surface of the rough ER where new peroxisomes are often seen being formed as outpouchings; alternatively, catalase may enter new peroxisomal vesicles after the latter have been formed and left the rough ER surface.

Recent *in vitro* studies indicate that catalase and uricase are synthesized by free polysomes and not by membrane-attached polysomes (Goldman and Blobel, 1978). It appears that nascent catalase translated *in vitro* may not contain a cleavable extended peptide sequence, since analysis of this product by electrophoresis on SDS polyacrylamide gels showed that it comigrated with nascent *in vivo* labeled catalase (Robbi and Lazarow, 1978). Also, catalase, in contrast to secretory proteins, is not segregated by microsomal vesicles *in vitro* (Goldman and Blobel, 1978). Hence both the *in vivo* and *in vitro* studies suggest that nascent catalase is not vectorially discharged into the ER lumen, and thus the sorting out of secretory proteins and peroxisomal proteins does not occur there. These studies with the peroxisomal proteins again emphasize that the site of synthesis plays a major role in separating different classes of proteins and in placing them in the correct intracellular position for subsequent transport to their correct destination. Histochemical evidence shows that peroxisomes may be formed from outpouchings of the rough ER (Novikoff and Shin, 1964), and the biochemical experiments described above indicate that catalase and uricase are synthesized on free polysomes. This suggests that the rough ER membrane can evaginate without including secretory proteins, and that during evagination catalase and uricase may be inserted into these preperoxisomal outpouchings. How these peroxisomal proteins are directed from the free polysomes to these vesicles and how they are transferred across the membrane awaits to be determined.

V. INTRACELLULAR PROTEIN MIGRATION

A. Movement from Rough ER to the Golgi Apparatus

From the lumen of the rough ER, the nascent secretory proteins move to another region of the ER devoid of polysomes (the smooth ER) and then pass from the ER to the Golgi apparatus where they are packaged into secretory vesicles. In rat liver, this progression from rough to smooth ER to the Golgi apparatus has been clearly shown to occur for newly made albumin and other plasma proteins. Within 5–10 min of intravenous injection of radioactive L-leucine into rats, maximal incorporation of radioactivity into proalbumin (a precursor of serum albumin) is observed in the rough ER. The radioactive proalbumin is then transferred to the smooth ER, reaching maximal level at this cellular site at about 15 min, followed by a rapid transfer of proalbumin to the Golgi apparatus region, and ending by the appearance of radioactive serum albumin in the blood by 15 to 20 min after the initial administration of radioactive L-leucine (Peters, 1962; Glaumann and Ericsson, 1960; Peters *et al.*, 1971; Geller *et al.*, 1972) (Fig. 5).

In the exocrine pancreas the time sequence is slightly different: the

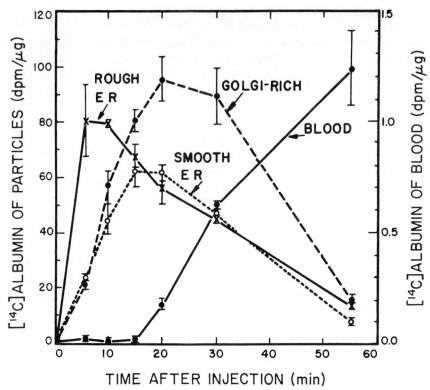

Fig. 5. Intracellular path of nascent albumin and its appearance in the blood. Rat albumin was labeled *in vivo* with L-[¹⁴C]leucine and the appearance of nascent albumin in various hepatic subfractions, obtained by cell fractionation, was measured. (From Peters *et al.*, 1971, with the permission of the *Journal of Biological Chemistry*.) Intracellularly nascent albumin exists as a precursor, proalbumin, which is converted into serum albumin in the Golgi complex (Ikehara *et al.*, 1976; Edwards *et al.*, 1976; Redman *et al.*, 1978).

secretory proteins pass from the rough ER to the transitional elements of the ER (which are akin to smooth ER of the hepatocyte), then to small peripheral vesicles which appear on the cis or proximal side of the Golgi apparatus, and after 30 min have elapsed they appear in the condensing vacuoles that are the precursors of zymogen granules (Jamieson and Palade, 1967b) (Figs. 1–4 and 6).

The transport of secretory protein from the rough ER lumen to the Golgi apparatus region occurs in the absence of protein synthesis (Jamieson and Palade, 1968a). This indicates that the intracellular movement of secretory protein is not motivated by a simple concentration gradient causing the movement of material from one cellular compartment to another and also indicates that the secretory cells contain, at least for a

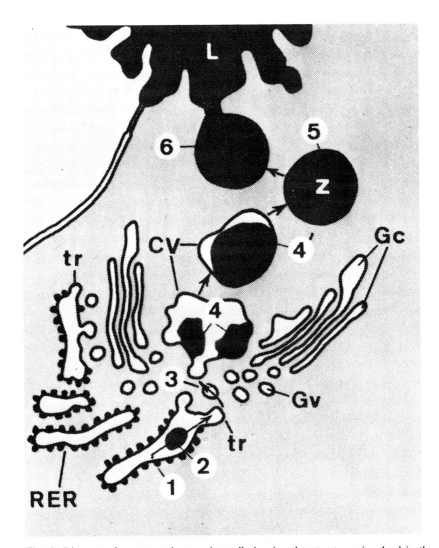

Fig. 6. Diagram of a pancreatic exocrine cell showing the structures involved in the synthesis and secretion of pancreatic enzymes. The numbers show the sequence of events which occur from the time that synthesis of these proteins occurs on membrane-attached polysomes of the rough ER, until they are discharged into the acinar lumen. Rough endoplasmic reticulum (RER); transitional elements of the rough ER (tr); smooth vesicles of the Golgi peripheral region (Gv); Golgi cisternae (Gc); condensing vacuoles (CV); zymogen granules (Z) and the acinar lumen (L). (From Jamieson, 1972, with the permission of Academic Press.)

short time, sufficient ancillary proteins that may be necessary to affect this movement.

The transfer of secretory proteins from the ER to the Golgi complex is, however, energy dependent (Jamieson and Palade, 1968b). In the absence of ATP synthesis, the pancreatic secretory proteins cannot leave the ER. When ATP synthesis is resumed, transport to the Golgi apparatus follows, suggesting an energy "lock-gate" exists at the level of the transitional elements of the ER, but this idea has not yet been extended to other secretory systems.

The transport of secretory protein from the rough ER via the smooth ER to the Golgi complex occurs entirely within membrane compartments, even though the ER and the Golgi apparatus may not be continuously connected by tubules. The connection between these two compartments is probably intermittent, effected by small closed vesicles composed of specific membranes which carry selected proteins from one compartment to another (Jamieson and Palade, 1967a,b). It is envisioned that to establish continuity between the various compartments, the ER membrane must be able to select, segregate, and enclose secretory proteins into small vesicles which then move to and specifically fuse with the accepting compartment. Studies of a variety of cell types show that the route of secretory proteins from rough ER to the Golgi complex described above is generally the same for all secretory cells examined, although minor variations occur between cell types (Palade, 1975). The biochemistry of these processes is not yet known and awaits further study.

B. Passage of Secretory Proteins through the Golgi Apparatus

That secretory proteins travel unidirectionally from rough ER via the smooth ER to the area of the Golgi apparatus and thence to secretory vesicles which are discharged by exocytosis is now reasonably well established. The concentration of the proteins into granules or secretory vesicles is thought to occur in the Golgi apparatus, a process that appears to occur in the trans or distal portion of the Golgi apparatus. According to this concept, secretory proteins emanating from the ER enter the cis side of the Golgi apparatus and move across the stacked Golgi cisternae, leaving the trans side as condensing vacuoles or immature secretory vesicles. The kinetic studies on which this proposal is based derive for the most part from the pioneering autoradiographic studies of Leblond and colleagues who showed that some radioactive sugars are added to thyroglobulin in the stacked Golgi cisternae prior to the appearance of radioactive thyroglobulin in the secretion granules (Leblond and Bennett, 1977). This view that secretory proteins pass through the stacked Golgi apparatus prior to entering secretion vesicles is supported by other studies that have used radiographic, cytochemical, and immunocytochemical techniques in a variety of tissues, such as in the parotid (Castle *et al.*, 1971), pituitary prolactin cells (Farquhar *et al.*, 1978), exocrine pancreas (Kraehenbuhl *et al.*, 1977) and the lacrimal gland (Hand and Oliver, 1977). The results of these experiments have located newly made secretory proteins in stacked Golgi cisternae and in immature and mature secretion granules, but these studies are limited in their resolution and have not adequately demonstrated the intracisternal movement of secretory proteins from the cis to the trans Golgi apparatus. Some indication that secretory proteins do travel from cis to trans Golgi elements derives from studies by Bergeron *et al.* (1978), which show that radioactive leucine-labeled plasma proteins first appear in an isolated cell fraction from rat liver that may be considered as deriving from cis Golgi elements and that later the radioactive plasma proteins appear in filled or partially filled secretory vesicles that may be derived from trans Golgi elements. Using the same methods for the isolation of hepatic Golgi cell fractions, the albumin precursor, proalbumin, has been shown to be present in cis Golgi elements and that the cell fraction composed of filled secretion vesicles contains both proalbumin and the processed product serum albumin (Redman *et al.*, 1978). This indicates that the cis Golgi elements contain a precursor of the protein seen in the secretion vesicles, but the movement of secretory proteins from cis to trans Golgi elements is not clearly demonstrated by any of these experiments.

C. GERL and Its Possible Role in Secretion

Novikoff (1976) described a close cytochemical and morphological relationship between lysosomes and parts of the Golgi apparatus. Cytochemically the enzyme acid phosphatase is located both in the smooth ER fraction usually seen at the *trans* side of the Golgi apparatus and in several different types of lysosomes, many of which appear to be produced in this area. This structural area has been termed GERL for Golgi–endoplasmic reticulum–lysosomes. First described in various neurons, GERL is also present in such secretory systems as the exocrine pancreas, the anterior pituitary gland, and hepatocytes. GERL is distinguished from other smooth-surfaced elements in the cell by the presence of acid phosphatase and by limiting membranes that are often thickened and sometimes coated. The cisternae of GERL often show a dense content. Anatomically GERL sometimes shows connections with the ER, but continuities between the stacked Golgi and GERL have not been seen. It is possible to discriminate cytochemically between the trans elements of the stacked Golgi apparatus and the neighboring GERL because the trans Golgi cisternae have thiamin pyrophosphatase but not acid phosphatase activity.

The possible involvement of GERL in secretion and more directly in packaging of secretory proteins is suggested by various observations that note that in several secretory systems (for example, guinea pig pancreas) acid phosphatase but not thiamin pyrophosphatase is present in immature secretion vesicles (Novikoff *et al.*, 1977). This observation has contributed to the view that the condensing vacuoles in the exocrine pancreas and immature secretory granules in other systems are formed from GERL and not from thiamin pyrophosphatase-staining Golgi apparatus.

There is, however, at present some disagreement as to the roles played by the Golgi apparatus and GERL in packaging of secretory proteins. As mentioned earlier, there are instances in which secretory products have been localized within the stacked Golgi cisternae. Some of these are based on the immunochemical and cytochemical observations of several pancreatic secretory proteins in bovine exocrine cells (Kraehenbuhl *et al.*, 1977), of B apoprotein of lipoproteins in Golgi hepatocytes (Alexander *et al.*, 1976), and of exportable peroxidases in the parotid (Herzog and Miller, 1970) and lacrimal gland cells (Hand and Oliver, 1977). These studies suggest the involvement of the stacked Golgi apparatus in the transport and packaging of secretory products, but do not exclude the participation of GERL in this process. Some of the confusion stems from disagreement on the extent of acid phosphatase location in the Golgi apparatus and has led some investigators to identify parts of the stacked Golgi cisternae as GERL, while others identify this same area as the trans or innermost element of Golgi apparatus.

Furthermore, depending on the system studied, the means by which packaging of proteins into secretion vesicles for exocytosis occurs appears to differ. As already mentioned, in the unstimulated guinea pig exocrine pancreas the condensing vacuoles (immature secretion vesicles) may arise from the fusion of small vesicles emanating from the transitional elements of the ER, thereby apparently bypassing the stacked Golgi cisternae (Jamieson and Palade, 1967a, 1968b), but in stimulated glands the pancreatic secretory proteins are detected in the stacked Golgi cisternae (Jamieson and Palade, 1971b). In many cells dilations from the innermost or trans Golgi elements are easily apparent, and these give rise to the secretion vesicles. In other systems, such as polymorphonuclear leukocytes (Bainton and Farquhar, 1966) and anterior pituitary gland cells (Smith and Farquhar, 1966), the secretion vesicles may derive not from the large dilation of the Golgi apparatus but from the fusion of several smaller vesicles that emanate from the Golgi sacules.

Not determined as yet is whether the trans element of the Golgi apparatus can be converted by vesicular membrane flow to GERL. Such a process would entail a loss of thiamin pyrophosphatase activity, a gain of acid phosphatase, and a change in structure. Novikoff *et al.* (1971) have presented arguments against this possible route, but some evidence from studies on peroxidase secretion in rat lacrimal glands suggests that this may occur. These studies suggest that peroxidase is transported through the stacked Golgi cisternae and is initially concentrated in the stacked Golgi apparatus. On the basis of cytochemical studies, transfer of peroxidase from the Golgi apparatus to immature secretion vesicles is thought to occur by conversion of the innermost (trans) Golgi saccules to GERL (Hand and Oliver, 1977). Thus, both the Golgi apparatus and GERL may be involved in the packaging of secretory products and the formation of secretory granules in this case.

Other possibilities that must be considered are that, since not all immature vesicles contain acid phosphatase (a marker for GERL), those vesicles which do not contain this marker enzyme are destined to be secreted and those which do contain acid phosphatase may be marked for intracellular degradation. In this respect it should be noted that very low density lipoprotein (VLDL) particles destined to be secreted are thought to be packaged by the Golgi apparatus (Alexander *et al.,* 1976), but these cells also contain another population of VLDL-containing vesicles which cytochemically appear to be part of GERL (Novikoff and Yam, 1978). The fate of this latter population of VLDL-containing vesicles has not yet been determined. Also some evidence suggests that more than one type of acid phosphatase exists in tissues, such as the liver, adrenal medulla, the prostate, and placenta, and thus the acid phosphatase noted in secretion vesicles may be distinct from that seen in GERL elements.

The difficulty in differentiating the functions of Golgi apparatus and GERL probably stems from their close functional and morphological relationship. The Golgi apparatus and GERL both appear to be involved in segregation of multiple products. For example, the newly synthesized exportable secretory proteins have to be separated from the lysosomal enzymes, and recycled secretory vesicles and autophagic vesicles also have to be distinguished and rerouted. Although these processes appear to be performed in the Golgi apparatus and GERL regions, we do not yet understand the mechanisms which govern them. More detailed studies, which may have to await more sensitive kinetic studies with defined proteins and membranes, are needed to unravel the interrelationships between the Golgi apparatus and GERL in secretion.

VI. STORAGE AND DISCHARGE

A. Concentration and Storage of Secretory Proteins

In most secretory tissues, the secretion vesicles are seen (by electron microscopy) to contain different densities of material. Autoradiographic studies with labeled secretory proteins show an increased number of grains associated with more mature granules, suggesting that secretory proteins reach the condensing vacuoles or immature granules in a dilute form and are progressively concentrated to the level seen in mature granules (termed zymogen granules in the exocrine pancreas) (Jamieson and Palade, 1967b; Caro and Palade, 1964). This concentration process occurs even when both protein synthesis and energy production are inhibited (Jamieson and Palade, 1971a). The mechanisms by which concentration occurs is not yet elucidated, although it has been suggested that in the exocrine pancreas, this occurs by interaction, within the Golgi apparatus, of high molecular weight sulfated compounds and the predominantly cationic secretory proteins, leading to the formation of an insoluble aggregate which will reduce the osmotic activity within the granule (Tartakoff *et al.,* 1974a; Reggio and Palade, 1978). The result of such a complex presumably will lead to an outflow of water and a concentration of secretory proteins within the vesicle. The attractiveness of this hypothesis is that the polysulfated compounds in the exocrine pancreas are known to be present in the stacks of the Golgi cisternae, in Golgi vesicles, in condensing vacuoles, and in zymogen granules (Berg and Young, 1971). Also, autoradiographic studies indicate that the primary site of $^{35}SO_4^{2-}$ incorporation into macromolecules is the Golgi apparatus (Young, 1973). Thus, these experiments place these polysulfated compounds in the correct cellular location to exert their effect on vesicular concentration.

Other types of interactions geared for concentrating proteins may occur in other cells, as may be the case for the concentration of insulin and proinsulin. The latter are acidic molecules and are thought to form insoluble paracrystals with divalent zinc within the secretion vesicles. The formation of insoluble insulin will lead, as described above, to a concentration of protein within the secretory vesicles (Steiner *et al.,* 1974b).

Mature secretion vesicles remain in the cell for varying periods of time. Most exocrine glands store secretory products in mature vesicles until the appropriate signal to release is received. This is advantageous, since in many cases a continuous discharge is not necessary; rather the cell must be able to secrete large amounts of secretion products quickly in response to specific physiological stimuli. In most cases *de novo* synthesis of the secretory proteins may not be rapid enough or may not produce sufficient secretory protein to satisfy the immediate demand. Such storage is evident in the apex of a resting exocrine pancreas cell which is filled with numerous stored zymogen granules. In some cells, however, such as the hepatocyte, secretory proteins are concentrated in mature secretory vesicles (see Fig. 7), but the vesicles are not stored within the cell for long periods of time. In the hepatocyte, as far as is known, the storage and discharge of the secretory vesicles is not hormonally controlled, and if a

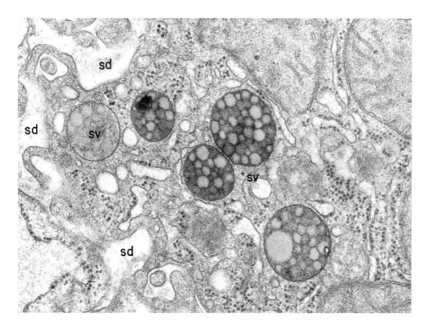

Fig. 7. Rat hepatocytes after the administration of colchicine. A part of the cell periphery at the sinusodal front of the cell is shown with a cluster of secretion vesicles marked with VLDL contents. Secretion vesicles, sv; space of Disse sd. (From Redman *et al.*, 1975, with the permission of the *Journal of Cell Biology*.)

greater amount of a circulating plasma protein is required, the rate of synthesis of these proteins is increased.

Concentration and storage of secretory proteins are not mandatory steps for secretion, and the extent to which these processes are involved in secretion varies in different secretory systems. A more detailed description of cells which do not markedly concentrate and store their secretory product will be given later when variations of the secretory process are discussed. Also, it should be noted that even cells which store the bulk of their secretory proteins in granules awaiting the hormonal stimulus to secrete may be capable in the interim of exporting small amounts of material. Thus, some exported proteins may be carried to the site of discharge by vesicles which contain less concentrated materials and which have not been stored for long periods of time.

B. Comparison of Protein Composition of Secretion Vesicles with That of Proteins Secreted

Greene *et al.* (1963) showed that the proportions of several secretory enzymes present in resting bovine pancreatic juice are identical to that found in isolated zymogen granules. Later Kraehenbuhl *et al.* (1977) extended this study and demonstrated by immunocytochemical techniques that trypsinogen, chymotrypsinogen A, carboxypeptidase A, RNase, and DNase could be detected in the Golgi cisternae and in all stored zymogen granules of all bovine pancreatic exocrine cells. These data suggest that the secretory proteins pass through the Golgi cisternae and are mixed and packaged together in zymogen granules. These studies also suggest that all of the secretory proteins from the exocrine pancreas are transported and discharged in parallel. This view is supported by kinetic studies which also show parallel processing and discharge of various pancreatic secretory proteins by the exocrine pancreas of the guinea pig (Tartakoff *et al.,* 1974a).

There are, however, some physiological conditions which indicate that certain secretagogues may selectively stimulate the secretion of certain digestive enzymes and not of others, and this has been taken to indicate that nonparallel secretion of pancreatic digestive enzymes may occur in the exocrine pancreas (Rothman, 1975). Whether this can be explained by regional differences in enzyme production by the exocrine gland, or by variations in the production of individual secretory proteins by different cells, or, less likely, by selective release of specialized granules containing a limited selection of secretory proteins is not yet known.

Whether all secretory vesicles contain a full complement of secretory proteins or whether some secretory vesicles contain a limited selection of exportable proteins appears to be answered for the bovine exocrine pancreas, but similar studies have not been performed with other tissues. Hepatic secretory vesicles appear by electron microscopy to contain both lipoprotein particles and other dense amorphous material, and the content of isolated hepatic secretion vesicles is similar to that of plasma proteins. However, careful immunocytochemical studies using several defined antibodies against various plasma proteins are needed to determine whether the hepatocyte, like the exocrine pancreas, contains secretory vesicles which carry a mixture of various exportable products.

C. Exocytosis

The final steps in secretion are (a) movement of the mature secretion vesicles to the sites of discharge, (b) recognition of mature secretory vesicles by specific areas of the plasma membrane, (c) attachment of the two membranes which probably leads to a reorganization of the membrane at the site of attachment, (d) fusion of the membranes such that the contents of the secretion vesicles are now open to the extracellular space and can be discharged, and (e) the secretion vesicle membranes remain fused with the plasma membrane, parts of which are later removed by endocytosis and reutilized for further packaging of secretory proteins or are degraded and used for the synthesis of new membranes (Fig. 8).

Studies in a variety of tissues firmly established that exocytosis is an energy-dependent process and that calcium ions are required (Douglas, 1974; Rubin, 1974; Kaliner and Austen, 1975). In glandular cells whose discharge is stimulated by specific hormones and exocytosis does not

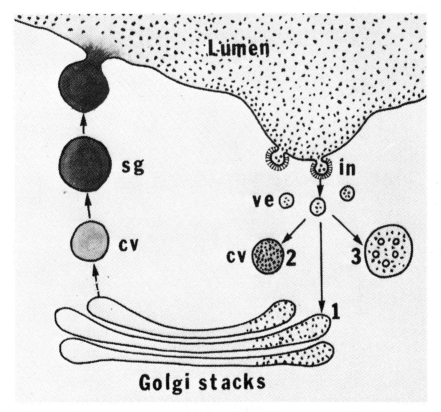

Fig. 8. A diagram showing the interactions between Golgi complex and the cell surface in secretory cells. On the left the secretory pathway depicting the route of secretory proteins from the stacked Golgi cisternae, to condensing vacuoles (cv) to mature secretion granules (sg) which fuse with the apical plasma membrane is shown. The right side shows the postulated pathways taken by retrieved membranes. Parts of the surface membrane, with a coat on their cytoplasmic side, invaginate (in), pinch off, and then lose the coats. These apical vesicles (ve) may then follow several routes: (1) they may move to the stacked Golgi with preference for the distended rims; (2) they may go to condensing vacuoles, or (3) they may be directed to lysosomes. [From Herzog and Farquhar, 1977, with permission from *Proceedings of the National Academy of Sciences* (U.S.A.).]

occur at a continuous rate but is intermittent and regulated by humoral agents, the stimulus–secretion coupling often involves the interplay of cyclic nucleotides and phosphorylation of one or more membrane proteins by specific protein kinases (Rassmussen *et al.,* 1975; Schramm and Selinger, 1975; Greengaard, 1978). During exocytosis, at least in the exocrine pancreas, a depolarization of the plasmalemma occurs (Mathews and Petersen, 1973). That calcium may be employed as a "second messenger" in triggering exocytosis is suggested by studies which show that many secretory cells require extracellular calcium, in addition to the relevant secretogogue, to stimulate maximal discharge of stored secretory products (Berridge, 1975). Extracellular calcium is not, however, always required for exocytosis as has been noted in the exocrine pancreas (Argent *et al.,* 1973; Robberecht and Christophe, 1971; Williams and Chandler, 1975). It is now a widely accepted concept that an increase in cytosolic Ca^{2+} triggers exocytosis and that this increase may come from intracellular calcium stores or may result from an influx of extracellular Ca^{2+} due to an increased plasma membrane permeability in response to the secretogogue. Still unanswered, however, is how the increase in cytoplasmic calcium concentration can lead to a stimulation of exocytosis. The exact sequence of events that occur after the secretogogue has interacted with its specific receptor on the plasma membrane and before discharge of secretion products has occurred has not yet been fully elucidated.

In some cells, for example, the hepatocyte stimulus–secretion coupling seems to be lacking. In these cells discharge of secretory products is not enhanced by humoral agents, and we do not know whether exocytosis involves the interplay of Ca^{2+}, cyclic nucleotides, and membrane protein phosphorylation. An indication that these processes may not occur or may be minor in these cells stems from studies on plasma cells which produce immunoglobulin but whose secretion is not regulated by secretogogues. In these cells secretion is not affected by the absence of extracellular Ca^{2+}, by the addition of cyclic nucleotide derivatives, or by agents which alter the cellular cAMP and cGMP levels (Tartakoff and Vassalli, 1977).

An important facet of exocytosis is that secretion vesicles must recognize areas of the plasma membrane with which to fuse. This process is very selective, and in actively secreting cells, such as the exocrine pancreas, the zymogen granules are concentrated toward the apex of the cell, and when exocytosis is triggered these vesicles fuse only with the apical plasmalemma, although at that time other membranes may be in the vicinity. The only other membrane fusion which has been noticed is the occasional preliminary fusing of two or three pancreatic zymogen granules which may lead to the exocytosis of several granules in tandem. In other cells, the specificity is just as marked. For instance, in the hepatocyte secretion vesicles are often seen both at the bile front and close to the plasma membrane which leads to the space of Disse. However, the secretion vesicles which contain albumin and other plasma proteins move only to, and fuse with, that part of the plasma membrane which leads to the space of Disse (Fig. 7). This specificity suggests the existence of recognition sites in both the secretion vesicles and the interacting portions of the plasma membrane that allows for specific fission of the two membranes prior to their fusion. Relevant in this regard are studies by Satir *et al.* (1973) using the freeze-fracture technique on *Tetrahymena,* an organism which actively discharges mucocysts by exocytosis. It was shown that a "fusion rosette" appears on the plasma membrane directly above the mature mucocysts. The resting mucocysts near the plasma membrane contain an annulus, or ring, at the anterior end of their membrane, but this ring is missing in mucocysts not close to the site of discharge. These matching membrane structures, the "fusion rosette" on the plasma membrane and the ring in the mature mucocysts, may contain the necessary recognition sites, and their appearance may signal membrane changes occurring in these organelles in preparation for membrane fusion (Satir, 1975).

Coated vesicles are often seen in secretory cells, and their involvement in endocytotic processes has been well demonstrated (Friend and Farquhar, 1967; Nagasawa *et al.,* 1971; Heuser and Reese, 1973). Coated vesicles may also, however, be involved in exocytosis. For example, coated secretory vesicles, containing casein, are seen in lactating mammary epithelial cells, and these coated vesicles are sometimes attached to the inner surface of the plasma membrane by "preexocytotic plaques" (Franke *et al.,* 1976).

Some agents, such as the antimicrotubular drugs colchicine and vinblastine, are known to inhibit exocytosis in a variety of secretory tissues (Roberts, 1974). In several secretory cells the block in secretion has been localized at the region of the Golgi apparatus where concentration, packaging, and exocytosis occurs (Olsen and Prockop, 1974; Redman *et al.,* 1975; Patzelt *et al.,* 1977). In the hepatocyte colchicine inhibits plasma protein secretion after secretion vesicles have been filled with secretory products but prior to their fusion with the plasma membrane (Redman *et al.,* 1975). The fact that colchicine acts on the hepatocyte, in which secretion is not thought to be governed by humoral agents, dissociates the effect of this antimicrotubular drug from the stimulus–secretion mechanism. In glandular cells where the release of stored material is stimulated by secretogogues, colchicine does not affect discharge of stored zymogen granules but does impede the replenishment of new mature secretion granules (Palzelt *et al.,* 1977).

The many published experiments on this topic indicate that colchicine, which binds to tubulin and inhibits microtubule polymerization, blocks the secretory pathway in the Golgi region, indicating that microtubules may function in secretory vesicle formation and/or exocytosis. However, despite much information gathered from a large number of studies on many different tissues, what role microtubules play in vesicle maturation and subsequent exocytosis is not clear. Whether colchicine acts by its antimicrotubular effect or some other way is still an open question.

A fibrillar network is often noticed around the area of discharge where the secretion vesicles meet the specific portions of the plasma membrane to which they will fuse. A fibrillar halo also is often seen around the discharging secretion vesicles (Palade, 1975). These structures may contain contractile proteins that play a role in moving and/or aligning the secretion vesicles to the region of fusion. The ground substance, probably modified and specialized in certain areas of the cell as described above, may be important in secretion and needs further experimental attention.

VII. PROCESSING OF SECRETORY PROTEINS DURING INTRACELLULAR TRANSPORT

A. Limited Proteolysis

As described earlier, most secretory proteins, with only one known exception (ovalbumin), are initially synthesized as precursor proteins that are proteolytically cleaved during vectorial discharge. Even with ovalbumin a small amount of processing occurs, since the initiator methionine at the amino terminus of the primary translation product is removed when the polypeptide is about 20 residues long and is replaced by an acetyl group on the amino-terminal glycine (Palmiter et al., 1978). Subsequent to vectorial discharge, some secretory proteins undergo further limited proteolysis at their amino termini. This is not, however, a general rule since some secretory proteins, such as albumin, insulin, and parathyroid hormone, exist as intermediate precursors (termed proalbumin, proinsulin and proparathyroid hormone, respectively), but a large number of other secretory proteins, for example, growth hormone and various of the pancreatic digestive enzymes, do not have detectable pro-intermediate forms (Steiner, 1976). The final proteolytic processing of many of these intermediate forms is performed in the Golgi area, and in the case of albumin, the cleavage of proalbumin into serum albumin occurs in mature secretory vesicles but not while proalbumin is in a less concentrated stage in immature secretory vesicles (Ikehara et al., 1976; Edwards et al., 1976; Redman *et al.,* 1978). Thus, this final proteolytic processing occurs at a very late stage of secretion. While in most cases the secretory proteins contain peptide extensions at the amino-terminal end, some proteins, such as proinsulin, proglucagon (Steiner *et al.,* 1974a), and procollagen (Fessler *et al.,* 1975; Byers *et al.,* 1975; Olsen *et al.,* 1976), also have peptide extensions at the carboxyl-terminal end.

Why some secretory proteins occur as intermediate precursors and others do not is still a puzzle. Whether the cleaved peptides have a function of their own, whether limited proteolysis is necessary to uncover an activity that does not exist in the intermediate precursor, or whether the amino-terminal extension is necessary for intracellular segregation and transport are some of the possibilities which have been entertained.

The processing of prosecretory proteins that occurs during intracellular transport necessitates the action of specific intracellular proteolytic enzymes housed within the membranous compartments involved in secretion. In the examples of proparathyroid hormone, proalbumin, and proinsulin, the cleavage is due to a trypsin-like enzyme, since an arginyl residue is present at the cleavage site. (See Chapter 5 by Steiner, this volume.)

Limited proteolysis is only one of the steps that secretory proteins undergo during intracellular transport prior to their discharge from the cell. Many other processing steps also occur. For example, it has been estimated that the polypeptide coded by procollagen mRNA undergoes at least seven posttranslational modifications before the protein is functional outside of the cell (Campbell and Blobel, 1976; Bornstein, 1974). These modifications include scission of both amino-terminal and carboxy-terminal peptides (Davidson *et al.,* 1975), hydroxylation of prolyl and lysyl residues of the polypeptide (Cardinale and Udenfriend, 1974), glycosylation (Harwood *et al.,* 1975), and disulfide formation. Not all secretory proteins undergo all of the co- and posttranslational modifications to which collagen is subjected, but all secretory proteins are known to be modified to a lesser or greater degree.

B. Other Modifications to Secretory Proteins Occurring in the Rough ER

In addition to signal peptidase, which cleaves the initial signal peptide extension, a variety of other enzymes which further modify secretory proteins are present in the rough ER. Some examples, such as the enzymes which hydroxylate the lysyl and prolyl residues in collagen, have already been noted. Another important early modification is disulfide bond formation (Anfinsen, 1973). This is important because it is estimated that the ER membrane is permeable to molecules of about 10 Å in diameter (Tedeschi *et al.*, 1963), and thus the early segregation of secretory polypeptides by the rough ER necessitates the conversion of nascent secretory proteins into a globular form too large to leave the ER. Thus, some of the modifications to secretory proteins, such as disulfide formation which occurs during or after vectorial discharge, may be needed to ensure that these proteins remain compartmentalized within the cisternae of the rough ER.

Other modifications of the amino acid residues are also known to occur in the ER. An interesting example is the processing of rat prothrombin which is synthesized from a precursor in a vitamin K-dependent step that involves the attachment of calcium binding groups to the protein precursor. In this process specific glutamyl residues of the prothrombin precursor are carboxylated to yield the novel amino acid residues, γ-carboxyglutamic acid, which actively bind calcium (Magnusson *et al.*, 1974; Suttie, 1974).

C. Glycosylation of Secretory Proteins

Secretory protein glycosylation is a multistep procedure to which many proteins are subjected. It begins early in the secretory process during protein translation and ends late while the proteins are being packaged in the Golgi apparatus. Thyroglobulin is a good example of a protein that is glycosylated during secretion; it contains both short and long glycosylated side chains with the proximal sugar, N-acetylglucosamine, attached to the amide nitrogen of the asparagyl residue of nascent thyroglobulin. The N-acetylglucosamine residues are usually followed by mannose, which are terminal in the short side chains, or which contain other sugars, often by branching, in the long side chains. The proximal sugars N-acetylglucosamine and mannose are added to the growing polypeptide chain during its passage through the ER membrane. The N-acetylglucosamine and mannose residues are probably added as a unit with dolichol oligosaccharide as a donor (Waechter and Lennarz, 1976; Baynes *et al.*, 1973). The remaining terminal sugars, galactose, fucose, and sialic acid, are added stepwise in the Golgi apparatus (Leblond and Bennett, 1977).

Proximal glycosylation of nascent secretory proteins is not needed to transfer the secretory protein across the ER membrane or subsequently to segregate these proteins within the ER. For example, several secretory proteins, such as albumin, are not detected in glycosylated forms, and two secretory proteins, the α subunit of the placental peptide of human chorionic gonadotropin and preplacental lactogen, which normally exist in glycosylated forms, have been shown to be sequestered *in vitro* by microsomal vesicles even though proximal glycosylation had been inhibited by the antibiotic tunicamycin. In these experiments, the signal sequences of these proteins were cleaved, showing that glycosylation and the signal peptidase action are independent (Bielinska *et al.*, 1978). Glycosylation of the asparaginyl residue on the nascent secretory protein requires that the latter be in contact with the specific glycosylation enzymes while still attached to the polysomes and that the nascent protein contain the determinant sequence -Asn-X-Thr(Ser) exposed to those enzymes during vectorial discharge. The tripeptide sequence containing asparagine is thought to be necessary, but not sufficient, for proximal glycosylation to occur (Marshall, 1974; Pless and Lennarz, 1977).

Terminal glycosylation of secretory proteins occurs in the Golgi apparatus (Schachter, 1974). In the case of thyroglobulin, autoradiography of thyroid follicular cells has shown that [^{3}H]fucose incorporation into protein occurs only in the Golgi apparatus and not in condensing vacuoles or in the secretion granules (Haddad *et al.*, 1971). The autoradiographic grains seen in those studies were distributed throughout the Golgi stack, suggesting that fucose is added as thyroglobulin passed through the Golgi apparatus. Biochemical, as well as autoradiographic, evidence suggests that galactose and sialic acid are also added to secretory proteins while they are within the Golgi apparatus and that these additions occur prior to the secretion of thyroglobulin into the colloid-containing lumen. Other secretory glycoproteins are believed to be glycosylated in a similar multistep procedure.

Most proteins secreted by eukaryotic cells are glycoproteins, but there are several exceptions. For example, albumin, the major protein secreted by the liver, is not a glycoprotein, and its precursor forms do not contain sugars. Among the pancreatic secretory proteins, trypsinogen, chymotrypsinogen, and carboxypeptidase are thought not to be glycoproteins, yet they appear in the Golgi cisternae together with such glycoproteins as DNase and RNase (Kraehenbuhl *et al.*, 1977). Proteins such as bovine pancreatic ribonuclease (Plummer and Hirs, 1964) and bovine lactalbumin (Hill and Brew, 1975) normally are secreted both as glycosylated and unglycosylated forms, and Weitzman and Scharff (1976) have described a stable mutant plasma cell that secretes unglycosylated IgG. These studies indicate that proteins need not be glycosylated to be secreted. However, this does not rule out the possibility that carbohydrate chains may play a role in the transport of specific proteins. Studies with tunicamycin in different secretory systems have given varied results. This antibiotic inhibits the secretion of IgA and IgE by myeloma tumor cells (Hickman *et al.*, 1977), but does not affect the secretion of transferrin and very low density lipoprotein by chick or rat hepatocytes (Struck *et al.*, 1978). It may be speculated that the presence of specific carbohydrates in some proteins may dictate whether they are to be secreted or whether they are to remain inside of the cell, perhaps inserted into intracellular or plasma membranes; yet other secretory proteins may require the carbohydrate moiety only for a function outside of the cell and may not require glycosylation in order to be secreted.

Several other processing modifications are known to occur within the secretory compartments. Two examples are the iodination of thyroxine and the phosphorylation of serine and threonine residues in a variety of secretory phosphoproteins.

Again, we must indicate a paucity of information as to the determinants which specify that a given secretory protein be directed to a given subcompartment within the multistep secretory process. Perhaps individual proteins contain specific signals in their amino acid sequence or in their carbohydrate moieties by which they recognize membrane regions containing the specific processing enzymes housed along the secretory pathway.

VIII. MEMBRANE REDISTRIBUTION DURING SECRETION

A. General Considerations

In the movement of secretory proteins from the rough ER to the area of the Golgi apparatus and subsequently during packaging, concentration, and exocytosis, there is extensive membrane translocation from one cellular compartment to another. How the cell maintains the proper equilibrium between the various membrane compartments and how the specificity of the various organellar membranes is preserved during the secretory process is not yet fully understood, but is the subject of extensive study.

Some points that need to be kept in mind when discussing this topic are the following. (a) Although many secretory vesicles fuse with the plasma membrane during exocytosis, there appears to be no long lived net gain in cell surface during active secretory periods in exocrine cells (Amsterdam *et al.*, 1969). The excess luminal membrane is removed and the characteristics of the endomembrane pool is retained in secretory cells. (b) *In vivo* membrane labeling studies show that membrane proteins are synthesized at lower rates than are secretory proteins (Meldolesi, 1974; Wallach *et al.*, 1975; Castle *et al.*, 1972). Therefore, the continual production of secretion vesicles for use in intracellular transport and exocytosis is thought not to be strictly dependent on *de novo* synthesis of complete membranes. Membrane recycling, which may occur either by reutilization of existing membranes or by the reassembly of preexisting membrane parts, is thought to contribute to membrane conservation during secretion (Steinman et al., 1976; Holtzman et al., 1977; Farquhar, 1978a). (c) During secretion there does not appear to be a general mixing of lipid and protein moieties of the various membranes, since each membrane compartment (although having many proteins and lipids in common) maintains a distinctively different chemical composition (Fleischer, 1974; Howell et al., 1978). This suggests that strict limitations on the fusion reactions are imposed during secretion. This is an especially important idea since current knowledge of membrane structure leads us to expect lateral mixing of membrane components following fusion, unless specific membrane components are restrained, perhaps by their attachment to part of the cytoskeletal system. How the secretory cell removes excess membrane from the recipient compartment and replenishes membrane to the donor compartment while retaining the individuality of the membranes is a key question that needs to be answered.

B. Membrane Biogenesis

The intracellular path of newly synthesized intrinsic plasma membrane proteins has been assumed but not definitely shown to be much like that of secretory proteins, i.e., rough ER → smooth ER → Golgi apparatus → secretory vesicles → plasma membrane. The most direct evidence for the involvement of the rough ER, smooth ER, and Golgi apparatus in plasma membrane synthesis derives from studies on the assembly of viral envelope proteins. The G protein of vesicular stomatitis virus (VSV), which is thought to be a transmembrane protein with parts of the molecule exposed to both the outside and the inside of the plasma membrane, is synthesized on the rough ER, receives its proximal sugars at this site, is terminally glycosylated (presumably in the Golgi apparatus), and then is deposited in the plasma membrane (Hunt and Summers, 1976; Knipe et al., 1977). On the other hand, the M protein of VSV, which differs from the G protein in being located on the inner surface of the plasma membrane and in not being a glycoprotein, follows a different intracellular route. It is synthesized on free polysomes rather than on the rough ER and appears more rapidly on the plasma membrane than does the G protein (Morrison and Lodish, 1975; Grubman et al., 1975; Knipe et al., 1977). This shows that different plasma membrane proteins follow different intracellular routes prior to their insertion into the plasma membrane. At present, the extent to which the intracellular pathway taken by viral surface membrane proteins can be compared to the intracellular pathway taken by cellular plasma membrane proteins during active secretion cannot be assessed.

Terminal glycosylation and sulfation of lipids and proteins are generally assumed to be activities of the Golgi apparatus (Schachter, 1974; Fleischer, 1977), and, because many plasma membrane components are thus modified, they are thought to pass through the Golgi apparatus prior to being deposited on the plasma membrane. Other than viral glycoproteins mentioned above, certain plasma membrane proteins, such as adenylate cyclase (Cheng and Farquhar, 1976) and the insulin receptor (Bergeron *et al.*, 1973), have been shown to be present in the Golgi apparatus, and their presence there has been taken as a sign that they are plasma membrane precursors. However, a precursor–product relationship between the insulin receptor and/or adenylate cyclase proteins in the Golgi apparatus and those in the plasma membrane has not been demonstrated. The receptors may be in the Golgi apparatus as a result of recycling from the plasma membrane or they may be permanent residents of the Golgi apparatus.

The aforementioned studies suggest that the Golgi apparatus plays an active role in the biogenesis of many plasma membrane proteins and the Golgi apparatus is known to be involved in the packaging of secretory protein, but whether and to what extent these two processes are coupled is undetermined.

Membrane flow as proposed by Morré and colleagues envisions membrane constituents synthesized in the rough ER, or in the nuclear envelope, moving via the smooth ER and the Golgi apparatus to the cell surface (Morré *et al.*, 1974a,b; Morré and Ovtracht, 1977). In order to account for both the similarities and differences seen in different membrane compartments, the membrane components are thought to be differentiated by modification, addition, and elimination during intracellular transit. This model proposes constant synthesis of membrane material coupled to a rapid "differentiation" of these components as they travel from compartment to compartment; the membrane compartments act not only as vehicles to carry secretory proteins between these compartments but are themselves carriers of cell membranes. This mechanism is thought to be operative not only in dividing and growing cells but also in secretory cells and to be the source by which the latter replaces membranes that may be degraded or lost during secretion, as occurs in mammary gland epithelia (Morré *et al.*, 1979).

C. Membrane Movements during Exocytosis

The major movement of membranes from one compartment to another occurs during exocytosis when secretory vesicles, filled with secretory proteins, fuse with the plasmalemma. Even though the secretory vesicle membrane and the plasma membrane share several components (Morré *et al.*, 1974a,b), the composition of these two types of membranes, in several examples noted, are markedly different (Meldolesi and Cova, 1972; Castle *et al.*, 1975; Hörtnagl, 1976). This raises the question of what happens to the secretion vesicle membranes when they fuse with the plasmalemma. The secretion vesicle components may mix with the existing plasma membrane or constraints may be placed on such mixing. It is known that the excess luminal membrane formed during exocytosis is removed, but how the intracellular membrane pool is maintained during active secretion has not as yet been properly answered. One view suggests that the excess luminal membrane is internalized, degraded, and reutilized during the formation of new membrane in order to satisfy the membrane needs of the donor compartment. An alternate view suggests that following exocytosis the vesicle membrane is recovered by internalization but that it is reutilized, without prior degradation, in successive cycles of packaging of secretory products and further exocytosis (for a review, see Holtzman *et al.*, 1977; Farquhar, 1978a).

Recent studies have expanded these suggestions, and some investigators think that luminal vesicles retrieved after exocytosis from the apical region of glandular cells have several options. They may (a) move to the stacked Golgi cisternae with preference for the distended rims of the Golgi apparatus, (b) they may move to condensing vacuoles, or (c) they may go to lysosomes via GERL (Fig. 8). These options are based on tracer studies performed on the rat lacrimal and parotid glands, in which endocytosed dextran was seen to reach the stacked Golgi apparatus, the condensing vacuoles, and lysosomes. It was also noticed in later studies that cationic ferritin, which binds to the cell membrane, was taken up by dissociated anterior pituitary cells and was then transferred to multiple stacked Golgi cisternae, deposited around immature secretion granules, and located within GERL and lysosomes. By contrast, anionic ferritin, which also is endocytosed by these cells, but which, unlike cationic ferritin, does not bind to the cell surface, was concentrated mostly in lysosomes and did not appear in organelles involved in protein secretion (Herzog and Farquhar, 1977; Farquhar, 1978b).

A unanimous view of membrane distribution events during secretion cannot be arrived at now. Nevertheless, most agree that the intracellular movement of secretory proteins is accomplished while the endomembranes retain a distinctive chemical composition. Whether this is accomplished by a flow of membrane components from rough ER to the Golgi apparatus to the plasma membrane, or by shuttling membrane vesicles which carry secretory proteins from donor to acceptor area, or by a combination of both processes is not known. However, in the process of

packaging secretory proteins and subsequent exocytosis, at least in the
lacrimal, parotid, and anterior pituitary cells, there appears to be a shuttling of vesicles between the Golgi complex and the plasmalemma. Studies
with cells, such as neurons, macrophages, and fibroblasts, also indicate
that membranes are conserved and recycled during exocytosis. The
membrane vesicles probably contain receptors that allow them to select
the proper proteins for transport and probably also contain receptors on
their surfaces which allow them to recognize and fuse to appropriate
donor areas. In this respect, it should be noted that the surface of vesicles
thought to be involved in membrane recycling are coated with a protein,
clathrin, which may serve to recognize specific membrane domains
(Pearse, 1976).

The Golgi apparatus in secretory cells is particularly active in membrane redistribution. Some of the membrane traffic through this organelle
is probably biosynthetic in nature and represents membrane flow from the
Golgi apparatus to the plasma membrane. Other membranes in this organelle may be returning from the plasma membrane to engage in a new
round of secretory protein packaging and yet other recycled membranes
may be rerouted, perhaps via GERL, to lysosomes. In addition, some of
the material in the Golgi apparatus may have arrived from the plasma
membrane to serve other purposes, such as the internalization of receptors. The extent to which these many paths are affected by or contribute
to protein secretion and the mechanisms by which these membranes are
separated and directed are not yet clear.

IX. VARIATIONS IN THE SECRETORY PATHWAY OF DIFFERENT TYPES OF CELLS

As already mentioned, not all secretory cells necessarily follow all of
the steps described for the pancreatic exocrine cells. In the hepatocyte
long-term storage and stimulus–secretion apparently do not occur, but the
newly made plasma proteins produced by the hepatocyte nevertheless
follow the general path of synthesis on membrane-attached polysomes,
segregation within the cisternae of the rough ER, transfer to the Golgi
region via the smooth ER, concentration into secretory vesicles, and discharge at the space of Disse (Peters, 1977). Endocrine cells follow the
same path plus a storage step, but in these cells exocytosis is not limited
to a specific domain of the plasmalemma as is true in exocrine cells
(Palade, 1975).

In many secretory cells, such as plasma cells (which secrete immunoglobulin) and fibroblasts (which produce extracellular fibrils and collage-

nous protein), secretion vesicles have been difficult to detect, which produced the former notion that discharge of secretory proteins occurs directly from the ER. Now, however, it is thought that the secretory process of these specialized cells are modified to accommodate specific needs. Since these cells are involved in continuous production and secretion, there is little or no need for them to concentrate and store the secretory products intracellularly. However, it has been determined that the secretory proteins of these cells do need to travel through the Golgi region in order to finish the processing of the proteins (Olsen and Prockop, 1974; Melchers, 1971; Tartakoff and Vassalli, 1977). In some of these cells, secretion vesicles have been identified, although at times they are not easily recognizable as such. These secretion vesicles move to the plasmalemma, fuse with the surface membrane, and discharge their products as do other secretory cells. There are some differences, however, between these cells and exocrine pancreas cells. For example, in plasma cells secretion occurs at a maximal rate and there is no evidence of short-term physiological control of discharge. Secretion by plasma cells is not easily influenced by changes in extracellular or intracellular calcium, and so far cyclic nucleotides have not been implicated as playing a role in secretion. In common with secretory cells, which concentrate and store secretory products, blocking respiration of plasma cells inhibits discharge of secretory products. Also, it has been shown that a carboxylic ionophore, monensin, causes a dilation of the Golgi elements, induces the formation of vacuoles in the Golgi area, and appears to inhibit the exit of secretory proteins from the Golgi apparatus (Tartakoff and Vassalli, 1978).

Palade (1975) has noted special cases that are variations on the general theme. Prominent among these are those cells which, instead of discharging their secretory products extracellularly, discharge proteins into endocytic vacuoles. For instance, release of lysosomal hydrolases by phagocytosing neutrophilic leukocytes, proceeds by fusion of the secretion granules with the phagocytic vacuole (Zucker-Franklin and Hirsch, 1964). Release of some of these enzymes to the outside may occur if the phagosome is open to the outside surface because of incomplete engulfment by the cell. These cells are also able, however, to release enzymes by direct extrusion of granule enzymes to the exterior when they encounter nonphagocytosable immune complexes. This has been shown *in vitro* to occur along the plasma membrane that is in contact with a surface-bound immune complex but not on the cell surface that is not adherent to this immune complex (Henson, 1971). In this latter situation, the secretion granules fuse with the cell surface that would normally be a vacuole, but in this *in vitro* situation it happens to be the exterior surface membrane. In contrast to the variation at the level of exocytosis, the synthesis and intracellular transport of secretory proteins by leukocytes occurs as for exocrine cells.

Macrophages also discharge secretory proteins into endocytic vacuoles (Cohn *et al.*, 1966). Unlike the neutrophilic and eosinophilic leukocytes, but like plasma cells and fibroblasts, these cells do not concentrate and store the secretory materials prior to discharge. It is thought that in macrophages small primary lysosomes (small secretion vesicles carrying dilute acid hydrolases) move from the Golgi complex to the endocytic vacuoles. The endocytic vacuoles may then fuse with secondary lysosomes providing a further source of hydrolases.

The shuttling or rerouting of small secretory vesicles or primary lysosomes to the endocytic vacuole highlights the discriminatory role that the Golgi apparatus plays in many cells that actively secrete proteins and produce lysosomes for autophagy. In this respect, the roles of GERL and the stacked Golgi cisternae become intermingled and confusing. However, this intracellular region obviously is capable of handling and packaging both secretory proteins and lysosomal proteins. It is important to note, moreover, that situations occur when exocrine cells, such as the anterior pituitary, prefer to degrade their secretory products rather than secrete them—perhaps because of an overproduction of secretory protein or a lowered demand for these proteins. Under these circumstances the secretion granules are discharged into secondary lysosomes (Farquhar, 1971).

X. CONCLUDING REMARKS

The salient features in protein secretion are the segregation of the proteins into membrane compartments and the specific interactions of these intracellular membranes that serve to move the proteins from their site of synthesis to the region of discharge. During intracellular transport, the proteins are modified at various cellular locations to prepare them for their extracellular functions.

The sequence of intracellular events has been elucidated for a large number of different secretory cells and for many different exportable proteins. However, we still do not understand the forces that transport the proteins from one cellular location to another. In particular, we have little or no understanding of the role, if any, that the cell ground substance may play in transport.

Intracellular transport of proteins is specific, and only selected proteins are transported and processed. It is thought that the membrane compartments which carry the secretory proteins may contain several receptors. Some of these receptors may reside intracisternally to enable membranes to choose the correct proteins for export and to discriminate between them for purposes of processing, packaging, and transport to the next compartment. Other receptors may be on the cytoplasmic surface of these membranes and may allow these membranes to recognize, bind, and fuse to the successive compartment. This view remains speculative, as does the nature of these receptors.

The movement of secretory proteins is thought to involve the controlled fission and fusion of successive membrane compartments, but we have little knowledge of the associated molecular events. Thus, further studies on membrane interrelationships are needed to understand and separate the biosynthetic and secretory functions of the various cellular membranes.

REFERENCES

Alexander, C. A., Hamilton, R. L., and Havel, R. J. (1976). Subcellular localization of B apoprotein of plasma lipoproteins in rat liver. *J. Cell Biol.* **69,** 241–263.

Amsterdam, A., Ohad, I., and Schramm, M. (1969). Dynamic changes in the ultrastructure of the acinar cell of the rat parotid gland during the secretory cycle. *J. Cell Biol.* **41,** 753–773.

Anfinsen, C. B. (1973). Principles that govern the folding of protein chains. *Science* **18,** 223–230.

Argent, B. E., Case, R. M., and Scratcherd, T. (1973). Amylase secretion of the perfused cat pancreas in relation to the secretion of calcium and other electrolytes and as influenced by the external ionic environment. *J. Physiol. (London)* **230,** 575–593.

Bainton, D. F., and Farquhar, M. G. (1966). Origin of granules in polymorphonuclear leukocytes. Two types derived from opposite faces of the Golgi complex in developing granulocytes. *J. Cell Biol.* **39,** 299–317.

Baynes, J. W., Hsu, A. F., and Heath, C. F. (1973). The role of mannosylphosphonyldihydropolyisoprenol in the synthesis of mammalian glycoproteins. *J. Biol. Chem.* **248,** 5693–5704.

Berg, N. B., and Young, R. W. (1971). Sulfate metabolism in pancreatic acinar cells. *J. Cell Biol.* **50,** 469–483.

Bergeron, J. J. M., Evans, W. H., and Geschwind, I. I. (1973). Insulin binding to rat liver Golgi fractions. *J. Cell Biol.* **59,** 771–775.

Bergeron, J. J. M., Borts, D., and Crug, J. (1978). Passage of serum-destined proteins through the Golgi apparatus of rat liver. An examination of heavy and light Golgi fractions. *J. Cell Biol.* **76,** 87–97.

Berridge, M. J. (1975). The interaction of cyclic nucleotides and calcium in the control of cellular activity. *Adv. Cyclic Nucleotide Res.* **6,** 2–98.

Bielinska, M., Grant, G. A., and Borine, I. (1978). Processing of placental peptide hormones synthesized in lysates containing membranes derived from tunicamycin-treated ascites tumor cells. *J. Biol. Chem.* **253,** 7117–7119.

Blobel, G. (1977). Synthesis and segregation of secretory proteins: The signal hypothesis. *In* "International Cell Biology" (B. R. Brinkley and K. R. Porter, eds.), pp. 318–325. Rockefeller Univ. Press, New York.

Bornstein, P. (1974). The biosynthesis of collagen. *Annu. Rev. Biochem.* **43**, 567–603.

Byers, P. H., Click, E. M., Harper, E., and Bornstein, P. (1975). Interchain disulfide bonds in procollagen are located in a large nontriple-helical COOH-terminal domain. *Proc. Natl. Acad. Sci. U.S.A.* **72**, 3009–3013.

Campbell, P. N., and Blobel, G. (1976). The role of organelles in the chemical modification of the primary translation products of secretory proteins. *FEBS Lett.* **72**, 215–226.

Cardinale, G. L., and Udenfriend, S. (1974). Prolyl hydroxylase. *Adv. Enzymol.* **41**, 245–300.

Caro, L. G., and Palade, G. E. (1964). Protein synthesis, storage and discharge in the pancreatic exocrine cell. *J. Cell Biol.* **20**, 473–495.

Castle, J. D., Jamieson, J. D., and Palade, G. E. (1972). Radioautographic analysis of the secretory process in the parotid acinar cell of the rabbit. *J. Cell Biol.* **53**, 290–311.

Castle, J. D., Jamieson, J. D., and Palade, G. E. (1975). Secretion granules of the rabbit parotid gland. Isolation, subfractionation and characterization of membrane and content sub-fractions. *J. Cell Biol.* **64**, 182–210.

Cheng, H., and Farquhar, M. G. (1976). Presence of adenylate cyclase activity in Golgi and other fractions from rat liver. II. Cytochemical localization within Golgi and ER membranes. *J. Cell Biol.* **76**, 237–244.

Cohn, Z. A., Fedorko, M. E., and Hirsch, J. B. (1966). The *in vitro* differentiation of mononuclear phagocytes. V. The formation of macrophage lysosomes. *J. Exp. Med.* **123**, 757–766.

Davidson, J. M., McEneany, L. S. G., and Bornstein, P. (1975). Intermediates in the limited proteolytic conversion of procollagen to collagen. *Biochemistry* **14**, 5188–5194.

deDuve, C., and Baudhuin, P. (1966). Peroxisomes (microbodies and related particles). *Phys. Rev.* **46**, 323–357.

DePierre, J. W., and Dallner, G. (1975). Structural aspects of the membrane of the endoplasmic reticulum. *Biochim. Biophys. Acta* **415**, 411–472.

Douglas, W. W. (1974). Involvement of calcium in exocytosis and the exocytosis-vesiculation sequence. *Biochem. Soc. Symp.* **39**, 1–28.

Edwards, K., Fleischer, B., Drybrugh, H., Fleischer, S., and Schreiber, G. (1976). The distribution of albumin precursor proteins and albumin in liver. *Biochem. Biophys. Res. Commun.* **72**, 310–318.

Farquhar, M. (1971). Processing of secretory products by cells of the anterior pituitary gland. *Mem. Soc. Endocrinol.* **19**, 79–122.

Farquhar, M. (1978a). Traffic of products and membranes through the Golgi complex. *In* "Transport of Macromolecules in Cellular Systems" (S. Silverstein, ed.), pp. 341–362. Dahlem Konferenzen, Berlin, West Germany.

Farquhar, M. G. (1978b). Recovery of surface membrane in anterior pituitary cells. *J. Cell Biol.* **77**, R35–R42.

Farquhar, M. G., Reid, J. J., and Daniell, L. W. (1978). Intracellular transport and packaging of prolactin. A quantitative electron microscope autoradiographic study of mammotrophs dissociated from rat pituitaries. *Endocrinology* **102**, 296–311.

Fessler, L. I., Morris, N. P., and Fessler, J. H. (1975). Procollagen: Biological scission of amino and carboxyl extension peptides. *Proc. Natl. Acad. Sci. U.S.A.* **72**, 4905–4909.

Fleischer, B. (1974). Isolation and characterization of Golgi apparatus and membranes from rat liver. *In* "Methods of Enzymology" (S. P. Colowick and M. V. Kaplan, eds.), Vol. 31A, pp. 180–191. Academic Press, New York.

Fleischer, B. (1977). Localization of some glycolipid glycosylating enzymes in the Golgi apparatus of rat kidney. *J. Supramol. Struct.* **249**, 5995–6003.

Franke, W. W., Luder, M. R., Kartembeck, J., Zerbon, H., and Keenan, T. (1976). Involvement of vesicle coat in casein secretion and surface regeneration. *J. Cell Biol.* **69**, 173–195.

Friend, D. S., and Farquhar, M. G. (1967). Function of coated vesicles during protein absorption in the vas deferens. *J. Cell Biol.* **35**, 357–376.

Gagnon, J., Palmiter, R. D., and Walsh, K. A. (1978). Comparison of the NH_2-terminal sequence of ovalbumin as synthesized *in vitro* and *in vivo*. *J. Biol. Chem.* **253**, 7464–7468.

Ganoza, M., and Williams, C. (1969). *In vitro* synthesis of different categories of specific proteins by membrane-bound and free ribosomes. *Proc. Natl. Acad. Sci. U.S.A.* **63**, 1370–1376.

Geller, D. M., Judah, J. D., and Nicholls, M. R. (1972). Intracellular distribution of serum albumin and its possible precursors in rat liver. *Biochem. J.* **127**, 865–874.

Glaumann, H., and Ericsson, J. L. E. (1970). Evidence for the participation of the Golgi apparatus in the intracellular transport of nascent albumin in the liver cell. *J. Cell Biol.* **47**, 555–567.

Goldman, B., and Blobel, G. (1978). Biogenesis of peroxisomes: Intracellular site of synthesis of catalase and uricase. *Proc. Natl. Acad. Sci. U.S.A.* **75**, 5066–5070.

Greene, L. J., Hirs, C. H. W., and Palade, G. E. (1963). On the protein composition of bovine pancreatic zymogen granules. *J. Biol. Chem.* **238**, 2054–2070.

Greengard, P. (1978). Phosphorylated proteins as physiological effectors. *Science* **199**, 146–152.

Grubman, M. J., Mayer, S. A., Banerjee, A. K., and Ehrenfeld, E. (1975). Sub-cellular localization of vesicular stomatitis virus messenger RNAs. *Biochem. Biophys. Res. Commun.* **62**, 531–538.

Haddad, A., Smith, M. D., Hercovics, A., Nadler, N. J., and Leblond, C. P. (1971). Radioautographic study of *in vivo* incorporation of fucose-^{3}H into thyroglobulin by rat follicular cells. *J. Cell Biol.* **49**, 856–882.

Hand, A. R., and Oliver, C. (1977). Relationship between the Golgi apparatus, GERL and secretory granules in acinar cells of the rat exorbital lacrimal gland. *J. Cell Biol.* **74**, 399–413.

Harwood, R., Grant, M. E., and Jackson, D. S. (1975). Studies on the glycosylation of hydroxylysine residues during collagen biosynthesis and the subcellular localization of collagen galactosyltransferase and collagen glucosyltransferase in tendon and cartilage cells. *Biochem. J.* **152**, 291–302.

Haugen, D. A., Armes, L. G., Yasunobu, K. T., and Coon, M. J. (1977). Amino terminal sequence of phenobarbitol-inducible cytochrome *P-450* from rabbit liver microsomes; Similarity to hydrophobic-terminal segments of preproteins. *Biochem. Biophys. Res. Commun.* **77**, 967–973.

Heuser, J. E., and Reese, T. S. (1973). Evidence for recycling of synaptic vesicle membrane during transmitter release at the frog neuromuscular junction. *J. Cell Biol.* **57**, 315–344.

Henson, P. (1971). The immunologic release of constituents from neutrophil leukocytes. I. The role of antibody and complement on nonphagocytosable surfaces on phagocytosable pacticles. *J. Immunol.* **107**, 1535–1546.

Herzog, V., and Miller, F. (1970). Die Lokalization endoginer Peroxydase in der Olandula Parotis der Ratte. *Z. Zellforsch. Mikrosk. Anat.* **107**, 403–420.

Herzog, V., and Farquhar, M. G. (1977). Luminal membrane retrieved after exocytosis reaches most Golgi cisternae in secretory cells. *Proc. Natl. Acad. Sci. U.S.A.* **74**, 5073–5077.

Hickman, S., Kulczycki, A. Jr., Lynch, R. D., and Kornfeld, S. (1977). Studies of the Mechanism of Tunicamycin Inhibition of IgA and IgE Secretion by Plasma Cells. *J. Biol. Chem.* **252**, 4402–4408.

Hicks, S. J., Drysdale, J. W., and Munro, H. N. (1969). Preferential synthesis of ferritin and albumin by different populations of liver polysomes. *Science* **164**, 584–585.

Hill, R. L., and Brew, K. (1975). Lactose Synthetase. *Adv. Enzym.* **43**, 411–490.

Holtzman, E., Schacher, S., Evans, J., and Teichberg, S. (1977). Origin and fate of the membrane of secretion granules and synaptic vesicles: Membrane circulation in neurons, gland cells and retinal photoreceptors, *In* "Cell Surface Reviews: The Synthesis Assembly and Turnover of Cell Surface Components" (G. Poste and G. Nicolson, eds.) pp. 166–246. North-Holland Publ. Amsterdam.

Hörtnagl, H. (1976). Membranes of the adrenal medulla. A comparison of membranes of chromaffin granules with those of endoplasmic reticulum. *Neuroscience* **1**, 9–18.

Howell, K. E., Ito, A., and Palade, G. E. (1978). Endoplasmic reticulum marker enzymes in Golgi fractions—What does it mean? *J. Cell Biol.* **79**, 581–589.

Hunt, L. A., and Summers, D. F. (1976). Glycosylation of vesicular stomatitis virus glycoprotein in virus-infected HeLa cells. *J. Virol.* **20**, 646–657.

Ikehara, Y., Oda, K., and Kato, K. (1976). Conversion of proalbumin into serum albumin in the secretory vesicles of rat liver. *Biochem. Biophys. Res. Commun.* **72**, 319–326.

Jamieson, J. D. (1972). Transport and discharge of exportable proteins in pancreatic exocrine cells. *In vitro* studies. *Curr. Top. Membr. Transp.* **3**, 273–338.

Jamieson, J. D., and Palade, G. E. (1967a). Intracellular transport of secretory proteins in the pancreatic exocrine cell. I. Role of peripheral elements of the Golgi complex. *J. Cell Biol.* **34**, 577–596.

Jamieson, J. D., and Palade, G. E. (1967b). Intracellular transport of secretory proteins in the pancreatic exocrine cell. II. Transport to condensing vacuoles and zymogen granules. *J. Cell Biol.* **34**, 597–615.

Jamieson, J. D., and Palade, G. E. (1968a). Intracellular transport of secretory proteins in the pancreatic exocrine cell. III. Dissociation of intracellular transport from protein synthesis. *J. Cell Biol.* **39**, 580–588.

Jamieson, J. D., and Palade, G. E. (1968b). Intracellular transport of secretory proteins in the pancreatic exocrine cell. IV. Metabolic requirements. *J. Cell Biol.* **39**, 589–603.

Jamieson, J. D., and Palade, G. E. (1971a). Condensing vacuole conversion and zymogen granule discharge in pancreatic exocrine cells: Metabolic studies. *J. Cell Biol.* **48**, 503–522.

Jamieson, J. D., and Palade, G. E. (1971b). Synthesis, intracellular transport and discharge of secretory proteins in stimulated pancreatic exocrine cells. *J. Cell Biol.* **50**, 135–158.

Kaliner, M., and Austen, K. F. (1975). Immunologic release of chemical mediators from human tissues. *Annu. Rev. Pharmacol.* **15**, 177–189.

Knipe, D. M., Baltimore, D., and Lodish, H. F. (1977). Localization of two cellular forms of the vesicular stomatitis viral glycoprotein. *J. Virol.* **21**, 1121–1127.

Kraehenbuhl, J. P., Racine, L., and Jamieson, J. D. (1977). Immunocytochemical localization of secretory proteins in bovine pancreatic exocrine cells. *J. Cell Biol.* **72**, 406–423.

Kreibich, G., and Sabatini, D. D. (1974). Selective release of content from microsomal vesicles without membrane disassembly. *J. Cell Biol.* **61**, 789–807.

Kreibich, G., Ulrich, B. L., and Sabatini, D. D. (1978). Proteins of rough microsomal membranes related to ribosome binding. *J. Cell Biol.* **77**, 464–487.

Lazarow, P. B., and deDuve, C. (1973). The synthesis and turnover of rat liver peroxisomes. V. Intracellular pathway of catalase synthesis. *J. Cell Biol.* **59**, 507–524.

Leblond, C. P., and Bennett, G. (1977). Role of the Golgi apparatus in terminal glycosylation.

In "International Cell Biology" (B. R. Brinkley and K. R. Porter, eds.), pp. 326–336. Rockefeller Univ. Press, New York.

Lin, J. J. C., Kamazawa, H., Ozola, J., and Wu, H. C. (1978). An *E. coli* mutant with an amino acid alteration within the signal sequence of outer membrane prolipoprotein. *Proc. Natl. Acad. Sci. U.S.A.* **75**, 4891–4895.

Magnusson, S., Sottrup-Jensen, L., Petersen, T. E., Morris, H. R., and Dell, A. (1974). Primary structure of the vitamin K-dependent part of prothrombin. *FEBS Lett.* **44**, 189–193.

Marshall, R. D. (1974). The nature and metabolism of the carbohydrate–peptide linkages of glycoproteins. *Biochem. Soc. Symp.* **40**, 17–26.

Mathews, E. K., and Petersen, O. H. (1973). Pancreatic acinar cells: Ionic dependence of the membrane potential and acetylcholine-induced depolarization. *J. Physiol. (London)* **231**, 283–295.

Melchers, F. (1971). Biosynthesis of the carbohydrate portion of immunoglobulin, radiochemical and chemical analysis of the carbohydrate moieties of two myeloma proteins purified from different subcellular fractions of plasma cells. *Biochemistry* **10**, 653–659.

Meldolesi, J. (1974). Dynamics of cytoplasmic membranes in guinea pig pancreatic acinar cells. I. Synthesis and turnover of membrane proteins. *J. Cell Biol.* **61**, 1–13.

Meldolesi, J., and Cova, D. (1972). Composition of cellular membranes in the pancreas of the guinea pig. IV. Polyacrylamide gel electrophoresis and amino acid composition of membrane proteins. *J. Cell Biol.* **55**, 1–18.

Milstein, C., Brownlee, G., Harrison, G., and Matthews, M. (1972). A possible precursor of immunoglobulin light chains. *Nature (London), New Biol.* **239**, 117–120.

Morré, D. J., and Ovtracht, L. (1977). Dynamics of the Golgi apparatus: Membrane differentiation and membrane flow. *Int. Rev. Cytol., Suppl.* **5**, 61–188.

Morré, D. J., Keenan, T. W., and Huang, C. M. (1974a). Membrane flow and differentiation: Origin of Golgi apparatus membranes from endoplasmic reticulum. *Adv. Cytopharmacol.* **2**, 107–125.

Morré, D. J., Yunghans, W. N., Keenan, T. W., and Vigil, E. (1974b). "Isolation of organelles and endo-membrane components from rat liver: Biochemical marker and quantitative morphometry." *Methodol. Dev. Biochem.* **4**, 195–236.

Morré, D. J., Kartenbeck, J., and Franke, W. (1979). Membrane flow and interconversions among endomembranes. *Biochim. Biophys. Acta* **559**, 71–152.

Morrison, T. G., and Lodish, H. F. (1975). Site of synthesis of membrane and nonmembrane proteins of vesicular stomatitis virus. *J. Biol. Chem.* **250**, 6955–6962.

Nagasawa, J., Douglas, W. W., and Schulz, R. A. (1971). Micropinocytotic origin of coated and smooth microvesicles ("synaptic vesicles") in neurosecretory terminals of posterior pituitary gland demonstrated by incorporation of horseradish peroxidase. *Nature (London)* **232**, 341–342.

Novikoff, A. B. (1976). The endoplasmic reticulum: A cytochemist's view (a review). *Proc. Natl. Acad. Sci. U.S.A.* **73**, 2781–2787.

Novikoff, A. B., and Shin, W. Y. (1964). The endoplasmic reticulum in the Golgi zone and its relations to microbodies, Golgi apparatus and autophagic vacuoles in rat liver cells. *J. Microsc. (Paris)* **3**, 187–206.

Novikoff, A. B., Mori, M., Quintana, N., and Yam, A. (1977). Studies of the secretory process in the mammalian exocrine pancreas. I. The condensing vacuole. *J. Cell Biol.* **75**, 148–165.

Novikoff, P. M., and Yam, A. (1978). Sites of lipoprotein particles in normal rat hepatocytes. *J. Cell Biol.* **76**, 1–11.

Novikoff, P. M., Novikoff, A. B., Quintana, N., and Hauw, J. J. (1971). Golgi apparatus, GERL and lysosomes of neurons in rat dorsal root ganglia studied by thick section and thin section cytochemistry. *J. Cell Biol.* **50**, 859–886.

Olsen, B. R., and Prockop, D. J. (1974). Ferritin conjugated antibodies used for labeling of organelles involved in the cellular synthesis and transport of procollagen. *Proc. Natl. Acad. Sci. U.S.A.* **71**, 2033–2037.

Olsen, B. R., Hoffman, H. P., and Prockop, D. J. (1976). Interchain disulfide bonds at the COOH-terminal end of procollagen synthesized by matrix-free cells from chick embryonic tendon and cartilage. *Arch. Biochem. Biophys.* **175**, 341–350.

Palade, G. (1975). Intracellular aspects of the process of protein synthesis. *Science* **189**, 347–358.

Palmiter, R. D., Gagnon, J., and Walsh, K. A. (1978). Ovalbumin: A secreted protein without a transient hydrophobic leader sequence. *Proc. Natl. Acad. Sci. U.S.A.* **75**, 94–98.

Patzelt, C., Brown, D., and Jeanrenaud, B. (1977). Inhibitory effect of colchicine on amylase secretion by rat parotid glands. Possible localization in the Golgi area. *J. Cell Biol.* **73**, 578–593.

Pearse, B. M., (1976). Clathrin: A unique protein associated with intracellular transfer of membrane coated vesicles. *Proc. Natl. Acad. Sci. U.S.A.* **73**, 1255–1259.

Peters, T. (1977). Intracellular albumin transport. *In* "Albumin Structure and Function" (V. M. Rosenoer, M. Dratz, and M. A. Rothschild, eds.), pp. 305–332. Pergamon, Oxford.

Peters, T., Fleischer, B., and Fleischer, S. (1971). The biosynthesis of rat serum albumin. *J. Biol. Chem.* **246**, 240–244.

Peters, T., Jr. (1962). The biosynthesis of rat serum albumin. *J. Biol. Chem.* **237**, 1186–1189.

Pless, D. D., and Lennarz, W. J. (1977). Enzymatic conversion of proteins to glycoproteins. *Proc. Natl. Acad. Sci. U.S.A.* **74**, 134–138.

Plummer, T. H., Jr., and Hirs, C. H. W. (1964). On the Structure of Bovine Pancreatic Ribonuclease B. *J. Biol. Chem.* **239**, 2530–2538.

Rasmussen, H., Jensen, P., Lake, W., Friedmann, N., and Goodman, D. B. P. (1975). Cyclic nucleotides and cellular calcium metabolism. *Adv. Cyclic Nucleotide Res.* **5**, 375–394.

Redman, C. M. (1967). Studies in the transfer of incomplete polypeptide chains across rat liver microsomal membranes *in vitro. J. Biol. Chem.* **242**, 761–768.

Redman, C. M. (1968). The synthesis of serum proteins on attached rather than free ribosomes of rat liver. *Biochem. Biophys. Res. Commun.* **31**, 845–850.

Redman, C. M. (1969). Biosynthesis of serum proteins and ferritin by free and attached ribosomes of rat liver. *J. Biol. Chem.* **244**, 4308–4315.

Redman, C. M., and Sabatini, D. D. (1966). Vectorial discharge of peptides released by puromycin from attached ribosomes. *Proc. Natl. Acad. Sci. U.S.A.* **56**, 608–615.

Redman, C. M., Siekevitz, P., and Palade, G. E. (1966). Synthesis and transfer of amylase in pigeon pancreatic microsomes. *J. Biol. Chem.* **241**, 1150–1158.

Redman, C. M., Grab, D. J., and Irukulla, R. (1972). The intracellular pathway of newly formed rat liver catalase. *Arch. Biochem. Biophys.* **152**, 496–501.

Redman, C. M., Banerjee, D., Howell, K., and Palade, G. E. (1975). Colchicine inhibition of plasma protein release from rat hepatocytes. *J. Cell Biol.* **66**, 42–59.

Redman, C. M., Banerjee, D., Manning, C., Huang, C. Y., and Green, K. (1978). *In vivo* effect of colchicine in hepatic protein synthesis and on the conversion of proalbumin to serum albumin. *J. Cell Biol.* **77**, 400–416.

Reggio, H. A., and Palade, G. E. (1978). Sulfated compounds in the zymogen granules of the guinea pig pancreas. *J. Cell Biol.* **77**, 288–314.

Robberecht, P., and Christophe, J. (1971). Secretion of hydrolases by perfused fragments of rat pancreas: Effect of calcium. *Am. J. Physiol.* **220**, 911–917.

Robbi, M., and Lazarow, P. B. (1978). Synthesis of catalse in two cell-free protein synthesizing systems and in rat liver. *Proc. Natl. Acad. Sci. U.S.A.* **75**, 4344–4348.

Roberts, K. (1974). Cytoplasmic microtubules and their functions. *Prog. Biophys. Mol. Biol.* **28**, 373–420.

Rothman, S. S. (1975). Protein transport in the pancreas. *Science* **190**, 747–753.

Rubin, R. P. (1974). "Calcium and the Secretory Process." Plenum, New York.

Sabatini, D. D., Tashiro, Y., and Palade, G. E. (1966). On the attachment of ribosomes to microsomal membranes. *J. Mol. Biol.* **19**, 503–524.

Satir, B. (1975). The final step in secretion. *Sci. Am.* **233**, 29–37.

Satir, B., Schooley, C., and Satir, P. (1973). Membrane fusion in a model system. *J. Cell Biol.* **56**, 153–176.

Schachter, H. (1974). The subcellular sites of glycosylation. *Biochem. Soc. Symp.* **40**, 57–71.

Schramm, M., and Selinger, Z. (1975). The functions of cyclic AMP and calcium as alternative second messengers in parotid gland and pancreas. *J. Cyclic Nucleotide Res.* **1**, 181–192.

Siekevitz, P., and Palade, G. E. (1960). A cytochemical study on the pancreas of the guinea pig. *J. Biophys. Biochem. Cytol.* **7**, 630–644.

Smith, P. E., and Farquhar, M. G. (1966). Lysosome function in the regulation of the secretory process in cells of the anterior pituitary gland. *J. Cell Biol.* **31**, 319–347.

Steiner, D. F. (1976). Peptide hormone precursors: Biosynthesis processing and significance. *In* "Peptide Hormones" (J. A. Parsons, ed.), pp. 49–64. Macmillan, New York.

Steiner, D. F., Kemmler, W., Tager, H. S., and Peterson, J. D. (1974a). Proteolytic processing in the biosynthesis of insulin and other proteins. *Fed. Proc., Fed. Am. Soc. Exp. Biol.* **33**, 2105–2115.

Steiner, D. F., Kemmler, W., Tager, H. S., and Rubenstein, A. H. (1974b). Molecular events taking place during intracellular transport of exportable protein. The conversion of peptide hormone precursors. *Adv. Cytopharmacol.* **2**, 195–205.

Steinman, R. M., Brodie, S. E., and Cohn, Z. A. (1976). Membrane flow during pinocytosis. A stereologic analysis. *J. Cell Biol.* **68**, 665–687.

Struck, D. K., Sierta, P. B., Lane, M. D., and Lennarz, W. J. (1978). Effect of Tunicamycin on the secretion of serum proteins by primary cultures of rat and chick hepatocytes. *J. Biol. Chem.* **253**, 5332–5337.

Suttie, J. W. (1974). Metabolism and properties of a liver precursor to prothrombin. *Vitam. Horm. (N.Y.)* **32**, 463–481.

Takagi, M., and Ogata, K. (1968). Direct evidence for albumin synthesis by membrane-bound polysomes in rat liver. *Biochem. Biophys. Res. Commun.* **33**, 55–60.

Tartakoff, A. M., and Vassalli, P. (1977). Plasma cell immunoglobulin secretion: Arrest is accompanied by alterations of the Golgi complex. *J. Exp. Med.* **146**, 1332–1345.

Tartakoff, A., and Vassalli, P. (1978). Comparative studies of intracellular transport of secretory proteins. *J. Cell Biol.* **79**, 694–707.

Tartakoff, A. M., Greene, L. J., and Palade, G. E. (1974a). Studies on the guinea pig pancreas: Fractionation and partial characterization of exocrine proteins. *J. Biol. Chem.* **249**, 7420–7431.

Tartakoff, A. M., Greene, L. J., Jamieson, J. D., and Palade, G. E. (1974b). Parallelism in the processing of pancreatic proteins. *Adv. Cytopharmacol.* **2**, 177–193.

Tedeschi, H., James, J. M., and Anthony, W. (1963). Photometric evidence for the osmotic behavior of rat liver microsomes. *J. Cell Biol.* **18**, 503–513.

Young, R. W. (1973). The role of the Golgi complex in sulfate metabolism. *J. Cell Biol.* **57,** 175–189.

Waechter, C. J., and Lennarz, W. J. (1976). The role of polyphenol-linked sugars in glyco-protein synthesis. *Annu. Rev. Biochem.* **45,** 95–112.

Wallach, D., Kirschner, N., and Schramm, M. (1975). Non-parallel transport of membrane proteins and content proteins during assembly of the secretory granules in rat parotid gland. *Biochim. Biophys. Acta* **375,** 87–105.

Weitzman, S., and Scharff, M. D. (1976). "Mouse myeloma mutants blocked in the assembly glycosylation and secretion of immunoglobulin." *J. Mol. Biol.* **102,** 237–252.

Williams, J. A., and Chandler, D. E. (1975). Ca^{++} and pancreatic amylase release. *Am. J. Physiol.* **228,** 1729–1732.

Zucker-Franklin, D., and Hirsch, J. G. (1964). Electron microscope studies on the degranu-lation of rabbit peritoneal leukocytes during phagocytosis. *J. Exp. Med.* **120,** 569–575.

Index

from reticulocytes, 26
termination, 2
termination pathway, 54
tRNA proofreading, 52
tRNA selection, 52
Protein topography
    ribosome, 16
Protein transport
    across membranes, 353
Protein-water relationships, 401
Pseudouridine, 4–5, 20
Putrescine
    and ribosome function, 81
    synthesis of, 80
Pyrimidine biosynthesis, 269

R

Red cell membrane
    major ectoproteins in, 362
    sialoglycoprotein of, 362
Regulatory enzymes, 207
Reversible enzyme reactions, 210
Ribonucleic acid (RNA)
    arrangement of TMV in, 410
    interferon-induced effects on, 93
    low-molecular weight, 329
        nucleocytoplasmic interrelations and,
            329
    nucleocytoplasmic interrelations and, 329
    protein granules of *Chironomus,* 139
Ribonucleoprotein (RNP)
    amino acid composition of, 150–151
    complexes, 137, 140
    containing mRNA, 160
    nuclear, 139
    nucleocytoplasmic interrelations and, 329
    particles, 330
        densities of, 144
        mRNA, 87
        and nuclear pore, 138
        in unfertilized eggs, 116
Ribosomal initiation complex
    mRNA binding to, 82
Ribosomal proteins, 13
    amino acid compositions, 20
    from different tissues, 21
    immunochemical comparison, 21
    number of, 20
    phosphorylated, 21
    phosphorylation of, 78

topography of, 21
Ribosomal RNA (rRNA) 5S, 20
    5.8S, 20
    interaction with mRNA, 42-43
    modified nucleosides of, 12
    precursors
        maturation, 139
    secondary structure of, 17
    sequence of 5S, 20
    sequencing, 12
    tertiary structure, 17
Ribosomal subunits
    assembly of large 415
    MET-tRNA binding, 71
    reconstitution of active, 13
    three-dimensional models of, 15
Ribosome(s)
    aminoacyl-tRNA binding to, 47
    assembly map, 14
    chemical properties of, 12
    of chloroplasts, 18
    composition of eukaryotic, 20
    of *Escherichia coli*, 11
    eukaryotic, 19, 21, 33
    general properties, 11
    interactions with mRNA, 43
    of mitochondria, 18
    model of, 404
    mRNA binding, 41
    physical properties of, 12
    prokaryotic, 33
    protein topography of, 16
    proteins of, 11, 13
    proteins of eukaryotic, 21
    role in initiation, 44
    sensitivity to antibiotics, 11
    structure of eukaryotic, 21
    synthesis of, 33
    three-dimensional structure of, 14
Ribosome assembly, 365, 413
    map, 14
Ribothymidine, 4–5
Ribulose-1,5-bisphosphatase carboxylase,
    370
RNP RNA
    characteristics of nuclear, 147

S

Secreted proteins
    precursors of, 187

191688

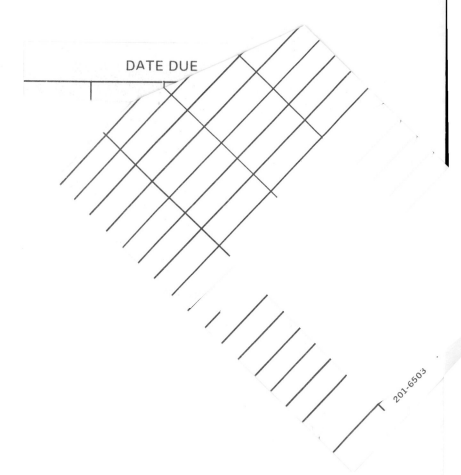

DATE DUE

201-6503